TODAY'S

CLASSROOM MANUAL

For Automotive Engine Performance

TODAY'S TECHNICIAN ™

CLASSROOM MANUAL

For Automotive Engine Performance

SEVENTH EDITION

Ken Pickerill

Southern Illinois University at Carbondale

CENGAGE
Learning·

Australia · Brazil · Japan · Korea · Mexico · Singapore · Spain · United Kingdom · United States

Today's Technician: Automotive Engine Performance, Seventh Edition

Ken Pickerill

SVP, GM Skills & Global Product Management: Jonathan Lau

Product Director: Matthew Seeley

Senior Product Manager: Katie McGuire

Senior Director, Development: Marah Bellegarde

Senior Product Development Manager: Larry Main

Senior Content Developer: Meaghan Tomaso

Product Assistant: Mara Ciacelli

Vice President, Marketing Services: Jennifer Ann Baker

Marketing Manager: Jonathan Sheehan

Senior Production Director: Wendy Troeger

Production Director: Andrew Crouth

Senior Content Project Manager: Cheri Plasse

Managing Art Director: Jack Pendleton

Cover image(s): Umberto Shtanzman/ Shutterstock.com

Library of Congress Control Number: 2016959775

Book Only ISBN: 978-1-3059-5826-5
Package ISBN: 978-1-3059-5828-9

Cengage Learning
20 Channel Center Street
Boston, MA 02210
USA

Cengage Learning is a leading provider of customized learning solutions with employees residing in nearly 40 different countries and sales in more than 125 countries around the world. Find your local representative at **www.cengage.com**

Cengage Learning products are represented in Canada by Nelson Education, Ltd.

To learn more about Cengage Learning, visit **www.cengage.com**

Purchase any of our products at your local college store or at our preferred online store **www.cengagebrain.com**

Notice to the Reader

Publisher does not warrant or guarantee any of the products described herein or perform any independent analysis in connection with any of the product information contained herein. Publisher does not assume, and expressly disclaims, any obligation to obtain and include information other than that provided to it by the manufacturer. The reader is expressly warned to consider and adopt all safety precautions that might be indicated by the activities described herein and to avoid all potential hazards. By following the instructions contained herein, the reader willingly assumes all risks in connection with such instructions. The publisher makes no representations or warranties of any kind, including but not limited to, the warranties of fitness for particular purpose or merchantability, nor are any such representations implied with respect to the material set forth herein, and the publisher takes no responsibility with respect to such material. The publisher shall not be liable for any special, consequential, or exemplary damages resulting, in whole or part, from the readers' use of, or reliance upon, this material.

Printed in the United States of America

Print Number: 02   Print Year: 2017

DEDICATION

I dedicate this book to the memory of my father, Miles K. Pickerill. Dad taught me love, honesty, and craftsmanship.

Also to Peggy, my wife of 40 years, and my two sons, Adam and Eric.

My thanks for their contributions to my career go out to:

The late Doug McNally, who gave me my first automotive job.

The late John Riddle, who guided me as a young apprentice many years ago.

Dennis Price, who always believed in my ability.

Ken Pickerill

CONTENTS

Thanks to the support the *Today's Technician™ Series* has received from those who teach automotive technology, Cengage Learning, the leader in automotive-related textbooks, is able to live up to its promise to provide new editions of the series every few years. By revising this series on a regular basis, we can respond to changes in the industry, changes in technology, changes in the certification process, and to the ever-changing needs of those who teach automotive technology.

The *Today's Technician™ Series* features textbooks and digital learning solutions that cover all mechanical and electrical systems of automobiles and light trucks. The individual titles correspond to the National Automotive Technicians Education Foundation (NATEF) and the National Institute for Automotive Service Excellence (ASE) certification areas and are specifically correlated to the 2013 standards for Automotive Service Technicians (AST) and Master Automotive Service Technicians (MAST).

The mission of NATEF is to improve the quality of automotive technician training programs. NATEF evaluates programs against standards developed by the automotive industry and recommends qualified programs for certification (accreditation) from ASE. NATEF's national standards reflect the skills that students must master to be successful. ASE certification through NATEF ensures that certified training programs meet or exceed industry-recognized, uniform standards of excellence. All titles in the *Today's Technician™ Series* include remedial skills and theories common to all of the certification areas, and advanced or specific subject areas that reflect the latest technological trends, such as this updated title on engine performance.

The technician of today and the future must know the underlying theory of all automotive systems and be able to service and maintain those systems. Dividing the material into two volumes, a Classroom Manual and a Shop Manual, provides the student with the information needed to begin a successful career as an automotive technician without interrupting the learning process by mixing cognitive and performance learning objectives into one volume.

The design of Delmar's *Today's Technician™ Series* was based on features that are known to promote improved student learning. The design was further enhanced by a careful study of survey results, in which the respondents were asked to value particular features. Some of these features can be found in other textbooks, while others are unique to this series.

Each Classroom Manual contains the principles of operation for each system and subsystem. The Classroom Manual also discusses design variations in key components used by the different vehicle manufacturers, and considers emerging technologies that will be standard or optional features in the near future. This volume is organized to build upon basic facts and theories. Its primary objective is to help the reader gain an understanding of how each system and subsystem operates. This understanding is necessary to diagnose the complex automobiles of today and tomorrow. Although the basics contained in the Classroom Manual provide the knowledge needed for diagnostics, diagnostic procedures appear only in the Shop Manual. An understanding of the underlying theories is also a requirement for competence in the skill areas covered in the Shop Manual.

A spiral-bound Shop Manual delivers hands-on learning experiences with step-by-step instructions for diagnostic and repair procedures. Photo Sequences are used to illustrate some of the common service procedures. Other common procedures are listed and are accompanied with fine line drawings and photos that let the reader visualize and conceptualize the finest details of the procedure. This volume explains the reasons for performing the procedures, as well as the circumstances when each particular service is appropriate.

The two volumes are designed to be used together and are arranged in corresponding chapters. Not only are the chapters in the volumes linked together, the contents of the chapters are also linked. The linked content is indicated by marginal callouts that refer the reader to the chapter and page where the same topic is addressed in the companion volume. This valuable feature saves users the time and trouble of searching the index or table of contents to locate supporting information in the other volume. Instructors will find this feature especially helpful when planning the presentation of material and when making reading assignments.

Both volumes contain clear and thoughtfully selected illustrations, many of which are original drawings or photos specially prepared for inclusion in this series. This means that the art is a vital part of each textbook and not merely inserted to increase the number of illustrations.

The page design of this series uses available margin space to deliver helpful information efficiently without interrupting the pedagogical lesson material. This information includes examples of concepts just introduced in the text, explanations or definitions of terms that are not defined in the text, examples of common trade jargon used to describe a part or operation, and unique applications of the system or service described in the text. Many textbooks also include this information but insert it in the main body of text; this tends to interrupt the reader's thought process. By placing this information to the side of the main text, students can read through the text uninterrupted and refer to the additional information when it is best for them.

Jack Erjavec,
Series Editor

HIGHLIGHTS OF THIS NEW EDITION—CLASSROOM MANUAL

We are very proud of the changes made to make the seventh edition of *Today's Technician™: Automotive Engine Performance*. The Classroom Manual has been updated to reflect current changes in automotive engine performance, while retaining some of the more pertinent information on older models that are still in wide use. Engine performance technology is often seen by students as an array of complicated subjects, all of which can be difficult to understand. This is why *Automotive Engine Performance* follows a natural progression of subjects, including basic theories, engine construction, electrical theory, and computer control, to bring all these topics together. More experienced students and working technicians can review the basic theories and then move on to more advanced areas in the latter half of the text.

This edition features an even further investigation of digital fuel injection (DFI) use and operation, pulse width modulated fuel pump control, and variable valve timing, along with expanded and updated coverage of hybrid vehicle technology. Coverage of ignition systems and fuel injection has been updated with coverage that reflects the vehicles in use today. This edition continues to present the latest in sensor technology and new developments in automotive engine performance. The Classroom Manual also has new ASE-style multiple-choice questions to help assess student understanding of the material.

HIGHLIGHTS OF THIS NEW EDITION—SHOP MANUAL

The *Automotive Engine Performance*, seventh edition, Shop Manual follows the same natural progression of the Classroom Manual but focuses on the skill set required to be a successful technician in today's shop. The Shop Manual starts with shop procedures and ASE certification, progresses through electrical testing, basic engine testing, and scan tool procedures, and finally through the most advanced diagnostics used today. The coverage includes new basic engine technology, such as variable valve timing, while retaining a thorough coverage of basic engine testing such as vacuum, compression, cylinder leakage, and engine balance that are the cornerstone of engine performance diagnosis. Students are then led through electrical troubleshooting. This coverage includes basic multimeter, test lamp, logic probe, and oscilloscope pattern reading and use. Electrical wiring and connector repair are also covered, and a review of battery starting and charging diagnostics is given. Engine performance ignition systems coverage reflects modern usage of coil-over-plug systems, while retaining some distributor ignition coverage. Fuel injection system coverage reflects port fuel and direct fuel injection. Engine control systems such as variable valve timing, fuel systems such as returnless fuel systems, pulse width modulated fuel pump control, and emission controls such as PCV, EGR, and evaporative emissions are covered in depth. A section on computer operation and networking is included, as well as a chapter on related systems to bring the operation of the whole vehicle together for the student. This broad coverage makes this edition suitable for a wide range of student capability, from beginning technician to a more seasoned veteran. Other important updates are updated photo sequences. Many job sheets have been added to provide complete coverage of MAST, AST, and MLR tasks.

CLASSROOM MANUAL

Features of the Classroom Manual include the following:

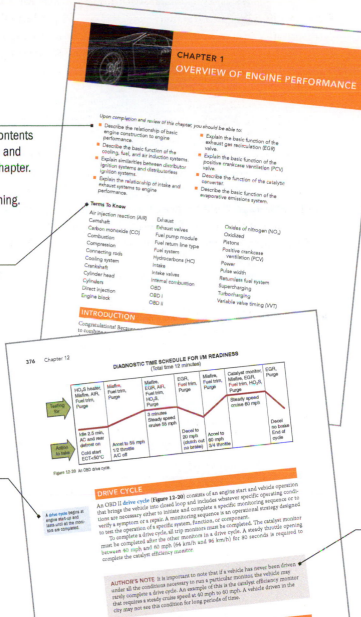

Cognitive Objectives

These objectives outline the chapter's contents and identify what students should know and be able to do upon completion of the chapter. Each topic is divided into small units to promote easier understanding and learning.

Terms to Know List

A list of key terms appears immediately after the Objectives. Students will see these terms discussed in the chapter. Definitions can also be found in the Glossary at the end of the manual.

Margin Notes

The most important terms to know are highlighted and defined in the margin. Common trade jargon also appears in the margin and gives some of the common terms used for components. This helps students understand and speak the language of the trade, especially when conversing with an experienced technician.

Cross-References to the Shop Manual

References to the appropriate page in the Shop Manual appear whenever necessary. Although the chapters of the two manuals are synchronized, material covered in other chapters of the Shop Manual may be fundamental to the topic discussed in the Classroom Manual.

Author's Note

This feature includes simple explanations, stories, or examples of complex topics. These are included to help students understand difficult concepts.

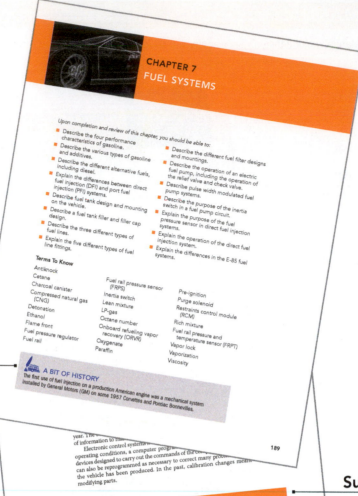

CHAPTER 7
FUEL SYSTEMS

Upon completion and review of this chapter, you should be able to:

- Describe the four performance characteristics of gasoline.
- Describe the various types of gasoline and additives.
- Describe the different alternative fuels, including diesel.
- Explain the differences between direct fuel injection (DFI) and port fuel injection (PFI) systems.
- Describe fuel tank design and mounting on the vehicle.
- Describe a fuel tank filler and filler cap design.
- Describe the three different types of fuel lines.
- Explain the five different types of fuel line fittings.

- Describe the different fuel filter designs and mountings.
- Describe the operation of an electric fuel pump, including the operation of the relief valve and check valve.
- Describe pulse width modulated fuel pump systems.
- Describe the purpose of the inertia switch in a fuel pump circuit.
- Explain the purpose of the fuel pressure sensor in direct fuel injection systems.
- Explain the operation of the direct fuel injection system.
- Explain the differences in the E-85 fuel systems.

Terms To Know

Antiknock
Cetane
Charcoal canister
Compressed natural gas (CNG)
Detonation
Ethanol
Flame front
Fuel pressure regulator
Fuel rail

Fuel rail pressure sensor (FRPS)
Inertia switch
Lean mixture
LP-gas
Octane number
Onboard refueling vapor recovery (ORVR)
Oxygenate
Paraffin

Pre-ignition
Purge solenoid
Restraints control module (RCM)
Rich mixture
Fuel rail pressure and temperature sensor (FRPT)
Vapor lock
Vaporization
Viscosity

A BIT OF HISTORY
The first use of fuel injection on a production American engine was a mechanical system installed by General Motors (GM) on some 1957 Corvettes and Pontiac Bonnevilles.

189

year. The v...
of information to ma...

Electronic control systems w...
operating conditions, a computer progra...
devices designed to carry out the commands of the co...
can also be reprogrammed as necessary to correct many prob...
the vehicle has been produced. In the past, calibration changes mean...
modifying parts.

SUMMARY

- The engine block contains the crankshaft, connecting rods, cylinders, and pistons.
- A gasoline or diesel engine is termed an *internal combustion engine.*
- The cylinder head fits over and seals the top of the cylinder.
- The four-stroke cycle consists of intake, compression, power, and exhaust.
- The intake system carries the air-fuel mixture in, and the exhaust system takes the burned fuel out.
- The AIR system is designed to reduce cold start emissions by burning hydrocarbons in the exhaust and heating the catalytic converter for quicker operation.
- The PCV system relieves crankcase pressure and removes unburned fuel and water from the crankcase.
- The EGR valve lowers combustion temperatures and thereby reduces NO_x production.
- The catalytic converter helps reduce emissions by oxidation and reduction reactions, which reduce HC, CO, and NO_x emission.

- The EVAP system prevents the evaporation of fuel and the resulting hydrocarbon emission.
- The ignition system is designed to deliver spark to the proper spark plug at the right time to start combustion.
- The cooling system is responsible for cooling the engine and heating the passenger compartment in winter.
- The fuel system must provide enough clean fuel at the right pressure.
- Supercharging and turbocharging raise engine horsepower by forcing air into the engine.
- OBD I vehicles were the first computer-controlled emission systems and were used from the late 1970s until 1996.
- OBD II systems are designed to monitor and test components while driving.
- Gasoline fuel injection methods include port fuel and direct.

REVIEW QUESTIONS

Short Answer Essays

1. Explain why the engine performance specialist is probably one of the most skilled technicians in the shop.
2. Explain why diesel and gasoline engines are classified as internal combustion engines.
3. Explain the purpose of the cylinder head.

4. Explain the four-stroke cycle.
5. Explain how a turbo- or supercharger raises engine power.
6. Explain why a PCV system is needed.
7. Explain how the EGR valve lowers combustion temperatures.

A Bit of History

This feature gives the student a sense of the evolution of the automobile. This feature not only contains nice-to-know information, but also should spark some interest in the subject matter.

Summary

Each chapter concludes with a summary of key points from the chapter. These key points help the reader review the chapter contents.

Review Questions

Short-answer essays, fill-in-the-blanks, and multiple-choice questions are found at the end of each chapter. These questions are designed to accurately assess the student's competence in the stated objectives at the beginning of the chapter.

SHOP MANUAL

To stress the importance of safe work habits, the Shop Manual also dedicates one full chapter to safety. Other important features of this manual include the following:

Performance-Based Objectives

These objectives define the contents of the chapter and what the student should have learned upon completion of the chapter.

Terms to Know List

Terms in this list can be found in the Glossary at the end of the manual.

Customer Care

This feature highlights those little things a technician can do or say to enhance customer relations.

Author's Note

This feature includes simple explanations, stories, or examples of complex topics. These are included to help students understand difficult concepts.

Basic Tools Lists

Each chapter begins with a list of the basic tools needed to perform the tasks included in the chapter.

Special Tools Lists

Whenever a special tool is required to complete a task, it is listed in the margin next to the procedure.

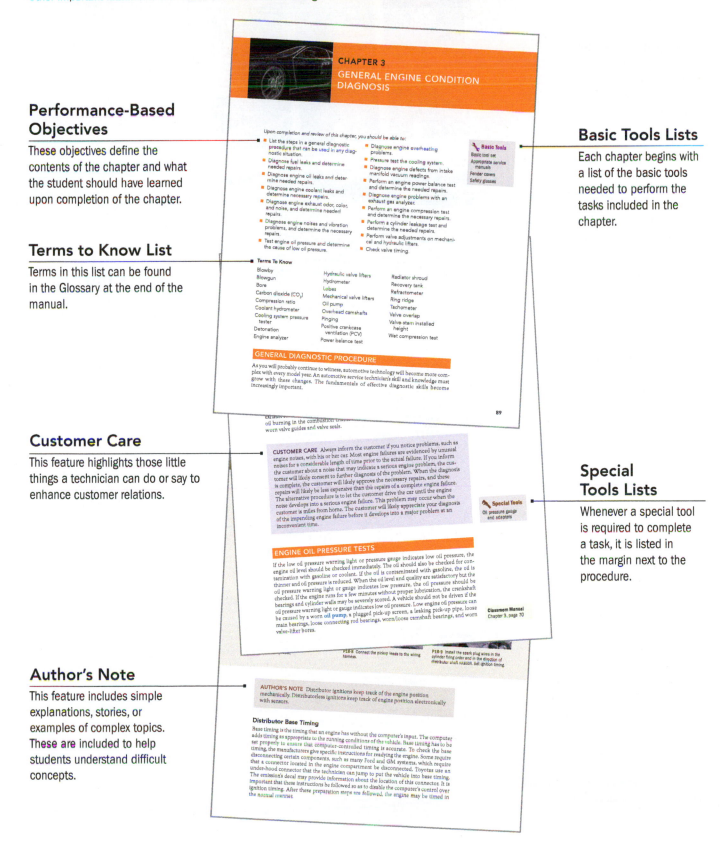

Photo Sequences

Many procedures are illustrated in detailed photo sequences. These detailed photographs show the students what to expect when they perform particular procedures. They also can provide the student a familiarity with a system or type of equipment, which the school might not have.

References to the Classroom Manual

References to the appropriate page in the Classroom Manual appear whenever necessary. Although the chapters of the two manuals are synchronized, material covered in other chapters of the Classroom Manual may be fundamental to the topic discussed in the Shop Manual.

Service Tips

Whenever a shortcut or special procedure is appropriate, it is described in the text. Generally, these tips describe common procedures used by experienced technicians.

Warnings and Cautions

Cautions appear throughout the text to alert readers to potentially hazardous materials or unsafe conditions. Warnings advise the students of things that can go wrong if instructions are not followed or if an incorrect part or tool is used.

Fuel System Diagnosis and Service 351

PHOTO SEQUENCE 13
Typical Procedure for Relieving Fuel Pressure and Servicing the Fuel Tank

P13-1 To begin, relieve any fuel pressure from the system.

P13-2 Disconnect the negative battery cable.

P13-3 If the tank has a significant amount of fuel inside, pump the fuel out of the tank with a gas caddy. (It may be difficult to insert a siphon if the tank has a check valve at the bottom of the fuel filler pipe.)

P13-4 Raise the vehicle on the lift.

P13-5 Disconnect electrical connections from the fuel pump, fuel sender, and vapor pressure sensor.

P13-6 Remove fuel filler tube and vapor and fuel lines from the fuel tank.

---

288   Chapter 6

Classroom Manual
Chapter 6, page 175

Prior to J1930 naming rules, a TP sensor was called a TPS by almost everyone.

Throttle Position Sensor (Potentiometer Type)

A malfunctioning throttle position (TP) sensor (**Figure 6-27**) may cause acceleration stumbles, engine stalling, and improper idle speed.

Using a scan tool, monitor the TP sensor for dropouts in the signal voltage. If the problem is not found, back-probe the sensor terminals to complete the meter connections. With the ignition switch on, connect a voltmeter from the 5-volt reference wire to ground (**Figure 6-28**). The voltage reading on this wire should be approximately 5 volts. Always refer to the vehicle manufacturer's specifications.

🔧 **SERVICE TIP** When testing the TP sensor voltage signal, use a DSO or graphing DMM because the gradual voltage increase on this wire is easier to monitor. If the sensor voltage increase is erratic, the reading fluctuates.

If the reference wire is not supplying the specified voltage, check the voltage on this wire at the computer terminal. If the voltage is within specifications at the computer but low at the sensor, repair the reference wire. When this voltage is low at the computer, check the voltage supply wires and ground wires on the computer also check the reference voltage with the sensor unplugged. Sometimes a defective sensor can ground the reference wire. If these checks are satisfactory, replace the computer.

With the ignition switch on, connect the voltmeter from the sensor ground wire to the battery ground. If the voltage drop across this circuit exceeds specifications, repair the ground wire from the sensor to the computer.

With the ignition switch on, connect a voltmeter from the sensor signal wire to ground. Slowly open the throttle and observe the voltmeter. The voltmeter reading should increase smoothly and gradually. Typical TP sensor voltage readings would be 0.5 volt to 1 volt with the throttle in the idle position and 4 to 5 volts at wide-open throttle. Always [...] the vehicle manufacturer's specifications. If the TP sensor does not have the speci[...] signal is erratic, replace the sensor.

[...] lowest value seen as 0 percent throttle. Make certain to consult [...] vehicle

---

176   Chapter 4

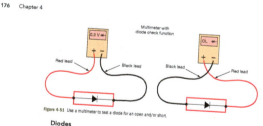

Figure 4-51 Use a multimeter to test a diode for an open and/or short.

Diodes

Regardless of the bias of the diode, it should allow current flow in one direction only. To test a diode, use the diode check feature of the multimeter. Disconnect the diode from the circuit. Connect the meter's leads across the diode (**Figure 4-51**). Observe the reading on the meter, and then reverse the meter's leads and observe the reading on the meter. Because of resistance in one direction should be very high or infinity, many meters read "OL" for out of limits and in the other direction a reading of 0.5 to 0.8 indicates a good silicon diode and 0.2 to 0.3 indicates a good germanium diode. The diode check function actually checks for the voltage drop across the diode. If any other readings were observed, the diode is bad. A diode that has low resistance in both directions is shorted. A diode that has high resistance or an infinity reading in both directions is open.

Problems may be encountered when checking a diode with a high-impedance digital ohmmeter. Since many diodes will not allow current flow through them unless the voltage is at least 0.6 volts, a digital meter may not be able to forward bias the diode. This will result in readings that indicate the diode is open, when in fact it may not be. Because of this problem, many multimeters are equipped with a diode testing feature. This test allows for increased voltage at the test leads. Some meters will display the voltage required to forward bias the diode. If the diode is open, the meter will display "OL" or another reading to indicate *infinity* or *out of range*. Some meters during a diode check will make a beep noise when there is continuity.

Diodes may also be tested with a voltmeter. Using the same logic as when testing with an ohmmeter, test the voltage drop across the diode. The meter should read low voltage in one direction and near source voltage in the other direction.

Classroom Manual
Chapter 4, page 101

⚠️ **Caution**
Battery electrolyte contains sulfuric acid, a strong, corrosive acid. Electrolyte is very damaging to vehicle paint, upholstery, and clothing. Do not allow the electrolyte to contact these items.

BATTERY DIAGNOSIS AND SERVICE

⚠️ **WARNING** Battery electrolyte is very harmful to human skin and eyes. Always wear face and eye protection, protective gloves, and protective clothing when handling batteries or electrolyte. If electrolyte contacts your skin or eyes, flush with clean water immediately and obtain medical help.

⚠️ **WARNING** Never smoke or allow other sources of ignition near a battery. Hydrogen gas discharged while charging the battery is explosive (Figure 4-52)! A battery explosion may cause personal injury or property damage. Even when the battery has not been charged for several hours, hydrogen gas still may be present under the battery cover.

Case Studies

Each chapter ends with a Case Study describing a particular vehicle problem and the logical steps a technician might use to solve the problem. These studies focus on system diagnosis skills and help students gain familiarity with the process.

ASE-Style Review Questions

Each chapter contains ASE-Style Review Questions that reflect the performance objectives listed at the beginning of the chapter. These questions can be used to review the chapter as well as to prepare for the ASE certification exam.

Diagnostic Charts

Some chapters include detailed diagnostic charts that list common problems and most probable causes. They also list a page reference in the Classroom Manual for the student to use to gain a better understanding of the system's operation and a page reference in the Shop Manual to refer the student to details on the procedure necessary for correcting the problem.

650　Chapter 12

CASE STUDY

A customer brought in his OBD II–equipped car into the dealership complaining of poor fuel economy. The customer stated that the gas mileage has been declining since the car was new. This is not normal. Gas mileage normally improves slightly as the engine is broken in. The customer had no other complaints. Because this is a very difficult problem to verify, the technician began the diagnostic process with a visual inspection and found nothing out of the ordinary. The car's MIL was not lit.

He then connected a scan tool to the DLC and reviewed the data. Comparing the input data being displayed to the normal range of values listed in the service information, he discovered that the upstream O_2 sensor was biased rich. That meant the PCM was seeing a lean condition and adding fuel to correct for this problem. To verify this, the technician watched the fuel trim. Sure enough, the long-term FT had moved to add more fuel. Normally this is caused by a vacuum leak or restricted fuel injectors. The latter probable cause seemed unlikely since the O_2 sensor showed that added fuel was being delivered. Therefore, the technician began to look for a possible cause of a vacuum leak.

The process continued for quite some time as he checked all of the vacuum hoses and components. Nothing appeared to be leaking. As he leaned over to check something in the back of the engine, he heard a slight "Pfftt." The noise had somewhat of a rhythm to it, and he focused his attention to it. It did not sound like a vacuum leak. As he increased the engine's speed, the noise pulses became closer together and soon became a constant noise. The noise appeared to be coming from the lower part of the engine.

Using a stethoscope, he was able to identify the source of the noise. It appeared that the gasket joining the exhaust manifold to the exhaust pipe was not seated properly. To verify this, he raised the car on a hoist and took a look. He found that the retaining bolts were loose. He took a quick look at the gasket and found it to be in reasonable shape, then tightened the bolts to specifications.

Not sure that the noise or the exhaust leak was related to the problem but suspected that it could be, he connected a lab scope to the oxygen sensor and watched its activity. The sensor's signal no longer showed a bias. It appears that the leak in the exhaust was pulling air into the exhaust between each pulse. This was adding oxygen to the exhaust stream causing the computer to think the mixture was lean.

ASE-STYLE REVIEW QUESTIONS

1. While discussing the adaptive learning of a PCM, *Technician A* says that when the battery is disconnected from the vehicle, the learning process resets. *Technician B* says that the vehicle must be driven under part throttle with moderate acceleration for the PCM to relearn the system.
Who is correct?
A. A only　　　　　C. Both A and B
B. B only　　　　　D. Neither A nor B

2. *Technician A* says that PO codes with definitions can be accessed through a generic scan tool.

Who is correct?
A. A only　　　　　C. Both A and B
B. B only　　　　　D. Neither A nor B

3. *Technician A* says that if the PCM is correcting for a lean condition, the long-term FT will indicate a negative number.
Technician B says that a rich condition causes the long-term FT to show a positive number.
Who is correct?
A. A only　　　　　C. Both A and B
B. B only　　　　　D. Neither A nor B

Onboard Diagnostic (OBD II) System Diagnosis and Service　　635

DTC PO108 MAP sensor circuit high voltage

Step	Action	Values	Yes	No
1	Was the Powertrain On-Board Diagnostic (OBD) System Check performed?	—	Go to Step 2	Go to Powertrain OBD System Check 2.4 L Powertrain OBD System Check 2.2 L
2	1. Install a scan tool 2. Engine at idle. Does the scan tool display the MAP voltage specified?	4.0V	Go to Step 3	Go to Step 4
3	1. Turn the ignition switch OFF. 2. Disconnect the MAP sensor electrical connector. 3. Turn the ignition switch ON. Does the scan tool display the MAP voltage specified?	1.0V	Go to Step 5	Go to Step 6
4	1. Turn the ignition switch ON, with the engine OFF, review freeze frame data. 2. Operate the vehicle within the freeze-frame conditions. Does the scan tool display the MAP voltage specified?	4.0V	Go to Step 3	Go to diagnostic aids
5	Probe the MAP sensor signal ground circuit with a test light connected to battery voltage. Does the test light illuminate?	—	Go to Step 7	Go to Step 11
6	Check the MAP sensor signal circuit for a short to voltage and repair as necessary. Was a repair necessary?	—	Go to Step 14	Go to Step 12
7	With a DVM connected to ground, probe the 5-volt reference circuit. Does the DVM display the specified voltage?	5V	Go to Step 8	Go to Step 9
8	Check the MAP sensor vacuum source for being plugged or leaking. Was a problem found?	—	Go to Step 10	Go to Step 13
9	Check the 5-volt reference circuit for a short to voltage and repair as necessary.	—	Go to Step 14	Go to Step 12
10	Repair the vacuum hose as necessary. Is the action complete?	—	Go to Step 14	
11	Check for an open in the MAP sensor ground circuit and repair as necessary. Was a repair necessary?	—	Go to Step 14	Go to Step 12
12	Replace the PCM. Is the action complete?	—	Go to Step 14	
13	Replace the MAP sensor. Is the action complete?	—	Go to Step 14	
14	1. Using the scan tool, clear the DTCs. 2. Start the engine and idle at normal temperature. 3. Operate the vehicle within the conditions for setting the DTC. Does the scan tool indicate that the diagnostic ran and passed?	—	Go to Step 15	Go to Step 2
15	Check for any additional DTCs. Are any DTCs displayed that have not been diagnosed?	—	Go to specific DTC charts	System OK

Figure 12-26　Diagnostic chart for a DTC—PO108.

Job Sheets

Located at the end of each chapter, the Job Sheets provide a format for students to perform procedures covered in the chapter. A reference to the ASE Task addressed by the procedure is included on the Job Sheet.

Name _____   Date _____

Diagnosing Related Systems   683

INSPECTING DRIVE BELT

JOB SHEET
58

Upon completion of this job sheet, you should be able to visually inspect a serpentine drive belt and check its tightness.

NATEF Correlation

This job sheet addresses the following **AST/MAST** task:

A.3.   Diagnose abnormal engine noises or vibration concerns; determine necessary action.

Tools and Materials
- Vehicle with a serpentine belt
- Service information for the above vehicle

Describe the vehicle being worked on:

Year _____ Make _____

Model _____ VIN _____

Procedure

1. Carefully inspect the belt and describe the general condition.

_____

_____

_____

_____

2. Check the tension of the belt. Belt tension and wear are generally assessed by looking at a wear indicator on the belt tensioner. (See Service Information.)
You found _____

3. Based on the above, what are your recommendations?

_____

_____

4. Describe the procedure for _____

ASE Challenge Questions

Each technical chapter ends with five ASE Challenge Questions. These are not mere review questions; rather, they test the students' ability to apply general knowledge to the contents of the chapter.

---

680   Chapter 13

ASE CHALLENGE QUESTIONS

1. *Technician A* says that improper tire sizes can cause the speedometer to be inaccurate. *Technician B* says that the PCM can be recalibrated to correct this condition.
Who is correct?
A. A only
B. B only
C. Both A and B
D. Neither A nor B

2. While discussing a vibration problem, *Technician A* says as a general rule, a U-joint problem is most noticeable in the 30 to 60 mph range. *Technician B* says that tire balance problems are most noticeable on acceleration and deceleration.
Who is correct?
A. A only
B. B only
C. Both A and B
D. Neither A nor B

3. *Technician A* says that a four-wheel alignment tells the technician whether the rear axle or wheels are square with the front wheels. *Technician B* says that total toe for all four wheels must be determined and rear toe adjusted where possible to bring the rear axle or wheels into square with the chassis.

Who is correct?
A. A only
B. B only
C. Both A and B
D. Neither A nor B

4. While discussing electronically controlled transmissions, *Technician A* says that line pressure is controlled by a pulse width modulated solenoid. *Technician B* says that the solenoid is controlled by the PCM.
Who is correct?
A. A only
B. B only
C. Both A and B
D. Neither A nor B

5. While discussing wheel alignment angles, *Technician A* says that camber angle changes while driving. *Technician B* says that the camber angle also changes when the vehicle is loaded and sags under the weight.
Who is correct?
A. A only
B. B only
C. Both A and B
D. Neither A nor B

SUPPLEMENTS

Instructor Resources

The *Today's Technician* series offers a robust set of instructor resources, available online at Cengage's Instructor Resource Center and on DVD. The following tools have been provided to meet any instructor's classroom preparation needs:

- An Instructor's Guide provides lecture outlines, teaching tips, and complete answers to end-of-chapter questions.
- PowerPoint presentations include images, videos, and animations that coincide with each chapter's content coverage.
- Cengage Learning Testing Powered by Cognero® delivers hundreds of test questions in a flexible, online system. You can choose to author, edit, and manage test bank content from multiple Cengage Learning solutions and deliver tests from your Learning Management System, or you can simply download editable Word documents from the DVD or Instructor Resource Center.
- An Image Gallery includes photos and illustrations from the text.
- The Job Sheets from the Shop Manual are provided in Word format.
- End-of-Chapter Review Questions are also provided in Word format, with a separate set of text rejoinders available for instructors' reference.
- To complete this powerful suite of planning tools, a pair of correlation guides map this edition's content to the NATEF tasks and to the previous edition.

REVIEWERS

The author and publisher would like to extend a special thanks to the following reviewers for their contributions to this text:

Terry L. Enyart
University of Northwestern Ohio
Lima, OH

Tim Mulready
University of Northwestern Ohio
Lima, OH

Chris McNally
Hudson Valley Community College
Troy, NY

Christopher D. Parrott
Vatterott College
Wichita, KS

CHAPTER 1
OVERVIEW OF ENGINE PERFORMANCE

Upon completion and review of this chapter, you should be able to:

- Describe the relationship of basic engine construction to engine performance.
- Describe the basic function of the cooling, fuel, and air induction systems.
- Explain similarities between distributor ignition systems and distributorless ignition systems.
- Explain the relationship of intake and exhaust systems to engine performance.
- Explain the basic function of the exhaust gas recirculation (EGR) valve.
- Explain the basic function of the positive crankcase ventilation (PCV) valve.
- Describe the function of the catalytic converter.
- Describe the basic function of the evaporative emissions system.

Terms To Know

Air injection reaction (AIR)	Exhaust	Oxides of nitrogen (NO$_x$)
Camshaft	Exhaust valves	Oxidized
Carbon monoxide (CO)	Fuel pump module	Pistons
Combustion	Fuel return line type	Positive crankcase ventilation (PCV)
Compression	Fuel system	
Connecting rods	Hydrocarbons (HC)	Power
Cooling system	Intake	Pulse width
Crankshaft	Intake valves	Returnless fuel system
Cylinder head	Internal combustion	Supercharging
Cylinders	OBD	Turbocharging
Direct injection	OBD I	Variable valve timing (VVT)
Engine block	OBD II	

INTRODUCTION

Congratulations! Because you are taking a course in engine performance, you are about to combine most of the knowledge you have gained to date and use it to start diagnosis of drivability problems. The engine performance specialist is usually one of the most highly talented people in the shop. It takes a good foundation in electricity and electronics. It requires knowledge of the engine and how it works along with basic transmission operation. The engine performance person also does most of the electrical diagnosis and scan tool work, so you will also need some knowledge of basic engine testing, computer operation, networking, body control modules, and so forth, which is included to help you in your career. We hope you find this text helpful in your pursuit of the most challenging and rewarding facets of automotive service.

BASIC ENGINE CONSTRUCTION

An automotive engine, whether gasoline or diesel, is classified as an **internal combustion** engine because the burning—or **combustion**—takes place inside the engine. Burning the air-fuel mixture inside the engine produces large amounts of pressure, which powers the engine. The engine must be strong enough to withstand the high temperatures and pressures of combustion.

The **engine block** is the largest single piece of the engine and is usually cast from iron or aluminum (**Figure 1-1**). The block is bored with large holes called **cylinders**, which are fitted with **pistons**. The pistons are connected to the **connecting rods**, which are bolted to the **crankshaft**. The crankshaft and connecting rods turn the up-and-down motion of the pistons into the rotary motion of the crankshaft (**Figure 1-2**). The engine block and crankshaft also contain lubrication and cooling passages.

The **cylinder head** fits over the top of the engine block and seals the area over the piston (**Figure 1-3**). The cylinder head contains the valves and passages necessary to allow the air-fuel mixture into the cylinder and the exhaust out (**Figure 1-4**). The valves are operated by a **camshaft** that opens the **intake valves** and **exhaust valves** at the correct time during the rotation of the engine.

Four-Stroke Cycle

The four-stroke cycle is described as **intake**, **compression**, **power**, and **exhaust** (**Figure 1-5**). During the intake stroke, the intake valve is opened and the piston is moving down, pulling the air-fuel mixture into the cylinder. The intake valve closes, and the piston

> One of the easiest methods to gain horsepower from an engine without using any additional parts is making it "breathe" with as little restriction as possible in the intake and exhaust systems.

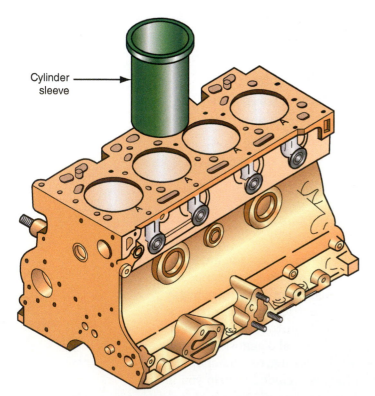

Figure 1-1 An aluminum engine block uses a steel cylinder liner for durability.

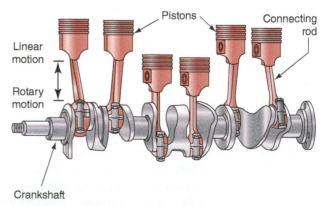

Figure 1-2 The reciprocating motion of the piston is converted to rotary motion by the crankshaft.

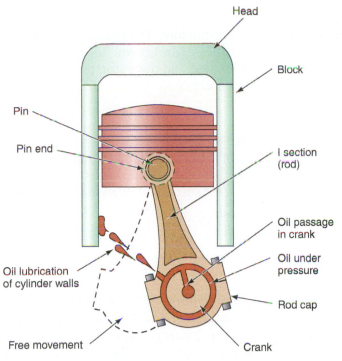

Figure 1-3 Piston, connecting rod, head, and block relationships.

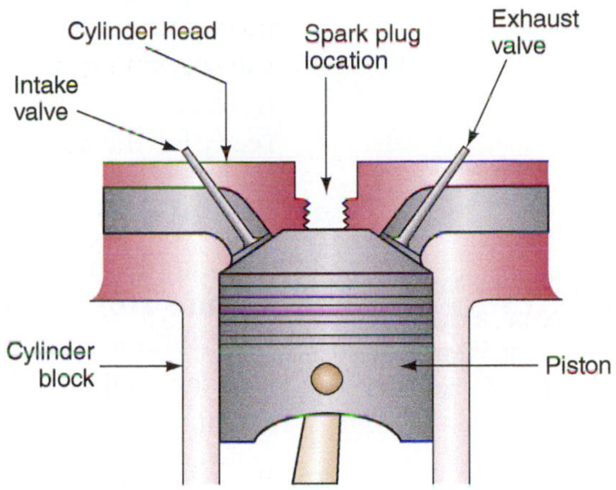

Figure 1-4 Placement of intake and exhaust valves.

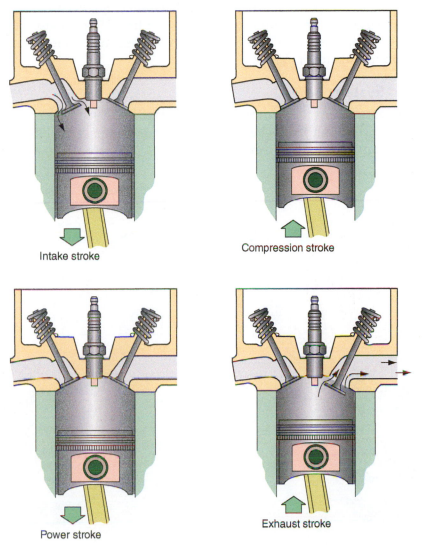

Figure 1-5 The four-stroke cycle.

moves upward, compressing and heating the air-fuel mixture. The mixture is ignited by the spark plug, and the rapidly expanding gases push the piston down. The piston goes to the bottom of the cylinder; then the exhaust valve opens and the piston moves back up, ready for the process to start over with another intake stroke.

Intake and Exhaust Manifolds

The intake manifold delivers the air-fuel mixture to the cylinder heads intake passage (**Figure 1-6**). There are many different designs of intake manifolds depending on the injection method, but all are designed to offer the least amount of restriction to airflow into the engine and exhaust out of the engine. The exhaust manifold carries the exhaust and its heat away from the engine (**Figure 1-7**). One of the easiest ways to gain power from an engine is to make it breathe easier through less restrictive exhaust and intake manifolds.

Engine Systems

Every engine also has to include the **cooling system** (**Figure 1-8**). The cooling system is responsible for carrying combustion heat away from the engine and maintaining a desired operating temperature as well as providing heat for the passenger compartment. A water pump circulates the coolant through the engine and radiator. The thermostat keeps the coolant at an operating temperature that is high enough for maximum cylinder efficiency. Engines are designed to operate at a temperature that promotes good combustion and low emissions, as well as ensuring unburned fuel and water from the combustion process are boiled off in the crankcase. On the other hand, an overheating engine can cause severe engine damage in a matter of minutes.

Lubrication System

Another critical system is the lubrication system (**Figure 1-9**). The oil pump provides pressurized oil throughout the engine via drilled passages in the head, block, and crankshaft. Lubrication helps prevent wear because engine parts are riding on a film of oil.

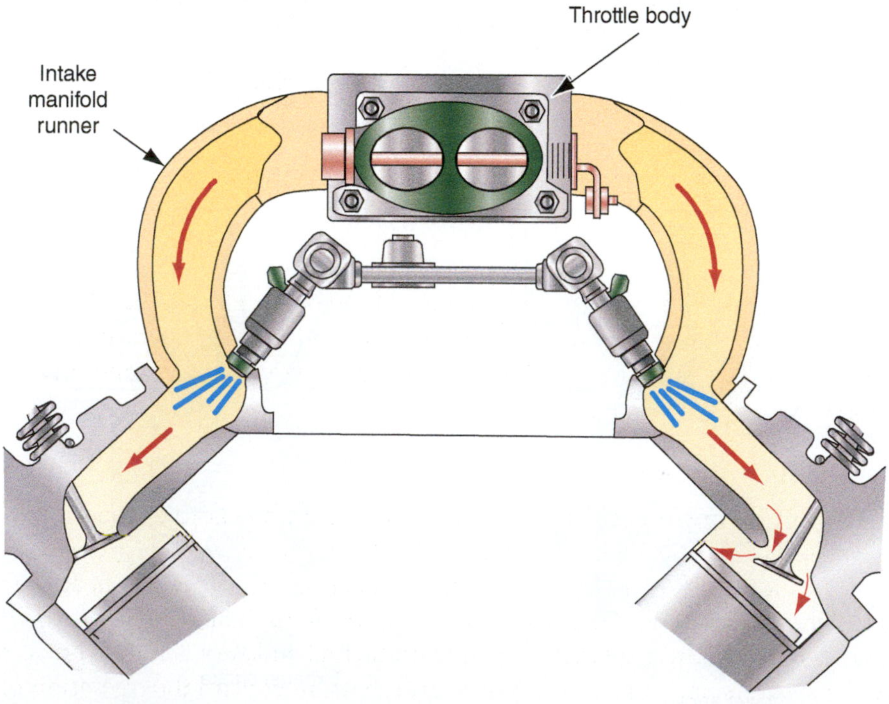

> During the combustion process, a significant amount of water is produced, mostly as steam.

> The cooling system carries excess heat from the engine by circulating coolant from the engine block through the radiator.

> The fuel system delivers filtered fuel from the tank to the injectors via an electric fuel pump, fuel lines, and a pressure regulator.

Figure 1-6 A fuel-injected engine's intake manifold.

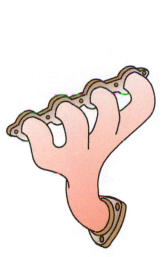

Figure 1-7 Exhaust manifold.

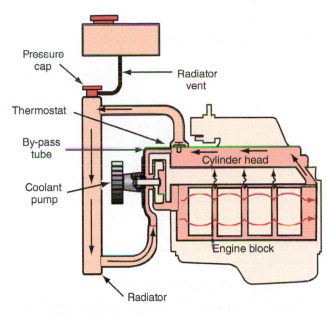

Figure 1-8 The engine's cooling system.

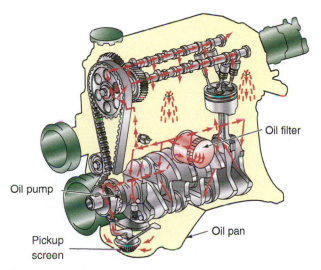

Figure 1-9 Pressurized oil is sent throughout the engine to prevent wear.

Obviously, if the metal parts of the engine touch, they wear rapidly and produce heavy damage to internal parts. The life of the engine depends on the operation of the cooling and lubrication systems. Make sure your customers realize the importance of maintaining these systems!

Fuel Systems

The **fuel system** (**Figure 1-10**) stores and delivers clean fuel at the proper pressure. The fuel is delivered to the fuel injector assembly. There are two principal types of fuel delivery systems. The older **fuel return line type** has a fuel pressure regulator near the injectors and a return line that sends unneeded fuel back to the fuel tank. In the **returnless fuel system**, the fuel pressure regulator is located in the fuel tank, and a return line is not used. Late model systems have a **fuel pump module** that can control the operation of the fuel pump according to the demand on the system.

Late model fuel systems do not use a return line, because the pressure regulator is located in the fuel tank, or pressure is regulated by the fuel pump driver module. Some older fuel injection systems used a return line with a pressure regulator mounted on the fuel rail.

The fuel pump module controls the speed and therefore the output pressure of the fuel pump according to the demand on the the fuel system. The module is controlled by the ECM/PCM.

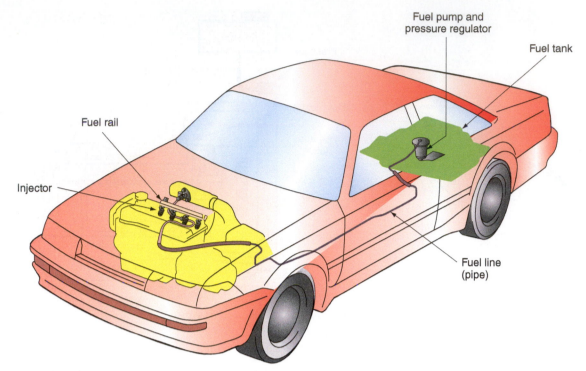

Figure 1-10 A typical fuel system.

Fuel pump and pressure regulator

Fuel tank

Fuel rail

Injector

Fuel line (pipe)

Pulse width of an injector refers to the length of time the injector is commanded on by the PCM.

There are two principal types of fuel injection in use today: port fuel injection and direct fuel injection. In port fuel injection (**Figure 1-11**), each cylinder has its own injector that sprays a fine atomized mist of fuel directly behind the intake valve. Under normal operating conditions, the computer tries to produce a mixture that is 14.7 parts air to 1 part fuel, but under some conditions such as a cold engine or acceleration, the mixture must be much richer. The air-fuel mixture is changed by the amount of time the fuel injectors are left on. The time the injectors are left on is called **pulse width**. The pulse width is longer when conditions require more fuel. The computer controls fuel delivery based on engine sensors such as coolant, intake air, manifold pressure, mass airflow, and throttle position sensors.

Direct injection is the process of injecting fuel directly into the combustion chamber instead of behind the intake valve or into the intake manifold.

Direct fuel injection has some major differences from port fuel injection (PFI), along with some similarities. The fuel is injected directly into the cylinder instead of behind the intake valve (**Figure 1-12**). The advantage of injecting fuel directly into the cylinder is the ability to deliver fuel without restriction from the intake valve and the ability to inject the fuel at any time during the intake cycle, nearly to top dead center (TDC). The fuel is injected at a much, much higher pressure than port fuel (over 2,000 psi). The high fuel pressure is necessary to prevent combustion pressures from going into the fuel system. Most of the direct injection engines also have a special piston that keeps the direct injected fuel near the spark plug so a very small amount of fuel (corresponding to a very lean mixture) can be ignited (**Figure 1-13**). The latent heat of evaporation helps cool the combustion chamber and allows for higher compression ratios without contributing to spark knock or oxides of nitrogen. Direct fuel injection provides more horsepower and torque, lower emissions, and better fuel economy. As different as these systems are, direct injection still uses most of the same sensors that port fuel injection does. Of course, we will learn more about both types of fuel injection later in the text.

The supercharger is belt driven from the engine, while exhaust gases leaving the engine drive the turbocharger.

Air Induction Systems

The fuel system needs to have a clean supply of fresh air. Remember, 14.7 times as much air as fuel is needed to run an internal combustion engine. Most engines have a system of ductwork to deliver cool air to the engine, and some engines have a mass air

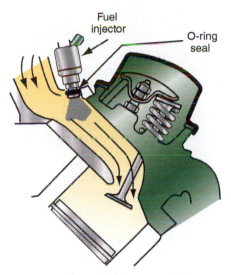

Figure 1-11 A port fuel injection system.

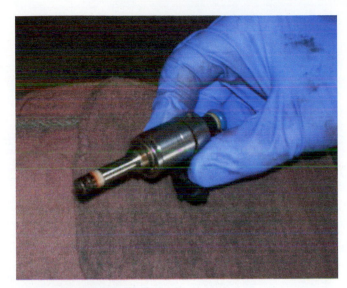

Figure 1-12 A direct fuel injector.

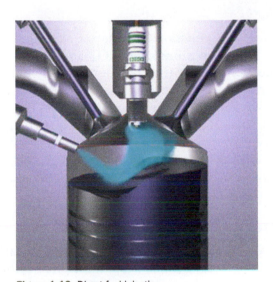

Figure 1-13 Direct fuel injection.

flow (MAF) sensor installed in the ducting to measure the amount of air entering the engine. The advent of more factory vehicles with **supercharging** and **turbocharging** (**Figure 1-14**) has renewed interest in this area of engine performance. Superchargers are belt driven from the engine, while turbochargers are driven from the force of the exhaust stream. By using large fuel injectors and high-pressure fuel pumps, getting enough fuel into the engine has not been a problem. Getting enough air into the engine to burn the fuel *is* a problem. That problem is overcome by forcing air into the intake manifold by use of a supercharger or turbocharger. Instead of a vacuum in the intake, the turbocharger or supercharger creates pressure inside the intake manifold. Getting high horsepower numbers from a small engine is possible with these induction systems. Of course, increased performance always comes with a cost, and you need to know the ins and outs before you recommend one of these vehicles to your customers. You will have to explain the need for high octane fuel and special oil change procedures to your customers.

Supercharging and turbocharging are both ways to increase engine power by forcing air into the combustion chamber.

Hydrocarbons are unburned fuel.

Carbon monoxide is partially burned fuel.

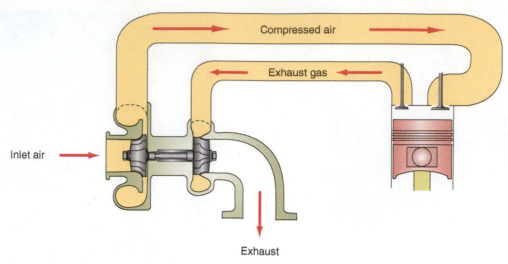

Figure 1-14 Turbocharging forces air into the combustion chamber.

EMISSION CONTROL SYSTEM

There are principally three emissions that have to be controlled on the automobile: **hydrocarbons (HC)**, **carbon monoxide (CO)**, and **oxides of nitrogen** (NO_x). The Environmental Protection Agency (EPA) has established guidelines for the emission of these gases. Great strides have been made since the 1960s in controlling these pollutants, but because more miles are being driven by more vehicles each year, there are still problems with these pollutants. HC is unburned fuel, CO is partially burned fuel, and NO_x is nitrogen that has been combined with oxygen in the combustion chamber. Nitrogen is generally inert, but can combine with oxygen due to high combustion temperatures and pressures. We will give you a quick overview of some of the most important systems that are used to control emissions (**Figure 1-15**).

Greenhouse Gases

Much concern over global warming has been raised over the last several years. One area that receives some of the most attention is the emission of carbon dioxide (CO_2). Motor vehicles with internal combustion engines emit CO_2 as a result of combustion. Vehicle emissions have been placed into categories based on their "footprint" and the levels of emissions and fuel mileage. Larger vehicles have larger footprints.

Positive Crankcase Ventilation

The **positive crankcase ventilation (PCV)** system controls "blow-by." Blow-by is the combustion gas that finds its way past the piston rings and into the crankcase. This would cause the crankcase to become pressurized, creating leaks and HC emissions. Additionally, the combustion gases contain unburned fuel and water that would eventually contaminate the oil. The PCV system removes the crankcase pressure and burned blow-by gases (**Figure 1-16**). The PCV is the earliest emission control system, which was first required on California cars in 1961.

Evaporative Emission System

As we have said before, hydrocarbon is an emission of unburned fuel from the exhaust. The evaporation of fuel is a hydrocarbon emission as well. Since the late 1960s, all fuel systems must be sealed with no vents directly to the atmosphere. The vapors are stored

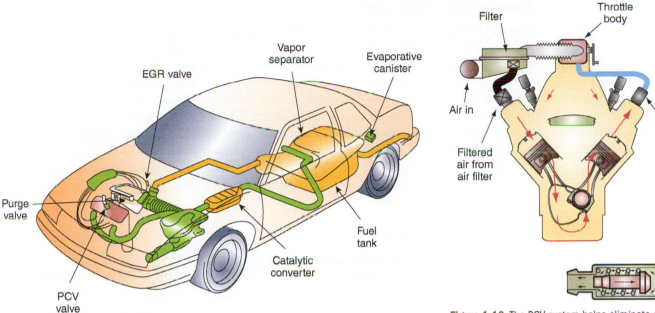

EGR valve

Vapor
separator

Evaporative
canister

Purge
valve

PCV
valve

Catalytic
converter

Fuel
tank

Figure 1-15 Overview of basic emission systems.

Filter

Throttle
body

Air in

Filtered
air from
air filter

PCV
valve

Figure 1-16 The PCV system helps eliminate crankcase pressure, along with water and unburned fuel.

in a charcoal canister until the engine can burn the stored vapors (**Figure 1-17**). The evaporative emission system on an OBD II vehicle must be able to monitor the system and detect a hole 0.020 inches in diameter since the 2001 model year.

Additionally, since 1998 the evaporative emission system has also had to store all the vapors produced while refueling with the advent of on-board refueling vapor recovery (ORVR).

The evaporation of
fuel is a hydrocarbon
emission.

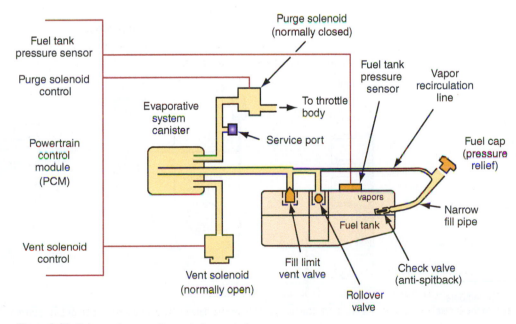

Fuel tank
pressure sensor

Purge solenoid
control

Powertrain
control
module
(PCM)

Vent solenoid
control

Purge solenoid
(normally closed)

Evaporative
system
canister

To throttle
body

Service port

Fuel tank
pressure
sensor

Vapor
recirculation
line

Fuel cap
(pressure
relief)

vapors

Fuel tank

Narrow
fill pipe

Fill limit
vent valve

Check valve
(anti-spitback)

Vent solenoid
(normally open)

Rollover
valve

Figure 1-17 Enhanced evaporative emission controls.

Exhaust Gas Recirculation

The exhaust gas recirculation (EGR) valve helps reduce oxides of nitrogen (NO_x) in the exhaust (**Figure 1-18**). Oxides of nitrogen are formed by high temperatures in the combustion chamber. The EGR valve meters a small amount of exhaust gas into the intake stream. Because this gas has already been burned, it is basically inert. The addition of the inert gas lowers the combustion chamber temperature. Fuel delivery is calculated by the powertrain control module (PCM) to include EGR. Without proper EGR flow, combustion temperatures are too high and can even cause engine damage. Many late model vehicles do not have EGR systems. These vehicles take care of EGR flow by the use of **variable valve timing (VVT)**. VVT can be used to manipulate valve opening and closing times so that exhaust gas is left in the intake manifold, which substitutes for EGR flow. More on these systems later.

> Variable valve timing uses a computer-controlled hydraulic actuator to provide optimal valve timing in most late model vehicles.

Air Injection System

The **air injection reaction (AIR)** system (**Figure 1-19**) adds pressurized air into the exhaust manifold. This burns some of the unburned and partially burned fuel in the exhaust manifold, which helps heat the catalytic converter and oxygen sensor, making them effective sooner. Control valves are used to keep air from the pump from entering the exhaust stream under some conditions (such as deceleration) to prevent backfire. Modern AIR pumps are electrically driven and operate only when the engine is cold. The AIR system is used on some vehicles for cold start emission control.

Catalytic Converter

> Oxidation is adding oxygen; reduction is removing oxygen.

The catalytic converter is designed to clean up vehicle exhaust by the oxidation and reduction of exhaust gases (**Figure 1-20**). A catalyst is anything that promotes a reaction with its presence but is not changed by the reaction. Problem exhaust gases are carbon monoxide, hydrocarbons, and oxides of nitrogen. Carbon monoxide and hydrocarbons are **oxidized** (burned) in the converter. Oxides of nitrogen are **reduced**, which means the nitrogen and oxygen are split apart.

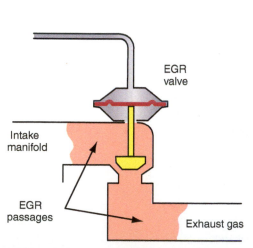

Figure 1-18 EGR cools down combustion temperatures by introducing some exhaust gas into the intake manifold.

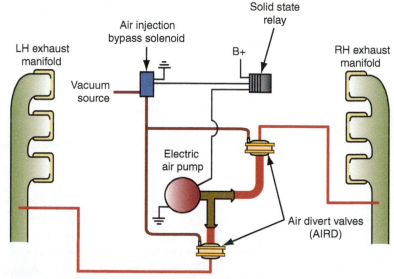

Figure 1-19 The air injection system helps speed up the operation of the catalytic converter when the engine is cold.

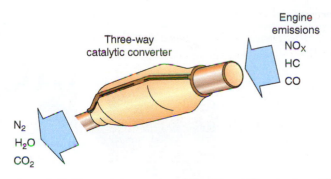

Figure 1-20 The catalytic converter oxidizes HC and CO and reduces NO_x.

Exhaust System

The exhaust system carries the toxic exhaust fumes away from the vehicle and makes the engine quiet. The catalytic converter is also part of the exhaust system. The exhaust system requires pipes, heat shields, mufflers, resonators, and hangers.

ELECTRICAL AND ELECTRONIC SYSTEMS

Automobiles have a variety of electrical systems and subsystems. Some of these systems include charging and starting, ignition, body control, powertrain control, and interior and exterior lighting, among others.

Ignition System

Of course, the gasoline engine cannot run without spark. The ignition systems of today look very different from the ignitions of early automobiles, but they still work basically the same. An ignition coil has two sets of windings: a primary and secondary (**Figure 1-21**). The primary winding is energized and is allowed to saturate. Then the primary is turned off. The magnetic field around the primary collapses, and high voltage is induced into the secondary, which is strong enough to jump the gap at the spark plug on compression. It does this on modern coil-over-plug (COP) ignition as well as on breaker contact points—equipped vehicles from 1960. The difference is that the system uses a transistor to switch the coil on and off instead of a set of mechanical points. Distributor ignition vehicles find the correct cylinder to fire because it is indexed mechanically when the rotor button points to the correct cylinder in the distributor (**Figure 1-22**). Coil-over-plug and distributorless ignition systems (**Figure 1-23**) find which cylinder needs to fire by using sensors and a computer. The computer's sensors allow very precise control of timing, resulting in increased fuel mileage and power. The computer knows or calculates the amount of air entering the engine, the gear the transmission is in, the engine load, the engine temperature, the temperature of the air entering the intake, and whether or not the engine is spark-knocking, among other details. You should start to get an idea of how the manufacturers managed to meet fuel economy standards and still have cars that are fun to drive.

Starting System

The purpose of the starting system is, of course, to start the vehicle. A small current from the ignition switch is used to close the contacts on the starter solenoid or relay, which sends full battery current to the starter motor (**Figure 1-24**). The starter motor turns the crankshaft, the ignition fires the plugs at the right time, and if successful in starting, the engine takes over on its own and the key is allowed to return to the run position. The starter has to be able to turn the engine at a high enough speed to allow the compression temperature to rise to a suitable level; otherwise, starting may not take place.

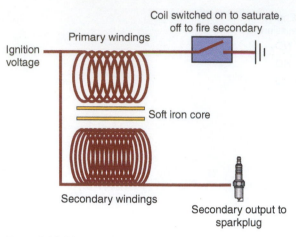

Figure 1-21 Diagram of coil internals showing primary and secondary windings.

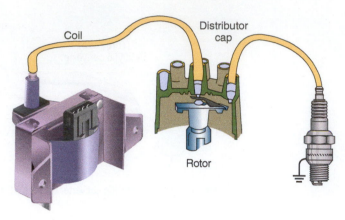

Figure 1-22 Typical distributor ignition.

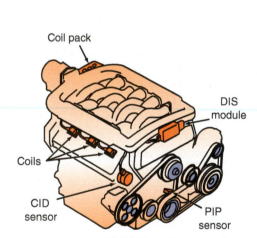

Figure 1-23 Ford's distributorless ignition components.

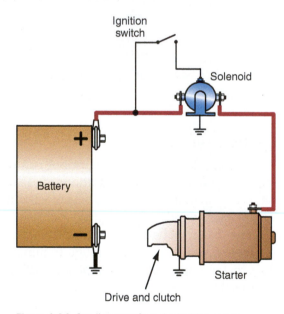

Figure 1-24 Small current from the ignition switch activates a solenoid, which connects battery power to the starter motor.

Imagine the problem of starting a cold engine on a winter day. Cold engines bleed off the needed heat, the battery is unable to work at full voltage because it is cold, and the oil is thick, making the engine harder to turn over.

Charging Systems

The charging system (**Figure 1-25**) exists to recharge the battery. Engine starting drains much of the battery power, and the charging system replenishes the battery for the next start. The charging system also powers accessories when the engine is running. The PCM has an input to the regulator over the vehicle network so the PCM can actually control the alternator current output. The PCM knows what the loads are on the alternator, as well as the ability to anticipate loads, such as when the air-conditioning is going to be turned on. The battery is there to help fill any temporary gaps the charging system may have, such as sitting still at a stoplight, air-conditioning or cooling fan

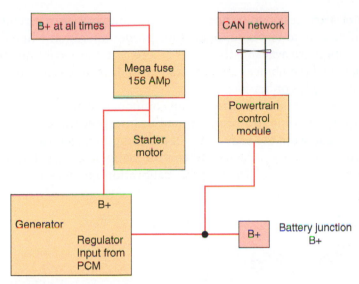

Figure 1-25 Modern charging system diagram.

turned on, and so on. If the alternator is charging the battery, it is said to be charging. If more current is flowing from the battery than from the alternator, it is said to be discharging.

Electronic Engine Controls

Most vehicles produced since the early 1980s have used some sort of electronic control system (**Figure 1-26**). Stringent gasoline mileage and emission laws required the accuracy of computers using a variety of sensors that have brought us to where we are today: good fuel economy and low emissions with good performance. Electronic systems have fewer moving parts to wear out or lose calibration than their mechanical counterparts. Fewer repairs are needed, and maintenance is reduced.

The first computer-controlled emission system has been termed **OBD** (onboard diagnostics) or **OBD I**. These systems were made from the late 1970s until the implementation of **OBD II** in 1996.

Modern vehicles have a network of several modules that can store many types of codes, such as those for the transmission, network, and body.

OBD stands for onboard diagnostics.

Inputs from sensors	Computer processing	Outputs using actuators
Mass airflow (amount of air entering the engine)	Computer takes inputs and processes them according to its programming and then determines the appropriate outputs. The computer also compensates fuel delivery for such conditions as a cold engine, wide-open throttle, or idle.	Fuel injectors
Crank sensor (engine speed)		Timing control
Cam sensor (camshaft position)		Exhaust gas recirculation (EGR)
Throttle position sensor		Idle control
Coolant temperature sensor		
Intake air temperature		
Oxygen sensor/air-fuel ratio		
Knock (spark knock) sensor		

Figure 1-26 Computer-controlled systems receive inputs from sensors and perform outputs based on their programming.

OBD II vehicles are designed to not only catch faults in sensor circuits and output devices, but also be able to test components for proper operation while driving. More codes and information are available than ever before, and these continue to grow every year. The technician who embraces and understands this technology can use this wealth of information to make his or her job easier.

Electronic control systems have three main components: input sensors to determine operating conditions, a computer programmed to interpret these results, and output devices designed to carry out the commands of the computer. Electronic control systems can also be reprogrammed as necessary to correct many problems that may arise after the vehicle has been produced. In the past, calibration changes meant changing or modifying parts.

SUMMARY

- The engine block contains the crankshaft, connecting rods, cylinders, and pistons.
- A gasoline or diesel engine is termed an *internal combustion engine.*
- The cylinder head fits over and seals the top of the cylinder.
- The four-stroke cycle consists of intake, compression, power, and exhaust.
- The intake system carries the air-fuel mixture in, and the exhaust system takes the burned fuel out.
- The AIR system is designed to reduce cold start emissions by burning hydrocarbons in the exhaust and heating the catalytic converter for quicker operation.
- The PCV system relieves crankcase pressure and removes unburned fuel and water from the crankcase.
- The EGR valve lowers combustion temperatures and thereby reduces NO_x production.
- The catalytic converter helps reduce emissions by oxidation and reduction reactions, which reduce HC, CO, and NO_x emission.

- The EVAP system prevents the evaporation of fuel and the resulting hydrocarbon emission.
- The ignition system is designed to deliver spark to the proper spark plug at the right time to start combustion.
- The cooling system is responsible for cooling the engine and heating the passenger compartment in winter.
- The fuel system must provide enough clean fuel at the right pressure.
- Supercharging and turbocharging raise engine horsepower by forcing air into the engine.
- OBD I vehicles were the first computer-controlled emission systems and were used from the late 1970s until 1996.
- OBD II systems are designed to monitor and test components while driving.
- Gasoline fuel injection methods include port fuel and direct.

REVIEW QUESTIONS

Short Answer Essays

1. Explain why the engine performance specialist is probably one of the most skilled technicians in the shop.

2. Explain why diesel and gasoline engines are classified as internal combustion engines.

3. Explain the purpose of the cylinder head.

4. Explain the four-stroke cycle.

5. Explain how a turbo- or supercharger raises engine power.

6. Explain why a PCV system is needed.

7. Explain how the EGR valve lowers combustion temperatures.

8. Explain the basic operation of the catalytic converter.

9. Explain what type of emission is controlled by the evaporative emissions and how it is accomplished.

10. Explain how the lubrication system prevents wear.

Fill-in-the-Blanks

1. The evaporation of fuel is a _____ emission.

2. _____ times as much air as fuel is needed to run an internal combustion engine.

3. An ignition coil has two sets of windings: a _____ and a _____.

4. One of the easiest ways to gain power from an engine is to make it _____ easier through less restrictive exhaust and intake manifolds.

5. A _____ _____ circulates coolant through the engine and radiator.

6. Lubrication helps prevent wear because engine parts are riding on a _____ of _____.

7. Two principal types of fuel injection in use today are _____ _____ injection and _____ _____ injection.

8. The air-fuel mixture is changed by the _____ of _____ the fuel injectors are on.

9. The _____ is the earliest emission control.

10. The _____ valve meters a small amount of exhaust gas into the intake stream.

Multiple Choice

1. The EGR valve is designed to reduce:
   A. Oxides of oxygen
   B. Carbon monoxide
   C. Carbon dioxide
   D. Oxides of nitrogen

2. *Technician A* says that the catalytic converter oxidizes hydrocarbons (HC).
   Technician B says that the catalytic converter oxidizes CO (carbon monoxide).
   Who is correct?
   A. Technician A
   B. Technician B
   C. Both technicians
   D. Neither technician

3. *Technician A* says that the evaporative emission system (EVAP) must detect a leak from a hole 0.020 inches in diameter beginning in the 2001 model year.
   Technician B says that the ORVR portion of the evaporative emission system is designed to capture fuel vapors produced during refueling.
   Who is correct?
   A. Technician A
   B. Technician B
   C. Both technicians
   D. Neither technician

4. All of the following about an alternator are true *except*:
   A. Recharge the battery after starting the engine.
   B. Sense electrical load with the diodes.
   C. PCM can control alternator output.
   D. Power accessories when the engine is running.

5. *Technician A* says that overheating can damage an engine in a matter of minutes.
   Technician B says engines are designed to operate at a temperature that promotes good combustion and low emissions, as well as ensuring unburned fuel and water from the combustion process are boiled off in the crankcase.
   Who is correct?
   A. Technician A
   B. Technician B
   C. Both technicians
   D. Neither technician

6. All of the following are true about supercharging *except*:
   A. Superchargers are driven by the exhaust flow.
   B. Superchargers require a belt.
   C. Superchargers create pressure inside the intake manifold.
   D. Supercharged engines require high octane fuel.

7. The proper sequence for the four-cycle engine is:
   A. Intake, exhaust, power, compression.
   B. Compression, power, intake, exhaust.
   C. Power, intake, compression, exhaust.
   D. Intake, compression, power, exhaust.

8. *Technician A* says that the PCV system was introduced in 1998.

   Technician B says the PCV system helps eliminate blow-by and oil contamination by unburned fuel. Who is correct?

   A. Technician A   C. Both technicians

   B. Technician B   D. Neither technician

9. Regarding **Figure 1-27**, *Technician A* says the solenoid is used so a small amount of current through the ignition switch can be used to control the much larger starter current.

   Technician B says current to start the vehicle travels directly through the ignition switch and then to the starter. Who is correct?

   A. Technician A   C. Both technicians

   B. Technician B   D. Neither technician

10. *Technician A* says that direct injection is an earlier version of port fuel injection.

    Technician B says that direct injection systems inject fuel directly into the intake manifold. Who is correct?

    A. Technician A   C. Both technicians

    B. Technician B   D. Neither technician

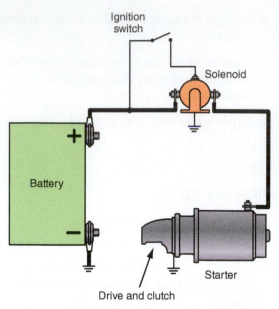

Figure 1-27 Starter circuit.

CHAPTER 2
BASIC THEORIES

Upon completion and review of this chapter, you should be able to:

- Define atoms and elements.
- Define compounds and molecules.
- Describe the parts of an atom.
- Explain Newton's laws of motion.
- Define work and force.
- List the most common types of energy and energy conversions.
- Define inertia and momentum.
- Define aerodynamics.
- Explain friction.
- Define mass, weight, and volume.
- Define power.
- Explain the compressibility of gases and the non-compressibility of liquids.
- Describe atmospheric pressure and vacuum.
- Explain venturi operation.
- Explain what happens during combustion.
- Describe acids and bases.
- Explain basic gear theory.

Terms To Know

Acid	Energy	Positive pressure
Atmospheric pressure	Friction	Power
Atom	Gear reduction	Pressure
Bar	Inertia	Proton
Base	Kilopascal (kPa)	Torque
Coefficient of drag (Cd)	Mass	Vacuum
Combustion	Molecule	Valence ring
Compound	Momentum	Venturi
Compressible	Negative pressure	Volume
Direct drive	Neutron	Weight
Electron	Non-compressible	Work
Element	Overdrive	

INTRODUCTION

An understanding of the basics is absolutely essential before attempting a study of complex systems and components. Basic engine theories such as force, work, torque, and power must be understood prior to a study of engine components, systems, and diagnosis in this book. To truly understand how an engine works, you must have a basic understanding of these principles of chemistry and physics. Other important concepts that you must understand are those related to electricity and electronics. These topics are covered in much detail in Chapter 4 of this textbook. If you have previously studied engine theory and physics, the information in this chapter may be used as a review. A thorough study of this chapter will provide all the necessary background information before you study the information in this book.

THE BASICS

Atoms and Elements

An **atom** may be defined as the smallest particle of an element in which all the chemical characteristics of the element are present. The atom has a nucleus, comprised of protons and neutrons. Electrons orbit the nucleus.

An **element** is defined as a liquid, solid, or gas that contains only one type of atom. For example, copper contains only copper atoms.

Compounds and Molecules

A **compound** is a liquid, solid, or gas that contains two or more types of atoms. Water is a compound that contains hydrogen and oxygen. A **molecule** is the smallest particle of a compound in which all the chemical characteristics of the compound are present.

Electrons, Protons, and Neutrons

A **proton** is a small, positively charged particle located at the center, or nucleus, of each atom.

An **electron** is a small, very light particle with a negative electrical charge. Electrons move in orbits around the nucleus of an atom.

A **neutron** does not have an electrical charge. These particles add weight to an atom. Neutrons are positioned in the nucleus of an atom.

Acids and Bases

An **acid** is a compound that breaks into hydrogen ions (+H) and another compound in a water solution. An ion is an atom or group of atoms with a positive or negative charge. Ions determine whether a compound is an acid or base. For example, H_2SO_4 is sulfuric acid, and in water it would break into +H and $-SO_4$ ions. Acids have a sour taste. For example, a lemon contains citric acid. Acids attack some metals and produce hydrogen gas. Acids turn litmus paper from blue to red. Acids also react with alkalis (bases) to form salts. For example, NaOH (or sodium hydroxide, a powerful base) could combine with HCl (or hydrochloric acid) to form table salt and water. A **base** (alkali) is a compound that produces hydroxide (−OH) and another compound in water. Bases feel slippery to the touch and turn litmus paper from red to blue. Soaps are usually weak bases.

Electron Movement

The outer ring on an atom is called a **valence ring**. The number of electrons on the valence ring determines the electrical characteristics of the element.

NEWTON'S LAWS OF MOTION

First Law

A body in motion remains in motion, and a body at rest remains at rest unless some outside force acts on it. When a car is parked on a level street, it remains stationary unless it is driven or pushed. If the gas pedal is depressed with the engine running and the transmission in drive, the engine delivers power to the drive wheels, and this force moves the car.

An **atom** is the smallest particle of an element that still has the properties of the element.

An **element** is a substance that contains only one type of atom.

A **compound** is made up of two or more types of atoms.

A **molecule** is the smallest part of a compound.

An **electron** is a small negatively charged particle that orbits an atom.

A **neutron** makes up part of the nucleus of the atom and does not have a charge.

Acids combine with **bases** to form salts.

 Caution

Always exercise extreme caution with strong bases and acids.

The **valence ring** determines the electrical and chemical properties of the atom.

Second Law

A body's acceleration is directly proportional to the force applied to it, and the body moves in a straight line away from the force. For example, if the engine power supplied to the drive wheels increases, the vehicle accelerates faster.

Third Law

For every action there is an equal and opposite reaction. A practical application of this law occurs when the wheel on a vehicle strikes a bump in the road surface. This action drives the wheel and suspension upward with a certain force, and a specific amount of energy is stored in the spring. After this action occurs, the spring forces the wheel and suspension downward with a force equal to the initial upward force.

WORK AND FORCE

When a force moves a certain mass a specific distance, **work** is produced. When work is accomplished, the mass may be lifted or slid on a surface (**Figure 2-1**). Since force is measured in pounds and distance is measured in feet, the measurement for work is foot-pounds (ft.-lb.). In the metric system, work is measured in Newton meters (N·m). If a force moves a 3,000-pound vehicle for 50 feet, 150,000 foot-pounds of work are produced. Mechanical force acts on an object to start, stop, or change the direction of the object. It is possible to apply force to an object and not move the object. Under this condition, no work is done. Work is only accomplished when an object is moved.

Work is defined as the result of applying a force.

 A BIT OF HISTORY

Around 1665, at the age of 23, Newton clarified the principles of mechanics, formulated the law of universal gravitation, separated white light into colors, and proposed a theory for the propagation of light, and invented differential and integral calculus. These principles help define the laws of physics and mechanical dynamics that are employed in almost every vehicle system today.

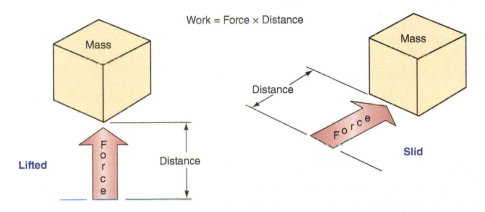

Work = Force × Distance

Figure 2-1 Work is accomplished when a mass is lifted or slid on a surface.

ENERGY

Energy may be defined as the ability to do work.

Energy may be defined as the ability to do work. When energy is released to do work, it is called kinetic energy. This type of energy may also be referred to as energy in motion. Stored energy may be called potential energy. Energy is available in one of these seven forms:

1. Chemical energy is contained in the molecules of atoms. In the automobile, chemical energy is contained in the molecules of gasoline as well as in the molecules of electrolyte in a battery.
2. Electrical energy is required to move electrons through an electric circuit. In the automobile, the battery is capable of producing electrical energy to start the vehicle, and the alternator produces electrical energy to power the electrical accessories and recharge the battery.
3. Mechanical energy is defined as the ability to move objects. In the automobile, the battery supplies electrical energy to the starting motor, and this motor converts the electrical energy to mechanical energy to crank the engine.
4. Thermal energy may be defined as energy produced by heat. When gasoline burns, thermal energy is released.
5. Radiant energy is defined as light energy. In the automobile, radiant energy is produced by the lights.
6. Nuclear energy is defined as the energy within atoms when they are split apart or combined. Nuclear power plants generate electricity with this principle. This type of energy is not used in the automobile.
7. Solar energy is defined as a light-source energy that can be converted to heat or electricity. The photovoltaic effect is applied to convert light energy directly into electrical energy. Solar cells, consisting largely of semiconductor materials, are the basic elements used for conversion.

ENERGY CONVERSION

Energy cannot really be destroyed or eliminated. It can only be changed or converted from one state, or form, to another. Energy conversion occurs when one form of energy is changed to another form. Since energy is not always in the desired form, it must be converted to a form we can use. The following are some of the most common automotive energy conversions.

Chemical to Thermal Energy Conversion

Chemical energy in gasoline or diesel fuel is converted to thermal energy when the fuel burns in the engine cylinders.

Thermal to Mechanical Energy Conversion

Mechanical energy is required to rotate the drive wheels and move the vehicle. The piston and crankshaft in the engine and the drivetrain are designed to convert the thermal energy produced by the burning fuel into mechanical energy (**Figure 2-2**).

Electrical to Mechanical Energy Conversion

The windshield wiper motor converts electrical energy from the battery or alternator to mechanical energy to drive the windshield wipers.

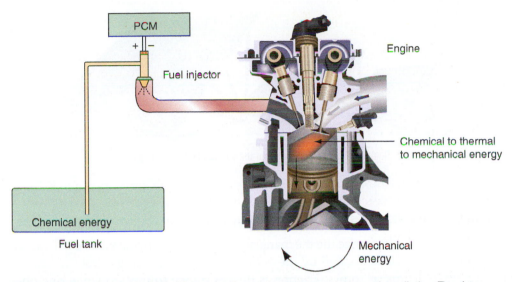

PCM

+ ‖ −

Fuel injector

Engine

Chemical to thermal
to mechanical energy

Chemical energy

Fuel tank

Mechanical
energy

Figure 2-2 Thermal energy in the fuel is converted to mechanical energy in the engine cylinders. The piston, crankshaft, and drivetrain deliver this mechanical energy to the drive wheels.

Mechanical to Electrical Energy Conversion

A windmill is an example of converting mechanical energy to electrical energy (**Figure 2-3**). A great example on the vehicle is the alternator. The alternator is driven by mechanical energy from the engine. The alternator converts this energy to electrical energy, which powers the electrical accessories on the vehicle.

Figure 2-3 Windmills convert mechanical energy to electrical energy.

Figure 2-4 Solar cells convert energy from the sun to electrical energy.

Mechanical to Thermal Energy Conversion

The brakes of a vehicle change the mechanical energy of the moving vehicle to heat energy, in the form of **friction**, created by the action of the brake linings rubbing against the rotating brake drums or rotors. Friction is the resistance to motion when one object is moved over another object.

> **Friction** is the resistance to motion when one object is moved over another object. Friction generates heat.

Solar to Electrical Energy Conversion

Solar energy can be converted to electrical energy using special semiconductors. One type of solar cell is a photoelectric cell or sensor (**Figure 2-4**). This converts the light signal into a voltage signal. It can be found on vehicles that have automatic headlamps that turn on when the sensor detects a specified amount of ambient light. When the voltage level reaches a specified point proportionate to the ambient light level, the headlamps will go on.

INERTIA

> **Inertia** is defined as the tendency of an object at rest to remain at rest, or the tendency of an object in motion to stay in motion.
>
> When a force overcomes static inertia and moves an object, the object gains **momentum**.

The **inertia** of an object at rest is called static inertia, whereas dynamic inertia refers to the inertia of an object in motion. Inertia exists in liquids, solids, and gases. When you push and move a parked vehicle, you overcome the static inertia of the vehicle. If you catch a ball in motion, you overcome the dynamic inertia of the ball.

MOMENTUM

Momentum is the product of an object's weight times its speed. Momentum is a type of mechanical energy. An object loses momentum if another force overcomes the dynamic inertia of the moving object.

FRICTION

Friction may occur in solids, liquids, and gases. When a car is driven down the road, friction occurs between the air and the car's surface. This friction opposes the momentum, or mechanical energy, of the moving vehicle. Since friction creates heat, some of the mechanical energy from the vehicle's momentum is changed to heat energy in the air and body components. The mechanical energy from the engine must overcome the vehicle inertia and the friction of the air striking the vehicle. Body design has a very dramatic effect on the amount of friction developed by the air striking the vehicle. The total resistance to motion caused by friction between a moving vehicle and the air is referred to as **coefficient of drag (Cd)**.

> **Coefficient of drag (Cd)** may also be called aerodynamic drag.

The study of Cd is very complicated and also very important. At 45 miles per hour (mph), 72 kilometers per hour (kph), half of the engine's mechanical energy is used to overcome air friction, or resistance. Therefore, reducing a vehicle's Cd can be a very effective method of improving fuel economy.

AERODYNAMICS

Engineers are always looking for ways to make the automobile more efficient. Aerodynamics is the study of how objects like automobiles move through the air (**Figure 2-5**). Fuel mileage as well as handling is adversely affected by boxy shapes and flat surfaces facing forward. Ideally, air will follow the contours of the vehicle as it moves down the road. Early in the design of a new vehicle, a model is tested in a wind tunnel to assess the way it is going to slice through the air. Even the bottom of the vehicle must be taken into account because if the air gets an opportunity to lift up under the vehicle, it can make handling difficult and dangerous, as well as increasing drag. Air dams and spoilers are used to redirect air around the vehicle instead of beneath.

MASS, WEIGHT, AND VOLUME

A lawn mower is much easier to push than a 2,500-pound vehicle because the lawn mower has very little inertia compared to the vehicle. A spaceship might weigh 100 tons here on earth where it is affected by the earth's gravitational pull, whereas in space with no gravitational pull the spaceship is weightless, but the mass remains the same. **Mass** is the measurement of an object's inertia. In outer space beyond the earth's gravity and atmosphere, the spaceship is almost weightless. **Weight** is the measurement of the earth's gravitational pull on the object. Here on earth, mass and weight are measured in pounds and ounces in the English system. In the metric system, mass and weight are measured in grams and kilograms.

   Volume is a measurement of size, and it is related to mass and weight. For example, a pound of gold and a pound of feathers both have the same weight, but the pound of feathers occupies a much larger volume. In the English system, volume is measured in cubic inches, cubic feet, cubic yards, or gallons. The measurement for volume in the metric system is cubic centimeters or liters.

> **Mass** is the measurement of an object's inertia.
>
> **Weight** is the measurement of the earth's gravitational pull on the object.
>
> **Volume** is the length, width, and height of a space occupied by an object.

TORQUE

Torque is a force that does work with a twisting, or turning force. When you pull a wrench to tighten a bolt, you supply torque to the bolt. This torque, or twisting force, is calculated by multiplying the force and the radius. For example, if you supply 10 pounds of force on

> **Torque** is a force that does work with a twisting, or turning, force.

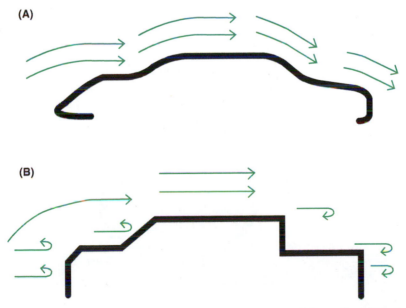

(A)

(B)

Figure 2-5 (A) Vehicle with low aerodynamic drag. (B) Vehicle with high aerodynamic drag.

the end of a 2-foot wrench to tighten a bolt, the torque is 10 times 2 = 20 foot-pounds (ft.-lb.) (**Figure 2-6**). The English measurement of torque is ft.-lb.; the metric measurement is the Newton-meter (N-m). If the bolt turns during torque application, work is done. When a bolt does not rotate during torque application, no work is accomplished.

GEAR THEORY

Referring to **Figure 2-7**, you can see that the driving gear is turning the driven gear. The radius of the driving gear is 1 foot and is turned with a force of 25 pounds. This means that the driving gear has a torque of 25 foot-pounds. If this torque is applied to a gear with a radius of 2 feet, the final output of the driven gear is 50 foot-pounds. This means that the torque at the driven gear has been doubled. Because the driven gear is twice as large as the driving gear, the ratio between the two gears would be referred to as 2:1. This relationship is also called a **gear reduction** because the speed of the driven gear will be less than that of the driving gear. The work that is performed is calculated by multiplying force times distance.

If the two gears were the same size (**Figure 2-8**), there would be no torque or speed change. This would be a 1:1 ratio, or sometimes called **direct drive**.

Finally, if the driving gear were larger than the driven gear, there would be a speed increase and a torque decrease (**Figure 2-9**). This ratio is expressed as a ratio of less than 1:1, such as a typical automotive **overdrive** ratio of around 0.8:1. Overdrive has little torque, so a vehicle in an overdrive gear range is likely to downshift quickly when climbing

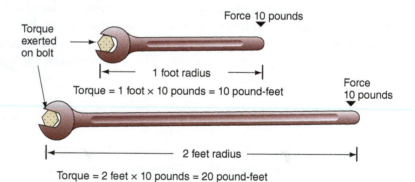

Figure 2-6 Force applied to a wrench produces torque. If the bolt turns, work is accomplished.

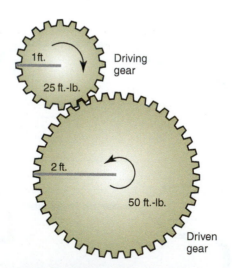

Figure 2-7 The driven gear will turn at half the speed but twice the torque because it is two times larger than the driving gear.

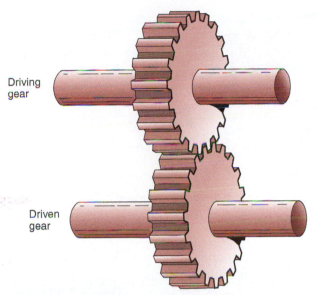

Driving gear

Driven gear

Figure 2-8 If both gears are the same size, the gear ratio is expressed as 1:1.

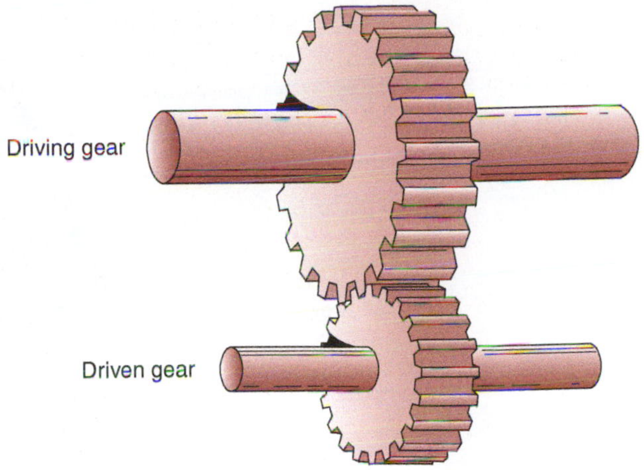

Driving gear

Driven gear

Figure 2-9 In an overdrive ratio, the driving gear is larger than the driven gear.

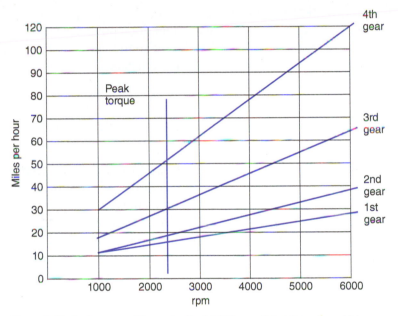

Figure 2-10 Peak torque of the engine is at 2,200 rpm. Using gears, the vehicle speed at peak torque can be changed to increase speed.

a hill or accelerating. Of course, the power through the transmission is still sent through a final drive or differential, which is generally a gear reduction of about 3:1. This means that there is not a true overdrive at the wheels, but the broader range of modern transmissions gives designers the opportunity to use a large gear reduction for good acceleration from a stop when more torque is needed and still be able to produce speed and good fuel mileage for the highway when large amounts of torque are not necessary. **Figure 2-10** shows the relationship between gear selection torque and speed.

> **Work** is calculated by multiplying force times distance.

POWER

James Watt, a Scotsman, is credited with being the first person to calculate **power**. He measured the amount of work that a horse could do in a specific amount of time. Watt calculated that a horse could move 330 pounds for 100 feet in 1 minute. If you multiply

> **Power** is a measurement for the rate, or speed, at which work is done.

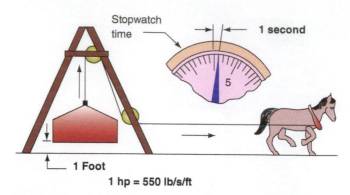

Figure 2-11 One horsepower is produced when 550 pounds are moved 1 foot in 1 second.

330 pounds by 100 feet, the answer is 33,000 foot-pounds of work. Watt determined that one horse could do 33,000 foot-pounds of work in 1 minute. Thus, 1 horsepower is equal to 33,000 foot-pounds per minute, or 550 foot-pounds per second (**Figure 2-11**). Two horsepower could do this same amount of work in one-half minute, or 4 horsepower would be capable of completing this work in one-quarter minute. If you push a 3,000 pound, 1,360 kilograms (kg), car for 11 feet (3.3 meters) in one quarter minute, you produce 4 horsepower. From this brief discussion about horsepower, we can understand that as power increases, the rate at which work is done also increases, or the time to do work decreases. Power is the measurement for the rate, or speed at which work is done.

PRINCIPLES INVOLVING LIQUIDS AND GASES

Molecular Energy

Molecular energy may be defined as the kinetic energy available in atoms and molecules because of the constant electron movement within these molecules.

Kinetic energy is energy released to do work.

Remember that kinetic energy refers to energy in motion. Given that electrons are constantly in motion around the nucleus in atoms or molecules, kinetic energy is present in all matter. Kinetic energy in atoms and molecules increases as the temperature increases. A decrease in temperature reduces this kinetic energy. Molecules in solids move slowly compared to those in liquids or gases. Gas molecules move quickly compared to liquid molecules. Because gas molecules are in constant motion, they spread out to fill all the space available. At higher temperatures, gas molecules spread out more, whereas lower temperatures cause gas molecules to move closer together. When pressure increases on a gas, so does temperature, and when temperature increases, so does pressure.

Temperature

Temperature affects all liquids, solids, and gases. The volume of any matter increases as the temperature increases. Conversely, the volume decreases in relation to a reduction in temperature. When the gases in an engine cylinder are burned, the sudden temperature increase causes rapid gas expansion, which pushes the piston downward and causes engine rotation (**Figure 2-12**).

Pressure and Compressibility

Pressure and temperature are directly related. If you increase one, you also increase the other.

lopascal is the **rement of** **the metric**

Because liquids and gases are both substances that flow, they may be classified as fluids. If a nail punctures an automotive tire, the air escapes until the pressure in the tire is equal to atmospheric pressure outside the tire. When the tire is repaired and inflated, air pressure is forced into the tire. If the tire is inflated to 32 pounds per square inch (psi), or 220 **kilopascals (kPa)**, this pressure is applied to every square inch on the inner tire surface.

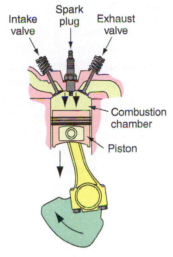

Figure 2-12 Hot expanding gases push the piston downward and rotate the crankshaft.

Pressure is always supplied equally to the entire surface of a container. Since air is a gas, the molecules have plenty of space between them. When the tire is inflated, the pressure in the tire increases, and the air molecules are squeezed closer together, or compressed. Under this condition, the air molecules cannot move as freely, but extra molecules of air can still be forced into the tire. Therefore, gases such as air are said to be **compressible**.

The air in the tire may be compared to a few balls on a billiard table without pockets. If a few more balls are placed on the table, the balls are closer together, but they can still move freely (**Figure 2-13**).

If the vehicle is driven at high speed, friction between the road surface and the tires heats the tires and the air in the tires. When air temperature increases, the pressure in the tire also increases. Conversely, a temperature decrease reduces pressure.

If 100 cubic feet (2.8 cubic meters) of air is forced into a large truck tire and the same amount of air is forced into a much smaller car tire, the pressure in the car tire is much greater.

Molecules in a liquid may be compared to a billiard table without pockets that is completely filled with balls. These balls can roll around, but no additional balls can be placed on the table because the balls cannot be compressed. Similarly, liquid molecules cannot be compressed (**Figure 2-14**).

Pressure may be defined as a force exerted on a given surface area.

Pressure and volume are inversely proportional. If volume is decreased, pressure is increased.

Compressible relates to the increased pressure of air or gas molecules as a result of being squeezed together.

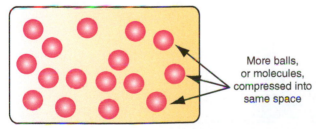

Balls are still free to move and bounce but with less space and more activity.

More balls, or molecules, compressed into same space

Figure 2-13 Gases can be compressed because there is empty space between the molecules.

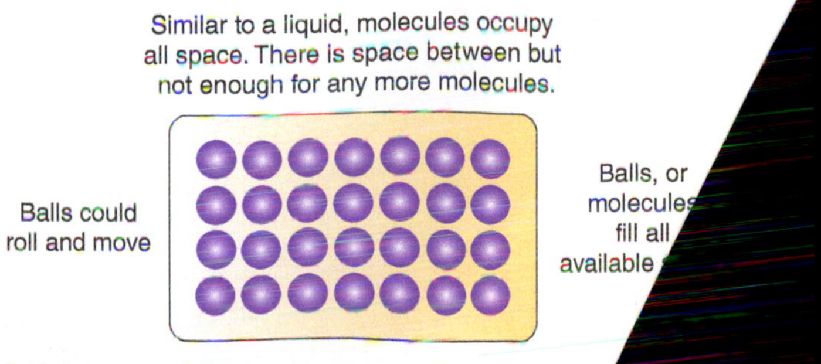

Similar to a liquid, molecules occupy all space. There is space between but not enough for any more molecules.

Balls could roll and move

Balls, or molecules fill all available

Figure 2-14 Liquids are non-compressible because there i space between molecules.

Liquids are not compressible.

Gases can be compressed; but liquids are **non-compressible**.

In nature, a high pressure always moves to a lower pressure.

Liquid Flow

If a tube is filled with billiard balls and the outlet is open, more balls may be added to the inlet. When each ball is moved into the inlet, a ball is forced from the outlet. If the outlet is closed, no more balls can be forced into the inlet (**Figure 2-15**).

The billiard balls in the tube may be compared to molecules of power steering fluid in the line between the power steering pump and steering gear. Because **non-compressible** fluid fills the line and gear chamber, the force developed by the pump pressure is transferred through the line to the gear chamber (**Figure 2-16**).

This pressure is applied equally to every square inch in the gear chamber. This pressure applied to the rack piston in the gear chamber helps move the rack piston. Since the rack is connected through steering linkages and arms to the front wheels, the force on the rack piston helps the driver move the front wheels to the left or right during a turn (**Figure 2-17**).

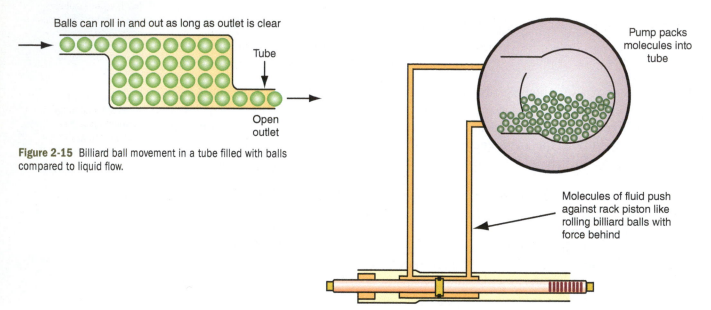

Balls can roll in and out as long as outlet is clear
Tube
Open outlet

Figure 2-15 Billiard ball movement in a tube filled with balls compared to liquid flow.

Pump packs molecules into tube
Molecules of fluid push against rack piston like rolling billiard balls with force behind

Figure 2-16 Power steering pump pressure supplied to the steering gear chamber.

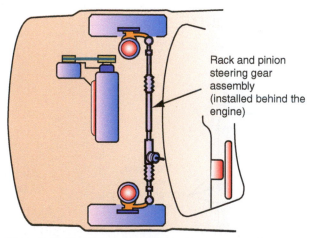

Rack and pinion steering gear assembly (installed behind the engine)

Figure 2-17 Steering gear, linkages, and arms connected to the front wheels.

ATMOSPHERIC PRESSURE

Because air is gaseous matter with mass and weight, it exerts pressure on the earth's surface. A 1-square-inch column of air extending from the earth's surface to the outer edge of the atmosphere weighs 14.7 pounds. Therefore, **atmospheric pressure** is 14.7 psi at sea level (**Figure 2-18**). The atmospheric pressure at sea level is 29.92 in. Hg (inches of mercury). Atmospheric pressure measured in the metric system at sea level is 100 kPa. You might also see a measurement called *bar*. A bar equals 100 kPa or 14.5 psi. Bar can also be used in terms of pressure measurements for oil and fuel. The relationship of the change of pressure and altitude is inversely proportionate. The lapse rate is 1 in. Hg per 1,000 feet increase in altitude. Water vapor is about 68 percent as dense as air; therefore, air that is saturated with moisture has seemingly less oxygen compared to the same volume of air with less moisture. Hot, dry air is less dense than cooler air and therefore has more space between the molecules. Cooler air is denser, with less space between the oxygen molecules.

Atmospheric Pressure and Temperature

When air becomes hotter, it expands. This hotter air is lighter than an equal volume of cooler air. This hotter, lighter air exerts less pressure on the earth's surface compared to cooler air. If the temperature decreases, air contracts and becomes heavier. Therefore, an equal volume of cooler air exerts more pressure on the earth's surface compared to hotter air. Cool air is said to be denser than warm air.

Atmospheric Pressure and Altitude

As you climb above sea level, atmospheric pressure decreases. At 5,000 feet (1,524 meters) above sea level, a one-square-inch column of air from the earth's surface to the outer edge of the atmosphere is 5,000 feet (1,524 meters) less than the same column of air at sea level. Therefore, the weight of this column of air is less at an elevation of 5,000 feet (1,524 meters) than at sea level. As altitude continues to increase, atmospheric pressure continues to decrease. At an altitude of several hundred miles above sea level,

Atmospheric pressure may be defined as the total weight of the earth's atmosphere. However, if the pressure is measured with a pressure gauge, the gauge would read zero psi. A pressure gauge measures only the difference in pressure between what surrounds the gauge and what is applied to the gauge. To avoid confusion, pressures measured by the gauge are listed as "psig." The g means gauge. True or absolute (such as atmospheric pressures) pressures are given as "psia."

One **bar** is equal to 14.50 psi.

An equal volume of hot air weighs less and exerts less pressure on the earth's surface compared to cold air.

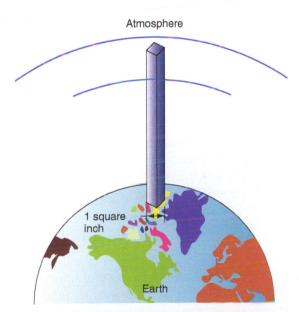

Atmosphere

1 square inch

Earth

Figure 2-18 A column of air 1-inch square extending from the earth's surface at sea level to the outer edge of the atmosphere weighs 14.7 pounds.

the earth's atmosphere ends. There is zero atmospheric pressure at that point. In addition to lower atmospheric pressure, the higher the altitude, the less oxygen there is proportionately in the air. Less density due to lower pressure means less oxygen per unit of measurement. The altitude also has an effect on the boiling point of liquids. As the altitude increases, the boiling point decreases.

VACUUM

Liquids, solids, and gases tend to move from an area of high pressure to a low-pressure area.

When air has a pressure higher than atmospheric pressure, a **positive pressure** exists. When air has a lower pressure than atmospheric, it has a **negative pressure**. *Vacuum* is the term commonly used to describe negative pressures. Vacuum is best described as any pressure less than atmospheric. A total or complete vacuum is the total or complete absence of air pressure. In nature, a high pressure always moves to a low pressure.

The understanding of pressure and vacuum is one of the keys to understanding how an engine works. It also explains how things work and why things may not work the way they are supposed to. The reaction of a high pressure to a lower pressure defines the action of many engine components. Whenever there is a high (positive) pressure and a lower (less positive) pressure, the higher pressure will move toward the lower in an attempt to equalize the pressures.

We all are quite familiar with this. A balloon expands as we blow in more air, simply because we are increasing the pressure inside the balloon. The high pressure inside is pushing out against the lower pressure outside the balloon.

Vacuum could be measured in pounds per square inch (psi), but inches of mercury (in. Hg) are most commonly used for this measurement. Let us assume that a plastic "U" tube is partially filled with mercury, and atmospheric pressure is allowed to enter one end of the tube. If a vacuum is supplied to the other end of the "U" tube, the mercury is forced downward by the atmospheric pressure. When this movement occurs, the mercury also moves upward on the side where the vacuum is supplied. If the mercury moves downward 10 inches, or 25.4 centimeters (cm), where the atmospheric pressure is supplied, and upward 10 inches (25.4 cm) where the vacuum is supplied, 20 in. Hg is supplied to the "U" tube. The highest possible, or perfect, vacuum is 29.9 in. Hg (**Figure 2-19**), which is 0 psi of pressure.

AUTHOR'S NOTE As altitude increases, atmospheric pressure decreases. As pressure decreases, the boiling point of a liquid also decreases. For example, water boils at 212°F (100°C) at sea level but will boil at an incrementally lower temperature as altitude increases. So if you are on a camping trip in the Rocky Mountains (say, 6,000 feet above sea level) and the water boils at a lower temperature (less than 212°F (100°C)), how long would you have to boil a three-minute egg before it has thoroughly cooked?

Vacuum and atmospheric pressure are used in several automotive systems. For example, atmospheric pressure is available outside the engine air intake. When a piston moves downward with an intake valve open, a vacuum is created in the cylinder above the piston. The air moves rapidly from the high pressure outside the air intake to the lower pressure in the cylinder (**Figure 2-20**).

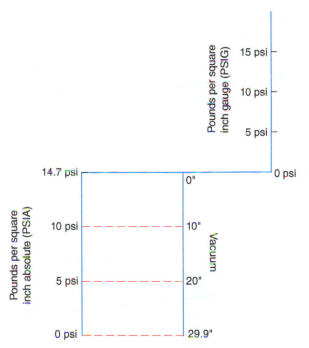

Figure 2-19 A comparison of pressure and vacuum readings.

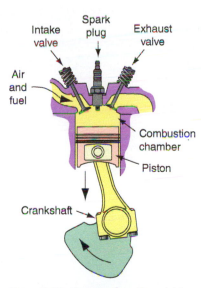

Figure 2-20 Air moves from the outside atmosphere, which is a high pressure, to the lower pressure in the cylinder during the engine's intake stroke.

Pumps use high and low pressure to move liquids or gases. For example, as a water pump rotates, it creates a high pressure at the pump outlet and a low pressure at the inlet. This pressure difference causes coolant to flow through the cooling system.

VENTURI PRINCIPLE

Pressures of a liquid or gas can be increased or decreased by applying more pressure onto it or by relieving some of the pressure on the liquid or gas. Pressure is also changed by putting a restriction in whatever the liquid or gas is flowing through. At the restriction, the pressure increases. After the restriction, the pressure is lower. A **venturi** is a restriction designed to increase the speed of the flow and decrease its pressure (**Figure 2-21**).

Let's assume we have a tube with air flowing through it. The air is moving from the high-pressure side to the low-pressure side. In the tube we have a restriction or venturi. As the moving air enters the venturi, the center column of air has no direct contact with the restriction. The air surrounding the center column is directed, by the venturi, toward the center column. This pushes the column through the venturi faster than it would on its own.

On the other side of the restriction, the size of the tube returns to normal. Although the speed of the air increased through the venturi, the volume of air entering into this larger area is less than the amount of air on the other side of the venturi. As a result, a lower pressure or vacuum is present at the outlet of the venturi.

> A **venturi** may be defined as a narrow area in a pipe through which a liquid or a gas is flowing.

COMBUSTION

An engine converts the energy found in fuel into heat energy, which is changed to mechanical energy that can be used to move a vehicle. The conversion of energy results from the mechanical design of the engine and through **combustion**. Combustion is a burning process. During this burning process, a chemical reaction takes place between the air and the fuel that entered the engine's cylinders. During this chemical reaction, heat energy is

> **Combustion** is the process whereby the proper mixture of fuel and oxygen is expose to a high temperatu spark and causes material or fuel t

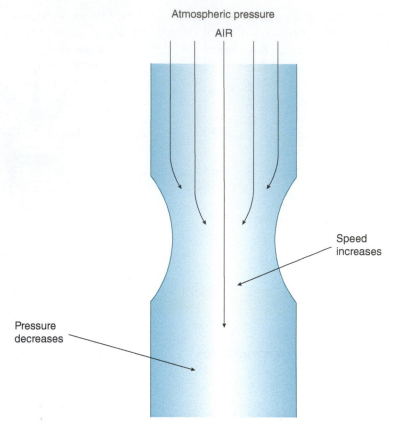

Figure 2-21 A venturi increases the speed of the incoming air and creates a vacuum below the venturi.

released. With the release of the heat, the pressure of the air in the cylinders increases. This high pressure pushes on the engine's piston, causing it to move. The movement of the piston is converted to move the vehicle.

The energy of the moving piston is dependent upon the amount of pressure there is on the piston. The amount of pressure on the piston depends on the amount of heat generated by the combustion process. Complete combustion of the air and fuel will provide for the maximum amount of heat from that amount of air and fuel.

To have complete combustion, four things must be present: (1) the correct amount of air must be mixed with (2) the correct amount of fuel in (3) a sealed container, and this mixture must be (4) shocked by the correct amount of heat at the correct time (**Figure 2-22**). Although other factors can affect combustion, these are absolutely the most important

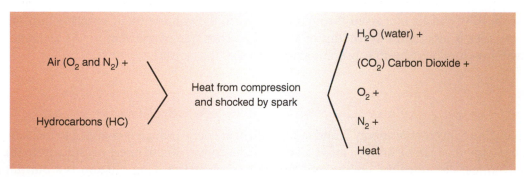

Figure 2-22 The ideal combustion process.

factors. As you will see, the engine has systems that attempt to meet these requirements and provide for complete combustion or maximum efficiency. When an engine doesn't run well, it is because one or more of these requirements have not been met.

All components of the engine are designed in an attempt to achieve total combustion. Although no engine can achieve complete combustion during all operating conditions, total combustion is the goal. Incomplete or inefficient combustion results from neither having a sealed cylinder nor the right amounts of air, fuel, heat, or any combination of these in the cylinder. The more incorrect any of these requirements are, the less complete the combustion will be.

The result of total combustion is the generation of great heat and the conversion of all of the cylinder's air and fuel into water and carbon dioxide. When total combustion does not take place, full conversion of the fuel and air also does not take place. This results in the release of pollutants from the engine.

SUMMARY

- An element contains only one type of atom.
- An atom is the smallest particle of an element.
- Compounds contain two or more types of atoms.
- A molecule is the smallest particle in a compound.
- Protons are positively charged particles at the center, or nucleus, of each atom.
- Neutrons have no electric charge, and they are located in the nucleus of most atoms.
- Electrons are negatively charged particles found in various orbits around the nucleus of an atom.
- Work is the result of applying a force.
- Force is measured in pounds and distance.
- Energy is the ability to do work, and there are seven basic types of energy.
- Inertia is the tendency of an object at rest to remain at rest, or the tendency of an object in motion to remain in motion.
- An object gains momentum when force overcomes static energy and moves the object.
- Friction is the resistance to motion when one object is moved over another object.
- Mass is a measurement of an object's inertia.

- Weight is a measurement of the earth's gravitational pull on an object.
- Volume is the length, width, and height of a space occupied by an object.
- Power is a measurement for the rate at which work is done.
- Torque is a twisting force that does work.
- Atmospheric pressure is the total weight of the earth's atmosphere.
- Vacuum is often used to describe a condition where air pressure is lower than atmospheric pressure.
- In nature, a high pressure always moves toward a lower pressure.
- Complete combustion takes place when there is the correct amount of air mixed with the correct amount of fuel in a sealed container, which is shocked by the correct amount of heat at the correct time.
- Acids react with metals to produce hydrogen.
- Acids and bases react to form salts.
- A gear reduction increases torque.

REVIEW QUESTIONS

Short-Answer Essays

1. Define an element and a compound.

2. Name the four factors that are necessary for complete combustion.

3. Define a molecule.

4. Describe three particles found in an atom, including the electrical charge and location of each particle.

5. Explain why a venturi decreases pressure.

6. What is atmospheric pressure?

7. What happens to air pressure as the temperature of the air changes?

8. Describe Newton's three laws of motion.

9. Describe the seven different forms of energy.

10. Describe four of the different types of energy conversion.

Fill-in-the-Blanks

1. At sea level, atmospheric pressure is _____ psi.

2. In nature, a _____ pressure always moves toward a(n) _____ pressure.

3. The atom has a nucleus, comprised of _____ and _____.

4. Work is calculated by multiplying _____ times _____.

5. Energy may be defined as the ability to do _____.

6. When one object is moved over another object, the resistance to motion is called _____.

7. Weight is the measurement of the earth's _____ _____ on an object.

8. Torque is a force that does work with a(n) _____ action.

9. Power is a measurement for the rate at which _____ is done.

10. Negative pressure may be called _____.

Multiple Choice

1. *Technician A* says that the valence ring of an atom is the innermost ring of electrons.
   Technician B says that the valence ring is the second ring of electrons.
   Who is correct?
   A. Technician A   C. Both technicians
   B. Technician B   D. Neither technician

2. To have complete combustion, we must have all of the following *except*:
   A. Correct amount of air.
   B. Correct amount of fuel.
   C. Unsealed container.
   D. Shocked by the correct amount of heat at the right time.

3. One horsepower, as calculated by Watt, is equivalent to:
   A. Moving 550 pounds 100 feet in 1 minute.
   B. Doing 33,000 foot-pounds of work in 1 hour.
   C. Doing 550 foot-pounds of work in 1 second.
   D. Both A and B.

4. *Technician A* states that the inertia of a vehicle at rest is called static inertia.
   Technician B says that the inertia of a vehicle in motion is called dynamic inertia.
   Who is correct?
   A. Technician A   C. Both technicians
   B. Technician B   D. Neither technician

5. *Technician A* says that torque is measured in the metric system as gram-centimeters.
   Technician B says torque is measured in the metric system as Newton meters.
   Who is correct?
   A. Technician A   C. Both technicians
   B. Technician B   D. Neither technician

6. *Technician A* says that the mass of an object is the same in space as it is on earth.
   Technician B says that mass is the measurement of an object's inertia.
   Who is correct?
   A. Technician A   C. Both technicians
   B. Technician B   D. Neither technician

7. A perfect vacuum is said to have been reached at:
   A. 18.8 in. Hg.   C. 29.9 in. Hg.
   B. 17.3 in. Hg.   D. 35.9 in. Hg.

8. *Technician A* says that energy released to do work is called kinetic energy.

   Technician B says that energy that is stored is called theoretical energy.

   Who is correct?

   A. Technician A

   B. Technician B

   C. Both technicians

   D. Neither technician

9. While discussing gases,

   Technician A says that when pressure increases so does temperature.

   Technician B says that when temperature increases, so does pressure.

   Who is correct?

   A. Technician A

   B. Technician B

   C. Both technicians

   D. Neither technician

10. All of the following are measurements of atmospheric pressure *except* _____

    A. bar.

    B. psi.

    C. kPa.

    D. kg-cm.

CHAPTER 3
ENGINE DESIGN AND OPERATION

Upon completion and review of this chapter, you should be able to:

- Define the methods used for engine classification.
- Describe the four strokes in the four-stroke cycle.
- Explain compression ratio.
- Explain the purpose of the camshaft, pushrods, and rocker arms.
- Explain volumetric efficiency.
- Describe the difference between an overhead cam engine and an overhead valve engine.
- Describe four different types of engine block designs.
- Briefly describe the different engine systems.
- Define *cylinder bore* and *stroke*.

- Explain how to calculate engine displacement.
- Describe three different methods of measuring engine efficiency.
- Name and describe the components of a typical lubricating system.
- Describe the purpose of a crankcase ventilation system.
- Explain oil service and viscosity ratings.
- List and describe the major components of the cooling system.
- Describe the operation of the cooling system.
- Describe the functions of the water pump, radiator, radiator cap, and thermostat in the cooling system.

Terms To Know

American Petroleum Institute (API)
Bore
Cams
Combustion chamber
Compression ratio
Coolant
Crude oil
Diesel engines
Displacement
Dual overhead camshaft (DOHC)

Efficiency
Energy-conserving oils
Expansion tank
Firing order
Glow plugs
Horsepower (hp)
Lobes
Mild hybrid
Oil pump pickup
Overhead camshaft (OHC)
Overhead valve (OHV)

Parallel hybrid
Recovery tank
Reverse flow cooling system
Rotary engine
Series hybrid
Society of Automotive Engineers (SAE)
Stroke
Surge tank
Wankel engine

INTRODUCTION

Modern engines are highly engineered power plants. These engines are designed to meet the performance and fuel-efficiency demands of the public. The heavy, cast-iron engine with its poor gas mileage has been replaced by compact, lightweight, and fuel-efficient engines. Modern engines are made of lightweight engine castings and stampings, noniron materials (e.g., aluminum, magnesium, fiber-reinforced plastics), and fewer and smaller fasteners to hold things together. These fasteners are made possible through computerized joint designs that optimize loading patterns. Each of these newer engine designs has its own distinct personality, based on construction materials, casting configurations, and design.

These modern engine-building techniques have changed how technicians do their job. Before these changes can be explained, it is important to explain the "basics" of engine design and operation.

ENGINE CLASSIFICATIONS

Today's automotive engines can be classified in several ways, depending on the following design features:

- *Operational cycles.* Most technicians will generally come in contact with only four-stroke cycle engines. However, a few older cars have used—and some cars in the future may use—a two-stroke engine.
- *Number of cylinders.* Current engine designs include 3-, 4-, 5-, 6-, 8-, 10-, and 12-cylinder engines.
- *Cylinder arrangement.* An engine can be flat (opposed), in-line, or V-type. Other more complicated designs have also been used.
- *Displacement.* The volume of the engine's cylinders added together.
- *Valvetrain type.* Engine valvetrains can be either the **overhead camshaft (OHC)** type or the camshaft in-block **overhead valve (OHV)** type. Some engines use separate camshafts for the intake and exhaust valves. These are based on the OHC design and are called **dual overhead camshaft (DOHC)** engines. V-type DOHC engines have four camshafts—two on each side.
- *Ignition type.* There are two types of ignition systems: spark and compression. Gasoline engines use a spark ignition system. In a spark ignition system, the air-fuel mixture is ignited by an electrical spark. Diesel engines, or compression ignition engines, have no spark plugs. An automotive diesel engine relies on the heat generated as air is compressed to ignite the air-fuel mixture for the power stroke.
- *Cooling systems.* There are both air-cooled and liquid-cooled engines in use. Nearly all of today's engines have liquid-cooling systems.
- *Fuel type.* Several types of fuel are currently used in automobile engines, including gasoline, natural gas, diesel, and propane. The most commonly used is gasoline. Recently, electric vehicles have been introduced for sale to the public. These vehicles do not rely on a fuel for power, but on electrical energy stored in a high capacity battery.

Some **overhead camshafts (OHC)** operate the valves directly, eliminating rocker arms and pushrods.

An **overhead valve (OHV)** uses rocker arms and pushrods to operate the valves.

A **dual overhead camshaft (DOHC)** engine uses one overhead camshaft for the intake valves and one for the exhaust valves.

An engine with two exhaust valves and two intake valves in each cylinder is commonly referred to as a four-valve engine. These are typically called by their total number of valves: for example, a four-valve, eight-cylinder engine may be called a 32-valve engine.

In 1998 (after more than 50 years), the only major manufacturer that still used air-cooled engines, Porsche, discontinued the air-cooled 911.

ENGINE LOCATION

The engine is typically placed in one of three locations. In the vast majority of vehicles, it is located at the front of the vehicle, forward of the passenger compartment. Front-mounted engines can be positioned either longitudinally or transversely with respect to the vehicle. The second engine location is a mid-mount position between the passenger compartment and rear suspension. Mid-mount engines are normally transversely mounted. The third, and least common, engine location is in the rear of the vehicle. The engines are typically opposed type engines. Each of these engine locations offers advantages and disadvantages.

CHARACTERISTICS OF FOUR-STROKE ENGINE DESIGN

Depending on the vehicle, an in-line, V-type, or horizontally opposed cylinder design can be used. The most popular designs are in-line and V-type engines.

In-Line Engines

In the in-line engine design (**Figure 3-1**), the cylinders are all placed in a single row. There is one crankshaft and one cylinder head for all of the cylinders. The block is cast so that all cylinders are located in an upright position.

In-line engine designs have certain advantages and disadvantages. They are easy to manufacture and service. However, because the cylinders are positioned vertically, the front of the vehicle must be higher. This affects the aerodynamic design of the car. Aerodynamic design refers to the ease with which the car can move through the air. When equipped with an in-line engine, the front of a vehicle cannot be made as low as it can be with other engine designs.

V-Type Engines

The V-type engine design has two rows of cylinders (**Figure 3-2**) located 60 to 90 degrees away from each other. A V-type engine uses one crankshaft that is connected to the pistons on both sides of the V. This type of engine has two cylinder heads, one over each row of cylinders.

One advantage of using a V configuration is that the engine is not as high or long as an in-line configuration. The front of a vehicle can now be made lower. This design improves the outside aerodynamics of the vehicle. If eight cylinders are needed for power, a V configuration makes the engine much shorter, lighter, and more compact. Many years ago, some vehicles had an in-line eight-cylinder engine. The engine was very long, and its long crankshaft also caused increased torsional vibrations in the engine.

Opposed Cylinder Engines

In this design, two rows of cylinders are located opposite the crankshaft (**Figure 3-3**). Opposed cylinder engines are used in applications where there is very little vertical room for the engine. For this reason, opposed cylinder designs are commonly used on vehicles that have the engine in the rear. The angle between the two cylinder heads is typically 180 degrees. One crankshaft is used with two cylinder heads.

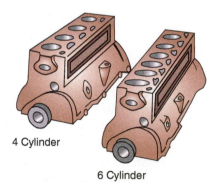

4 Cylinder

6 Cylinder

Figure 3-1 In-line engine designs.

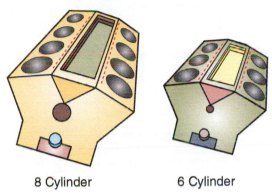

8 Cylinder   6 Cylinder

Figure 3-2 V-type engine designs.

Figure 3-3 Horizontally opposed cylinder engine designs.

FOUR-STROKE GASOLINE ENGINES

In a passenger car or truck, the engine provides the rotating power to drive the wheels through the transmission and driving axles. All automobile engines, both gasoline and diesel, are classified as internal combustion because the combustion or burning takes place inside the engine. These systems require an air-fuel mixture that arrives in the combustion chamber at the correct time and an engine constructed to withstand the temperatures and pressures created by the burning of fuel.

The combustion process is the result of several engine systems working independently *and* in concert to bring together the precise amount of air, fuel, heat, pressure, and spark at precisely the right time to create power. The internal combustion engine can be compared to an air pump. The engine operates, in part, on the principles of thermodynamics. It is responsible for compressing the incoming air-fuel charge to raise its pressure and, in turn, raise its temperature close to the self-ignition temperature. The ignition system is responsible for supplying the required amount of spark to adequately ignite the air-fuel mixture at the perfect time under various pressures. The air intake and fuel systems are designed to deliver varying ratios of mixture to meet varying operating requirements. Of course, each cylinder should receive the same amounts of mixture, pressure, and spark if the engine is to run smoothly and efficiently. The construction of the engine and its different parts is engineered and manufactured to withstand a harsh operating environment, both internal and external to the **combustion chamber**.

Most vehicles today use a four-stroke, reciprocating gasoline engine. As the name implies, four strokes of the piston are required to complete one engine cycle (**Figure 3-4**). A stroke is the full up or down travel of the piston in the cylinder bore. The piston is connected to the crankshaft by a connecting rod and moves up and down within the cylinder (reciprocating motion). The reciprocating motion of the piston is converted to a rotary motion via the crankshaft (**Figure 3-5**).

At the point of ignition in the end of the compression stroke, the air-fuel mixture burns rapidly, beginning the power stroke. The virtual immediate expansion of gases as heat energy is released from the gasoline due to ignition creating a sudden rise in temperature and pressure, forcing the piston downward and rotating the crankshaft. The next cylinder in the firing order immediately repeats this process, and so on for each cylinder. It takes two complete crankshaft revolutions, or 720 degrees, and one revolution of the camshaft to complete one engine cycle, firing all cylinders once. The camshaft rotates at half the crankshaft speed. For example, a four-cylinder engine will fire a cylinder every

The **combustion chamber** is the space between the top of the piston and cylinder

Some common names for various types of combustion chambers are hemi, wedge, Fast Burn heads, Vortec, and high-swirl, to name a few. Some direct fuel injection engines use a specially designed piston to aid in the combustion process.

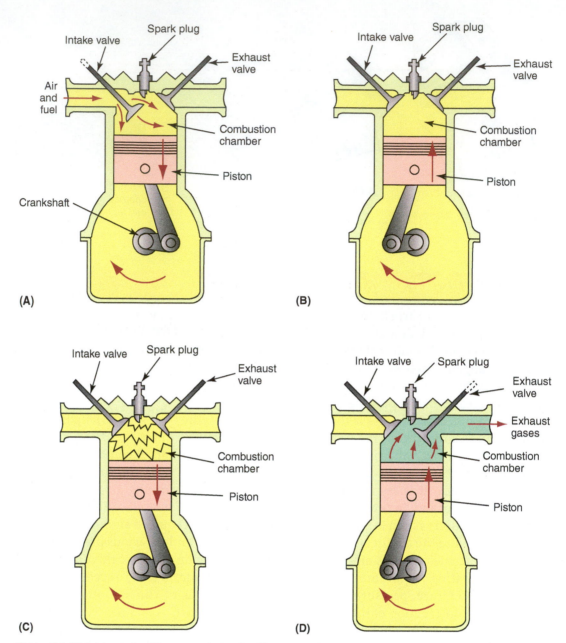

Figure 3-4 (A) Intake stroke, (B) compression stroke, (C) power stroke, and (D) exhaust stroke.

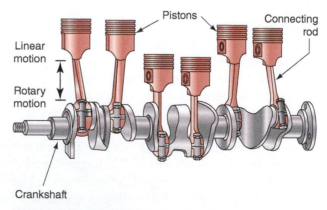

Figure 3-5 The reciprocating motion of the pistons is converted to rotary motion by the crankshaft.

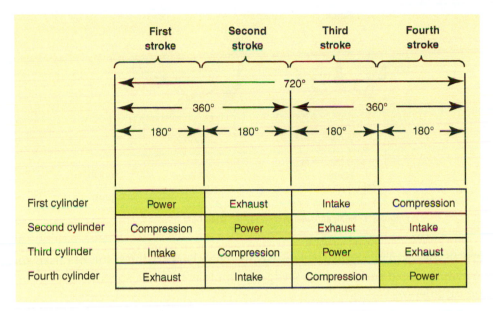

	First stroke	Second stroke	Third stroke	Fourth stroke
First cylinder	Power	Exhaust	Intake	Compression
Second cylinder	Compression	Power	Exhaust	Intake
Third cylinder	Intake	Compression	Power	Exhaust
Fourth cylinder	Exhaust	Intake	Compression	Power

Figure 3-6 A four-cylinder engine has one cylinder on a power stroke every 180 degrees of crankshaft rotation.

180 degrees of crankshaft rotation. **Figure 3-6** maps the crankshaft position, measured in degrees, and the cylinder event for one complete engine cycle in a four-stroke, four-cylinder engine.

A flywheel (manual transmission) or flexplate-torque converter (automatic transmission) is connected to the end of the crankshaft. The flywheel is a heavy mass that stores some of the engine's energy (inertia) to even out the power pulses as each cylinder fires. It does this by helping keep the crankshaft moving beyond bottom dead center (BDC) after the power stroke occurs for each cylinder, in addition to maintaining momentum to assist the piston on the compression stroke. In addition, the flywheel couples the transmission to the engine for the transfer of power created by the engine to be delivered to the drive wheels.

The combustion chamber is the space in which the air-fuel mixture is compressed and ignited to create a controlled explosion. It is not really an explosion by definition, but rather a rapid, calculated, and controlled burn. The term helps describe the sudden violent release of heat energy from a precise fuel mixture. Made up of the area between the valve surface side of the cylinder head and the top of the piston as it approaches top dead center (TDC), the combustion chamber is where the actual production of power begins. The design of the combustion chamber is determined by the size and shape of the cylinder head and piston top. Certain designs are engineered to create a type of turbulence within the combustion chamber to move the air-fuel mixture in a way that promotes a more efficient burn. Various designs in use today include the wedge, hemispherical (hemi), pent-roof, D shape, and heart shape. The two most common designs are the wedge and hemi (**Figure 3-7**).

The camshaft rotates at one-half the speed of the crankshaft.

Wedge Type

The wedge-type combustion chamber has the spark plug positioned off-center in the widest part of the wedge. The intake and exhaust valves are usually positioned next to each other. The air-fuel mixture is compressed into an area called the quench area as the piston travels through the compression stroke to TDC. The quench area promotes a thorough blending of the air and fuel before combustion. This also causes a turbulence or movement of the mixture within the cylinder, causing a more complete burn at lower and

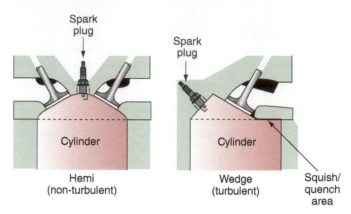

Figure 3-7 Hemi and wedge combustion chambers.

mid-cruise speeds. The high-swirl combustion chambers are the result of raised surfaces on the cylinder head surface of the wedge design that creates a specific amount of combustion turbulence.

Hemispherical Type

The main characteristic of the hemispherical combustion chamber is that it is rounded. When the piston is at TDC, the combustion chamber resembles a half-ball, hence the name. This design allows the intake and exhaust valves, typically located across from each other with the spark plug placed between them, to be angled in the cylinder head. This cross-flow arrangement of the valves helps create a less restricted airflow entering and exiting the cylinder. The central location of the spark helps ignite and maintain a centralized burn of the flame front. This type of combustion chamber is also referred to as a non-turbulent design. A variation of this style uses a domed piston that creates a tighter quench area that also creates some turbulence.

Other Types

A design that is popular with the four-valve-per-cylinder engines is the V-shaped, pent-roof combustion chamber. This and other newer types are designed to improve combustion efficiency and emissions.

The concept of these types of combustion chambers is to enhance the air-fuel mixture movement during the combustion process. The turbulence causes the burning air-fuel molecules to blend with and help quickly ignite the unburned molecules, resulting in a more complete combustion.

Although the combustion must occur in a sealed cylinder, the cylinder must also have some means of allowing heat, fuel, and air into it. There must also be a means to allow the burnt air-fuel mixture out so that a fresh mixture can enter and the engine continues to run. To accommodate these requirements, engines are fitted with valves (**Figure 3-8**).

There are at least two valves at the top of each cylinder. The air-fuel mixture enters the combustion chamber through an intake valve (on port fuel-injected vehicles) and leaves (after having been burned) through an exhaust valve. The valves are accurately machined plugs that fit into machined openings. A valve is said to be seated or closed when it rests in its opening, thus creating a "seal." When the valve is pushed off its seat, it opens.

A rotating camshaft, driven by the crankshaft, opens and closes the intake and exhaust valves. **Cams** are raised sections of a shaft that have high spots called **lobes**. As the camshaft rotates, the lobes rotate and push the valve open by pushing it away from its seat. Once the cam lobe rotates out of the way, the valve, forced by a spring, closes. The camshaft can be located either in the cylinder block or in the cylinder head.

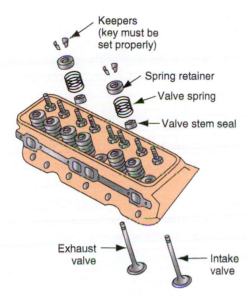

Figure 3-8 Parts of a cylinder head.

When the action of the valves and the spark plug is properly timed to the movement of the piston, the combustion cycle takes place in four strokes of the piston: the intake stroke, the compression stroke, the power stroke, and the exhaust stroke.

The up-and-down movement of the piston on all four strokes is converted to a rotary motion by the crankshaft. It takes two full revolutions of the crankshaft and one revolution of the camshaft to complete the four-stroke cycle.

Intake Stroke

The first stroke of the cycle is the intake stroke. As the piston moves away from TDC, the intake valve opens. The downward movement of the piston increases the volume of the cylinder above it. This reduces the pressure in the cylinder. This reduced pressure, commonly referred to as engine vacuum, causes the atmospheric pressure to push a mixture of air and fuel through the open intake valve. (Some engines are equipped with a super-or turbocharger that pushes more air past the valve.) As the piston reaches the bottom of its stroke, the reduction in pressure stops. This causes the intake of air-fuel mixture to slow down. It does not stop, because of the weight and movement of the air-fuel mixture. It continues to enter the cylinder until the intake valve closes. The intake valve closes after the piston has reached BDC. This delayed closing of the valve increases the volumetric efficiency of the cylinder by packing as much air and fuel into it as possible.

Shop Manual
Chapter 3, page 110

Compression Stroke

The compression stroke begins as the piston starts to move from BDC. The intake valve closes, trapping the air-fuel mixture in the cylinder. At TDC, the piston and cylinder walls form a combustion chamber in which the fuel will be burned. The upward movement of the piston compresses the air-fuel mixture to one-eighth of its volume (in an 8:1 compression ratio engine), raising its pressure to over 100 pounds per square inch (psi). The sudden rise in pressure and temperature increases fuel volatility, thereby bringing the air-fuel mixture almost to the point of self-ignition. Just prior to TDC, the spark plug ignites the air-fuel mixture and the fuel begins to burn. The displacement of the piston is the amount of cylinder volume change that takes place within the cylinder as the piston travels from TDC to BDC. The volume at TDC is the clearance volume of the cylinder. If the volume measured in the cylinder with the piston at BDC is eight times the clearance volume, it is an 8:1 compression-ratio engine.

Shop Manual
Chapter 3, page 116

Power Stroke

The power stroke begins as the piston reaches TDC. The burning fuel rapidly expands, creating a very high pressure against the top of the piston. This drives the piston down toward BDC. The downward movement of the piston is transmitted through the connecting rod to the crankshaft.

Exhaust Stroke

The exhaust valve opens just before the piston reaches BDC on the power stroke. Pressure within the cylinder causes the exhaust gas to rush past the open valve and into the exhaust system. Movement of the piston from BDC pushes most of the remaining exhaust gas from the cylinder. As the piston nears TDC, the exhaust valve begins to close as the intake valve starts to open. The exhaust stroke completes the four-stroke cycle. The opening of the intake valve begins the cycle again. This cycle occurs in each cylinder and is repeated over and over, as long as the engine is running.

> The period of time that the exhaust and intake valves are both open during the exhaust stroke is called valve overlap.

VALVE AND CAMSHAFT PLACEMENT CONFIGURATIONS

Two basic valve and camshaft placement configurations of the four-stroke gasoline engines are used in automobiles (**Figure 3-9**).

 A BIT OF HISTORY

Automotive technology and component integrity have certainly come a long way over the last 60 years. By definition, according to a parts and labor guide of the early 1950s, a "*complete tune-up*" included a carburetor and distributor overhaul *and* removal of the cylinder head(s) to grind and reface the valves and valve seats!

Overhead Valve

As the name implies, the intake and exhaust valves in an overhead valve engine are mounted in the cylinder head and are operated by a camshaft located in the cylinder block. This arrangement requires the use of valve lifters, pushrods, and rocker arms to transfer camshaft rotation to valve movement. The intake and exhaust manifolds are attached to the cylinder head.

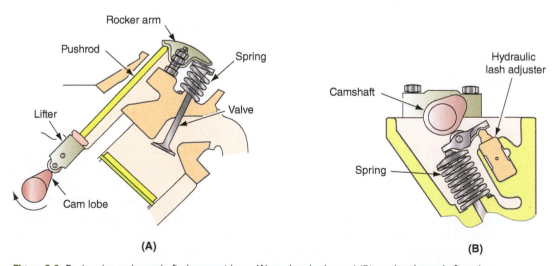

(A)

(B)

Figure 3-9 Basic valve and camshaft placement in an (A) overhead valve and (B) overhead camshaft engine.

Overhead Cam

An overhead cam (OHC) engine also has the intake and exhaust valves located in the cylinder head. But as the name implies, the cam is located in the cylinder head. In an overhead cam engine, the valves are operated directly by the camshaft or through cam followers or tappets, typically eliminating the pushrods and rocker arms. The OHC camshaft(s) can be driven by a belt or chain. A timing belt is quiet but generally has to be replaced at a mileage interval. The timing chain generally lasts longer but can make more noise during operation.

> Overhead cam (OHC) engines can use a belt or a chain to drive the camshaft(s).

VALVE AND CAMSHAFT OPERATION

In OHV engines with the camshaft in the block (**Figure 3-10**), the valves are operated by valve lifters and pushrods that are actuated by the camshaft. In OHC engines, the cam lobes operate the valves directly, and there is no need for pushrods or lifters.

> **Shop Manual**
> Chapter 3, page 120

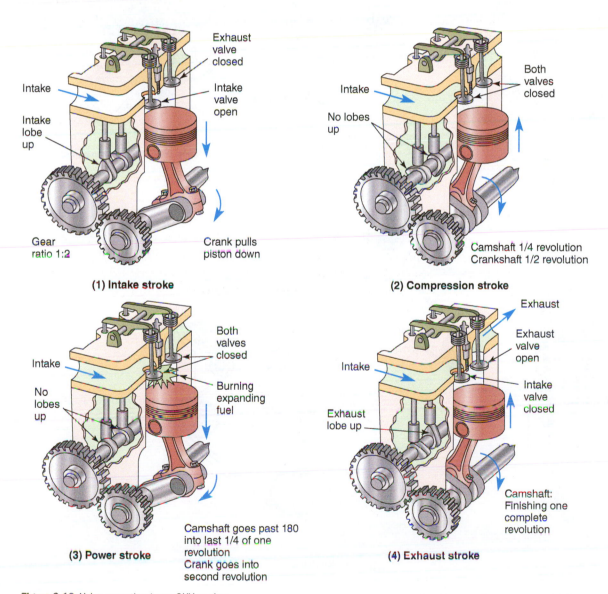

(1) Intake stroke

Exhaust valve closed
Intake
Intake valve open
Intake lobe up
Gear ratio 1:2
Crank pulls piston down

(2) Compression stroke

Both valves closed
Intake
No lobes up
Camshaft 1/4 revolution
Crankshaft 1/2 revolution

(3) Power stroke

Intake
No lobes up
Both valves closed
Burning expanding fuel
Camshaft goes past 180 into last 1/4 of one revolution
Crank goes into second revolution

(4) Exhaust stroke

Exhaust
Exhaust valve open
Intake
Intake valve closed
Exhaust lobe up
Camshaft: Finishing one complete revolution

Figure 3-10 Valve operation in an OHV engine.

The camshaft rotates at half crankshaft speed.

Cam lobes are oval shaped. The placement of the lobe on the shaft determines when the valve will open. Design of the lobe determines how far the valve will open and how long it will remain open in relation to piston movement.

Timing Gears, Belts, and Chains

The camshaft is driven by the crankshaft by way of gears, a timing chain, or a timing belt (**Figure 3-11**).

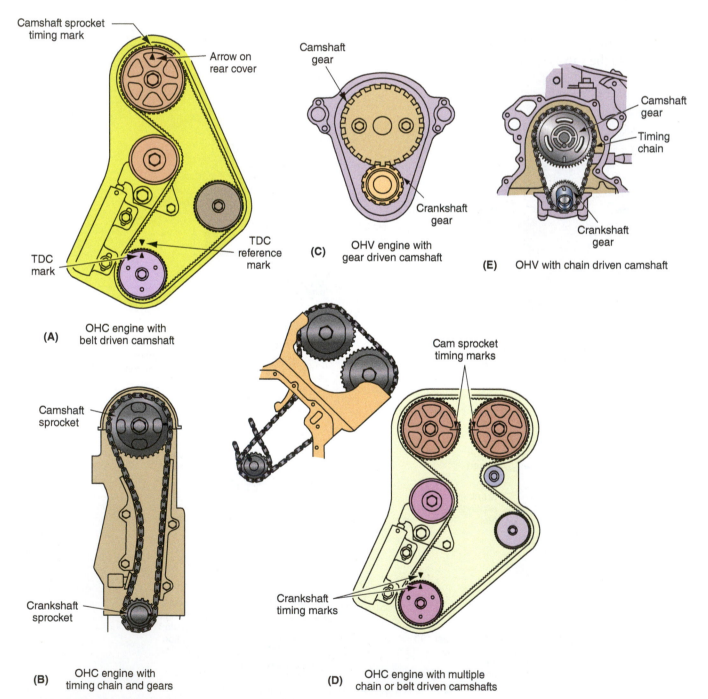

Camshaft sprocket timing mark

Arrow on rear cover

TDC reference mark

TDC mark

(A)   OHC engine with belt driven camshaft

Camshaft gear

Crankshaft gear

(C)   OHV engine with gear driven camshaft

Camshaft gear

Timing chain

Crankshaft gear

(E)   OHV with chain driven camshaft

Camshaft sprocket

Crankshaft sprocket

(B)   OHC engine with timing chain and gears

Cam sprocket timing marks

Crankshaft timing marks

(D)   OHC engine with multiple chain or belt driven camshafts

Figure 3-11 The different timing drive mechanisms used today.

Routine maintenance is essential to prevent valvetrain problems, especially with regard to timing belts. Certain engines have a recommended mileage interval in which the belt should be replaced. Some timing chains—and all timing belts—use a tensioner to push against the chain or belt to keep it tight, compensating for stretch and normal wear. Maintaining proper tension helps ensure the belt or chain will stay tight against the gears or sprockets to prevent slipping and maintain precise valve timing. Some tensioners are adjustable, while others are self-adjusting. If a timing belt or chain becomes excessively worn, it is possible that a change in valve timing accuracy will result. In addition, if the timing belt becomes damaged and breaks while the engine is running, the camshaft will stop suddenly while the crankshaft continues for a brief moment before the engine quits. Depending on engine design, damage may result to the valves, cylinder head, and pistons due to the valves coming in sudden contact with the tops of the pistons. Some engines have a small but adequate clearance between the open valve and the piston at TDC. Should a timing belt failure occur, this will likely prevent damage if the crankshaft and camshaft happen to come to rest in that particular position. If the timing chain or belt jumps from its correct position on the timing gear or if the tension is not correct, drivability problems or engine damage may occur. This is because the valves open and close at the wrong time relative to crank and piston position, resulting in incorrect airflow into and out of the cylinder. While timing gears seemingly require no maintenance, they can wear or break just like any other part. Today's advanced electronic engine management systems can actually mask some drivability symptoms caused by these or other mechanical problems in an effort to maintain emission control.

Valve Timing and Overlap

Volumetric efficiency is a measure of how much of the cylinder volume is filled during various operating conditions. The timed flow rate of a given volume is the basis for understanding this concept. For example, if a given volume of air at a given pressure flows through a fixed opening, it is a time-related event. If that same volume were under pressure, such as with a turbocharged or supercharged engine, more air would flow through the fixed opening in the same period of time. This is true up to the point where additional pressure no longer yields an increased amount of flow. This is because the area through which the air is flowing stays the same. Therefore, it is the size of the fixed opening that limits the flow. This explains why valvetrain-related problems such as a worn cam lobe, carbon-restricted intake valve, lifter adjustment, or faulty lifters can reduce compression in the cylinder. Even a restricted air filter could cause a major problem with volumetric efficiency. As you can see, volumetric efficiency can be reduced by problems with getting air into the cylinder or simply by the amount of time the air has to get into the cylinder, such as a higher rpm. Volumetric efficiency is also dependent on the exhaust flow leaving the cylinder. If exhaust cannot be moved out, then the intake charge has nowhere to go. Intake valve problems mean the time the valve is open is reduced, thereby creating less time for the air-fuel charge to flow into the chamber and resulting in less volume in the cylinder. To summarize, it can be simply stated that the ratio of ambient air density to that in the cylinder when filled is that particular engine's volumetric efficiency at a given rpm.

For the intake and exhaust valves to open and close properly, they must be timed precisely to the crankshaft position. Refer to **Figure 3-12** to aid in illustrating the relationship of the valves to the four strokes of the piston. As the piston approaches the intake stroke, the intake valve begins to open at about 20 degrees before the piston reaches TDC. The piston moves past TDC, enters the down stroke, and passes BDC at about 50 degrees when the intake valve closes. The movement of the incoming air continues although the piston has begun its upward travel. The compression stroke begins while the intake valve is closing, at about 50 degrees after BDC. The compression stroke continues toward TDC. Ignition occurs just before the piston reaches TDC to ensure that peak combustion

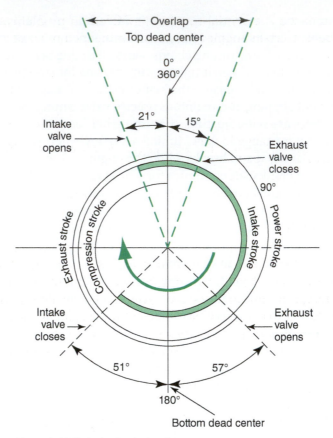

Figure 3-12 Typical valve timing diagram.

pressure occurs when the piston is about 15 degrees after TDC. This maximizes the power from ignition because the combustion pressure is effective for approximately 90 degrees of the total crankshaft rotation. That translates to about 15 percent of the crankshaft movement that is used to create power. The combustion continues to push the crankshaft to about 60 degrees before BDC, when the exhaust valve opens. The exhaust stroke continues to occur as the crankshaft rotates to about 20 degrees after TDC. Depending on the engine design, for about 30 or more degrees both the intake and exhaust valves are open. This period of time when both valves are open is called valve overlap. Overlap helps control the volume of the fresh air-fuel charge being drawn into the cylinder. The movement of the outgoing exhaust gases creates a lower pressure in the cylinder that acts on the intake charge. This action is called scavenging. This scavenging effect allows the fresh mixture to begin to flow into the cylinder even before the piston begins the intake stroke. If anything affects the valve timing or the valve opening, the flow will be reduced or interfered with, reducing mixture intake, airflow, volumetric efficiency, and compression.

Multivalve Engines

Volumetric efficiency relates to the concept of more air input equals more power output. Multivalve engines allow more volume to enter the cylinder in a specific period of time. The velocity of the intake air is actually higher with smaller multiple valve ports as compared to a single larger passage. Multivalve engines today come in a variety of arrangements and number of valves. A dual-valve engine is a bit of a misnomer in that there are actually four valves per cylinder—two, or dual, intake valves and two exhaust valves. Other combinations use three or five valves per cylinder. With the three-valve system, there are two intake valves and one exhaust, while the five-valve arrangement uses three for the intake and two for exhaust. In many of the multivalve applications, two camshafts are used and are referred to as dual overhead camshafts (DOHC). Any multivalve engine will require a more complex camshaft operation, which requires more components that add cost to the engine.

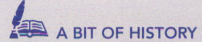

A BIT OF HISTORY

Among the first multivalve engines were the French Peugeot four-valve head in 1913 and a Pierce Arrow built in 1918.

Variable Cam Timing

As with any camshaft design, a good compromise is engineered into the valve timing to optimize high-end performance while maintaining good, low speed and idle quality. High-performance camshafts have a high duration (the amount of time the valves stay open) to enhance high rpm performance. However, the idle quality is erratic at best. To achieve the best of the entire performance spectrum, some engines use variable camshaft timing. By controlling camshaft and valve operation for best idle quality, low-speed performance, and higher horsepower output at high speeds, the incoming air-fuel charge can be adjusted specifically and precisely for the varying conditions and demands. Engine speed and load determine when the valves will open and for how long. Several types of variable cam timing designs are in use today. On some, camshaft rotational timing is controlled hydraulically through commands from the powertrain control module (PCM). Others use computer-controlled solenoids to control rocker arms to vary the intake and exhaust valve timing by alternating between separate cam lobes featuring two different lift and duration profiles.

Variable Displacement Engines

At light loads, such as during steady cruising, only a small percentage of available power is needed. Also, at these light engine loads, the throttle is almost completely closed, which limits the engine's ability to breathe, and in turn the cylinders do not get enough air for an efficient combustion. With some of the cylinders deactivated, the throttle is opened further, which contributes to cylinder fill. In a way, four cylinders are better than eight under a light load. Engineers have developed ways to shut down some of the cylinders in an engine but still have them available for action quickly when needed. So far, V-8 and V-6 engines have been fitted with the technology. Every other cylinder in the firing order is shut down when cylinder deactivation is available (**Figure 3-13**). Special engine mounts, computer programming, variable valve timing technology, and a partial cylinder fill keep the engine from running rough as you might suspect. The cylinders are generally deactivated by closing the exhaust valve on the exhaust stroke. This keeps the exhaust charge in the cylinder. This charge provides some cushion for the piston on the way up. Also, if there is no charge at all in the cylinder, oil would be pulled up through the piston ring gaps when the piston was on the way down the cylinder due to the vacuum created. The fuel injectors are also turned off, but the spark at the plug is still fired. The reasoning for this is the fact that the plug could cool down and foul out during periods of inactivity. Generally, there are two ways used to shut down cylinders: lifter deactivation and rocker arm deactivation. General Motors' Active Fuel Management and Chrysler's Multiple Displacement System (MDS) use a special lifter (**Figure 3-14**) that collapses when oil pressure sent from a PCM-activated solenoid acts on a pin in the lifter. The lifter collapses, so the push-rod is no longer lifted and the valve stays closed. A spring located on the lifter takes up the lash created and prevents the valvetrain from being noisy. Honda

Figure 3-13 This V-8 engine is running on four cylinders under certain conditions to greatly improve fuel economy.

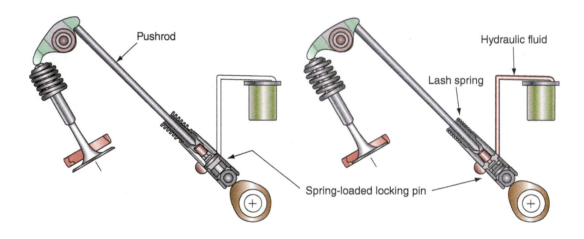

Normal operation
No oil supply to special locking pin, so lifter works as one unit. Normal combustion occurs

Variable displacement active
Oil supply unlatches locking pin, interior of lifter can no longer lift the pushrod. Pushrod is held in place by lash spring

Figure 3-14 Special valve lifters used in variable displacement engines from GM and Chrysler.

uses a version called variable cylinder management (VCM). This system uses the rocker arms and cam followers that are locked together by a synchro pin. When the cylinders are deactivated, the pin is pulled out and the valves are no longer opened by the rocker arms. When all cylinders are needed, the pins are reinserted and the cam follower and rocker arms are reconnected.

AUTHOR'S NOTE One development that has been researched over a long period of time is the idea of cutting back on the number of cylinders when in periods of low load. The first, such engines were actually experimented with during World War II. In 1981, Cadillac developed a V-8 engine called the V-8, 6, 4. The engine was used in that year and discontinued. At the time, computer and engine technologies were just not up to the task.

GASOLINE ENGINE SYSTEMS

The operation of an engine relies on several other systems. The efficiency of these systems affects the overall operation of the engine.

Air-Fuel System

This system makes sure the engine receives the right amount of both air and fuel needed for efficient operation. For many years, air and fuel were mixed in a carburetor, which supplied the resulting mixture to the cylinder. Today, automobiles have a fuel injection system that replaces the carburetor but performs the same basic function.

Ignition System

This system delivers a spark to ignite the compressed air-fuel mixture in the cylinder at the end of the compression stroke. The **firing order** of the cylinders is determined by the engine's manufacturer and can be found in the vehicle's service manual. Typical firing orders are illustrated in **Figure 3-15**.

Lubrication System

This system supplies oil to the various moving parts in the engine. The oil lubricates all parts that slide in or on other parts, such as the piston, bearings, crankshaft, and valve stems. The oil enables the parts to move easily so that little power is lost and wear is kept to a minimum. The lubrication system also helps transfer heat from one part to another for cooling. Engine oil also helps seal the engine, such as the small clearances in the piston rings.

Cooling System

This system is also extremely important. Coolant circulates in jackets around the cylinder and in the cylinder head. This removes part of the heat produced by combustion and prevents the engine from being damaged by overheating.

> The **firing order** of an engine is the order in which the pistons reach just a few degrees before the start of the power stroke, and the spark plug firing must be synchronized with this firing order.

> Lubricating oil cleans, cools, lubricates, and seals engine components.

Shop Manual
Chapter 3, page 101

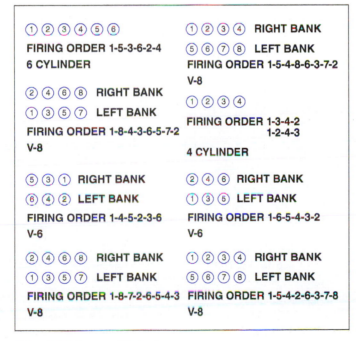

Figure 3-15 Common firing orders.

Exhaust System

This system removes the burned gases from the combustion chamber and limits the noise produced by the engine.

Emission Control System

Several control devices, which are designed to reduce the amount of pollutants released by the engine, have been added to the engine. Engine design changes, such as reshaped combustion chambers and altered valve timing, have also been part of the manufacturers' attempt to reduce emission levels.

ENGINE MEASUREMENT AND PERFORMANCE

An engine's bore is the diameter of the cylinder.

Some of the engine measurements and performance characteristics that a technician should be familiar with follow.

Bore and Stroke

The engine stroke is twice the crankshaft throw. The throw is the amount the connecting rod journal is offset from the crankshaft centerline.

The **bore** of a cylinder is simply its diameter measured in inches (in.) or millimeters (mm). The **stroke** is the distance of the piston travel between TDC and BDC. Between them, bore and stroke determine the displacement of the cylinders (**Figure 3-16**). When the bore of the engine is larger than its stroke, it is said to be oversquare. When the stroke is larger than the bore, the engine is said to be undersquare. Generally, an oversquare engine will provide for high engine speeds, such as for automobile use. An undersquare or long-stroke engine will deliver good low-speed power, such as an engine for a truck or tractor.

Displacement

The displacement of the engine is the volume of the cylinder between BDC and TDC multiplied by the number of cylinders.

Displacement is the volume of a cylinder between the TDC and BDC positions of the piston. It is usually measured in cubic inches, cubic centimeters, or liters (**Figure 3-17**). The total displacement of an engine (including all cylinders) is a rough indicator of its power output. Displacement can be increased by increasing the bore to a larger

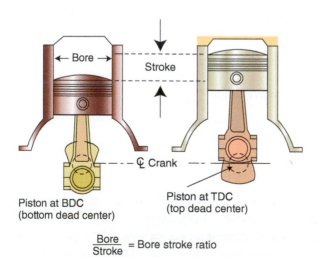

Figure 3-16 The bore and stroke of an engine.

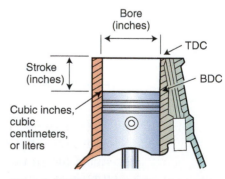

Figure 3-17 Displacement is the volume of a cylinder between TDC and BDC.

diameter or by increasing the length of the stroke. Total displacement is the sum of displacements for all cylinders in an engine. Cubic inch displacement (CID) may be calculated as follows.

$$CID = pi \times R^2 \times L \times N$$

where pi = 3.1416

$$R = \text{bore radius or} \frac{\text{bore diameter}}{2}$$

L = length of stroke

N = number of cylinders

Example: Calculate the CID of a six-cylinder engine with a 3.7 in. bore and 3.4 in. stroke.

$$CID = 3.1416 \times 1.85^2 \times 3.4 \times 6$$

$$CID = 219.66$$

Today's engines are described by their metric displacement. Cubic centimeters and liters are determined by using metric measurements in the displacement formula.
Example: Calculate the metric displacement of a four-cylinder engine with a 78.9 mm stroke and a 100 mm bore.

$$\text{Displacement} = 3.1416 \times \left(\frac{100}{2}\right)^2 \times 78.9 \times 4$$

$$\text{Displacement} = 2,479 \text{ cubic centimeters (cc)} = 2.5 \text{ liters (L)}$$

Larger, heavier vehicles are provided with large displacement engines. Large displacement engines produce more torque than smaller displacement engines. They also consume more fuel. Smaller, lighter vehicles can be adequately powered by lower displacement engines that use less fuel.

Compression Ratio

The **compression ratio** of an engine expresses how much the air and fuel mixture will be compressed during the compression stroke. The compression ratio is defined as the ratio of the volume in the cylinder above the piston when the piston is at BDC to the volume in the cylinder above the piston when the piston is at TDC (**Figure 3-18**). The formula for calculating the compression ratio is as follows.

$$\frac{\text{volume above the piston at BDC}}{\text{volume above the piston at TDC}}$$

or

$$\frac{\text{total cylinder volume}}{\text{total combustion chamber volume}}$$

In many engines, the top of the piston is even or level with the top of the cylinder block at TDC. The combustion chamber is in the cavity in the cylinder head and cylinder bore above the piston. This is modified slightly by the shape of the top of the piston. The volume of the combustion chamber must be added to each volume in the formula to achieve an accurate calculation of compression ratio.

When people refer to an engine's size, they are normally talking about the engine's displacement.

The compression ratio is the ratio of the cylinder volume at bottom dead center compared to the volume at TDC.

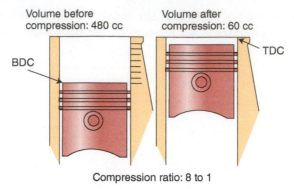

Volume before
compression: 480 cc

Volume after
compression: 60 cc

TDC

BDC

Compression ratio: 8 to 1

Figure 3-18 Compression ratio measures the amount the air and fuel mixture will be compressed.

Example: Calculate the compression ratio if the total piston displacement is 45 cubic in. and the combustion chamber volume is 5.5 cubic in.

$$\frac{45 + 5.5}{5.5} = \text{compression ratio}$$

$$9.1 \text{ to } 1 = \text{compression ratio}$$

(Be sure to add the combustion chamber volume to the piston displacement to achieve the total cylinder volume.)

The higher the compression ratio, the more power an engine theoretically can produce. Also, as the compression ratio increases, the heat produced by the compression stroke also increases. Gasoline with a low octane rating burns fast and may explode rather than burn when introduced to a high compression ratio. This can cause detonation. The higher a gasoline's octane rating, the less likely it is to explode.

As the compression ratio increases, the octane rating of the gasoline also should be increased to prevent abnormal combustion.

ENGINE EFFICIENCY

Engine efficiency is
the measure of the
energy an engine
requires verses the
energy produced.

Engine **efficiency** is a measure of the relationship between the amount of energy put into the engine and the amount of available energy from the engine. Engine efficiency is expressed in a percentage. The formula for determining efficiency is.

$$\text{efficiency} = \frac{\text{output energy}}{\text{input energy}} \times 100$$

Other aspects of the engine are expressed in efficiencies. These include mechanical efficiency, volumetric efficiency, and thermal efficiency. They are expressed as a ratio of input (actual) to output (maximum or theoretical). Efficiencies are expressed as percentages. They are always less than 100 percent. The difference between the efficiency and 100 percent is the percentage lost during the process. For example, if 100 units of energy were put into the engine and 28 units came back, the efficiency would be equal to 28 percent. This would mean that 72 percent of the energy received was wasted or lost.

Thermal Efficiency

An engine converts the chemical energy of the fuel into heat energy, which is then converted into mechanical energy by the pistons and connecting rods of the engine. Thermal energy is the relationship between the engine power output and the heat energy available in the fuel. A percentage is used to express this relationship.

As the cooling system and the lubrication system cool the engine components, these systems carry away much of the heat energy in the fuel. However, this cooling action is necessary to prevent the destruction of engine parts. Part of the heat energy in the fuel is dissipated through the exhaust system. Some of the heat energy in the fuel is consumed to overcome the internal friction in the engine and powertrain. From the total energy in the fuel, these average losses may occur:

- 35 percent loss to the cooling and lubrication system
- 35 percent loss to the exhaust gas
- 5 percent loss to engine friction
- 10 percent loss to powertrain friction

After these losses, only 15 percent of the thermal energy in the fuel is left to drive the vehicle. In recent years, engine components have been designed to reduce engine friction, and improved lubricants reduce engine and powertrain friction. However, the average engine thermal efficiency still remains below 30 percent. Some engine manufacturers are designing ceramic engine components that are capable of operating at higher temperatures than metal and aluminum components. The use of ceramic engine components increases engine thermal efficiency.

Mechanical Efficiency

Mechanical efficiency is the relationship between the engine power delivered and the power that would be delivered if the engine operated without any power loss. The mechanical efficiency of an engine is expressed as a percentage.

Volumetric Efficiency

As discussed earlier, volumetric efficiency is the relationship between the amount and density of air actually taken into the cylinder on the intake stroke compared to the amount of air required to fill the cylinder at atmospheric pressure. Volumetric efficiency is also expressed as a percentage. Intake manifold design, valve timing, and exhaust system design affect the volumetric efficiency of an engine.

TORQUE AND HORSEPOWER

Torque is a turning or twisting force. The engine's crankshaft rotates with a torque that is transmitted through the drivetrain to turn the drive wheels of the vehicle. **Horsepower (hp)** is the rate at which torque is produced.

Engines produce power by turning a crankshaft in a circular motion. To convert terms of force applied in a straight line to a force applied in a circular motion, the formula is.

$$torque = force \times radius$$

Example: A 10-pound force applied to a wrench 1 foot long will produce 10 pound-feet (lb.-ft.) of torque. Imagine that the 1-foot-long wrench is connected to a shaft. If 1 pound of force is applied to the end of the wrench, 1 pound-foot of torque is produced. Ten pounds of force applied to a wrench 2 feet long will produce 20 pound-feet of torque (**Figure 3-19**).

The technically correct torque measurement is stated in pound-feet (lb.-ft.). However, it is rather common to state torque in terms of foot-pounds (ft.-lb.). In the metric or SI system, torque is stated in Newton meters (N•m) or kilogram-meters (kg-m).

If the torque output of an engine at a given speed (rpm) is known, hp can be calculated by the following formula.

$$hp = (torque \times rpm) \div 5,252$$

Brake **horsepower (hp)** may be defined as the usable power at the engine's crankshaft.

Friction horsepower is the power required to overcome the internal friction of the engine.

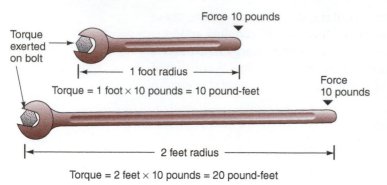

Figure 3-19 Force applied to a wrench produces torque.

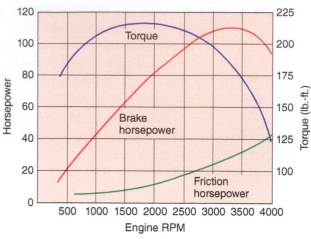

Figure 3-20 The relationship between horsepower and torque.

An engine produces different amounts of torque based on the rotational speed of the crankshaft and other factors. A mathematical representation, or graph, of the relationship between the horsepower and torque of one engine is shown in **Figure 3-20**.

This graph shows that torque begins to decrease when the engine's speed reaches about 1,700 rpm. Brake horsepower increases steadily until about 3,500 rpm. Then it drops. The third line on the graph indicates the horsepower needed to overcome the friction or resistance created by the internal parts of the engine rubbing against each other.

OTHER ENGINE DESIGNS

The gasoline-powered, internal combustion piston engine has been the primary automotive power plant for many years and will probably remain so for years to come. Present-day social requirements and new technological developments, however, have necessitated searches for ways to modify or replace this time-proven workhorse. This portion of the chapter takes a brief look at the most likely contenders and how they work.

Atkinson Cycle Engine

The Atkinson cycle engine is being used in some of the hybrid technology vehicles on the market today. Basically, the Atkinson cycle as currently used involves using an engine with a long stroke and a unique arrangement on the intake valve. It is a four-stroke engine with the normal intake-compression-power-exhaust strokes. The Atkinson cycle involves leaving the intake valve open on the intake stroke past BDC. This way, the start of the compression stroke is delayed until the intake valve is closed. This affects the compression ratio of the engine. The variable valve timing of the engine allows it to respond to conditions as needed. The advantage of the Atkinson cycle is the long power stroke that squeezes every bit of energy out of the fuel burned during the short compression stroke. A conventional four-stroke engine has a power and compression stroke that lasts for the same amount of time. With the long stroke of the Atkinson cycle, this would result in an engine that had a compression ratio too high for readily available pump gasoline. The engine is not known for high power unless supercharged, which is known as a Miller cycle engine.

Two-Stroke Gasoline Engines

In the past, several imported vehicles have used two-stroke engines. As the name implies, this engine requires only two strokes of the piston to complete all four

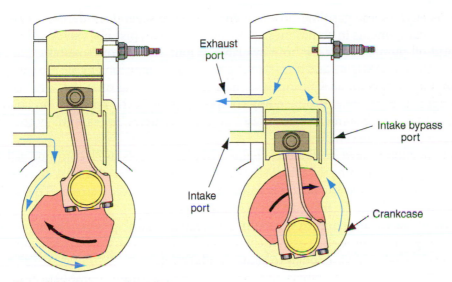

Figure 3-21 The two-stroke engine.

operations: intake, compression, power, and exhaust. As shown in **Figure 3-21**, this is accomplished as follows:

1. Movement of the piston from BDC to TDC completes both intake and compression.
2. When the piston nears TDC, the compressed air-fuel mixture is ignited, causing an expansion of the gases. Note that the reed valve is closed and the piston is blocking the intake port.
3. Expanding gases in the cylinder force the piston down, rotating the crankshaft.
4. With the piston at BDC, the intake and exhaust ports are both open, allowing exhaust gases to leave the cylinder and the air-fuel mixture to enter.

Although the two-stroke-cycle engine is simple in design and lightweight because it lacks a valvetrain, it has not been widely used in automobiles. They tend to be less fuel efficient and have dirtier exhaust than four-stroke engines. Oil is often in the exhaust stream because these engines require constant oil delivery to the cylinders to keep the piston lubricated. Some of these engines require that a certain amount of oil be mixed with the fuel.

In recent years, however, thanks to a revolutionary pneumatic fuel injection system, there has been increased interest in the two-stroke engine. The injection system, which works something like a spray paint gun, uses compressed air to flow highly atomized fuel directly into the top of the combustion chamber. The system becomes the long sought-after answer to the fuel economy and emissions problems of the conventional two-stroke engine. This fuel injection system is the basis for orbital two-stroke direct injection piston engine, which may be used in cars in the future.

Diesel Engines

Diesel engines represent tested, proven technology with a long history of success. Invented by Dr. Rudolph Diesel, a German engineer, and first marketed in 1897, the diesel engine is now the dominant power plant in heavy-duty trucks, construction equipment, farm equipment, buses, and marine applications.

Many manufacturers are using diesel engines in their light-duty trucks because diesel engines are capable of delivering large amounts of torque. Additionally, a few manufacturers are using diesel engines in their cars because of the high fuel mileage potential.

Diesel engines are engines that ignite the air-fuel mixture by high compression, not spark.

Diesel engines and gasoline-powered engines share several similarities. They have a number of components in common, such as the crankshaft, pistons, valves, camshaft, and water and oil pumps. They are both available in four-stroke combustion cycle models. However, the diesel engine and four-stroke compression-ignition engine are easily recognized by the absence of an ignition system. Instead of relying on a spark for ignition, a diesel engine uses the heat produced by compressing air in the combustion chamber to ignite the fuel. The systems used in diesel-powered vehicles are essentially the same as those used in gasoline vehicles.

Fuel injection is used in all diesel engines. Modern diesels use a common rail system similar to a gasoline engine. Injectors spray fuel directly into the cylinders at a pressure as high as 26,000 psi, just as the piston is completing its compression stroke. The heat of the compressed air ignites the fuel and begins the power stroke.

Glow plugs are used only to warm the combustion chamber when the engine is cold. Cold starting can be difficult in extremely cold conditions because even high compression ratios cannot heat cold air enough to cause combustion.

Diesel combustion chambers are different from gasoline combustion chambers because diesel fuel burns differently. Common rail diesels generally use an open combustion chamber. The open combustion chamber has the combustion chamber located directly inside the piston. Diesel fuel is injected directly into the center of the chamber. The shape of the chamber and the quench area produces turbulence. On the power stroke, fuel is injected into the small chamber at the top of the piston.

The high injection pressure of a common rail diesel ensures that the fuel charge is very finely atomized for maximum efficiency.

Glow plugs are used to preheat the combustion chamber in diesel engines.

Rotary Engine

The **rotary engine**, or **Wankel engine**, is somewhat similar to the standard piston engine in that it is a four-cycle, spark ignition, internal combustion engine. Its mechanical design, however, is quite different. For one thing, the rotary engine uses a rotating motion rather than a reciprocating motion. In addition, it uses ports rather than valves for controlling the intake of the air-fuel mixture and the exhaust of the combusted charge.

The heart of a rotary engine is a roughly triangular rotor that "walks" around a smaller, rigidly mounted gear. The rotor is connected to the crankshaft through additional gears in such a manner that for every rotation of the rotor the crankshaft revolves three times. The tips of the triangular rotor move within the housing and are in constant contact with the housing walls. As the rotor moves, the volume between each side of the rotor and the housing walls continually changes.

Referring to **Figure 3-22**, when the rotor is in position A, the intake port is uncovered and the air-fuel mixture is entering the upper chamber. As the rotor moves to position B, the intake port closes and the upper chamber reaches its maximum volume. When full compression has reached position C, the two spark plugs fire, one after the other, to start the power stroke. At position D, rotor side A uncovers the exhaust port and exhaust begins. This cycle continues until position A is reached where the chamber volume is at minimum and the intake cycle starts once again.

The fact that the rotating combustion chamber engine is small and light for the amount of power it produces makes it attractive for use in automobiles. Using this small, lightweight engine can provide the same performance as a larger engine. But the rotary engine, at present, cannot compete with the piston gasoline on durability, exhaust emission control, and economy.

Rotary (or **Wankel**) **engine** components rotate on a shaft; whereas piston engines are referred to as having a reciprocating motion.

Miller-Cycle Engines

This engine design is a modification of the four-stroke-cycle engine. During the intake stroke, a supercharger feeds highly compressed air to an intercooler. This cooled, but compressed, air is fed to the cylinders. During the compression stroke, the intake valve remains open for longer than normal. This prevents compression from occurring until

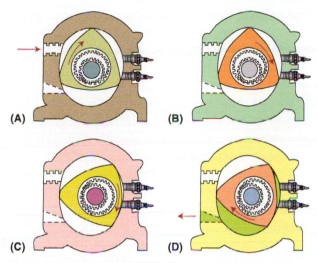

(A)   (B)

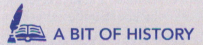

(C)   (D)

Figure 3-22 Rotary operational cycles.

the piston has moved one-fifth of its upward travel. Then the valve closes and compression occurs. The shortened compression stroke keeps the compression ratios and cylinder temperatures low. The power stroke begins as soon as the piston is ready to move down its bore and continues until it reaches BDC. This longer power stroke provides more torque and increased efficiency. The exhaust stroke is much the same as that in a four-stroke-cycle engine.

A BIT OF HISTORY

In the 1940s, Ralph H. Miller, an American, designed an engine with a compressor that forced more air-fuel mixture into the cylinders during the compression stroke. Today, modified versions of his design can be found in marine and industrial applications.

Electric Motors

In the early days of the automobile, electric cars outnumbered gasoline cars. An electric motor is quiet, has little or no emissions, and has few moving parts. It starts well in the cold, is simple to maintain, and does not burn petroleum products to operate. The main disadvantage is limited range, so they are suitable only for commuting. Several electric cars have recently been introduced. The Nissan Leaf and Ford Focus Electric are totally electric cars that have been released. The Leaf and Focus are different in that they do not have the option to use a gasoline motor; their operation is totally electric. The vehicles are plugged into a household power and recharged when the vehicle is not in use. In some regions, power is cheaper "off peak" at night, which can save money on the electrical power required to recharge the batteries.

Hybrid Electric Vehicles

Hybrid electric vehicles (HEVs) have been available for some time. The HEV uses both a small gasoline engine and an electric motor. Power for the electric motor is stored in a battery pack, which is charged by a generator powered by the gasoline engine. Depending on the type of hybrid vehicle, the gasoline engine may be a source of power to help drive the wheels, or in some cases the gasoline engine may just be used to drive a generator to charge the battery and provide electrical power to the electric motor. Additionally, the braking system is typically regenerative, so power is generated while stopping during light braking (during heavy braking, a normal hydraulic braking system is used).

Hybrid Types

Hybrid vehicles are classified as either mild, parallel, or series. In a **mild hybrid**, which is also known as a CAS (combined alternator starter) type, a gas or diesel engine runs to charge the batteries for the electric motor. The electric motor runs only when the small internal combustion motor needs assistance during acceleration or hill pulling.

The **series hybrid** uses a small internal combustion engine to charge the batteries when they become discharged. The electric motor provides power to the drive wheels. The series hybrid (**Figure 3-23**) uses a complex controller to provide seamless acceleration from the electric motor. The Chevrolet Volt is a series hybrid that uses the gasoline engine only to charge the battery. The gasoline engine is not used to provide power to the drive wheels at all.

The **parallel hybrid** uses both the electric motor and the internal combustion engine to power the drive wheels. The electric motor in the parallel system is small and uses principally the internal combustion engine except during times of low load, such as in town driving or freeway cruising. The Toyota Prius is a parallel hybrid.

Fuel Cells

Although commonly used in spacecraft and submarines, emerging technological advances are making the fuel cell a contender for automotive use. In the simplest terms, the fuel cell is a component that converts hydrogen to electricity. The electricity that is converted from hydrogen (derived from methanol or gasoline) will be used to power an electric motor that powers the car. The fuel cell is perhaps the answer to the problems associated with a conventional electric vehicle. The advantage of the fuel cell vehicle is that the batteries are lighter and are charged by the fuel cell, as compared to other electric vehicles, which have to be plugged in. The fuel cells do not store electricity, but rather convert it to be used as required to drive the electric motors. When power requirements exceed the fuel cell's capacity, power is provided by storage batteries. The hydrogen will likely come from methanol. Methanol can be handled and stored much the same way gasoline is. Currently under development for automotive application, fuel cell technology as an alternative to gasoline-powered engines will be the focal point for some time.

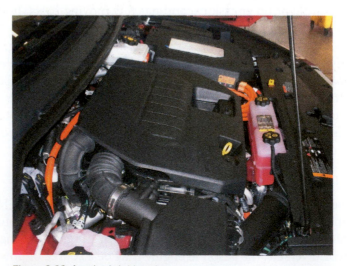

Figure 3-23 A series hybrid engine.

ENGINE LUBRICATION

An engine's lubricating system performs several important tasks. It holds an adequate supply of oil to cool, clean, lubricate, and seal the engine. It also removes contaminants from the oil and delivers oil to all necessary areas of the engine.

Engine Oil

Engine oil is a clean or refined form of **crude oil**. Crude oil, when taken out of the ground, is dirty and does not work well as a lubricant for engines. Crude oil must be refined to meet industry standards. Engine oil is just one of the many products that come from crude oil. Engine oil is specially formulated so that it has the following properties:

- Prompt circulation through the engine's lubrication system
- The ability to lubricate without foaming
- The ability to reduce friction and wear
- The ability to prevent the formation of rust and corrosion
- The ability to cool the engine parts it flows on
- The ability to keep internal engine parts clean
- The ability to suspend and carry small particles to the oil filter

To provide these properties, engine oil contains many additives. Because of these additives, choosing the correct oil for each engine application can be a difficult task. However, the **American Petroleum Institute (API)** has developed service ratings for motor oil that greatly simplify oil selection.

Performance requirements, test methods, and limits for motor oil are cooperatively established by vehicle and engine manufacturers and technical societies such as the American Society for Testing and Materials (ASTM) and the **Society of Automotive Engineers (SAE)**.

The API uses two types of marks: the API Service Symbol and the API Certification Mark (**Figure 3-24**). The API Service Symbol indicates the performance level and grade of the oil. The API Certification Mark, introduced in 1993, helps consumers identify

Shop Manual
Chapter 3, page 98

Engine oil is often called motor oil.

Crude oil is unrefined, uncleaned oil as it is pumped from the ground.

The **American Petroleum Institute (API)** is a technical society that has developed service ratings for motor oil.

The API circle and markings on oil containers is commonly called the API doughnut.

The **Society of Automotive Engineers (SAE)** established a classification system for oil viscosity ratings.

Figure 3-24 Identification labels on oil containers.

engine oil that is suitable for use in gasoline engines. Motor oils licensed to use the API Certification Mark meet the latest standards of the International Lubricant Standardization and Approval Committee (ILSAC), which comprises U.S. and Japanese automobile manufacturers.

The API Service Symbol has been around for many years and is well recognized. It is divided into three parts:

The top portion describes the performance level of the oil.
The center section displays the SAE viscosity grade of the oil.
The lower section indicates whether the oil has energy-conserving properties.

The top portion of the API Service Symbol indicates the oil's performance for gasoline or diesel engines. The API classifies engine oil as an S (Service) class for passenger cars and light- and medium-duty trucks with gasoline engines and as a C (Commercial) class for heavy-duty diesel applications. As a coincidence (not a true definition), think of the S-class oils for a spark-ignited engine and the C-class oils for compression-ignited engines. The second letter following both the S and C categories is generally assigned alphabetically in order of development.

The following service rating examples for gasoline engines are:

SG—Obsolete (was introduced in 1993)
SH—Obsolete (was introduced in 1996)
SJ—Obsolete (was introduced in 1997)
SL—Obsolete (was introduced in 2001)
SM—Obsolete (was introduced in 2004)
SN—Introduced in 2010 for current and older engines

It is important to note that each gasoline engine category above exceeds the performance properties of all the previous categories and can be used in place of the lower one(s). SN-rated oil, for example, can be used for any previous category. It is expected that the next API change will move the S rating one step higher. The tests that the oil must pass will be more stringent, so the letter rating will go up to reflect that new standard.

The following are the service ratings for diesel engines:

CE—Obsolete—Introduced in 1987 for high-speed, four-stroke, naturally aspirated, and turbocharged engines.
CF—Current—Introduced in 1994. The CF-rated oil can be used in place of CD-rated oils.
CF-4—Current—Introduced in 1990 for high-speed, four-stroke, naturally aspirated, and turbocharged engines.
CG-4—Current—Introduced in 1995 for severe-duty, high-speed, four-stroke engines using fuel with less than 0.5 percent weight sulfur. The CG-4 rated oil can be used in place of CD, CE, and CF-4 oils.
CH-4—Current—Introduced in 1999 for high-speed, four-stroke engines designed to meet 1998 emission standards using fuel with up to 0.5 percent weight sulfur. The CH-4 rated oil can be used in place of CD, CE, CF-4, and CG-4 oils.
CI-4—Current—Introduced in 2002 for four-stroke engines meeting 2004 emission standards. Can be used in place of CD, CE, CF-4, and CH-4 oils.
CJ-4—Introduced for 2007 and later diesels to be compatible with the diesel particulate filter installed on these engines to limit particulate emissions. This oil has limited amounts of sulfates and phosphorus because they are incompatible with the particulate filter.

The center section of the API Service Symbol indicates the SAE viscosity grade of the oil. The SAE classifications are the internationally accepted standard for defining viscosity. However, this standard provides no information specific to the quality of the oil. Although single-grade oils are still used in certain applications, the multigrade oils are most widely used in today's automotive applications.

Friction Modifiers

Oil refiners add a number of different additives to their oil that improve oil performance in the engine. Friction modifiers in the engine oil reduce friction between moving parts that contact each other. The oil film between two moving surfaces is actually sheared by the moving components. During this oil-shearing process, there is a certain amount of friction in the oil. This friction within the oil is referred to as viscous friction, and this type of friction is reduced by friction modifiers in the oil.

Antifoaming Agents

Oil foaming may be caused by the rotating action of the crankshaft and connecting rods. Because oil foam contains oil and air, excessive oil foaming causes air in the lubrication system, which partially destroys the oil film on moving components in contact with each other. This action allows these moving components to contact each other, causing excessive friction and wear. Antifoaming agents added to the engine oil reduce oil foaming in the engine. Oil foaming may be caused by overfilling the crankcase; therefore, it is important to maintain the oil at the proper level on the dipstick.

Corrosion and Rust Inhibitors

Because the engine experiences severe temperature changes, condensation may form water in the engine. Condensation on engine components can cause rust. When rust inhibitors are added to the oil, they help disperse moisture from the metal surfaces. Certain acids may be formed in small quantities during the combustion process, and these acids get past the rings into the engine oil. Acids cause corrosion on engine components as they circulate through the lubrication system with the oil. Acids in the oil are neutralized by corrosion inhibitors added to the oil.

Extreme Pressure Resistance

When an engine is operating under heavy load, the pistons and connecting rods produce an extremely high downward force on the crankshaft journals. Under this condition, the film of oil in the connecting rod and main bearings may be subjected to pressures up to 1,000 psi (6,900 kPa). This high pressure tends to squeeze the oil out of the crankshaft bearings. If the oil is squeezed out of the bearings and the bearing insert contacts the crankshaft, bearing damage results. Extreme pressure resistance additives help prevent the oil from being squeezed out of the bearings.

Detergents and Dispersants

Carbon is a normal by-product of the combustion process. This carbon collects on the combustion chamber surfaces. However, some of this carbon collects on the top ring lands and compression rings. Detergents added to the oil help clean these carbon deposits from the piston and rings, and this action keeps the rings working freely in the ring grooves.

A dispersant added to the engine oil helps prevent carbon particles from sticking together to form larger particles. Oil passages may be plugged by large carbon particles. The smaller carbon particles are removed by the oil filter, while some larger particles may remain in the oil pan. These carbon particles in the oil pan are removed when the oil is changed.

Oxidation Inhibitors

The rotation of the crankshaft causes oil agitation in the crankcase. Air combines with high-temperature oil when the oil is agitated. During this combination of air and high-temperature oil, the oxygen in the air combines with the oil and forms a sticky substance

Viscous friction refers to the resistance of an oil film to shearing apart.

Oils marked "Energy Conserving" can increase fuel economy by more than 1.5 percent. Those oils marked "Energy Conserving II" can increase fuel economy by more than 2.7 percent.

that is similar to tar and contains corrosive compounds. This tarlike material may plug oil passages or destroy the oil film on some of the engine components. The corrosive compounds may attack and erode bearing materials. Oxidation inhibitors added to the oil help prevent oxidation.

Viscosity

Shop Manual
Chapter 3, page 97

In addition to the additives and performance rating of the motor oil, viscosity is another important aspect of oil selection. Viscosity is merely the resistance of the oil to flow. Thin oil flows freely and is less resistant to flow, whereas thick oil has a higher resistance to flow and flows more slowly. Testing standards are established to measure the specific amount of oil that flows in a given period at specific testing temperatures, both hot and cold. The multigrade oils have additives called viscosity index improvers that allow the oil to flow as evenly when cold as when hot. In other words, with a multigrade oil there is a less noticeable difference in oil flow characteristics between temperature extremes. Engines operate in a wide range of temperatures and conditions: climate, stop-and-go city driving, prolonged highway speeds, pulling loads, and so on. Engines must function reliably in the harshest of operating environments. Consider today's engine designs, some of which have complex valvetrains, block and piston arrangements, tight clearances, and engines that run hotter in a smaller engine compartment. Generally, the largest percentage of critical engine wear occurs within the first 15 seconds of starting a cold engine. The properties of the oil must allow it to flow properly to provide lubrication and not break down, regardless of the conditions.

The SAE has established a classification system that standardizes oil viscosity ratings. This system is a numeric rating in which the higher viscosities, or thicker oils, receive the higher numbers.

For example, an SAE 40 oil is thicker and flows slower than an SAE 5 oil. Obviously, a cold engine needs to have the oil flow quickly and freely. Because heat can cause the oil to flow easier, thicker oil is needed for high temperatures. Straight-viscosity oils will have a single number or rating, such as SAE 40. There are no viscosity index improvers added to extend the flow properties. The multigrade has two numbers, indicating a cold flow rate and a hot flow rate. The numbers are assigned as a result of the oil's performance during the cold- and hot-flow rate tests performed. For example, SAE 5W-30 has the flow characteristics of 5 "weight" oil when cold and the hot flow rate of a 30 "weight." The number before the *W* indicates the cold rating. The *W* actually stands for "winter," not "weight" as some may assume. The oil testing standards are being made more stringent, and the motor oils are constantly evolving with new compounds and additives to keep pace with engineering advances. Oil change intervals have been extended due to these improvements. A longer interval between oil changes not only saves money for the consumer, but also makes sense environmentally. Engine life will be prolonged by using the proper rated oil recommended specifically by the manufacturer in addition to changing the oil and oil filter at the recommended intervals.

The SAE classification and the API rating are usually indicated on the oil container (**Figure 3-25**). Selecting oils that specifically meet or exceed the manufacturer's recommendations and changing the oil on a regular basis will allow the owner to achieve the maximum service life from an engine.

Engine oils can be classified as **energy-conserving (fuel-saving) oils**. These are designed to reduce friction, which in turn reduces fuel consumption. Friction modifiers and other additives are used to achieve this. Energy-conserving oil is identified as such on the top of the oil can. Some oils are specially formulated for turbocharged engines. This oil is identified on the container with a marking such as "Turbo Formula."

Figure 3-25 API designation and SAE rating.

Synthetic and Recycled Oils

Synthetic oils are recommended by a few engine manufacturers. These oils are manufactured from crude oil base stock and chemically altered to provide improved stability over a wide temperature range and improved viscous friction qualities. When synthetic oil is used in an engine, there may be a slight improvement in fuel economy and cold weather cranking speed.

The introduction of synthetic motor oils dates back to World War II. It is often described as the oil of the future. They offer a variety of advantages over natural oils including better fuel economy, stability over a wide range of temperatures and operating conditions, and longevity. The impurities found in conventional oils cause problems of oxidation when hot and limit cold-temperature performance. Less oxidation means fewer engine deposits. Synthetic oils are more costly than natural oil.

Engine oil that is based on petroleum and synthetic products is available. These oils combine some of the better attributes of synthetics with the lower-cost petroleum-based oils. Typically, these oils are 80 percent petroleum and 20 percent synthetic.

Used oil that has been processed to remove impurities and restored to original standards is called recycled oil. Many oil recycling plants have been built in North America, and the recycled oil from these plants should be equivalent to new refined oil. Most states have strict laws regarding the dumping of used oil.

GM Dexos

Recently, GM has introduced proprietary engine oil for use in its 2010 and later vehicles, called Dexos. Dexos is a semisynthetic blend that meets GM specifications for thermal degradation, friction reduction, resistance to foaming (important for variable valve timing components), oxidation, and deposit formation. GM feels this oil will allow for longer oil change intervals without excessive engine wear. GM has also released a list of engine oils that meet the same specifications as Dexos so that compatible engine oils are available on the open market. In addition, GM has released Dexos 2 for diesel engines.

AUTHOR'S NOTE The EPA estimates that the waste oil from a single oil change can ruin a million gallons of drinking water. Encourage your customers and friends who change their own oil to recycle. Recycling not only helps the environment but also helps cut down on our dependence on foreign oil.

LUBRICATING SYSTEMS

The main components of a typical lubricating system are described here.

Oil Pump

The oil pump is the heart of the lubricating system. Just as the heart in a human body circulates blood through veins, an engine's oil pump circulates oil through passages in the engine.

The oil pump on most late-model vehicles is driven by the crankshaft and is located in the front cover, behind the crankshaft pulley. Many of these pumps are variable displacement style to save fuel. The pump's displacement can be varied to cut down on the load the oil pump puts on the engine. The oil pump's purpose is to supply oil to the various moving parts in the engine. To make sure the parts are lubricated, an adequate amount of oil must be delivered to the parts. To get the oil to move through the engine and into the various parts, the oil must be pressurized. Variable displacement oil pumps are designed to move the correct volume of oil and pressurize it to a certain amount. The amount of oil flow through the engine depends on the volume of oil required, the pressure of the oil, and the clearance (or space) the oil must flow through (**Figure 3-26**). Some older vehicles on the road may have a positive displacement oil pump. The positive displacement pump always has the same load on the engine for a given RPM, so they require a pressure relief valve to limit oil pressure at higher RPMs. (See the discussion of pressure relief valves below.)

> The oil pump is generally driven by the crankshaft on most late-model engine designs.

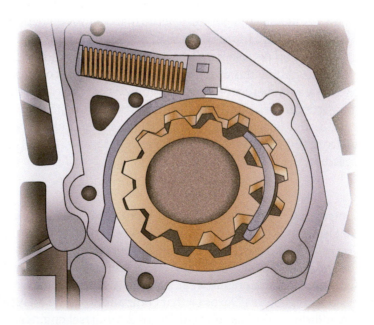

Figure 3-26 Oil pump and drive assembly.

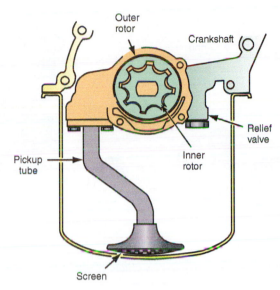

Figure 3-27 Oil pump and pickup assembly.

Oil Pump Pickup

The **oil pump pickup** is a line from the oil pump to the oil stored in the oil pan (**Figure 3-27**). It usually contains a filter screen, which is submerged in the oil at all times. The screen serves to keep large particles from reaching the oil pump.

The oil pump pickup is in the bottom of the oil pan.

Oil Pan or Sump

The oil pan attaches to the crankcase or block. It serves as the reservoir for the engine's oil. It is designed to hold the amount of oil that is needed to lubricate the engine when it is running, plus a reserve. The oil pan helps cool the oil through its contact with the outside air.

The anti-drain back valve ensures some oil will stay in the overhead cam area.

Pressure Relief Valves

There are two types of oil pumps: a variable displacement pump, much like those found in transmissions, and a positive displacement pump may still be found in some older model engines.

Engines using variable displacement pumps incorporate the regulator in the design of the pump itself. The regulator functions to change the displacement of the pump. Because the oil pump does not always need full pressure in operation, the load on the engine to drive the oil pump can be reduced in times of low need. This helps to save gasoline and improve performance.

Some engines use a positive displacement pump, and an oil pressure relief valve (**Figure 3-28**) is included in the system to prevent excessively high system pressures from occurring as engine speed is increased. Once oil pressure exceeds a preset limit, the spring-loaded pressure relief valve opens and allows the excess oil to bypass the rest of the system and return directly to the pump.

Shop Manual
Chapter 3, page 98

Oil Filter

Under pressure from the oil pump, oil flows through a filter (**Figure 3-29**) to remove any impurities that might have become suspended in the oil. This prevents impurities from circulating through the engine, which can cause premature wear. Filtering also increases the usable life of the oil. Oil filters are equipped with a bypass valve to allow oil to bypass the filter if the filter is clogged.

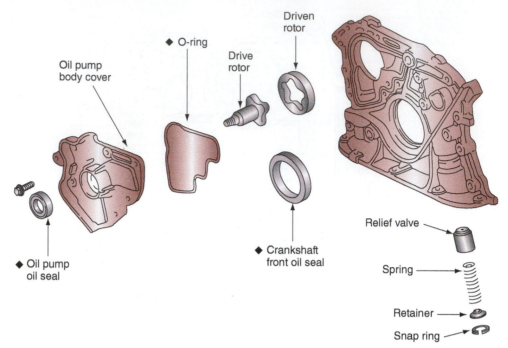

Figure 3-28 Oil pressure relief valve.

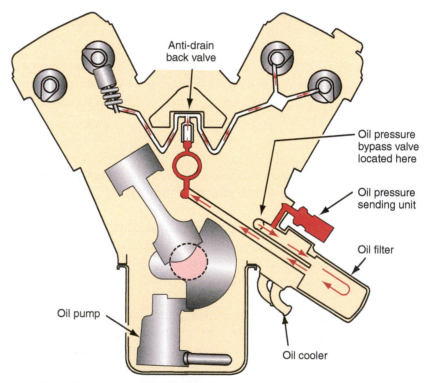

Figure 3-29 Oil passes from the filter through the anti-drain back valve to the lifters and the camshaft journals.

Engine Oil Passages or Galleries

From the filter, the oil flows into the engine oil galleries (**Figure 3-30**). These galleries consist of interconnecting passages that have been drilled completely through the engine block during manufacturing. The outside ends of the passages are blocked off so the oil

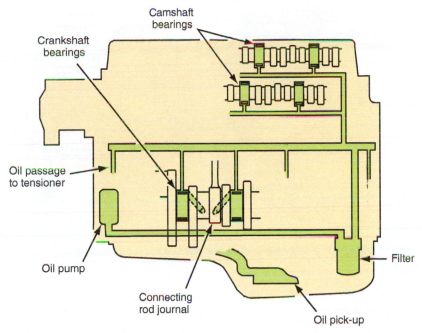

Camshaft bearings

Crankshaft bearings

Oil passage to tensioner

Oil pump

Connecting rod journal

Oil pick-up

Filter

Figure 3-30 Oil passes from the main oil gallery to the main bearings and connecting rod journals.

can be routed through these galleries to various parts of the engine. The crankshaft also contains oil passages (oilways) to route the oil from the main bearings to the connecting rod bearing surfaces.

Engine Bearings

Oil is delivered to the engine bearings by an oil gallery. An oil hole is therefore machined in the bearing for alignment with the oil gallery in the engine block. Many engine bearings are manufactured with an oil groove to help distribute the oil over the surface of the bearing. Once the oil has been used by the bearing, it flows out of the oil clearance space and is replenished with a fresh supply of oil under pressure from the oil pump.

Crankcase Ventilation

Crankcase ventilation is necessary because pressure can build in the crankcase due to combustion pressure. This pressure passes by the piston rings. Piston rings do not provide a total seal of the combustion area, and some combustion gases are able to reach the oil pan. These gases contaminate the oil and apply unwanted pressure on gaskets and seals.

The positive crankcase ventilation (PCV) system removes gases from the crankcase and delivers them back into the intake manifold and the cylinders. In the PCV system, a clean air hose is connected from the air cleaner to the rocker arm cover. A small filter is mounted on the end of this clean air hose inside the air cleaner. The PCV valve hose is connected from the rocker arm cover to one end of the PCV valve, and the other end of this valve is connected to the intake manifold. Inside the PCV, a spring-loaded, tapered valve controls the flow of crankcase vapors in relation to engine vacuum and load. With the engine idling, the PCV valve is nearly closed. The valve moves toward the open position as the throttle is opened and engine load increases. Intake manifold vacuum moves clean air from the air cleaner into the crankcase, and the crankcase gases are moved through the PCV valve into the intake manifold (**Figure 3-31**).

A dirty oil filter allows unfiltered oil to circulate when the bypass valve closes.

The true purpose of the PCV system is often overlooked because it is typically considered an emission control device. The PCV is an emission control device, but most important, it relieves the engine's crankcase of pressure and vapors.

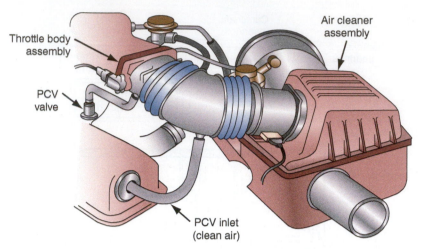

Figure 3-31 The PCV valve relieves crankcase pressure and burns blowby gases in the engine.

Oil Pressure Indicator

The driver can monitor oil pressure by looking at a gauge. The gauge indicates the engine oil pressure at all times, or it can be a warning light that will come on whenever the engine is running with insufficient oil pressure. The warning light is the most common oil pressure indicator.

Oil Seals and Gaskets

Oil seals and gaskets are used throughout the engine to prevent both external and internal oil leaks. The most common materials used for sealing are synthetic rubber, soft plastics, fiber, and cork. In critical areas, these materials might be bonded to a metal. It is important to note that because of today's engines with computer-controlled management systems, a seemingly innocent gasket or seal leak can actually affect drivability. A damaged or leaking seal or gasket can allow extra air to be drawn into the crankcase through the PCV system. This results in unmetered air entering the intake manifold. Although minor, depending on the type of system the computer uses to read air intake, the engine can idle rough, idle high, or have other symptoms.

Dipstick

The dipstick is used to measure the level of oil in the oil pan. The end of the stick is marked to indicate when the engine oil level is correct. It also has a mark to indicate the need to add oil to the system (**Figure 3-32**).

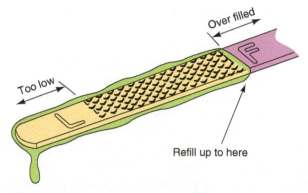

Figure 3-32 The dipstick indicates the oil level in the crankcase.

Oil Coolers

Some engines, such as diesel engines or turbocharged engines use an external oil cooler. These typically look like a small radiator mounted near the front of the engine. Heat is removed from the oil as air flows through the cooler. Normal maximum engine oil temperature is considered to be 250°F (121°C). Hot oil mixed with oxygen breaks down (oxidizes) and forms carbon and varnish. The higher the temperature, the faster these deposits build. An oil cooler helps keep the oil at its normal operating temperature.

COOLING SYSTEMS

During the combustion process, temperatures can reach 4,500°F (2,468°C). While the engine is idling or operating at moderate speeds, the average combustion chamber temperature is 2,000°F (1,080°C). These temperatures in the cylinder are high enough to melt aluminum pistons, distort the cylinder walls, warp the cylinder head, and destroy the lubricating oil. Coolant circulation through the jackets in the cylinder head and block must absorb heat and lower component temperatures to a safe operating level (**Figure 3-33**).

Heat may be transferred by conduction, convection, or radiation in an engine cooling system. However, most of the heat is transferred by conduction from the hot engine components to the lower-temperature coolant. Coolant temperature also affects combustion and engine efficiency. Some computer system input sensors also depend on correct coolant temperature to produce a normal signal.

> Heat transfer by conduction means the heat transfers to a cooler object.
>
> Heat is transferred by convection when heated atoms or molecules of fluid become less dense and lighter than the cooler atoms or molecules. The heated atoms or molecules rise upward, while the cooler portion of the fluid sinks downward.

Liquid-Cooled System

By far, the most popular and efficient method of engine cooling is the liquid-cooled system (**Figure 3-34**). In this system, heat is removed from around the combustion chambers by a heat-absorbing liquid (coolant) circulating inside the engine. This liquid is pumped through the engine and after absorbing the heat of combustion, flows into the radiator where the heat is transferred to the atmosphere. The cooled liquid is then returned to the engine to repeat the cycle. These systems are designed to keep engine temperatures within a range where they provide peak performance.

The engine's thermostat controls the temperature and amount of coolant entering the radiator. While the engine is cold, the thermostat remains closed, allowing coolant

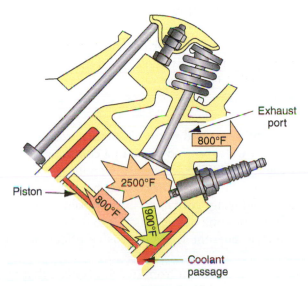

Figure 3-33 Heat is transferred from the cylinders to the coolant circulating through the coolant jackets.

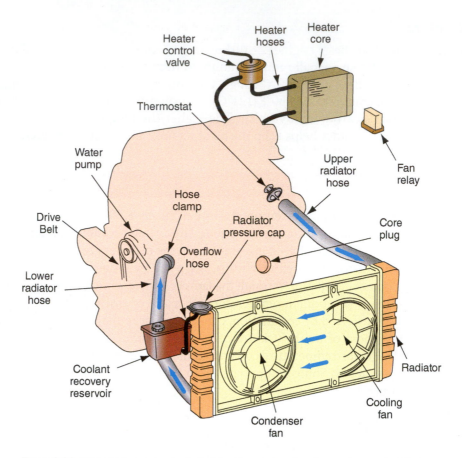

Figure 3-34 Major components of a typical liquid-cooled engine. Arrows show coolant flow.

Heat is transferred by radiation when heat waves travel through the atmosphere and strike another object, causing a warming effect on that object.

to circulate only inside the engine. This allows the engine to warm up uniformly and eliminates hot spots. When the coolant reaches the opening temperature of the thermostat, the thermostat begins to open and allows the flow of coolant to the radiator. The hotter the coolant becomes, the more the thermostat opens, allowing more coolant to flow through the radiator. Once the coolant has passed through the radiator and has given up its heat, it reenters the water pump. Here it is again pushed through the passages surrounding the combustion chambers to pick up heat and start the cycle once again.

The thermostat also maintains proper coolant temperatures to keep the emission control systems working properly. Many vehicles fail the emissions test because they are operating at a temperature that is too low.

Shop Manual
Chapter 3, page 100

Coolant

Ethylene glycol is the basic liquid in automotive antifreeze.

Coolant is a mixture of antifreeze and water. The antifreeze contains ethylene glycol and anticorrosion chemicals. If water is used alone in a cooling system, the water promotes rust, corrosion, and electrolysis. The water in the cooling system may become slightly acidic because of minerals and metals in the cooling system. Metals such as brass, copper, aluminum, and cast-iron alloy are contained in the cooling system. A small amount of electric current may flow from one of these metals to the other through the acid formed in the water. This electric current has a corrosive effect on cooling system metal components.

The antifreeze protects the coolant from freezing. Given that water expands when it is frozen, the coolant jackets in the block and head may be cracked by frozen coolant. The radiator and heater core are also damaged by frozen coolant.

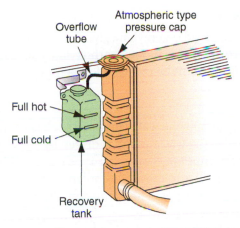

Figure 3-35 Coolant recovery (degas bottle) tank with coolant levels marked on the tank.

If the coolant boils, the liquid is no longer in contact with the cylinder walls and combustion chamber surfaces in the head. Under this condition, the heat is no longer transferred from these surfaces to the coolant, and the cylinder walls and combustion chamber surfaces become overheated. This action can score or collapse pistons and crack or warp cylinder heads.

A mixture of 50 percent antifreeze and water has a boiling point of approximately 230°F (103°C) at atmospheric pressure compared to water, which has a boiling point of 212°F (100°C) at sea level. This mixture of antifreeze and water provides extra protection from boiling. The same mixture of 50 percent antifreeze and water provides freezing protection to approximately −40°F (−35.5°C).

Most cooling systems today use an **expansion** or **recovery tank** (**Figure 3-35**). Cooling systems with expansion tanks are called closed-cooling systems. They are designed to catch and hold any coolant that passes through the pressure cap when the engine is hot. As the engine warms up, the coolant expands. This eventually causes the pressure cap to release, and the coolant passes to the expansion tank. When the engine is shut down, the coolant begins to shrink. Eventually, the vacuum spring inside the pressure cap opens and the coolant in the expansion tank is drawn back into the cooling system.

Recently, certain cooling systems have moved the radiator pressure cap to the coolant reservoir called the **surge tank** (**Figure 3-36**). The surge tank is a part of the cooling system's pressurized system. Unlike the recovery tank system, there is no direct access to the radiator for adding or checking coolant. Use caution when servicing a hot cooling system, and treat the pressure cap on the surge tank as if the cap were on the radiator. The surge tank can hold pressure just like the radiator. The cap on the tank will have a warning on it advising of potential pressure.

The coolant recovery system is an emission control system that prevents spillage of coolant, which is a hazardous material.

The **expansion** or **recovery tank** catches and holds coolant that passes through the pressure cap.

The **surge tank** is actually a pressurized part of the cooling system.

Figure 3-36 A cooling system surge tank.

At 60 mph, an average water pump can push as much as 20 or more gallons of coolant per minute through the engine and radiator.

In most automotive applications, the water pump is driven by the crankshaft.

Extended Life Coolant

Antifreeze formulations are based on ethylene glycol, with the corrosion protection based on carboxylates. The corrosion protection of carboxylates does not tend to diminish over time, so coolant is not scheduled to be changed for as much as 150,000 miles or 10 years.

Water Pump

The heart of the cooling system is the water pump. Its job is to move the coolant through the cooling system. Typically, the water pump is driven by the crankshaft through pulleys and a serpentine belt. The pumps are centrifugal-type pumps (**Figure 3-37**), with a rotating paddle wheel–type of impeller to move the coolant. The shaft is mounted in the water pump housing and rotates on bearings. The pump contains a seal to keep the coolant from passing through it. At the drive end (the exposed end), a pulley is mounted to accept the belt. The pulley is driven by the crankshaft. The pump housing usually includes the mounting point for the lower radiator hose.

When the engine is started, the pump impeller pushes the water from its pumping cavity into the engine block. When the engine is cold, the thermostat is closed. This stops the coolant from reaching the top of the radiator. In order for the water pump to circulate the coolant through the engine during warm-up, a bypass passage is added below the thermostat, which leads back to the water pump. This passage must be kept free to eliminate hot spots in the engine during warm-up. It also allows hot coolant to pass through the valve, which will open the thermostat when it reaches the proper temperature.

Radiator

Shop Manual
Chapter 3, page 101

The radiator is basically a heat exchanger, transferring heat from the engine to the air passing through it. The radiator itself is a series of tubes and fins that expose the heat from the coolant to as much surface area as possible. This maximizes the potential of heat being transferred to the passing air. The inlet tank in the radiator contains a baffle to distribute and de-aerate the coolant.

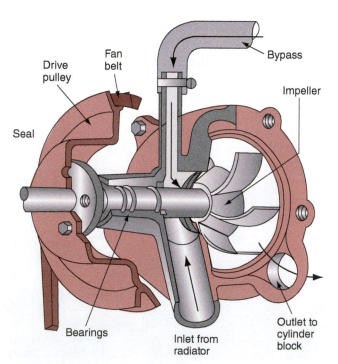

Figure 3-37 Impeller-type water pump.

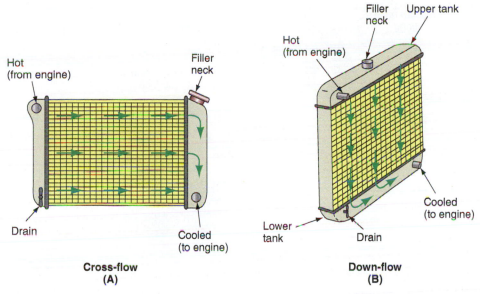

Figure 3-38 Radiator coolant flow.

Factors influencing the efficiency of the radiator are the basic design of the radiator, the area and thickness of the radiator core, the amount of coolant going through the radiator, and the temperature of the cooling air.

A radiator's efficiency can be greatly increased by increasing the difference between the temperature of the coolant and the outside air flowing through it. This can be done by only raising the temperature of the coolant. Doing this permits the use of a smaller radiator or the use of the same size radiator to cool a larger engine. This is the primary reason manufacturers have been specifying higher start-to-open temperatures for thermostats and higher pressure ratings for radiator pressure caps.

The radiator is usually based on one of these two designs: cross-flow or down-flow. In a cross-flow radiator (**Figure 3-38A**), coolant enters on one side, travels through tubes, and collects on the opposite side. In a down-flow radiator (**Figure 3-38B**), coolant enters the top of the radiator and is drawn downward by gravity. Cross-flow radiators are seen most often in large-engine or late-model cars because all the coolant flows through the fan airstream, which provides maximum cooling.

The two types of radiator core construction are honeycomb or cellular type and tube and fin. There are also two radiator core construction materials that are used: the copper/brass, soft solder coolers, and the vacuum-brazed, aluminum cores cinched to nylon tanks.

The aluminum type is thin, lightweight, and less costly. (The initial price of aluminum is less than copper.) For these reasons, aluminum core radiators are being used in new vehicles.

Most radiators feature petcocks or plugs that allow a technician to drain coolant from the system. Coolant is added to the system at the radiator cap or the recovery tank depending on the type of system being used.

Cooling System Pressure Caps

Atmospheric pressure is 14.7 psi at sea level, and this pressure decreases as elevation increases above sea level. At elevations below sea level, atmospheric pressure increases. Water boils at 212°F (100°C) at sea level, and this boiling point decreases at higher elevations with reduced atmospheric pressure. The cooling system pressure cap contains a spring-loaded lower sealing gasket that contacts the lower sealing surface in the radiator

Shop Manual
Chapter 3, page 103

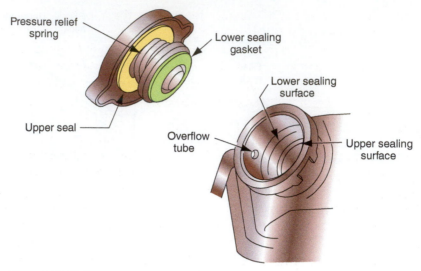

Figure 3-39 Radiator pressure cap assembly.

filler neck (**Figure 3-39**). The radiator pressure cap, along with antifreeze, allows the coolant to reach a high temperature without the danger of boiling.

On many cooling system caps, the spring and sealing gasket are designed to maintain 18 psi (124 kPa) in the cooling system. Although the cooling system pressure remains below 18 psi (124 kPa), the spring keeps the sealing gasket closed. For each 1 psi (6.8 kPa) of pressure increase, the boiling point of the water increases approximately 3°F (1.6°C). Therefore, the boiling point of water at 18 psi (124 kPa) is about 222°F (106°C). The 50 percent antifreeze and water solution increases the coolant boiling point to approximately 266°F (134°C). A pressurized cooling system provides additional protection against coolant boiling.

If the coolant overheats and the cooling system pressure exceeds 18 psi (124 kPa), this pressure forces the sealing gasket open against the spring tension. This action allows coolant to escape from the filler neck through the overflow tube into the coolant recovery system.

Older vehicles used a 15-pound pressure cap.

Water Outlet

Shop Manual
Chapter 3, page 102

The water outlet is the connection between the engine and the upper radiator hose through which hot coolant from the engine is pumped back into the radiator. The water outlet has been called a gooseneck, elbow, inlet, outlet, or thermostat housing. Generally, it covers and seals the thermostat and, in some cases, includes the thermostat bypass.

Hoses

As already mentioned, the primary function of a cooling system hose is to carry the coolant between different elements of the system. Nearly all automobiles have an upper and lower radiator hose and two heater hoses. Some may also have a bypass hose. The majority of cooling system hoses are made of butyl or neoprene rubber. Many hoses are wire reinforced, and some are used to connect metal tubing between the engine and the radiator. Normally, radiator hoses are designed with expansion bends to protect radiator connections from excessive engine motion and vibration.

The upper radiator hose is subjected to the roughest service life of any hose in the cooling system. It must absorb more engine motion than any of the other hoses. It is exposed to the coolant at its hottest stage, and it is insulated by the hood during hot soak periods. These conditions make the upper hose the most likely to fail.

Nearly all original equipment radiator hose is of the molded, curved design. Aftermarket products may be of this type or of a wire inserted flex type. This flex-type hose allows greater vehicle coverage per part number, but may not be designed for some cars that require radical bends and shapes. Lower radiator hoses are normally wire reinforced to prevent collapse due to the suction of the water pump. Hose clamps are used to secure the hoses to the inlets and outlets of the cooling system.

Thermostat

An automotive thermostat works somewhat like a typical home thermostat. It attempts to control the engine's operating temperature by routing the coolant either to the radiator or through the bypass or sometimes by a combination of both.

Shop Manual
Chapter 3, page 107

A mechanical thermostat is composed of a specially formulated wax and powdered metal pellet, which is tightly contained in a heat-conducting copper cup equipped with a piston inside a rubber boot. Heat causes the wax pellet to expand, forcing the piston outward, which opens the valve of the thermostat. The pellet senses temperature changes and opens and closes the valve to control coolant temperature and flow. Today's thermostats are also designed to slow down coolant flow when they are open. This helps prevent overheating, which can result from the coolant moving too quickly through the engine to absorb enough heat.

While the thermostat might be situated in several locations, there are two common locations: one is at the front of the engine on top of the engine block (**Figure 3-40**) and the other location is near the water pump inlet (**Figure 3-41**). Thermostats near the water pump inlet can have less thermal cycling than a thermostat located at the engine coolant outlet.

The heater element fits into a recess where it will be exposed to hot coolant. The top of the thermostat is then covered by the thermostat housing, which holds it in place and provides a connection to the radiator hose.

Two basic types of mechanical thermostats are on the market today. Both function in the same manner, but each has distinct differences.

The reverse poppet thermostat (**Figure 3-42A**) opens against the flow of the coolant from the water pump. The coolant, under water pump pressure, is used to help the reverse poppet thermostat stay closed when it is cool. The valve is self-aligning and also self-cleaning.

The balanced sleeve thermostat (**Figure 3-42B**) eliminates pressure shocks by allowing pressurized coolant to circulate around all of its working parts. By reducing pressure shocks, it provides coolant temperature stability during the most difficult operating conditions.

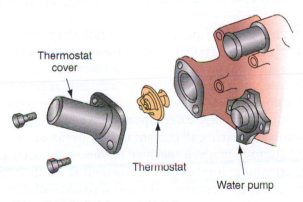

Figure 3-40 Typical thermostat location.

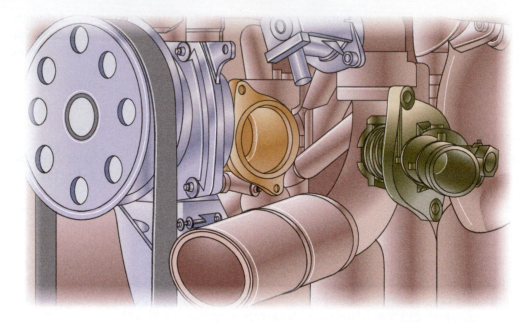

Figure 3-41 Many engines use a thermostat in the water pump inlet.

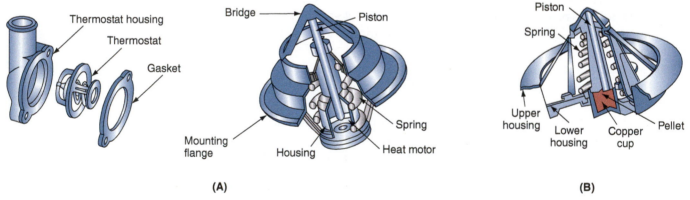

(A)   **(B)**

Figure 3-42 Types of thermostats: (A) reverse poppet, (B) balanced sleeve.

Electronic Thermostats

The thermostat rating is the coolest temperature at which the engine will operate after warm-up.

Some late-model vehicles use an electronic thermostat to maintain optimal engine temperature for engine performance and fuel mileage. The thermostat opening is controlled by the ECM, which also contains the programming for the thermostat opening program.

The thermostat permits fast warm-up of the engine after it has been started. Slow warm-up causes moisture condensation in the combustion chambers, which finds its way into the crankcase and causes sludge formation. The thermostat keeps the coolant above a designated minimum temperature required for efficient engine performance. Most engines are equipped with a coolant bypass, either outside the engine block or built into the casting. Some thermostat models are equipped with a bypass valve that shuts off the engine bypass after warm-up, forcing all coolant to flow to the radiator.

Serpentine belts are also called ribbed V-belts.

Thermostats must start to open at a specified temperature, normally 3°F (1°C) above or below its temperature rating. It must be fully opened at about 20°F (10°C) above the "start to open" temperature. It must also permit the passage of a specified amount of coolant when fully open and leak no more than a specified amount when fully closed.

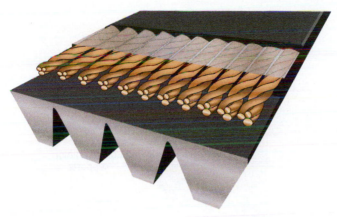

Figure 3-43 Multi-V-ribbed (serpentine) belt construction.

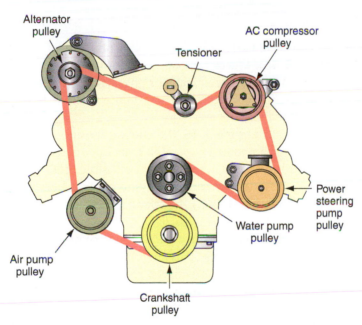

Figure 3-44 A single belt may drive many components.

Belt Drives

Belt drives (**Figure 3-43**) have been used for many years. Serpentine belts are used to drive water pumps, power steering pumps, air-conditioning compressors, alternators, and emission control pumps (**Figure 3-44**). The flexibility of belt drives has allowed many inexpensive improvements in the modern automobile.

The belt system is popular for a number of reasons. It is very inexpensive and quiet when compared to chains and gears. It is also very easy to repair. Because the belts are flexible, they will absorb some shock loads and cushion shaft bearings from excessive loads.

Serpentine belts are kept under constant tension by an idler pulley or tensioner. This pulley is designed to compensate for belt stretch and to keep a proper amount of tension on the belt.

Fans and Fan Clutches

The efficiency of the cooling system is based on the amount of heat that can be removed from the system and transferred to the air. The system needs air. At highway speeds, the ram air through the radiator should be sufficient to maintain proper cooling. At low speeds and idle, the system needs additional air. This air is delivered by a fan.

Shop Manual
Chapter 3, page 107

Shop Manual
Chapter 3, page 108

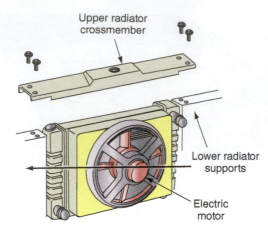

Upper radiator
crossmember

Lower radiator
supports

Electric
motor

Figure 3-45 Electrically driven cooling fan.

Vehicles generally use one of two methods of cooling fan control: the fan clutch or the electric fan motor (**Figure 3-45**). The electric fan and motor are 46 mounted to the radiator shroud and are not connected mechanically or physically to the engine. The 12-volt, motor-driven fan is electrically controlled by an engine coolant temperature switch, air-conditioner switch, or the PCM.

Following the schematic in **Figure 3-46**, the cooling fan motor is connected to the 12-volt battery supply through a normally open (NO) set of contacts (points) in the cooling fan relay. Protection for this circuit is provided by a fusible link (F/L). During normal operation, with the air conditioner off and the engine coolant below a predetermined temperature of approximately 215°F (102°C), the relay contacts are open and the fan motor does not operate.

Should the engine coolant temperature exceed approximately 230°F (110°C), the engine coolant temperature switch closes. This energizes the fan relay coil, which in turn closes the relay contacts. The contacts provide 12 volts to the fan motor if the ignition switch is in the on position. The 12-volt supply for the relay coil circuit is independent of the 12-volt supply for the fan motor circuit. The coil circuit is from the on terminal of the ignition switch, through a fuse in the fuse panel, and to ground through the relay coil and the temperature sensor.

There are many variations of electric cooling fan operation. Some provide a cooldown period whereby the fan continues to operate after the engine has been stopped and the ignition switch is turned off. The fan stops only when the engine coolant falls to a predetermined safe temperature, usually about 210°F (99°C). In some systems, the fan does not start when the air-conditioner select switch is turned on unless the high side of the air-conditioning (A/C) system is above a predetermined safe temperature.

Most vehicles control the cooling fan by completing the ground through the engine control computer. Check the service manual to see how an electric cooling fan is controlled before working with it.

Another way of controlling fan noise and horsepower loss is by using a fan clutch (**Figure 3-47**). This unit connects the fan to the drive pulley (usually on the water pump shaft). The clutch slips at high speeds; therefore, it is not turning at full engine speed.

Cooling fan clutches are temperature controlled. The obvious advantage of the temperature-controlled unit is that it knows when air is needed to cool the system. A silicone fluid couples the fan blade drive plate to a driven disc via a series of annular grooves in the two pieces. The fluid fills these grooves and drives the fan until the differential torque between the fan and the drive disc causes the fluid to shear or slip. The unit has a bimetallic element that senses the air temperature behind the radiator (**Figure 3-48**). This

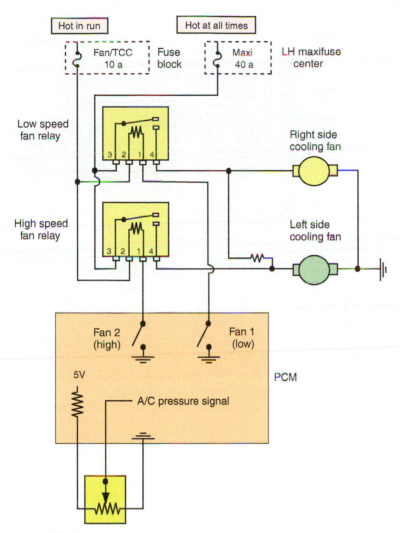

Figure 3-46 Electric cooling fan circuit with two cooling fans.

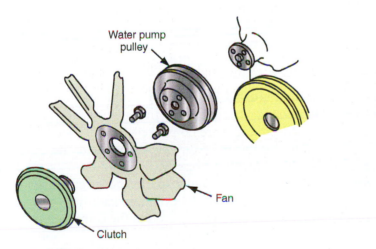

Figure 3-47 Viscous-drive fan clutch mounting.

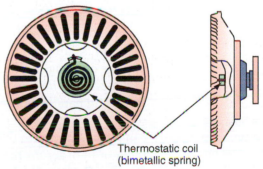

Figure 3-48 Thermostatic coil on a viscous-drive fan clutch.

bimetal is calibrated to open and close a valve in the clutch that dispenses the silicone at particular temperatures. When the fluid is returned to its reservoir, the unit freewheels the fan until more cooling is required. Tests have proven that maximum cooling is necessary less than 10 percent of the time. Therefore, a temperature-controlled fan clutch saves

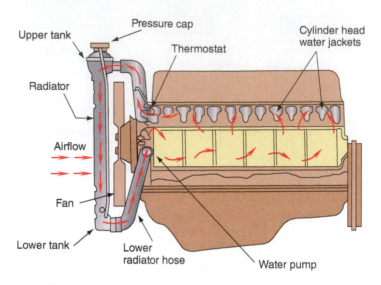

Figure 3-49 Coolant circulates through the engine.

The bimetallic element is called a thermostatic coil.

fuel and reduces noise 90 percent of the time. The shear characteristics of the silicone fluid gradually deteriorate over time, with an average loss in drive efficiency of about 100 rpm to 200 rpm per year depending on the mileage driven. Eventually, slippage reaches the point where the fan slips too much to effectively cool the engine, causing the engine to run hot.

Water Jackets

Hollow passages in the block and cylinder heads surround the areas closest to the cylinders and combustion chambers (**Figure 3-49**). Coolant flow through the block and head can be in one of the following ways:

Series flow system. In this system (**Figure 3-50A**), the coolant flows around all the cylinders on each bank as it flows to the rear of the block. Large main coolant passages at the rear of the block direct the coolant through the head gasket and to the head.

Parallel flow system. In this system (**Figure 3-50B**), coolant flows into the block under pressure and then crosses the gasket to the head through main coolant passage openings beside each cylinder.

Series-parallel flow system. This system (**Figure 3-50C**) uses a combination of series and parallel flow systems. The cooling passages inside the engine are designed so that the whole system can be drained and there are no pockets in which steam can form. Any steam that develops must be able to go directly to the top of the radiator in order for the cooling system to function properly.

Reverse flow cooling system. This system is on a few high-performance engines. In it, the coolant flows from the water pump through the cylinder heads first, and then it flows through the block. With the coolant flowing around the combustion chamber first, chamber temperatures are better maintained.

In a reverse flow cooling system, immediately after the coolant leaves the water pump, it flows through the cylinder heads.

Core Plugs

Included in the water jackets are soft (core) plugs and a block drain plug. The soft plugs and drain are usually removed during engine teardown. New ones are installed during reassembly. Core plugs are prone to rust and corrosion, and therefore will

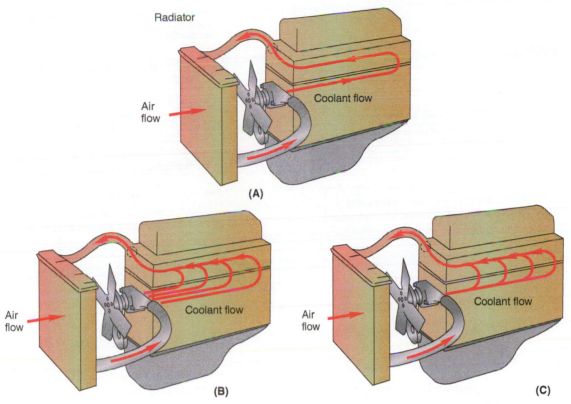

Figure 3-50 Different coolant flow designs: (A) series flow, (B) parallel flow, and (C) series-parallel flow.

weep coolant or rust through completely. When this happens, the core plugs should be replaced.

> Core plugs are often erroneously called *freeze plugs*.

Temperature Indicator

The engine temperature sensor (ECT) is screwed into the water jacket and sends a voltage signal to the PCM. Older vehicles used several ECTs. Many times there would be a separate sensor for the cooling fan, hot lamp, gauge, and the computer's engine controls. Modern vehicles use the engine ECT signal processed by the PCM to control all of these functions (**Figure 3-51**).

> The warmer coolant at the bottom of the cylinders helps reduce cylinder taper, which in turn reduces frictional horsepower loss. Lower frictional horsepower loss increases brake horsepower.

Heater System

A hot liquid passenger compartment heater is part of the engine's cooling system. Heated coolant flows from the engine through heater hoses and a heater control valve to a smaller heater core, or radiator, located in a hollow container on either side of the bulkhead. Air is directed or blown over the hot heater core, and the heated air flows into the passenger compartment. Movable doors can be controlled to blend cool air with heated air for more or less heat.

Transmission/Engine Oil Cooler

Vehicles with automatic transmissions have a cooler mounted in the radiator. The radiator cooler is a sealed metal coil that is usually located in the radiator's outlet tank. Metal or rubber hoses carry hot automatic transmission fluid to the heat cooler. The coolant passing over the sealed heat exchanger cools the fluid, which is then returned

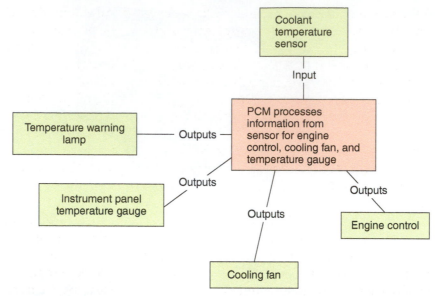

Figure 3-51 The coolant temperature sender on a modern vehicle is used by the PCM to control several functions.

to the transmission. Cooling the transmission fluid is essential to the efficiency and durability of an automatic transmission. Some transmission coolers are located outside the radiator as well. Many engines also use an engine oil cooler that is part of the radiator cooling tank.

Air-Cooled System

As mentioned at the beginning of this chapter, a few engines (such as the Porsche 911 and 930 models) use a cooling system that relies on air rather than a liquid to transfer heat from the engine to the atmosphere. Cylinders and heads have fins that are enclosed in a shroud to control airflow. The fins expose more of the surface area to airflow for better heat dissipation. Ducts and shrouds direct the airflow over the engine components, especially over the hotter cylinder head area. A belt-driven or electric blower provides for airflow. Fresh air is taken in and heated air is expelled into the atmosphere. A thermostat connected to a control valve or door regulates airflow to control engine temperature.

SUMMARY

- Automotive engines are classified by several different design features such as operational cycles, number of cylinders, cylinder arrangement, valvetrain type, valve arrangement, ignition type, cooling system, and fuel system.

- The basis of automotive gasoline engine operation is the four-stroke cycle. This includes the intake stroke, compression stroke, power stroke, and exhaust stroke. The four strokes require two full crankshaft revolutions.

- The most popular engine designs are the in-line (in which all the cylinders are placed in a single row) and V-type (which features two rows of cylinders). Opposed cylinder engines use two rows of cylinders located opposite the crankshaft.

- The two basic valve and camshaft placement configurations currently in use in four-stroke engines are the overhead valve and overhead cam. A third type, the flathead or side valve, was once popular but is no longer in use.

- Bore is the diameter of a cylinder, and stroke is the length of piston travel between top dead center and bottom dead center. Together these two measurements determine the displacement of the cylinder, which is the volume the cylinder holds between the TDC and BDC positions of the piston.
- Compression ratio is a measure of how much the air and fuel have been compressed. The higher the compression ratio, the more power an engine can produce. The compression ratio of an engine must be suited to the fuel available. As compression ratio increases, the octane rating of the fuel must increase to prevent abnormal engine combustion.
- Horsepower is the rate at which torque is produced by an engine. The torque is then transmitted through the drivetrain to turn the driving wheels of the vehicle.
- Instead of relying on a spark for ignition, diesel engines use the heat produced by compression air in the combustion chamber to ignite the fuel.
- In addition to the diesel, other automotive engines that may figure prominently in the future include the rotary or Wankel, Miller-cycle, electric, and HEVs.
- An engine's lubrication system has several important purposes: hold an adequate supply of oil to cool, clean, lubricate, and seal the engine; remove contaminants from the oil; and deliver oil to all necessary areas of the engine.
- Engine oil additives include friction modifiers, antifoaming agents, corrosion and rust inhibitors, extreme pressure resistance, detergents and dispersants, and oxidation inhibitors.
- Society of Automotive Engineers (SAE) oil viscosity ratings compare oils in relation to their ability to flow.
- American Petroleum Institute (API) oil classification ratings indicate the type of service for which the oil is suitable.
- Synthetic oil is manufactured in a laboratory. This oil has improved viscous friction qualities, but it is more expensive than refined oil.
- Recycled oil is used oil that has been recycled and restored to original specifications.
- The main components of a typical lubrication system are an oil pump, oil pump pickup, oil pan, pressure relief valve, oil filter, engine oil passages, engine bearings, crankcase ventilation, oil pressure indicator, oil seals and gaskets, dipstick, and oil coolers.
- The purpose of the oil pump is to supply oil to the various moving parts in the engine.
- Because the faster the pump turns the greater the pressure becomes, a pressure-regulating valve is installed to control the maximum oil pressure.

- All automotive vehicles are equipped with either an oil pressure gauge or a low-pressure indicator light. The gauges are either mechanically or electrically operated.
- All oil leaving the oil pump is directed to the oil filter. The filter is a disposable metal container filled with a special type of treated paper or other filter substance that catches and holds the oil's impurities.
- The PCV system removes vapors from the crankcase and directs them into the intake manifold.
- Two basic types of cooling systems are used by automotive manufacturers: liquid-cooled systems and air-cooled systems. The most popular and efficient method of engine cooling is the liquid-cooled system.
- Heat may be transferred by conduction, convection, or radiation.
- A mixture of 50 percent ethylene glycol and water in the cooling system reduces acid formation, provides antifreeze protection, and increases the coolant boiling point.
- The function of the water pump is to move the coolant efficiently through the system. The radiator transfers heat from the engine to the air passing through it. The radiator is usually one of two designs: cross-flow or down-flow. The thermostat attempts to control the engine's operating temperature by routing the coolant either to the radiator or through the bypass, or sometimes a combination of both.
- The radiator pressure cap pressurizes the cooling system and raises the coolant boiling point.
- Serpentine belts are used to drive water pumps, power steering pumps, air-conditioning compressors, alternators, and emission control pumps. The fan delivers additional air to the radiator to maintain proper cooling at low speeds and idle. Since fan air is usually only necessary at idle and low-speed operation, various design concepts are used to limit the fan's operation at higher speeds.
- The hollow passages in the block and cylinder heads through which coolant flows may be arranged as a series flow system, a parallel flow system, a series-parallel, or reverse flow system.
- Electric-drive cooling fans operate only when they are required to reduce the engine temperature.
- A viscous fan clutch is temperature operated, and it drives the fan faster when it is required to reduce coolant temperature.
- An engine temperature warning light is operated by a temperature switch in the cooling system.

REVIEW QUESTIONS

Short-Answer Essays

1. Explain how compression ratio is calculated.

2. Describe two different methods of camshaft drive in an overhead camshaft (OHC) engine.

3. Explain the advantage of a pressurized cooling system.

4. Describe the thermostat's purpose and operation.

5. Explain the reason for adding friction modifiers to engine oil.

6. Describe the difference between a 5W-40 oil and a 10W-30 oil.

7. Describe the engine component that drives most late-model oil pumps.

8. What is the name of the component in the lubrication system that prevents excessively high system pressures from occurring as engine speed increases?

9. Name the four strokes of a four-stroke cycle engine.

10. Explain what volumetric efficiency is and what can affect it.

Fill-in-the Blanks

1. In most automotive applications, the water pump is driven by the _____.

2. The stroke of an engine is _____ the crank throw.

3. The exhaust valve opens just _____ the piston reaches BDC on the power stroke.

4. In a four-stroke engine, the camshaft is rotating at _____ the crankshaft speed.

5. Lubricating oil cleans, cools, lubricates, and _____ engine components.

6. Engine displacement may be increased by increasing the bore diameter or the _____.

7. In a four-stroke engine, the point at which both the intake and exhaust valves are open is called _____.

8. For every pound per square inch (psi) of cooling system pressure, the boiling point of the coolant is increased approximately _____ degrees.

9. In a reverse flow cooling system, immediately after the coolant leaves the water pump, it flows through the _____.

10. An automatic transmission cooler is usually mounted in the _____.

Multiple Choice

1. All of the following are examples of engine designs *except*:
   A. Horizontally opposed.
   B. Inclined cylinder.
   C. V type.
   D. In-line.

2. *Technician A* says that a hemispherical combustion chamber is non-turbulent.
   Technician B says that a wedge combustion chamber is non-turbulent.
   Who is correct?
   A. Technician A　　C. Both technicians
   B. Technician B　　D. Neither technician

3. Two technicians are discussing the intake stroke,
   Technician A says that the intake valve is open during the intake stroke.
   Technician B says that the piston is moving from top dead center to bottom dead center.
   Who is correct?
   A. Technician A　　C. Both technicians
   B. Technician B　　D. Neither technician

4. *Technician A* says that the time that both valves are open is called overlap.
   Technician B says that valve overlap creates a scavenging effect.
   Who is correct?
   A. Technician A　　C. Both technicians
   B. Technician B　　D. Neither technician

5. *Technician A* says that the Atkinson cycle engine is a two-stroke engine.
   Technician B says that the Atkinson cycle engine delays the compression stroke by varying the time the exhaust valve is closed.
   Who is correct?
   A. Technician A　　C. Both technicians
   B. Technician B　　D. Neither technician

6. Two technicians are discussing 10W-30 motor oil. *Technician A* says that the *W* stands for "winter." *Technician B* says that the oil will have the viscosity of a 30 "weight" oil when warm. Who is correct?

   A. Technician A

   B. Technician B

   C. Both technicians

   D. Neither technician

7. *Technician A* says that the pressure relief valve limits oil pressure to a preset value. *Technician B* says oil filters have a bypass valve that allows oil to bypass the oil filter in case it becomes clogged. Who is correct?

   A. Technician A

   B. Technician B

   C. Both technicians

   D. Neither technician

8. All of the following are true of modern radiators *except*:

   A. Radiators are pressurized to 15 to 18 psi.

   B. Most radiators are diagonal flow.

   C. Radiators can be made of plastic and aluminum.

   D. A 50/50 mix of antifreeze and water is best.

9. *Technician A* says that the thermostat should be fully open about 10°F (5°C) above the rating of the thermostat. *Technician B* says that radiators are pressurized to help raise the boiling point of the coolant. Who is correct?

   A. Technician A

   B. Technician B

   C. Both technicians

   D. Neither technician

10. The three measures of engine efficiency are:

   A. Mechanical, volumetric, isometric.

   B. Mechanical, theoretical, thermal.

   C. Exocentric, theoretical, volumetric.

   D. Mechanical, volumetric, thermal.

CHAPTER 4
ELECTRICITY AND ELECTRONICS

Upon completion and review of this chapter, you should be able to:

- Explain the basic principles of electricity.
- Define the terms *voltage*, *current*, and *resistance*.
- Name the various electrical components and their uses in electrical circuits.
- Use the Ohm's law formula to calculate volts, amperes, or ohms in a circuit.
- Explain the differences between series, parallel, and series-parallel circuits.
- Define an electromagnet.
- Explain electromagnetic induction.
- Describe the operation of a diode in forward and reverse bias.
- Explain the operation of a transistor.
- Describe briefly how an integrated chip is manufactured.
- Explain the purposes of the battery in the vehicle electrical system.
- Describe the design of a cell group in an automotive battery.

- Describe the chemical changes that occur while the battery is charging and discharging.
- Explain four different battery ratings.
- Describe the operation of the neutral safety switch, starter relay, and theft deterrent computer.
- Explain the current flow through the solenoid windings and starting motor while the starter is engaging.
- Describe the current flow through the solenoid windings and starting motor while the engine is cranking.
- Explain the starter drive operation when the engine is cranking and while the starting motor is disengaging.
- Explain how a voltage regulator limits the AC generator voltage.
- Describe how the voltage is induced in the AC generator stator windings, and explain the current flow from the stator windings through the charging circuit when the engine is running.

Terms To Know

Actuator	Diode	Output driver
Alternating current (AC)	Direct current (DC)	Parallel circuit
Ammeter	Electricity	Peak inverse voltage (PIV)
Ampere	Electrolyte	Potentiometers
Analog	Field of flux	Power sources
Cell group	Integrated circuit (IC)	Reference voltage (Vref)
Conductors	Loads	sensors
Controllers	Ohm	Relay
Current	Ohmmeter	Rheostats
Digital	Ohm's law	Semiconductor

Series circuit

Series-parallel circuit

Solenoids

Thermistor

Transistor

Voltage

Voltage-generating devices

Voltmeter

Volts

Zener diode

INTRODUCTION

Many complicated electronic systems such as electronic fuel injection and distributorless ignition systems are described in this text. A study of electric and electronic fundamentals is necessary before you begin to learn about these systems. If you have studied the fundamentals of electricity before, the information in this chapter may be used as a review. To understand the operation of many of the components that affect engine performance, you must have a good understanding of electricity and electronics.

There is often confusion concerning the terms *electrical* and *electronic*. In this book, *electricity* and *electrical systems* will refer to wiring and electrical parts. *Electronics* will mean computers and modules used to control engine and vehicle systems.

A basic understanding of electrical principles is important for proper diagnosis of any system that is monitored, controlled, or operated by electricity. Although the subject is normally covered in a separate course, a quick overview of electricity and its principles is presented here. This chapter also covers the battery, starting motor, and AC generator, which are very important components of an automobile. The battery supplies the voltage and current to operate the starting motor to crank the engine. If the electrical accessories are turned on when the engine is not running, the battery supplies the power to operate these accessories. When the engine is running and the charging system cannot meet the power requirements of the electrical system, some power is supplied by the battery.

 A BIT OF HISTORY

Electricity was discovered by the Greeks over 2,500 years ago. They noticed when amber was rubbed with other materials, it was charged with an unknown force that had the power to attract objects, such as dry leaves and feathers. The Greeks called amber "elektron." The word *electric* was derived from the word and meant "to be like amber."

BASIC ELECTRICITY

All things are made up of atoms. An atom is the smallest particle of something. Atoms are very small and cannot be seen with your eye. This may be the reason many technicians struggle to understand electricity. The basics of electricity focus on atoms. Understanding the structure of the atom is the first step in understanding how electricity works. The following principles describe atoms, which are the building blocks of all materials.

- In the center of every atom is a nucleus.
- The nucleus contains positively charged particles called *protons* and particles called *neutrons* that have no charge.
- Negatively charged particles called *electrons* orbit every nucleus.
- Every type of atom has a different number of protons and electrons, but each atom has an equal number of protons and electrons. Therefore, the total electrical charge of an atom is zero, or neutral.

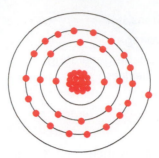

Figure 4-1 Basic structure of a copper atom.

In all atoms, the electrons are arranged in different orbits called *shells*. Each shell contains a specific number of electrons. The outer shell of electrons is called a *valence ring*. The number of electrons on the valence ring determines the electrical characteristics of the element. For example, a copper atom, which is a good conductor of electricity, has 29 electrons and 29 protons (**Figure 4-1**). The 29 electrons are arranged in shells. The outer shell has only one electron. This outer shell needs 32 electrons to be completely full. This means that the one electron in the outer shell is loosely tied to the atom and can be easily removed.

The looseness or tightness of the electrons in orbit around the nucleus of an atom explains the behavior of electricity. Electricity is caused by the flow of electrons from one atom to another. The release of energy as one electron leaves the orbit of one atom and jumps into the orbit of another is **electricity**. The key behind creating electricity is to give a reason for the electrons to move.

There is a natural attraction of electrons to protons. Electrons have a negative charge and are attracted to a proton with a positive charge. When an electron leaves the orbit of an atom, the atom then has a positive charge. An electron moves from one atom to another because the atom next to it appears to be more positive than the one it is orbiting. An electrical power source provides for a more positive charge, and to allow for a continuous flow of electricity, it supplies free electrons. To have a continuous flow of electricity, three things must be present: an excess of electrons in one place, a lack of electrons in another place, and a path between the two places.

An AC generator and battery are used in an automobile's electrical system; these are based on a chemical reaction and on magnetism. A car's battery is a source of chemical energy. A chemical reaction in the battery provides for an excess of electrons and a lack of electrons in another place. Batteries have two terminals—a positive and a negative. Basically, the negative terminal is the outlet for the electrons and the positive terminal is the inlet for the electrons to get to the protons. The chemical reaction in a battery causes a lack of electrons at the positive (+) terminal and an excess at the negative (−) terminal. This creates an electrical imbalance, causing the electrons to flow through the path provided by a wire. A simple example of this process is shown in the battery and light arrangement in **Figure 4-2**.

The chemical process in the battery continues to provide electrons (power) until the chemicals become weak. At that time, either the battery has run out of electrons or all of the protons are matched with an electron. When this happens, there is no longer a reason for the electrons to want to move to the positive side of the battery. It no longer looks more positive. Fortunately, the vehicle's charging system restores the battery's supply of electrons. This allows the chemical reaction in the battery to continue indefinitely.

Electricity and magnetism are interrelated. One can be used to produce the other. Moving a wire (a conductor) through an already existing magnetic field (such as a permanent magnet) can produce electricity. This process of producing electricity through

For current to flow, there must be a potential difference and a path for the electrons to flow; the positive post of the battery has a lack of electrons, while the negative post has an excess of electrons. The path they flow on is called a *circuit*.

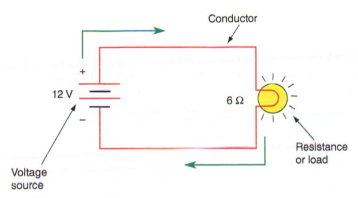

Figure 4-2 A simple light circuit consists of a voltage source, conductors, and a resistance or load.

magnetism is called *induction*. In a generator, a coil of wire is moved through a magnetic field. In an AC generator, a magnetic field is moved through a coil of wire. In both cases, electricity is produced. The amount of electricity that is produced depends on a number of factors, including the strength of the magnetic field, the number of wires that pass through the field, and the speed at which the wire moves through the magnetic field.

Measuring Electricity

Electrical **current** is a term used to describe the movement or flow of electricity. The greater the number of electrons flowing past a given point in a given amount of time, the more current the circuit has. This current, like the flow of water or any other substance, can be measured. **Voltage** is electrical pressure. Voltage is the force developed by the attraction of the electrons to the protons. The more positive one side of the circuit is, the more voltage is present in the circuit. Voltage does not flow; rather it is the pressure that causes current flow. When any substance flows, it meets resistance. The resistance to electrical flow can be measured.

Electrical Flow (Current)

The unit for measuring electrical current is the **ampere**, usually called an *amp*. The instrument used to measure electrical current flow in a circuit is called an **ammeter**.

In the flow of electricity, millions of electrons are moving past any given point at the speed of light. The electrical charge of any one electron is extremely small. It takes millions of electrons to make a charge that can be measured.

There are two types of electrical flow or current: **direct current (DC)** and **alternating current (AC)**. In direct current, the electrons flow in one direction only. The example of the battery and light shown earlier is based upon direct current. In alternating current, the electrons change direction at a fixed rate. Most automobile circuits operate on DC current, while the current in homes and buildings is AC.

Resistance

Every atom, the electrons resist being moved out of their shell. The amount of resistance depends on the type of atom. As explained earlier, in some atoms (such as those in copper) there is very little resistance to electron flow because the outer electron is loosely held. In other substances, there is more resistance to flow because the outer electrons are tightly held.

The resistance to current flow produces heat. This heat can be measured to determine the amount of resistance. A unit of measured resistance is called an *ohm*. Resistance can be measured by an instrument called an *ohmmeter*.

The flow of 6.28 billion, billion electrons past a given point in 1 second equals 1 ampere. That's 6.28 followed by 16 zeros!

Current is measured in amperes, electrical voltage is measured in volts, and electrical resistance is measured in ohms.

Shop Manual
Chapter 4, page 145

Current is the amount of electrons (electricity).

The ampere is named after André Ampere, who worked with magnetism and current flow in the 1700s.

Pressure

In electrical flow, some force is needed to move the electrons between atoms. This force is the pressure that exists between a positive and a negative point within an electrical circuit. This force, also called *electromotive force (EMF)*, is measured in units called ***volts***. One volt is the amount of pressure (force) required to move 1 ampere of current through a resistance of 1 ohm. An instrument called a ***voltmeter*** measures voltage.

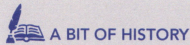

A BIT OF HISTORY

George S. Ohm was a German scientist in the 1800s who discovered that all electrical quantities are proportional to each other and therefore have a mathematical relationship.

Voltage Drop

As current passes through a resistance, heat is generated. With this generation of heat comes a loss in voltage. The voltage that is converted to heat by a resistance is called the *voltage drop*. In all circuits, all of the voltage provided by the source of power is dropped across the circuit. When a circuit has only one resistance, source voltage is dropped across that resistance. If a circuit has more than one resistance, the voltage drop across each resistance depends upon the resistance value of each resistor.

For example, if a 12-volt circuit has two 3-ohm resistors, each resistor will drop 6 volts. If we add a 6-ohm resistor to the circuit, each 3-ohm resistor will drop 3 volts and the 6-ohm resistor will drop 6 volts.

Ohm's Law

To understand the relationship between current, voltage, and resistance in a circuit, you need to know the basic law of electricity, **Ohm's law**. This law states that it takes 1 volt of electrical pressure to push 1 ampere of electrical current through 1 ohm of resistance. As such, the law provides a mathematical formula for determining the amount of current, voltage, or resistance in a circuit when two of these are known. The basic formula is Voltage = Current × Resistance.

Although the basic premise of this formula is calculating unknown values in an electrical circuit (**Figure 4-3**), it also helps define the behaviors of electrical circuits. A knowledge of these behaviors is important to an automotive technician.

If voltage does not change but there is a change in the resistance of the circuit, the current will change. If resistance increases, current decreases. If resistance decreases, current will increase. If voltage changes, so must the current or resistance. If the resistance stays the same and current decreases, so will voltage. Likewise, if current increases, so will the voltage.

> Current can be thought of as the amount of electricity.

> Ohm's law can be expressed as:
> $E = I \times R$
> $I = E \div R$
> $R = E \div I$

Circuits

When electrons are able to flow along a path (wire) between two points, an electrical circuit is formed. An electrical circuit is considered complete when there is a path that connects the positive and negative terminals of the electrical power source. Somewhere in the circuit there must be a load or resistance to control the amount of current in the circuit. Most automotive electrical circuits use the chassis as the path to the negative side of the battery. Electrical components have a lead that connects them to the chassis. These are called the *chassis ground connections*. In a complete circuit, the flow of electricity can be controlled and applied to do useful work, such as light a headlamp or turn

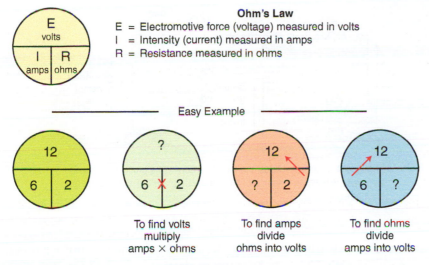

Ohm's Law

E = Electromotive force (voltage) measured in volts
I = Intensity (current) measured in amps
R = Resistance measured in ohms

Easy Example

To find volts
multiply
amps × ohms

To find amps
divide
ohms into volts

To find ohms
divide
amps into volts

Figure 4-3 Ohm's law.

over a starter motor. Components that use electrical power put a load on the circuit and consume electrical energy.

The amount of current that flows in a circuit is determined by the resistance in that circuit. As resistance goes up, the current goes down. The energy used by a load is measured in volts. Amperage stays constant in a circuit, but the voltage is dropped as it powers a load. Measuring voltage drop determines the amount of energy consumed by the load.

A complete electrical circuit exists when electrons flow along a path between two points. In a complete circuit, resistance must be low enough to allow the available voltage to push electrons between the two points. Most automotive circuits contain four basic parts.

1. **Power sources** such as a battery or AC generator that provide the energy needed to create electron flow
2. **Conductors** such as copper wires and the vehicle's frame that provide a path for current flow
3. **Loads**, which are devices such as lightbulbs, electric motors, or resistors that use electricity to perform work
4. **Controllers** such as switches or relays that direct the flow of electrons

A complete circuit must have a complete path from the power source to the load and back to the source. With the many circuits on an automobile, this would require hundreds of wires connected to both sides of the battery. To avoid this, automobiles are equipped with power distribution centers or fuse blocks that distribute battery voltage to various circuits. The positive side of the battery is connected to the fuse block, and power is distributed from there.

As a common return circuit, auto manufacturers use a wiring style that involves using the vehicle's metal frame as part of the return circuit. The load is often grounded directly to the metal frame. The metal frame then acts as the return wire in the circuit. Current passes from the battery, through the load, and into the frame. The frame is connected to the negative terminal of the battery through the battery's ground wire. This completes the circuit (**Figure 4-4**).

An electrical component such as an AC generator is often mounted directly to the engine block, transmission case, or frame. This direct mounting effectively grounds the component without the use of a separate ground wire. However, in other cases, a separate

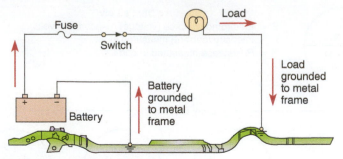

Figure 4-4 In most vehicles, the metal frame, engine block, or transmission case is used as a source of ground to complete the circuit back to the battery.

ground wire must be run from the component to the frame or another metal part to ensure a sound return path. The increased use of plastics and other nonmetallic materials in body panels and engine parts has made electrical grounding more difficult. To ensure good grounding back to the battery, some manufacturers now use a network of common grounding terminals and wires.

TYPES OF CIRCUITS

A series circuit has only one path for current to flow.

In a **series circuit**, the resistances are connected one after the other in the circuit, and the same current flows through all resistances (**Figure 4-5**). These facts may be stated about a series circuit:

1. The total resistance is the sum of all resistors in the circuit.
2. The same amount of current flows through the entire circuit.
3. Part of the source voltage is dropped across each resistor.
4. The sum of the voltage drops across each resistor equals the source voltage.

A parallel circuit has more than one path for current to flow back to the battery.

In a **parallel circuit**, each resistance is connected across the circuit, and each resistance is a separate path for current flow (**Figure 4-6**). These facts may be stated about a parallel circuit:

1. The current flow through each resistor depends on the ohms in that resistor.
2. The total current is the sum of the current flow through each branch.
3. Full source voltage is supplied to each branch, and full source voltage is dropped across each resistor.
4. The total resistance is less than the value of the smallest resistive branch in the circuit.

The series-parallel circuit has characteristics of both circuits.

In a **series-parallel circuit**, a resistor (or resistors) is connected in series with some parallel resistors in the circuit (**Figure 4-7**). Series-parallel circuits are used in many automotive applications. For example, in the instrument panel lamp circuit, the variable resistor in the headlamp switch is connected in series with the parallel instrument panel lamps.

Circuit Components

Automotive electrical circuits contain a number of different types of electrical devices. The more common components are outlined in the following sections.

Resistors are used to limit current flow (and therefore voltage) in circuits where full current flow and voltage are not needed or where too much voltage may cause damage. Resistors are devices specially constructed to introduce a measured amount of electrical

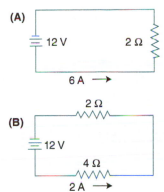

(A)

12 V   2 Ω

6 A →

(B)

2 Ω

12 V

4 Ω

2 A →

Figure 4-5 In a series circuit, the same amount of current flows through the entire circuit.

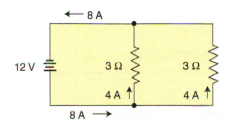

← 8 A

12 V   3 Ω   3 Ω

4 A ↑   4 A ↑

8 A →

Figure 4-6 A simple parallel circuit.

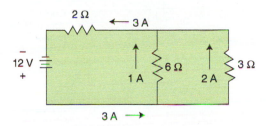

2 Ω   ← 3 A

12 V   6 Ω   3 Ω
+   1 A   2 A

3 A →

Figure 4-7 In a series-parallel circuit, the sum of the currents through the parallel legs must equal the current through the series part of the circuit.

resistance into a circuit (**Figure 4-8**). In addition, some other components use resistance to produce heat and even light. An electric window defroster is a specialized type of resistor that produces heat. Electric lights are resistors that become so hot that they produce light.

Resistors in common use in automotive circuits are of three types: fixed value, stepped or tapped, and variable.

Fixed value resistors are designed to have only one rating that should not change. These resistors are used to control voltage such as in an automotive ignition system.

Tapped or stepped resistors are designed to have two or more fixed values available by connecting wires to the several taps of the resistor. Heat motor resistor packs, which provide for different fan speeds, are an example of this type of resistor.

Variable resistors are designed to have a range of resistances available through two or more taps and a control. Two examples of this type of resistor are rheostats and potentiometers. **Rheostats** have two connections (**Figure 4-9**), one to the fixed end of a resistor and one to a sliding contact with the resistor. Turning the control moves the sliding contact away from or toward the fixed end tap, increasing or decreasing the resistance. **Potentiometers** have three connections (**Figure 4-10**), one at each end of the resistance and one connected to a sliding contact with the resistor. Turning the control moves the sliding contact away from one end of the resistance but toward the other end.

Shop Manual
Chapter 4, page 165

Rheostats have one connection fixed to the end of a resistor and one to a sliding contact.

Potentiometers are often used as position sensors.

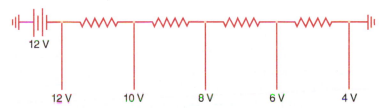

12 V

12 V   10 V   8 V   6 V   4 V

Figure 4-8 The use of a stepped resistor assembly is a common application of fixed resistors.

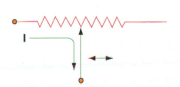

Figure 4-9 A rheostat.

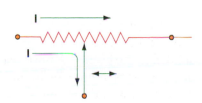

Figure 4-10 A potentiometer.

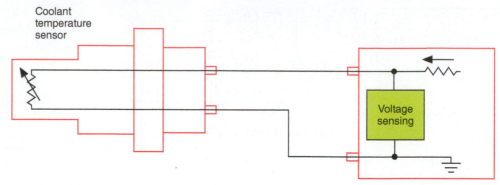

Coolant
temperature
sensor

Voltage
sensing

Figure 4-11 A thermistor is used to measure temperature. The sensing unit measures the change in resistance and translates this into a temperature valve.

Another type of variable resistor is the **thermistor** (**Figure 4-11**). This resistor is designed to change in values as its temperature changes. Although most resistors are carefully constructed to maintain their rating within a few ohms through a range of temperatures, the thermistor is designed to change its rating. Thermistors are used to provide compensating voltage in components or to determine temperature. As a temperature sender, the thermistor is supplied a voltage from the internal circuitry of the ECM. The ECM measures the voltage drop across the thermistor and interprets this voltage as a temperature. This temperature value from the ECM is sent out over the vehicle network for use by various modules in the automobile.

There are two types of thermistors: negative temperature coefficient and positive temperature coefficient. These names describe how the resistance of the thermistor changes with a fluctuation in temperature.

Circuit Protective Devices

When overloads or shorts in a circuit cause too much current to flow, the wiring in the circuit heats up, the insulation melts, and a fire can result unless the circuit has some kind of protective device. Fuses, fuse links, maxi-fuses, and circuit breakers are designed to provide protection from high current (**Figure 4-12**). These protection devices open the circuit when high current is present. As a result, the circuit no longer works, but the wiring and the components are saved from damage.

Switches

A switch of some type usually controls electrical circuits. Switches have two functions. They turn the circuit on or off, and they direct the flow of current in a circuit. Switches can be under the control of the driver or can be self-operating through a condition of the circuit, the vehicle, or the environment.

Contacts in a switch can be of several types, each named for the job they do or the sequence in which they work. A hinged-pawl switch is the simplest type of switch. It either makes or breaks the current in a single conductor or circuit. It is a single-pole, single-throw (SPST) switch. The throw refers to the number of output circuits, and the pole refers to the number of input circuits made by the switch (**Figure 4-13**).

Another type of SPST switch is the momentary contact switch. The spring-loaded contact on this switch keeps it from making the circuit except when pressure is being applied to the button. A horn is a switch of this type. Because the spring holds the contacts open, the switch has a further designation: normally open. In the case where the contacts are held closed except when the button is pressed, the switch is designated normally closed.

Shop Manual
Chapter 4, page 174

Figure 4-12 Under-hood fuse block.

Figure 4-13 Basic switches.

Single-pole, double-throw switches have one wire in and two wires out. This type of switch allows the driver to select between two circuits, such as high-beam or low-beam headlights.

Switches can be designed with a great number of poles and throws. The transmission neutral start switch may have two poles and six throws and is referred to as a multiple-pole, multiple-throw (MPMT) switch. It contains two movable wipers that move in unison across two sets of terminals. The dotted line shows that the wipers are mechanically linked or ganged. The switch closes a circuit to the starter in either P (park) or N (neutral) and to the backup lights in R (reverse).

Most switches are combinations of hinged-pawl and push-pull switches with different numbers of poles and throws. However, some special switches are required to satisfy the circuits of modern automobiles.

A temperature-sensitive switch usually contains a bimetallic element heated either electrically or by some component where the switch is used as a sensor. When engine coolant is at or below normal operating temperature, the engine coolant temperature sensor is in its normally open condition (**Figure 4-14**). If the coolant exceeds the temperature limit, the bimetallic element bends the two contacts together and the switch is closed to the indicator or the instrument panel. Other applications for heat-sensitive switches are time-delay switches and flashers.

Relays

A **relay** (**Figure 4-15**) is an electric switch that allows a small amount of current to control a much larger one. It consists of a control circuit. When the control circuit switch is open, no current flows to the coil so the windings are de-energized. When the switch is closed, the coil is energized, turning the soft iron core into an electromagnet and drawing the armature down. This closes the power circuit contacts connecting power to the load circuit. When the control switch is opened, the current stops flowing in the coil, the electromagnet disappears, and the armature is released, which breaks the power circuit contacts.

Solenoids

Solenoids are also electromagnets with movable cores used to translate electrical current flow into mechanical movement. The movement of the core causes something else to move, such as a lever. They can also close electrical contacts, acting as a relay at the same time.

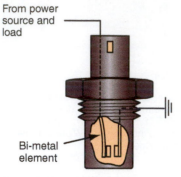

Figure 4-14 A temperature-sensitive switch.

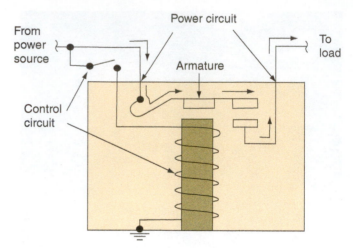

Figure 4-15 A typical relay.

ELECTROMAGNETISM BASICS

Electricity and magnetism are related. One can be used to create the other. Current flowing through a wire creates a magnetic field around the wire. Moving a wire through a magnetic field creates current flow in the wire.

Many automotive components such as AC generators, ignition coils, starter solenoids, and magnetic pulse generators operate using principles of electromagnetism.

Although almost everyone has seen magnets at work, a simple review of magnetic principles is in order to ensure a clear understanding of electromagnetism.

A substance is said to be a magnet if it has the property of magnetism—the ability to attract such substances as iron, steel, nickel, or cobalt. These are called *magnetic materials*.

A magnet has two points of maximum attraction, one at each end of the magnet. These points are called *poles*, with one being designated the north pole and the other the south pole. When two magnets are brought together, opposite poles attract while similar poles repel each other.

A magnetic field, called a ***field of flux***, exists around every magnet (**Figure 4-16**). The field consists of invisible lines along which a magnetic force acts. These lines emerge from the north pole and enter the south pole, returning to the north pole through the magnet itself. All lines of force leave the magnet at right angles to the magnet. None of the lines cross each other. All lines are complete.

Magnets can occur naturally in the form of a mineral called *magnetite*. Artificial magnets can also be made by inserting a bar of magnetic material inside a coil of insulated wire and passing a heavy direct current through the coil. This principle is very important in understanding certain automotive electrical components. Another way of creating a magnet is by stroking the magnetic material with a bar magnet. Both methods force the randomly arranged molecules of the magnetic material to align themselves along north and south poles.

Artificial magnets can be either temporary or permanent. Temporary magnets are usually made of soft iron. They are easy to magnetize but quickly lose their magnetism when the magnetizing force is removed. Permanent magnets are difficult to magnetize, but once magnetized they retain this property for very long periods.

Four metals (iron, steel, nickel, and cobalt) have magnetic qualities. When one of these metals is not magnetized, the molecules in the metal are randomly arranged and the

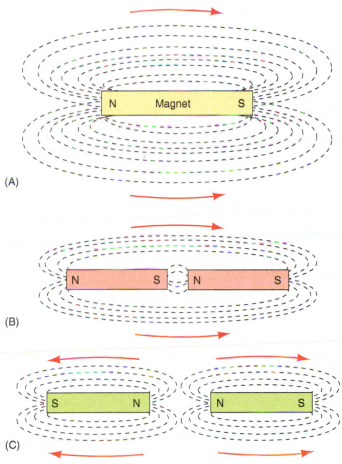

(A)

(B)

(C)

Figure 4-16 Magnetic principles: (A) field of flux around a magnet, (B) unlike poles attract each other, and (C) like poles repel.

magnetic strength of each molecule does not add together. If a metal is magnetized, the molecules are aligned so their magnetic strength adds together.

✒️📖 A BIT OF HISTORY

The force of a magnet was first discovered over 2,000 years ago by the Greeks. They found that a type of rock, magnetite, was attracted to iron. During the Dark Ages, the strange powers of the magnetite were believed to be caused by evil spirits.

When current flows through a conductor, an invisible field of force surrounds the wire. This magnetic field is concentric to the conductor, and an increase in current flow results in a stronger magnetic field. The direction of the magnetic field around the conductor is determined by the direction of current flow through the conductor. If the current flow is reversed, the magnetic field is also reversed. If an iron core is placed in the center of the coil, the magnetic lines of force are strengthened because iron is a better conductor for lines of force compared to air.

The magnetic strength of an electromagnet is determined mainly by the number of turns and the current flow through the winding. To calculate the strength of an electromagnet, multiply the number of turns and the current flow through the winding. Thus, we have the formula: number of turns × amperes = ampere turns of magnetic strength.

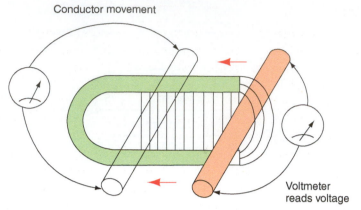

Conductor movement

Voltmeter
reads voltage

Figure 4-17 Moving a conductor through a magnetic field induces a voltage in the conductor.

Induced Voltage

Now that we have explained how current can be used to generate a magnetic field, it is time to examine the opposite effect of how magnetic fields can produce electricity. Consider a straight piece of conducting wire with the terminals of a voltmeter attached to both ends. If the wire is moved across a magnetic field, the voltmeter registers a small voltage reading (**Figure 4-17**). A voltage has been induced in the wire.

It is important to remember that the conducting wire must cut across the flux lines to induce a voltage. Moving the wire parallel to the lines of flux does not induce voltage. The wire need not be the moving component in this setup. Holding the conducting wire still and moving the magnetic field at right angles to it also induces voltage in the wire.

The wire or conductor becomes a source of electricity and has a polarity, or distinct positive and negative ends. However, this polarity can be switched depending on the relative direction of movement between the wire and magnetic field. This is why an AC generator produces alternating current.

When voltage is induced in a conductor and the conductor is connected to a complete circuit, current flows through the circuit. This process of inducing a voltage in a conductor with a moving magnetic field is referred to as electromagnetic induction. These requirements are necessary to induce a voltage in a conductor:

1. A conductor or conductors
2. A magnetic field
3. Relative motion

During electromagnetic induction, the amount of voltage induced in the conductor is determined by these factors:

1. The strength of the magnetic field
2. The number of conductors
3. The speed of motion

> AC generators produce AC volts, but the AC voltage is converted to DC before it is released into the rest of the electrical system.

BASICS OF ELECTRONICS

Computerized engine controls and other features of today's cars would not be possible if it were not for electronics. For purposes of clarity, let us define electronics as the technology of controlling electricity. Electronics has become a special technology beyond electricity. Transistors, diodes, semiconductors, integrated circuits, and solid-state devices are all considered to be part of electronics rather than just electrical devices. But keep in mind that all the basic laws of electricity apply to electronic controls.

Semiconductors

A **semiconductor** is a material or device that can function as either a conductor or an insulator, depending on how its structure is arranged. Semiconductor materials have less resistance than an insulator but more resistance than a conductor. Some common semiconductor materials include silicon (Si) and germanium (Ge).

In semiconductor applications, materials have a crystal structure. This means that their atoms do not lose and gain electrons as the atoms in conductors do. Instead, the atoms in these semiconductor materials share outer electrons with each other. In this type of atomic structure, the electrons are tightly held and the element is stable.

Because the electrons are not free, crystals cannot conduct current. These materials are called *electrically inert materials*. To function as semiconductors, a small amount of trace element must be added. The addition of these traces, called *impurities*, allows the material to function as a semiconductor. The type of impurity added determines what type of semiconductor will be produced.

The **diode** is the simplest semiconductor device. A diode allows current to flow in one direction but not in the opposite direction. Therefore, it can function as a switch, acting as either conductor or insulator depending on the direction of current flow.

Silicon has four valence electrons. When two pieces of silicon are joined together, the outer rings on the atoms join together, and this places eight electrons on the valence rings. This joining process between two pieces of silicon is called *covalent bonding*. A silicon crystal is the result of this action. When one side of a silicon crystal is mixed (called *doping*) with phosphorus, a negative-type material, an excess of electrons is formed. Doping the opposite side of the silicon crystal with boron creates a positive-type material with a lack of electrons.

If the positive polarity from a voltage source is connected to the positive side of a diode and the negative polarity is connected to the negative side of the diode, the current flows through the diode. This type of connection to a diode is called *forward bias*.

When the positive polarity from a voltage source is connected to the negative side of a diode and the negative polarity is connected to the positive side of the diode, the diode blocks current flow. *Reverse bias* is the term applied to this type of diode connection. A diode may be defined as a one-way electronic control device.

A diode connected with reverse bias is capable of blocking a certain amount of voltage. The maximum voltage that a diode blocks is called *peak inverse voltage (PIV)*. If the PIV is exceeded, the diode may break down in the reverse direction. This action ruins the diode.

A variation of the diode is the **Zener diode**. This device functions like a standard diode until a certain voltage is reached. A Zener diode is doped more heavily in the manufacturing process. This type of diode breaks down at a specific voltage in the reverse direction. For example, an 8-volt Zener diode breaks down and conducts current when an 8-volt reverse-bias voltage is supplied to the diode. If a reverse-bias voltage of 7.99 volts is supplied to the diode, it does not conduct current. A Zener diode is not damaged by a normal amount of reverse current flow.

A **transistor** is an electronic device produced by joining three sections of semiconductor materials. Like the diode, it is very useful as a switching device, functioning as either a conductor or an insulator.

A transistor is manufactured by joining three positive and/or negative materials. These positive and negative materials may be arranged as NPN or PNP. The terminals on a transistor are referred to as the emitter, base, and collector. In an NPN transistor, the emitter is negative, the base is positive, and the collector is negative (**Figure 4-18**).

The base material is sandwiched between the emitter and the collector. The base material is much thinner than the emitter and collector materials. If positive circuit polarity is connected to the positive emitter on the PNP transistor and the negative base is connected

A conductor is an element with one, two, or three valence electrons. These electrons can be moved easily to other atoms in the conductor. This type of element is classified as a good conductor.

A **semiconductor** is an element with four valence electrons. Semiconductors have unusual characteristics when they are combined with other elements and are used to manufacture diodes and transistors. Silicon is one of the most common semiconductors used in diodes and transistors.

Shop Manual
Chapter 4, page 176

Insulators are elements with five or more valence electrons in the valence ring. The electrons do not move easily from one atom to another. Therefore, this type of element is called an *insulator*.

A **transistor** is a switching device that can function as either a conductor or an insulator.

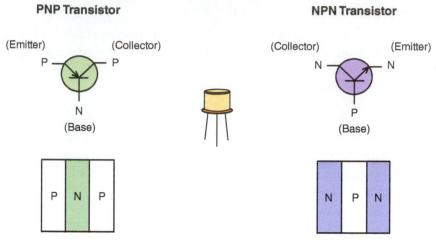

Figure 4-18 NPN and PNP transistors.

to negative circuit polarity, the emitter base is forward biased. This type of connection results in current flow through the emitter base circuit. If the emitter base is forward biased and the collector terminal is also connected to the negative circuit polarity, a higher current flows from the emitter through the base material and out the collector terminal.

The relationship between the lower emitter base current and the higher emitter collector current is called the *gain* in a transistor. If a set of contacts in the base circuit is opened, current flow in the emitter base circuit is stopped. Under this condition, the current flow is also stopped in the emitter collector circuit. A very low current in the emitter base circuit may be used to switch off a much higher current in the emitter collector circuit.

The emitter base circuit of a PNP transistor is reverse biased when positive circuit polarity is connected to the negative base material and negative circuit polarity is connected to the positive emitter material. If the emitter base circuit is reverse biased, the emitter base current flow is stopped. Under this condition, the emitter collector current is also stopped.

One transistor or diode is limited in its ability to do complex tasks. However, when many semiconductors are combined into a circuit, they can perform complex functions.

An **integrated circuit (IC)** is simply a large number of diodes, transistors, and other electronic components, such as resistors and capacitors, all mounted on a single piece of semiconductor material. This type of circuit has a tremendous size advantage. It is extremely small. Circuitry that used to take up entire rooms can now fit into a pocket. The principles of semiconductor operation remain the same in integrated circuits—only the size has changed.

In the IC manufacturing process, an etching and photographic procedure is used to form thousands of components on a silicon slice (**Figure 4-19**). A single chip may contain millions of transistors and other components. As more transistors are added to a computer chip, the computer is capable of performing more instructions per second. Currently, we have computers that handle more than 1 billion instructions per second. Computer response time and the control of output functions have been increasing dramatically.

While the cost of electronics per vehicle has increased gradually, the price of chips has decreased consistently. Therefore, the car manufacturers can actually purchase much smarter and smaller chips for their electronic dollar. This is one reason why the number of electronic systems on cars has increased and will continue to increase.

Because of the small size of ICs, electronics are no longer confined to simple tasks such as rectifying AC generator current. Enough transistors, diodes, and other solid-state components can be installed in a car to make logic decisions and issue commands to other areas of the engine. This is the foundation of computerized control systems.

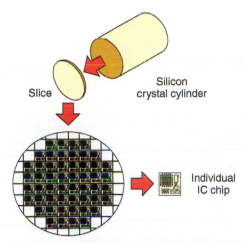

Figure 4-19 An integrated circuit etched on a silicon slice.

The computer has taken over many of the tasks in cars and trucks formerly performed by vacuum, electromechanical, or mechanical devices. When properly programmed, they can carry out explicit instructions with blinding speed and almost flawless consistency.

A typical electronic control system is made up of sensors, actuators, and related wiring that is tied to a computer.

The vehicle of today has dozens of computers, all connected by a vehicle network.

Sensors

All sensors perform the same basic function. They detect a mechanical condition (movement or position), chemical state, or temperature condition and change it into an electrical signal used by the computer to make decisions. The computer makes decisions based on information it receives from sensors. Each sensor used in a particular system has a specific job to do. Together these sensors provide enough information to help the computer form a complete picture of vehicle operation. Even though there are a variety of different sensor designs, they all fall under one of two operating categories: reference voltage sensors or voltage-generating sensors.

Shop Manual
Chapter 4, page 160

Reference voltage (Vref) sensors provide input to the computer by modifying or controlling a constant, predetermined voltage signal (**Figure 4-20**). This signal, which can have a reference value from 5 volts to 9 volts, is sent to each sensor by a reference voltage regulator located inside the processor. The term *processor* is used to describe the actual metal box that houses the computer and its related components. Because the computer knows that a certain voltage value has been sent out, it can indirectly interpret factors, such as motion, temperature, and component position, based on what comes back. For example, consider the operation of the throttle position sensor (TP sensor). During acceleration (from idle to wide-open throttle), the computer monitors throttle plate movement based on the changing reference voltage signal returned by the TP sensor. (The TP sensor is a type of variable resistor known as a rotary potentiometer that changes circuit resistance based on throttle shaft rotation.) As TP sensor resistance varies, the computer is programmed to respond in a specific manner (e.g., increase fuel delivery or alter spark timing) to each corresponding voltage change.

Reference voltage sensors modify a fixed voltage that is sent to the ECM as an input.

Most sensors presently in use are variable resistors or potentiometers. They modify a voltage to or from the computer, indicating a constantly changing status that can be calculated, compensated for, and modified. That is, most sensors simply control a voltage signal from the computer. When varying internal resistance of the sensor allows more or less voltage to ground, the computer senses a voltage change on a monitored signal line. The monitored signal line may be the output signal from the computer to the sensor

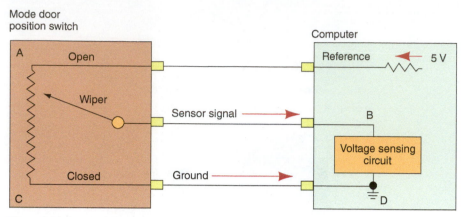

Figure 4-20 A potentiometer senses voltage in response to the movement of the wiper and sends a signal back to the voltage-sensing circuit of the computer.

(one- and two-wire sensors) or the computer may use a separate return line from the sensor to monitor voltage changes (three-wire sensors).

While most sensors are variable resistance/reference voltage, there is another category of sensors—the **voltage-generating devices** (**Figure 4-21**). These sensors include components like the magnetic reluctance sensor, oxygen sensor (zirconium dioxide), and knock sensor (piezoelectric), which are capable of producing their own input voltage signal. This varying voltage signal, when received by the computer, enables the computer to monitor and adjust for changes in the computerized engine control system.

In addition to variable resistors, two other commonly used reference voltage sensors are switches and thermistors. Switches provide the necessary voltage information to the computer so that vehicles can maintain the proper performance and drivability. Thermistors are special types of resistors that convert temperature into a voltage. Regardless of the type of sensors used in electronic control systems, the computer is incapable of functioning properly without input signal voltage from sensors.

Communication Signals

Analog voltages vary in strength.

Digital voltage is a specific value when on or off.

Most input sensors are designed to produce a voltage signal that varies within a given range (from high to low, including all points in between). A signal of this type is called an **analog** signal. Unfortunately, the computer does not understand analog signals. It can read only a **digital** binary signal, which is a signal that has only two values—on or off (**Figure 4-22**).

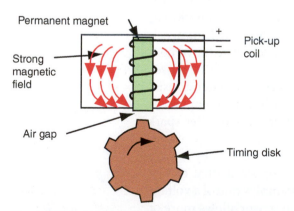

Figure 4-21 In a magnetic reluctance sensor a strong magnetic field is produced in the pick-up coil as the teeth align with the core. The field weakens as the teeth pass the core.

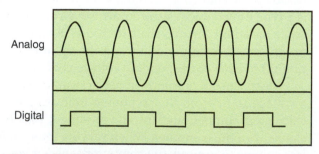

Figure 4-22 Analog signals are constantly variable, whereas digital signals are either on or off, or high and low.

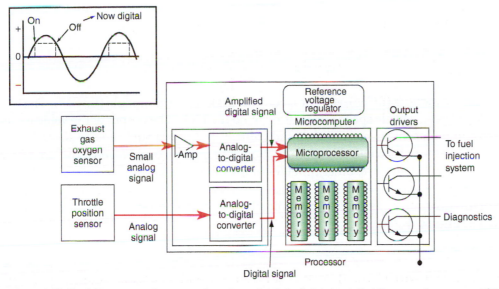

Figure 4-23 The A/D converter changes analog voltage signals to digital signals that the computer can recognize.

To overcome this communication problem, all analog voltage signals are converted to a digital format by a device known as an analog-to-digital converter (A/D converter). The A/D converter is located in a section of the processor called the *input signals*. However, some sensors like the Hall-effect switch produce a digital or square wave signal that can go directly to the computer as input (**Figure 4-23**).

A computer's memory holds the programs and other data, such as vehicle calibrations, that the microprocessor refers to in performing calculations. To the computer, the program is a set of instructions or procedures that it must follow. Included in the program is information that tells the microprocessor when to retrieve input (based on temperature, time, etc.), how to process the input, and what to do with it once it has been processed.

> Computers rely on sensors that produce analog and digital signals.

Actuators

After the computer has assimilated the information and the tools used by it to process this information, it sends output signals to control devices called *actuators*. These actuators, which are solenoids, switches, relays, or motors, physically act or carry out a decision the computer has made.

Actually, an **actuator** is an electromechanical device that converts an electrical current into mechanical action. This mechanical action can then be used to open and close valves, control vacuum to other components, or open and close switches. When the microcomputer receives an input signal indicating a change in one or more of the operating conditions, the microcomputer determines the best strategy for handling the conditions. The microcomputer then controls a set of actuators to achieve a desired effect or strategy goal. For the computer to control an actuator, it must rely on a component called an ***output driver***.

Output drivers are also located in the processor (along with the input conditioners, microprocessor, and memory) and operate by the digital commands issued by the microcomputer. Basically, the output driver is nothing more than an electronic on/off switch that the computer uses to control the ground circuit of a specific actuator. Four output drivers in one component is called a *quad driver*.

> A solenoid is a commonly used actuator.

> An **output driver** is a type of switch that is controlled by a microprocessor.

Communication between Modules in the Vehicle Network

Communication between modules is accomplished by the vehicle network. The vehicle network allows modules to share information. An example of this sharing can be illustrated by looking at some common modules like the engine control module (ECM), the

> An actuator converts an electrical signal into a mechanical action.

Figure 4-24 Symbol for network wiring.

body control module (BCM), and the electronic brake control module (EBCM). The three modules all need vehicle speed information to do their job. The output of the vehicle speed sensor could be sent to each module individually, but this would require several feet of wiring. If the modules could share information, one module could receive and process the signal and share it with the other modules in the network. Networking allows the manufacturer to save wiring, which also means fewer associated wiring problems. In reading schematic diagrams, network communications wiring is often marked with a symbol shown in **Figure 4-24**.

BATTERY, STARTING, AND CHARGING SYSTEMS

Of all the electrical functions found in an automobile, the battery, starting, and charging systems are commonly grouped together in their own category. For example, when performing a diagnosis in this area, many refer to it as a battery, starting, and charging system test.

Battery

It has often been stated that the battery is the heart of the electrical system. Nothing can happen without it. The battery serves three basic purposes. The first is to crank the engine. Because energy is stored in the battery, it must provide enough power to turn the electrical starter motor. In addition, the battery must be able to supply ample voltage to power the ignition system properly so it can build adequate spark to easily start the engine during cranking in a wide range of operating conditions.

The second function of the battery is to supply voltage when the AC generator is unable to do so. When the engine is running, the primary source for electrical power is the AC generator. When the AC generator cannot keep up with the demand or load, the battery acts as a reserve. If the AC generator fails, the battery can operate the vehicle until the battery's reserve capacity is used up. Theoretically, if everything on the vehicle is operating properly, the battery will always be in some state of recharge while the engine is running.

The third purpose of the battery is to stabilize the voltage of the vehicle. The battery provides a reference for the charging system output and helps absorb the small, normal voltage surges that occasionally occur. If the battery were to be disconnected while the engine was running, the reference would suddenly be lost. With no reference, the AC generator would be out of control because it would not have any regulated voltage output. Uncontrolled, some AC generators are capable of producing 200 volts. Usually, in today's 12-volt automotive systems, anything over 16 volts will cause damage to electrical components—especially computer systems. Never disconnect the battery with the engine running!

Starting Motor

The purpose of the starting motor is to turn the crankshaft fast enough to quickly build compression. As a rule of thumb, approximately 200 rpm is necessary. The compression increases cylinder pressure and temperature to aid in quicker ignition. The starting motor must be able to deliver proper cranking speed regardless of engine or ambient temperature.

Charging System

The AC generator must provide voltage and current to maintain the proper battery charge and power all the vehicle's electrical accessories while the engine is running. When functioning properly, the AC generator is capable of maintaining enough power output to meet or exceed the vehicle's total electrical load requirements.

Shop Manual
Chapter 4, page 188

BATTERY CONSTRUCTION

Case

The battery case holds and protects the battery components and electrolyte (**Figure 4-25**). Separating walls in the case form a separate reservoir for each cell. Many batteries have a translucent plastic case that makes the electrolyte level inside the battery visible. Very few batteries made today have removable cell caps. Most have sealed caps that are designed to capture much of the liquid that would otherwise have boiled out from charging. Some battery cases are too dark to see the electrolyte level. Unless the battery case is cracked or damaged where a sign of leakage is evident, there is usually no reason to suspect low electrolyte levels unless a problem caused it to boil away. Years ago, battery cases were made with a space between the floor of the battery and the bottom of the cell plates to allow sediment from the cell plates to collect. Due to the physical size restriction of the battery, the water level was close to the top of the plates. It was commonplace to add water regularly. Over a period of time, sediment from the cell plates would build up high enough on the case floor to actually come in contact with the cell plates. When this happened, the battery was soon destined to fail. Batteries are now designed using individual cell envelopes or pouches that isolate the cell plates from one another and also trap cell sediment. This allows the plates to be placed on the floor of the battery case. The cell plates can be made slightly taller, and there is more room on top of the cell plates for additional water. This makes the battery low maintenance or maintenance free.

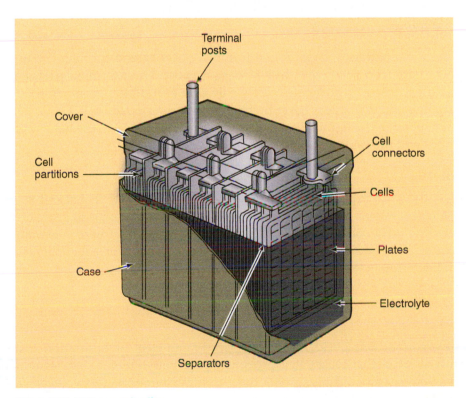

Figure 4-25 Battery construction.

Cover

The battery cover is permanently sealed to the top of the case. The cover has openings for the terminal posts and vent holes for the venting of gases. Some batteries have the terminals extending through the side of the case, and other batteries have terminals in the sides and top. Some covers have removable vent caps that may be removed to add water or test the electrolyte. These vent caps may be strip-type or box-type.

Plates

Battery plates contain a grid or coarse screen. In some batteries, the plate grids were made from lead mixed with antimony, which stiffened the lead. Many batteries manufactured today have plate grids made from lead and calcium, nickel and cadmium, or strontium. The active plate materials are pasted on the grids. The material on the positive plates is lead peroxide, and the negative plate material is sponge lead. A tab on the top of each plate grid allows the plate to be connected to the cell connector.

Cell Group

A **cell group** contains a number of alternately spaced negative and positive plates. There is a negative plate positioned on both sides of each cell group, and thus each group has an odd number of plates. For example, a cell group may have five positive plates and six negative plates.

Thin, porous separators are positioned between the negative and positive plates. These separators must be porous to allow electrolyte to contact the plates. Separators may be made from fiberglass sheets or plastic envelopes that fit over the plates. The separators prevent the plates from touching each other and shorting together electrically.

Cell connectors are lead burned to the positive and negative plate tabs in each cell group. These cell connectors also extend through the partitions between the cells, and the negative plates are connected to the positive plates in the next cell. The negative plates in one end cell and the positive plates in the opposite end cell are connected to the terminal posts that extend through the top or sides of the battery.

Terminal Posts

On top-terminal batteries, large, round terminal posts extend through the top of the battery. The positive post is larger than the negative post. These terminal posts are sealed into the battery cover to prevent electrolyte leakage around the posts. POS or NEG is usually stamped on the battery cover beside the proper terminal post. The battery cable ends are clamped to the terminal posts. On side-terminal batteries, the battery terminals are threaded and the cables are bolted to the terminals.

Electrolyte

An **electrolyte** is a liquid chemical that promotes voltage production in a battery.

The **electrolyte** is a mixture of approximately 64 percent water (H_2O) and 36 percent sulfuric acid (H_2SO_4). The electrolyte reacts with the plate materials to produce voltage. The electrolyte must contain enough sulfuric acid so it does not freeze in extremely cold temperatures. If the battery contained a higher percentage of sulfuric acid, this acid would attack and deteriorate the plate grids too quickly.

BATTERY OPERATION

If one negative and one positive plate are placed in an electrolyte solution, a chemical reaction takes place that causes an approximately 2.1-volt difference between the plates. Since a 12-volt battery contains six cells, the total fully charged battery voltage is approximately 12.6 volts. If an electrical load such as a lightbulb is connected to these two plates,

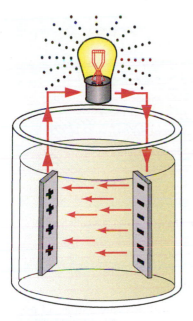

Discharging

Figure 4-26 Current flows from
one battery plate to the other when
a conductor with some electrical
resistance is connected between the
plates.

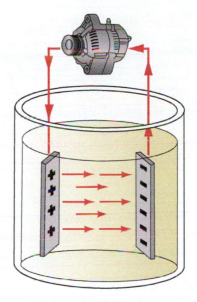

Charging

Figure 4-27 When the engine is
running, current flows from the
alternator through the battery plates.

the higher voltage on one plate forces current through the light to the other plate with the
lower voltage. This action is called *battery discharging* (**Figure 4-26**). A battery cell with
two plates will not supply a great deal of current flow. When more plates are added, the
current flow capability increases.

If a voltage source such as the AC generator in the charging system is connected to
the battery plates, the AC generator forces current flow in one battery plate and out of the
other battery plate (**Figure 4-27**). This action is called *battery charging*. The AC generator
voltage must be higher than the battery voltage to allow the AC generator to charge the
battery.

A charged battery has lead peroxide (PbO_2) on the positive plates and sponge lead
(Pb) on the negative plates. The electrolyte contains water (H_2O) and sulfuric acid (H_2SO_4)
(**Figure 4-28**).

As the battery discharges, these chemical changes occur:

1. The H_2SO_4 breaks up in the electrolyte, and the SO_4 goes to both plates where it
   joins with the Pb to form lead sulfate ($PbSO_4$) on both plates.
2. The O_2 on the positive plates joins with the hydrogen (H) in the electrolyte to form
   H_2O (**Figure 4-29**). The percentage of water in the electrolyte increases as the
   battery discharges.

In a discharged battery, both sets of plates are coated with lead sulfate ($PbSO_4$), and
the electrolyte contains a high percentage of H_2O (**Figure 4-30**).

When charging the battery, these chemical changes occur in the battery:

1. The SO_4 comes off both plates and joins with the H in the electrolyte to form
   H_2SO_4 (**Figure 4-31**).
2. The H_2O breaks up and the O goes to the positive plates where it joins with the
   Pb to form PbO_2.

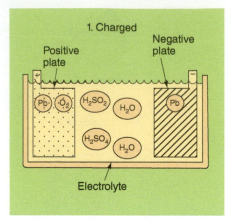

Figure 4-28 Plate materials and electrolyte content in a charged battery.

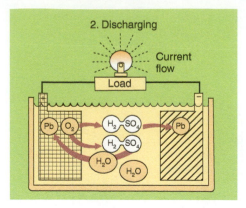

Figure 4-29 Chemical action in a battery while discharging.

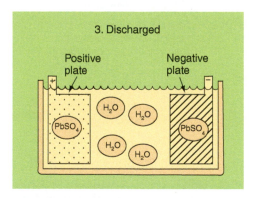

Figure 4-30 Plate materials and electrolyte content in a discharged battery.

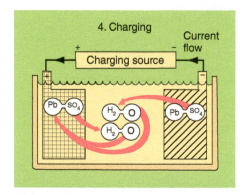

Figure 4-31 Chemical action in a battery while charging.

While a battery is charging, H gas escapes at the negative plates and O gas escapes at the positive plates. These two gases combine to form H_2O. The SO_4 is always in the electrolyte or on the plates; thus it does not escape from the battery. Since H_2O is the only chemical that escapes from the battery, it is the only chemical that should be added to the battery.

BATTERY CLASSIFICATIONS

Low-Maintenance Batteries

Low-maintenance batteries are designed to reduce internal heat and water loss. The addition of water may be required at 15,000 miles (24,000 km).

Maintenance-Free Batteries

Maintenance-free batteries are designed to reduce internal heat and water loss, so the addition of water is not required during the life of the battery. The battery cover is vented to allow gases to escape, but there are no vent caps to add water or test the electrolyte.

Shop Manual
Chapter 4, page 177

BATTERY RATINGS

Many different battery rating methods have been used by industry. What follows are the most common.

Cold Cranking Amperes

The cold cranking amperes (CCA) rating indicates the amperes a battery will deliver at 0°F (−18°C) for 30 seconds while maintaining a voltage of at least 1.2 volts per cell or 7.2 volts in the complete battery. Many automotive batteries have a CCA rating of 350 to 600 or more.

The most common way of rating a battery is by cold cranking amps.

Reserve Capacity Rating

The reserve capacity rating indicates in minutes the length of time a fully charged battery at 80°F (27°C) will deliver 25 amperes while the voltage remains above 1.75 volts per cell or 10.5 volts in the complete battery. Reserve capacity ratings are usually between 55 and 115 minutes.

Ampere-Hour Rating

The ampere-hour rating indicates the amount of current a battery will deliver for 20 hours with the voltage remaining above 1.75 volts per cell or 10.5 volts for the complete battery. If a battery delivers 5 amperes for 20 hours with the voltage above the specified value, the battery is rated at 5 × 20 = 100 ampere-hours. This is an outdated method of rating batteries and has been superseded by cold cranking amps (CCA).

Power (Watt) Rating

The watt rating is determined by multiplying the available battery voltage and the current flow delivered at 0°F (−18°C). Watt or power battery ratings are usually between 2,000 and 4,000 watts.

Shop Manual
Chapter 4, page 184

STARTING THE ENGINE

Any motor converts electrical energy into a rotating mechanical energy. When two like magnetic poles face each other, they tend to push away. In a motor, one of the magnetic poles is stationary in the starter's case. The other pole is fixed to a shaft that is fitted in the center of the case. The shaft is allowed to move. When the magnetic poles of the windings in the case are the same as the magnetic poles of the shaft, they push apart. This pushing causes the shaft to rotate. By connecting the shaft of the starter motor to the flywheel of the engine, the engine cranks as the shaft or armature spins.

All motors are made up of the same basic parts: a case, armature, brushes, and field windings (**Figure 4-32**).

A steel shaft is the basis of an armature, and it runs through the center of the armature assembly. A laminated iron core is pressed onto this shaft. Heavy insulated windings are mounted in the armature core slots.

A commutator is mounted on an insulating sleeve near the front of the armature. The commutator contains a series of copper bars that are insulated from each other. The armature windings are soldered to the commutator bars. In most starting motors, four brushes are mounted in holders, and springs keep the brushes in contact with the commutator bars. In many starting motors, two of the brushes are connected to ground on the starter case, and the other two powered brushes are connected to the field coils.

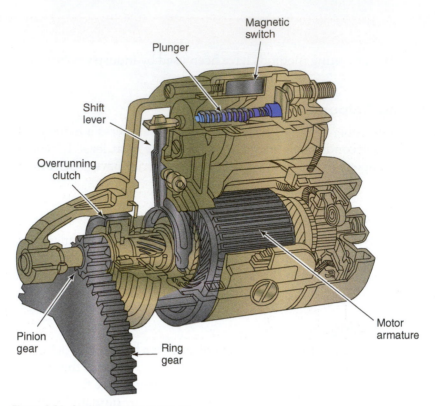

Figure 4-32 Starting motor components.

Most starter motors have two positive and two negative brushes.

The starter drive is mounted on splines on the back of the armature shaft. A pinion gear on the drive is pulled into mesh with the flywheel ring gear by a shift lever. Most starter drives are the overrunning clutch-type. Bushings in the starter end housings support the armature shaft.

Field Windings

Many starting motors have two positive brushes and two ground brushes.

Some starting motors contain four insulated field windings mounted on steel pole shoes that are bolted to the starter case. The field coils in many starting motors are connected through an insulated terminal in the starter case to the solenoid terminal. In modern vehicle starting motors, the field windings are replaced with strong permanent magnets.

Solenoid

In many starting motors, the solenoid is mounted on the starter housing. A plunger is mounted in a bore at the center of the solenoid, and two windings surround the plunger bore. The rear of the plunger is connected to the pivoted shift lever and the lower end of this lever is mounted in the drive collar. A heavy copper disc is mounted in front of the plunger. The large starting motor terminal and the battery cable terminal are mounted in front of the heavy copper disc. A plunger return spring holds the plunger toward the rear of the starting motor, and a smaller disc return spring holds the disc away from the terminals. A small terminal in the front of the solenoid is connected to the solenoid windings.

Some mistakenly call starting circuit relays solenoids. They are not! A solenoid is something that performs a mechanical act due to the control of its magnetic field.

Some starting circuits are fitted with a relay or magnetic switch that completes the battery positive circuit to the starter when the ignition key is moved to the start position. These systems may or may not have a solenoid mounted to the starter.

As the contacts in a relay or solenoid open and close, some arcing may occur across the contacts. This causes voltage spikes to be present in the electrical system. Some manufacturers install clamping diodes at the solenoid, relay, or both to prevent these voltage spikes.

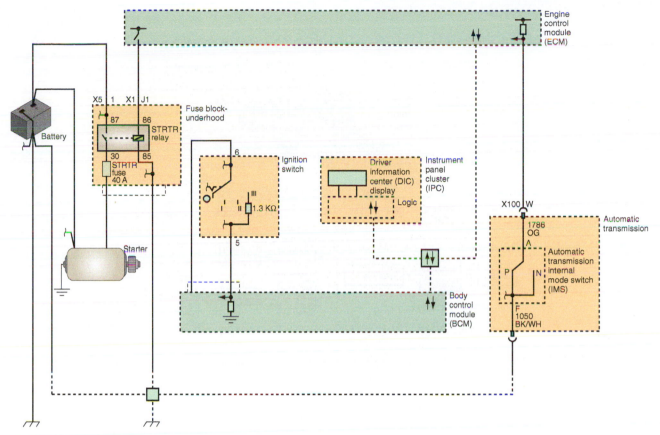

Figure 4-33 Starting motor circuits are integrated into the vehicle network on late-model vehicles

Transmission Range/Mode Switch Operation

Starter motor circuits can use the vehicle network to operate, such as the example shown in **Figure 4-33**. The ignition switch is used as an input to the BCM. When the ignition is turned to start, a 5-volt signal is sent to pin 2. Since the BCM, ECM, and IPC are connected via the vehicle network, the signal is shared with the ECM. Note that the transmission mode switch connects pin X1 or X4 (depending on the engine used) to ground, which tells the PCM that the transmission is in either park or neutral. When the ignition switch is activated and the transmission is in park, then the ECM applies voltage to the start relay, which engages the starter.

The theft-deterrent module inputs are also fed into the vehicle network. If any of conditions are met to activate the theft deterrent, then the ECM prevents vehicle starting.

Starting Motor Operation while Engaging

Current flows through the solenoid terminal and through the hold-in winding to ground. Current also flows through the pull-in winding and the starting motor to ground. This current flow is not high enough to operate the starting motor. However, the combined magnetic strength of both solenoid windings pulls the plunger ahead, and the plunger drives the disc against the battery cable as well as the starting motor terminals. This solenoid disc action completes the circuit between the battery and the starting motor, and a very high current now flows from the battery positive cable through the solenoid disc, field windings, insulated brushes, armature windings, and ground brushes (**Figure 4-34**). This current flow returns to the battery negative terminal through the ground return circuit. Forward movement of the solenoid plunger also pulls the drive into mesh with the flywheel ring gear.

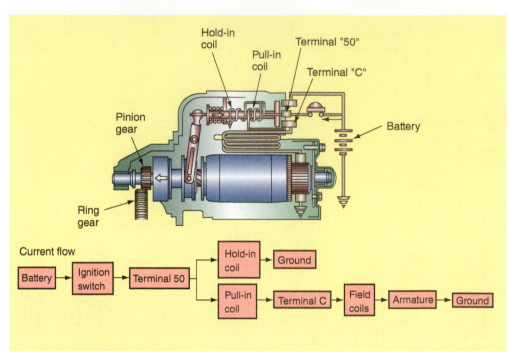

Figure 4-34 Starting motor and solenoid current flow while engaging.

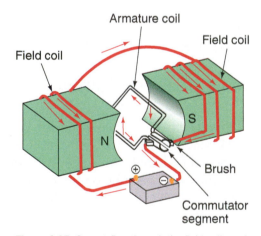

Figure 4-35 Current flow through the field coils and armature windings.

Starting Motor Operation while Engaged

Since the starting motor field coils and armature windings have very low resistance, the average current flow through this circuit is 100 to 200 amperes. This high current flow creates very strong magnetic fields between the field coils and around the armature windings (**Figure 4-35**). The interaction of the magnetic field between the field coils and the magnetic field around the armature windings creates a strong armature turning force to crank the engine (**Figure 4-36**). The starting motor is designed to deliver very high horsepower for short time periods, such as 15 seconds. Cranking the engine continually for more than 15 seconds may damage the starting motor.

Once the solenoid disc contacts the terminals, equal voltage is supplied to both ends of the solenoid pull-in winding, and the current flow in this winding is stopped. Current continues to flow through the hold-in winding, and the magnetic field around this winding holds the plunger ahead (**Figure 4-37**).

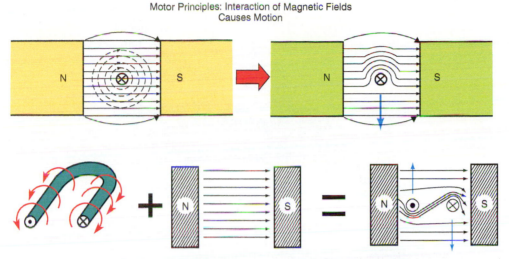

Motor Principles: Interaction of Magnetic Fields
Causes Motion

Figure 4-36 Interaction of the magnetic fields between the field coils and armature windings causes the armature to rotate.

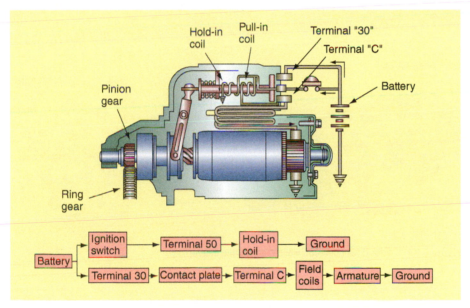

Figure 4-37 Current flow while the starting motor is engaged.

Starting Motor Operation while Disengaging

When the driver releases the ignition switch from the start to the on position, the circuit is opened between the ignition switch and the solenoid windings. When current stops flowing through the hold-in winding, the plunger spring moves the plunger rearward. This action pushes the drive out of mesh with the flywheel ring gear and allows the disc return spring to move the disc away from the solenoid terminals (**Figure 4-38**).

Starter Drive Operation

The starter drive clutch housing is splined to the armature shaft and must rotate with the shaft. Four spring-loaded steel rollers are mounted in tapered grooves in the drive housing. A pinion gear is mounted in the end of the drive housing, and the steel rollers contact a machined steel surface on this gear.

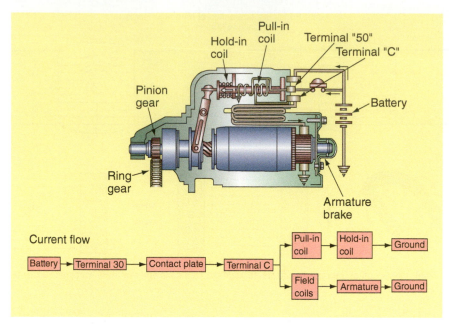

Figure 4-38 Starting motor operation while disengaging.

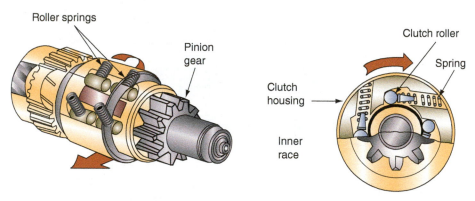

Figure 4-39 Starter drive operation while cranking the engine.

The solenoid shift lever moves the drive into mesh with the flywheel ring gear before the armature begins turning. When the solenoid disc contacts the terminals and the armature starts turning, the steel rollers are wedged into the narrow part of the tapered grooves where they are jammed between the drive housing and the pinion gear machined surface (**Figure 4-39**). Under this condition, the pinion gear must rotate with the drive housing to crank the engine.

Once the engine starts, the ring gear drives the pinion gear faster than the armature. This action moves the steel rollers against the spring tension, and the rollers are now positioned in the wider part of the tapered grooves (**Figure 4-40**). Under this condition, the starter drive overruns so the flywheel ring gear does not drive the armature at high speed.

This overrunning action of the drive is very important in protecting the armature from excessive speed. The average flywheel ring gear-to-pinion gear ratio is 15:1. If the drive did not overrun and the engine was running at 1,000 rpm, the armature speed would be 15,000 rpm. At this rpm, centrifugal force would throw the armature windings from their slots in the core and destroy the starting motor.

Prior to the implementation of SAE J1930, generators were commonly called *alternators*.

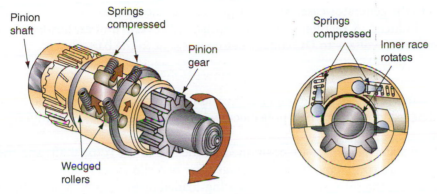

Figure 4-40 Starter drive operation while overrunning.

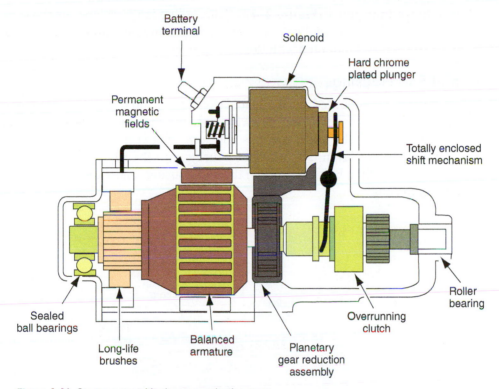

Figure 4-41 Starter motor with planetary reduction gears.

Gear Reduction Starting Motors

Many starting motors have a gear reduction between the armature and the drive. The reduction gears may be conventional or planetary (**Figure 4-41**). Planetary gear-sets use a combination of external and internal gears. The reduction gears allow a smaller, more compact starting motor to provide the same cranking power as a larger starting motor without reduction gears.

Shop Manual
Chapter 4, page 186

AC GENERATORS

AC generators keep a vehicle's battery charged by sending electrical energy, which was converted from mechanical energy, to the battery. The AC generator's inner shaft is rotated by the engine via a drive belt. This inner shaft is called the *rotor* and is a rotating magnetic field. The rotor and its magnetic field spin inside the stator windings, which are fastened to the generator case. As the magnetic field moves past the stator windings, AC voltage is induced. A rectifier circuit, comprised of several diodes, inside the generator converts the AC voltage to DC voltage before it is sent to the battery.

Rotor

A steel shaft is mounted in the center of the rotor assembly, and an insulated field winding is positioned on this shaft. Steel pole pieces with interlaced fingers are mounted over the field coil.

Two insulated slip rings are positioned on the end of the rotor shaft, and the ends of the field coil are connected to the slip rings. The two slip rings are insulated from each other and from the rotor shaft. A small spring-loaded brush contacts each slip ring.

The drive end of the rotor shaft is supported on a ball bearing in the drive end housing, and the slip ring end of the shaft is mounted on a needle bearing in the slip ring end frame. A pulley is bolted or pressed on the drive end of the rotor shaft, and a cooling fan is positioned behind the pulley (**Figure 4-42**). The cooling fan circulates air through the AC generator to cool the internal components. The cooling fan can be internal to the alternator or external and mounted with the drive pulley.

Stator and End Housings

The stator windings are mounted in a laminated iron frame, and this frame surrounds the rotor assembly. Many stators contain three windings, and the ends of these windings are connected to the diodes. The stator frame is positioned between the two end housings, and through bolts extend from the drive end housing into the slip ring end housing to hold the assembly together. Bolt openings in the end housings allow the AC generator to be mounted on the engine.

> In an AC generator, the ends of the rotor field windings are connected to the slip rings.

> **Shop Manual**
> Chapter 4, page 189

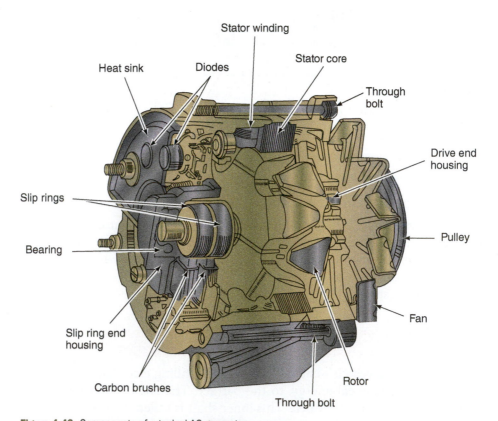

Figure 4-42 Components of a typical AC generator.

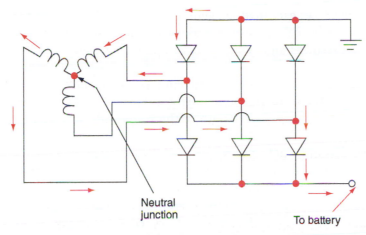

Neutral
junction

To battery

Figure 4-43 A wye-wound stator wired to six diodes for rectification.

Diodes

The purpose of diodes is to convert AC to DC. As explained earlier, a diode is an electrical one-way check valve causing current to flow in one direction only. Each winding of the stator produces a single-phase current. A three-phase current is produced with three windings. A set of diodes is used for each phase.

Many AC generators contain three negative diodes and three positive diodes (**Figure 4-43**). The positive diodes are mounted in an insulated heat sink to dissipate heat from the diodes. These diodes are connected to the insulated battery terminal in the end housing. Three negative diodes are connected to the end housing.

Shop Manual
Chapter 4, page 189

Voltage Regulators

Vehicles are equipped with integrated circuit (IC) regulators (**Figure 4-44**). Voltage regulators are mounted inside the AC generator or are part of and controlled directly by the ECM.

The voltage regulator is always connected in the AC generator field circuit, regardless of the type of voltage regulator. The purpose of the voltage regulator is to limit the AC generator voltage to protect the electrical accessories and battery on the vehicle. If the

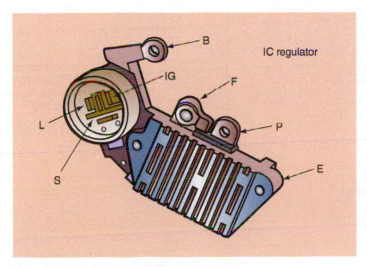

Figure 4-44 Integrated circuit (IC) voltage regulators.

voltage regulator allows excessive AC generator voltage, this voltage forces high current flow through the electrical accessories and battery. This high current flow may burn out the electrical accessories and cause excessive battery gassing, which may damage the battery.

The voltage regulator controls field current through the field winding, and this action regulates the magnetic strength around the rotor. The induced voltage in the stator windings is limited by the magnetic strength of the rotor and the speed at which the rotor spins.

Voltage regulation is controlled by pulse width modulation of the AC generator field. In fact, an actual physical voltage regulator may not exist as part of the charging system. Instead, the charge rate is controlled by the ECM and/or a combination of processors that sense system voltage and issue an output command to the charging system, specifically the AC generator. By managing the on and off time or duty cycle of the AC generator field, a precise charge rate can be maintained and small corrections can be made quickly and accurately.

AC Generator Operation

When the engine starts, the rotor magnetic field induces an alternating current (AC) voltage in the stator windings. This AC voltage is changed to a direct current (DC) voltage by the diodes. Current flows from the diodes through the AC generator battery terminal to the battery and electrical accessories. This current flows through the ground return on the vehicle to the AC generator end housing and negative diodes. The generator circuit from this late-model Ford vehicle has a battery-monitoring sensor that reports to the BCM on the state of the charge and health of the battery over a local interconnect network (LIN) (more on vehicle networks later in the text) (**Figure 4-45**). The battery monitor allows the alternator charge to be brought down when feasible, thus taking some of the load off the engine that is normally used to turn the alternator. The charging level of the generator is controlled by the voltage regulator, but in this example, the BCM can set the regular charging points at the most optimum level (**Figure 4-46**). Should the alternator fail, the BCM will command the instrument panel module to illuminate the charge warning lamp on the dash.

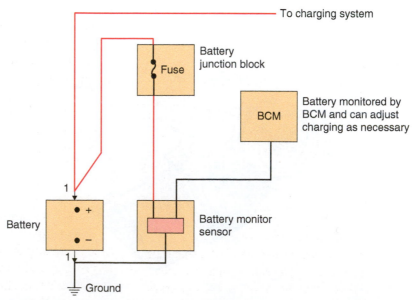

Figure 4-45 The generator circuit from this vehicle has a battery-monitoring sensor that reports to the BCM on the state of the charge and health of the battery.

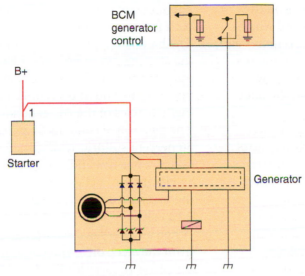

Figure 4-46 In this example, the BCM can set the regular charging points at the most optimum level.

SUMMARY

- Electricity is the release of energy as one electron leaves the orbit of an atom and jumps into the orbit of another.
- Two power or energy sources are used in an automobile's electrical system. These are based on a chemical reaction and on magnetic principles.
- A car's battery is a source of chemical energy. A chemical reaction in the battery provides for an excess of electrons and a lack of electrons in another place.
- The flow of electricity is called current and is measured in amperes. There are two types of electrical flow: direct current (DC) and alternating current (AC).
- Resistance to current flow produces heat. The amount of resistance is measured in ohms.
- Voltage is electrical pressure, and it is measured in volts.
- For electrical flow to occur, there must be an excess of electrons in one place, a lack of electrons in another, and a path between the two places.
- Ohm's law states that it takes 1 volt of electrical pressure to push 1 ampere of electrical current through 1 ohm of resistance.
- If voltage does not change but there is a change in the resistance of the circuit, the current will change. If resistance increases, current decreases. If resistance decreases, current will increase. If voltage changes, current or resistance changes. If the resistance stays the same and current decreases, so will voltage. Likewise, if current increases, voltage increases.

- The mathematical relationship between current, resistance, and voltage is expressed in Ohm's law, E = IR, where voltage is measured in volts, current in amperes, and resistance in ohms.
- Resistors in common use in automotive circuits are of three types: fixed value, stepped or tapped, and variable.
- Fixed value resistors are designed to have only one rating, which should not change. These resistors are used to control voltage, such as in an automotive ignition system.
- A variable resistor has a value than does not remain fixed. The variable resistor can be adjusted or varies according to conditions such as temperature or light.
- Rheostats have two connections—one to the fixed end of a resistor and one to a sliding contact with the resistor.
- Potentiometers have three connections, one at each end of the resistance and one connected to a sliding contact with the resistor. Turning the control moves the sliding contact away from one end of the resistance but toward the other end.
- A thermistor is a type of variable resistor that changes voltage in relationship to temperature. A negative temperature coefficient (NTC) thermistor's resistance increases as the temperature it is sensing decreases.
- Electrical schematics are diagrams with electrical symbols that show the parts and how electrical current flows through the vehicle's electrical circuits. They are used in troubleshooting.

- The strength of an electromagnet depends on the number of current-carrying conductors and what is in the core of the coil. Inducing a voltage requires a magnetic field producing lines of force, conductors that can be moved, and movement between the conductors and the magnetic field so that the lines of force are cut.
- Fuses, fuse links, maxi-fuses, and circuit breakers protect circuits against overloads. Switches control on/off and direct current flow in a circuit. A relay is an electric switch. A solenoid is an electromagnet that translates current flow into mechanical movement. Resistors limit current flow.
- A semiconductor is a material or device that can function as either a conductor or an insulator depending on how its structure is arranged.
- An integrated circuit is simply a large number of diodes, transistors, and other electronic components, such as resistors and capacitors, all mounted on a single piece of semiconductor material.
- The diode allows current to flow in one direction but not in the opposite direction.
- Transistors are used as switching devices.
- Computers are electronic decision-making centers. Input devices called sensors feed information to the computer. The computer processes this information and sends signals to controlling devices.
- A typical electronic control system is made up of sensors, actuators, microcomputer, and related wiring.
- Most input sensors are variable resistance/reference types, switches, and thermistors.
- All sensors detect a mechanical condition, chemical state, or temperature condition and change it into an electrical signal that can be used by the computer to make decisions.
- All sensors are either reference voltage sensors or voltage-generating sensors.
- Most input sensors are designed to produce a voltage signal that varies within a given range called an analog signal.
- A computer does not understand analog signals. It can read only a digital binary signal, which is a signal that has only two values—on or off.
- After the computer has assimilated the information and the tools used by it to process this information, it sends output signals to control devices called actuators, which are solenoids, switches, relays, or motors that physically act or carry out a decision the computer has made.
- In a series circuit, the same current flows through each resistance.

- Each resistance in a parallel circuit is a separate path for current flow.
- The battery supplies voltage and current to operate the starting motor while cranking the engine to operate the electrical accessories if the engine is not running and to overcome any charging inefficiencies of the AC generator.
- The electrolyte contains 64 percent water and 36 percent sulfuric acid.
- When a positive and a negative plate are placed in an electrolyte, a chemical reaction takes place and the pressure difference between the plates is about 2.13 volts.
- When a battery is discharging, the sulfate comes from the electrolyte and combines with the lead on the plates to form lead sulfate. The oxygen comes off the positive plates and combines with the hydrogen in the electrolyte to form water.
- The battery voltage depends on the chemical condition of the plates.
- When a battery is charging, the sulfate comes off both plates and combines with the hydrogen in the electrolyte to form sulfuric acid. The oxygen breaks away from the hydrogen in the electrolyte and combines with the lead on the positive plates to form lead peroxide.
- Batteries may be rated in cold-cranking amperes, reserve capacity, ampere-hours, or watts.
- A neutral safety switch is connected in series in the circuit between the ignition switch and the solenoid windings. This switch is closed if the gear selector is in neutral or park, and it is open in other selector positions.
- When the solenoid plunger forces the disc against the terminals, a very high current flows from the battery through the starting motor.
- The rotating magnetic field on the rotor induces an AC voltage in the AC generator stator windings, and this AC voltage is changed to a DC voltage by the diodes.
- Many AC generators have an integrated circuit (IC) regulator, while other AC generators are regulated by the PCM.
- Voltage regulators control field current and magnetic strength around the rotor to limit the voltage in the stator windings.
- If the AC generator is allowed to produce excessive voltage, this voltage forces high current through the battery and electrical accessories on the vehicle.

REVIEW QUESTIONS

Short-Answer Essays

1. Name the two energy sources used in automobile electrical systems.

2. For electrical flow to occur, what must be present?

3. Describe the chemical changes that occur while a battery is discharging.

4. What is the difference between voltage and current?

5. Explain the design of a starting motor armature.

6. State Ohm's law.

7. Describe the differences between a rheostat and a potentiometer.

8. What is the difference between a fixed resistor and a variable resistor?

9. What types of sensors are typically used in an automotive computer system?

10. Explain how the voltage is induced in the AC generator stator windings, and describe the type of voltage induced.

Fill-in-the-Blanks

1. As current passes through a resistance _____ is generated.

2. Current is measured in _____, electrical voltage is measured in _____, and electrical resistance is measured in _____.

3. Most batteries have a(n) _____ plastic case that makes the electrolyte level inside the battery visible.

4. _____, _____, _____, and _____ protect circuits against overloads.

5. _____ control on/off and direct current flow in a circuit.

6. The strength of an _____ depends on the number of turns and the current flow through the windings.

7. In an AC generator, the ends of the rotor field winding are connected to the _____.

8. A(n) _____ is a material or device that can function as either a conductor or an insulator, depending on how its structure is arranged.

9. A computer does not understand analog signals. It can read only a digital binary signal, which is a signal that has only two values: _____ or _____.

10. The AC generator voltage regulator is connected in the _____ circuit.

Multiple Choice

1. *Technician A* says electrical pressure is termed *voltage.*
   Technician B says amperage can be described as the amount of electricity flowing in a circuit.
   Who is correct?
   A. Technician A only   C. Both technicians
   B. Technician B only   D. Neither technician

2. All of the following are versions of Ohm's law *except*:
   A. $E = I \times R$ C. $R = E/I$
   B. $I = E/R$ D. $E = I/R$

3. *Technician A* says that in a series circuit the voltage is the same at all points.
   Technician B says that in a parallel circuit the amperage is the same at all points.
   Who is correct?
   A. Technician A only   C. Both technicians
   B. Technician B only   D. Neither technician

4. Which of the following statements are true?
   A. If current goes up and voltage stays the same, then resistance increases.
   B. If current goes down and voltage stays the same, then resistance goes down.
   C. If current goes down and voltage stays the same, then resistance increases.
   D. If current goes up and voltage goes down, then resistance stays the same.

5. *Technician A* says that the voltage regulator can be located inside the AC generator.

    Technician B says that the AC generator voltage regulator can be located inside the ECM.

    Who is correct?

    A. Technician A only   C. Both technicians

    B. Technician B only   D. Neither technician

6. Two technicians are discussing thermistors,

    Technician A says that thermistors change resistance as temperature changes.

    Technician B says thermistors are often used as computer input sensors.

    Who is correct?

    A. Technician A only   C. Both technicians

    B. Technician B only   D. Neither technician

7. *Technician A* says that a solenoid is an electromagnet with a fixed core.

    Technician B says a relay uses a small current to control a larger current.

    Who is correct?

    A. Technician A only   C. Both technicians

    B. Technician B only   D. Neither technician

8. During electromagnetic induction, the amount of current induced in the conductor is determined by all of these factors *except* the:

    A. number of conductors.

    B. strength of the magnet.

    C. shape of the magnet.

    D. speed of motion between magnet and conductor.

9. *Technician A* says that the transistor is the simplest semiconductor device.

    Technician B says that the transistor can act as an insulator or a conductor depending on the direction of current flow.

    Who is correct?

    A. Technician A only   C. Both technicians

    B. Technician B only   D. Neither technician

10. *Technician A* says that the potentiometer is a variable resistor that changes value according to temperature.

    Technician B says a diode allows current to flow in one direction but not the other.

    Who is correct?

    A. Technician A only   C. Both technicians

    B. Technician B only   D. Neither technician

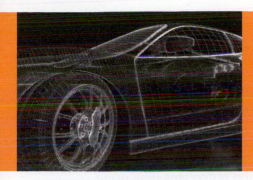

CHAPTER 5
INTAKE AND EXHAUST SYSTEMS

Upon completion and review of this chapter, you should be able to:

- Describe air cleaner purpose and operation.
- Describe different types of air cleaner elements.
- Describe port fuel injection (PFI) and direct fuel injection (DFI) in relation to intake manifold design.
- Describe how intake manifold vacuum is created, and explain the relation of throttle opening to intake manifold vacuum.
- Explain the difference between tuned exhaust headers and a conventional exhaust manifold.
- Explain how carbon monoxide (CO), unburned hydrocarbons (HC), and oxides of nitrogen (NO_x) are formed during the combustion process.
- Explain a three-way catalytic converter.
- Define diesel particulate emissions and their control.

- Explain the use of diesel exhaust fluid.
- Explain basic turbocharger operation.
- Describe how the turbocharger boost pressure is controlled.
- Explain two ways in which the turbocharger bearings are cooled.
- Describe the results of inadequate turbocharger bearing cooling.
- List three items that cause premature turbocharger failure.
- Explain basic supercharger operation.
- Describe the difference in compression ratio in a turbocharged or supercharged engine compared to a naturally aspirated engine.
- List nine components that are strengthened in a supercharged engine compared to a naturally aspirated engine.

Terms To Know

Air cleaner ducts
Backpressure
Catalytic converter
Diesel oxidation converter (DOC)
Diesel particulate filter (DPF)
Direct fuel injection (DFI)
Emission control system
Exhaust headers
Helmholtz resonator

Intercooler
Intermediate pipe
Monolithic-type catalytic converter
Muffler
Negative pressure
Oxidation catalyst
Oxidize
Particulate
Powertrain control module (PCM)

Reducing catalyst
Resonator
Reverse-flow muffler
Roots-type
Selective catalyst reduction (SCR)
Supercharger
Tailpipe
Turbo lag
Turbocharger
Waste-gate

Volumetric efficiency refers to the ability of an engine to fill the cylinders with the air and fuel mixture during the intake stroke.

INTRODUCTION

The intake and exhaust systems are extremely important to engine operation. An air intake system must provide an adequate supply of clean air to the engine cylinders to maintain volumetric efficiency. If any component in the air intake system offers excessive air restriction, volumetric efficiency is reduced. The air cleaner element must remove all dust and abrasives from the incoming air. If the air cleaner allows any dust and abrasives to enter the engine, scoring of the cylinder walls, pistons, and piston rings will occur.

The exhaust system must deliver exhaust from the cylinder head exhaust ports to the atmosphere with a minimum amount of restriction and noise. If any exhaust system component offers excessive restriction to exhaust flow, volumetric efficiency is reduced. Exhaust leaks in any of the exhaust system components provide objectionable noise. Catalytic converters in the exhaust system are responsible for reducing tailpipe emission levels.

CAFE ratings requirements are based on the "footprint" (wheelbase times track width) of the vehicle. Smaller vehicles must meet a higher CAFE rating than large vehicles.

Fuel and emissions regulations called corporate average fuel economy (CAFE) ratings have pushed the manufacturers to develop new technologies to increase fuel mileage while providing good performance. Many changes have been necessary to meet regulations, and the intake and exhaust systems have been included in the list of improvements made to vehicles over the past several years.

Turbocharged or supercharged engines were once seen primarily as an option for sports cars, but can now be found on many regular production vehicles. A turbocharger or supercharger provides the best of two worlds. A turbocharger or supercharger may be used on a relatively small engine that provides adequate fuel economy when driven at normal cruising speeds. However, the turbocharger or supercharger increases engine power to provide the faster acceleration desired by the driver.

IMPORTANCE OF INTAKE AND EXHAUST SYSTEMS

The engine needs almost fifteen times as much air as fuel to operate.

An internal combustion engine requires air to operate. This air supply is drawn into the engine by the vacuum created during the intake stroke of the pistons (**Figure 5-1**). The air is mixed with fuel in port fuel-injected engines and is delivered to the combustion chambers. Direct fuel injection systems deliver only air to the combustion chambers. Controlling the flow of air and the air-fuel mixture is the job of the induction system.

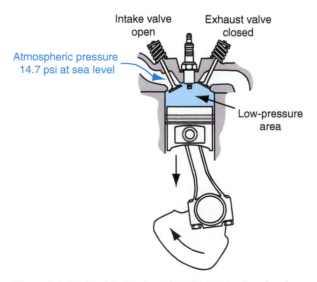

Figure 5-1 Intake air being drawn into the combustion chamber.

The air intake system channels cool air from outside the engine compartment to the throttle body assembly. Cooler air is denser than warm air, so it contains more oxygen. Engines today use advanced intake designs that take advantage of natural pressure pulsations of the engine, called "tuning" of the intake manifold. Advanced materials such as high-temperature plastic have also been used to cut weight and cost. Air cleaners are located away from the top of the engine to accommodate aerodynamic body designs. These systems also measure the amount of air and its density going into the engine so the powertrain control module (PCM) can add the correct amount of fuel for a precise burn.

Sensors measure airflow, temperature, and density. Air injection systems provide air to the exhaust stream to **oxidize** unburned hydrocarbons in the exhaust on some engines when started cold. These components allow the air induction system to perform the following functions:

- Provide the air that the engine needs to operate
- Filter the air to protect the engine from wear
- Monitor airflow temperature and density for more efficient combustion and a reduction of hydrocarbon (HC) and carbon monoxide (CO) emissions
- Operate with the positive crankcase ventilation (PCV) system to burn the crankcase fumes in the engine
- Provide clean air for some air injection systems

Before each cylinder can be refilled with a fresh charge of air-fuel mixture, the cylinders must be free of the burnt gases resulting from the last four-stroke cycle. This is the primary purpose of the exhaust system. The exhaust system collects these gases, quiets exhaust noise, and carries the gases out at the rear of the vehicle.

Shop Manual
Chapter 5, page 206

AIR CLEANERS

The primary function of the air filter is to prevent airborne contaminants and abrasives from entering into the engine with the air-fuel mixture. Without proper filtration, these contaminants can cause serious damage and appreciably shorten engine life. All incoming air should pass through the filter element before entering the engine.

Most air filter assemblies are located inside the air cleaner housing (**Figure 5-2**). In most cases, the air cleaner unit is connected to the throttle body assembly by an air transfer tube or duct.

Figure 5-2 A typical air cleaner on a fuel-injected vehicle.

> **⚠ Caution**
>
> Some technicians use compressed shop air to clean an air filter. While this cleaning saves the customer money for a new filter, care must be taken not to damage the filter. Holes can be made in the filter that, although small, are large enough to pass debris, thereby rendering the filter useless.

The air cleaner assembly also helps muffle noise caused by the airflow through the throttle plates. The air cleaner also provides filtered air to the PCV system and provides engine compartment fire protection in the event of backfire.

Most air cleaners on fuel-injected engines contain an intake air temperature and mass airflow sensors that send signals to the computer in relation to air intake temperature and quantity or mass. The computer uses this signal and other inputs to control the air-fuel mixture.

Many air filters have a pleated paper element with a wire or expanded metal screen on the outside of the element to provide protection against accidental paper damage during shipping and handling. A fine mesh screen is mounted on the inside of the pleated paper. The paper element and wire screens are molded into a heat-resistant plastisol on the top and bottom of the element. This heat-resistant plastisol has large sealing beads on both sides of the element. Air filter elements are available in many different shapes and sizes to fit various air cleaners. Some pleated-paper air filter elements contain oil-wetted resin-impregnated paper to provide longer element life and improved element efficiency. Some aftermarket manufacturers have developed reusable air filters that can be cleaned and are advertised to increase the amount of air drawn into the engine.

Some heavy-duty air cleaners have an oil-wetted polyurethane cover placed over the paper element. The polyurethane traps larger dirt particles, and the smaller particles pass through the polyurethane where they are caught in the paper element. The polyurethane cover may be removed, cleaned, oil-wetted, and reinstalled.

Air Filter Ducts

There are many different air cleaner designs depending on the type of fuel system and the available under-hood space (**Figure 5-3**). Many air cleaner assemblies have a snorkel attached to the air cleaner body and the air intake at the end of the snorkel. Most air cleaner assemblies have a plastic duct connected to the snorkel, and the outer end of this duct is positioned in a cool air location such as in front of or beside the radiator (**Figure 5-4**). A flexible coupling is positioned between the plastic duct and the snorkel. Some **air cleaner ducts** contain a large specially shaped resonator chamber that silences the airflow into the air cleaner (**Figure 5-5**).

These air cleaner ducts are designed to reduce the amount of engine noise transmitted into the passenger compartment (Figure 5-5). Most people wonder why there are so many

> **Air cleaner ducts** are tubes or hoses that direct air to the filter housing.

Figure 5-3 Air cleaner assembly and mass airflow meter.

— MAF sensor
— Air cleaner
— Air ducting to intake

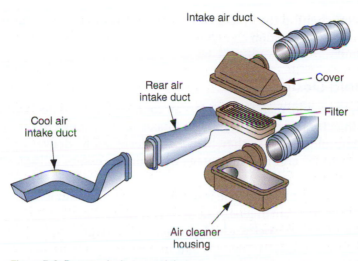

Figure 5-4 Remote air cleaner and ducts.

Figure 5-5 A Helmholtz resonator reduces engine noise.

tubes and containers under the hood going to the air cleaner. These are called **Helmholtz resonators** because they are designed to counteract what are called Helmholtz oscillations. A column of air drawn into or across a container, such as an automotive intake manifold, can produce oscillations that can result in a droning sound. These sounds occur because the air cleaner, hose, and intake have a specific frequency that will set up a vibration. (Blowing across the top of a soda pop bottle produces a similar effect.) These oscillations are always changing, due in part to the opening and closing of the intake valves and the throttle opening. The purpose of the resonator is to produce a counter-oscillation that cancels out the vibrations and therefore the sound produced.

INTAKE MANIFOLDS

The intake manifold distributes the clean air or air-fuel mixture as evenly as possible to each cylinder of the engine. Many intakes are made in two sections, a metal lower intake and a plastic composition upper intake. In port fuel injection (PFI), fuel is

Because the port fuel and direct fuel intake manifolds do not have to carry air-fuel mixtures that will condense out on the walls of the intake, the intake manifold design can be optimized for performance, fuel economy, and weight.

introduced directly behind the intake valves, while **direct fuel injection (DFI)** injects fuel directly into the combustion chamber itself, so DFI does not use the intake manifold to deliver any fuel.

Intake Manifold Design

On PFI systems, the intake manifold delivers air to the cylinders. Fuel is introduced by the individual fuel injectors directly behind the intake valves. They are often referred to as tuned intake manifolds because they have been designed to deliver equal amounts of airflow to each cylinder and take advantage of natural pressure pulsations in the intake manifold (**Figure 5-6**).

Each cylinder's intake runner tube is designed to be a certain length. All the runners for each cylinder are uniform in length as well. The effect the tubes have on incoming air causes a slight increase in velocity, which gives the air charge a type of ram effect. As air continues to enter the intake runner, a slight increase of pressure occurs. This increase takes place during the time the intake valve is closed. The intake air is packed inside the runner because it has nowhere to go. The moment the intake valve opens, the tightly packed air enters the combustion chamber with slightly more force behind it.

For years, intake manifolds were made mainly of either cast iron or aluminum. Most intake manifolds produced today leverage the advancements made in plastics and carbon-based technology. The use of this material is helpful in reducing heat-related problems. Another obvious advantage is weight reduction. The composite manifolds may also include an integral throttle body unit designed for certain applications. When performing service operations that require the removal and installation of a plastic or carbon-fiber style intake manifold, it is important to follow the manufacturer's procedures. Although tightening sequence and torque specifications are important with all intake manifolds, they are especially important with any composite-type manifold. Service information procedure must be used to prevent a cracked intake manifold.

Some intake manifolds have an upper half and a lower half. The lower half of the intake manifold is bolted to the cylinder heads and the upper half is bolted to the lower half (**Figure 5-7**).

On an in-line engine, the intake manifold-to-cylinder head area is a straight machined surface, whereas the intake manifold-to-cylinder head mounting surface on a V-type engine fits in the V between the cylinder heads. The intake manifold must distribute air and fuel, or air, as equally as possible to all the cylinders. The intake manifold air passages are designed to make their lengths as equal as possible. For example, in the manifold from

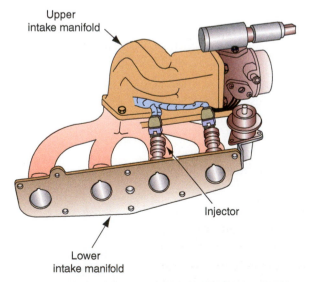

Figure 5-6 This type of tuned intake manifold, used with port fuel injection, gives a more equal amount of air to each cylinder.

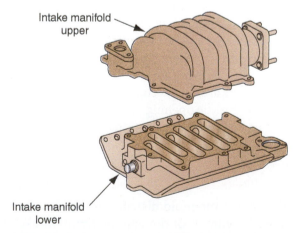

Figure 5-7 Intake manifold: the upper and lower halves.

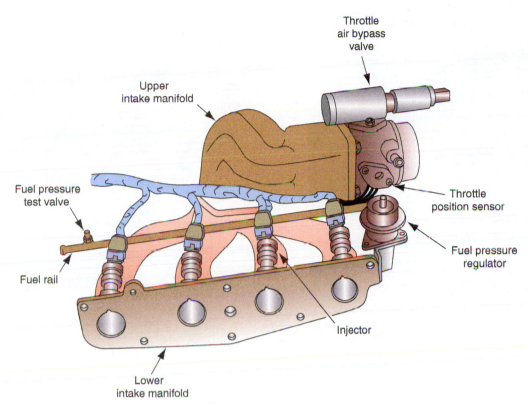

Throttle
air bypass
valve

Upper
intake manifold

Fuel pressure
test valve

Fuel rail

Throttle
position sensor

Fuel pressure
regulator

Injector

Lower
intake manifold

Figure 5-8 Intake manifold from a four-cylinder engine with curved runners in the lower half and a plenum area in the upper half.

a four-cylinder engine, four curved air passages split off from the upper plenum and connect with passages in the lower half of the intake, which is bolted to the cylinder head (**Figure 5-8**). The curved passages in the upper plenum and lower half of the intake are designed with similar lengths. Some intake manifolds have water passages cast into them to allow coolant to flow through.

An intake manifold vacuum leak can cause problems, because the air has not been measured by the computer through the mass airflow (MAF) sensor.

A series of bolts retains the intake manifold to the cylinder head, or heads, and a gasket is positioned between these two components. All intake manifold gaskets and mounting surfaces must be in satisfactory condition to prevent air leaks.

The intake manifold may be designed with longer curved air passages to improve airflow and volumetric efficiency (**Figure 5-9**). This type of intake may be referred to as a tuned intake. Some engines have small heater hoses connected from the cooling system to the throttle body assembly to prevent frost formation around the throttles.

Shop Manual
Chapter 5, page 209

Intake Manifold Tuning Valve

Some high-performance engines have two intake runners for each cylinder. There is a short set and a long set. Each of these sets is designed for different operating conditions—the longer set for low-speed, high-torque operation, and the shorter set for high-speed operation. The switching from one set of intake runners to the next is normally controlled by the PCM (**Figure 5-10**). The PCM relies on input from a variety of sensors to determine the optimum time to switch from one set of runners to the other. The intake manifold tuning valve (IMTV) is located away from the engine's intake valves. The tuning valve changes the volume of the intake to improve performance and emissions. The tuning valve can be vacuum or electrically operated. The tuning valve opens at higher RPM and gives less restriction. The tuning valve does not have to be monitored for operation.

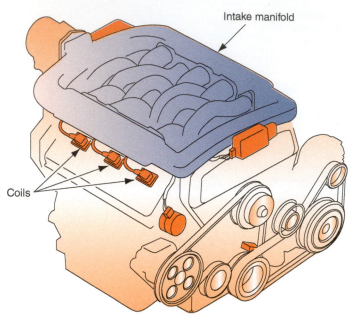

Figure 5-9 Tuned intake manifold with long intake runners on a port fuel-injected engine.

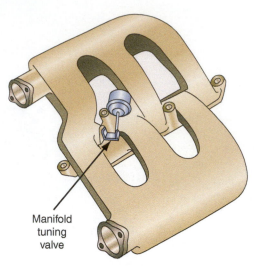

Figure 5-10 An intake manifold tuning valve (IMTV).

Intake Manifold Runner Control

The intake manifold runner control (IMRC) is located near the intake valves (**Figure 5-11**). Typically, there are two passages to each cylinder. At higher RPM, the butterfly opens and aids airflow into the cylinders, thus improving the performance. This keeps the airflow velocity high at low speeds, while allowing the necessary airflow into the engine at higher speeds.

Intake Manifold Sensors

Intake manifolds and related intake systems contain a variety of computer input sensors. These sensors include the engine coolant temperature sensor (ECT), intake air temperature (IAT) sensor, mass airflow (MAF) sensor, manifold absolute pressure sensor (MAP), and knock sensor. The ECT sensor may be mounted in a coolant passage in the intake manifold. This sensor sends a signal to the computer in relation to the coolant temperature. An IAT sensor is mounted in one of the airflow passages in the intake manifold. This sensor sends a signal to the computer in relation to the air temperature in the intake manifold. On some engines, the IAT sensor is mounted in the air cleaner (**Figure 5-12**). The IAT sensor is also commonly found as part of the MAF sensor assembly.

The MAF sensor transmits a signal to the computer in relation to the total mass of air entering the engine. This sensor is located in the air intake hose between the air cleaner and the throttle body assembly or directly in the throttle body assembly (**Figure 5-13**). A detailed explanation of these sensors is included later in the appropriate chapters.

Shop Manual
Chapter 5, page 211

Vacuum

Vacuum is measured in relation to atmospheric pressure. Atmospheric pressure is the pressure exerted on every object on earth and is caused by the weight of the surrounding air. At sea level, the pressure exerted by the atmosphere is 14.7 psi. The normal measure of vacuum is in inches of mercury, but the metric measure of kilopascals (kPa) is becoming increasingly popular. Using kilopascals eliminates the confusion caused by switching from psi and inches of mercury. Another unit of measure for vacuum and pressure is bar, pressure at sea level is around 1 bar, or 100 kilopascals.

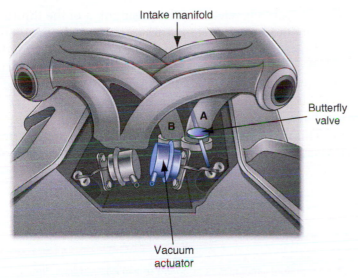

Intake manifold

Butterfly
valve

B   A

Vacuum
actuator

Figure 5-11 An intake manifold runner control (IMRC). It can be vacuum or electrically actuated by the PCM.

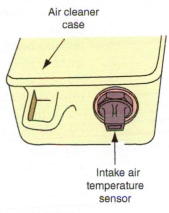

Air cleaner
case

Intake air
temperature
sensor

Figure 5-12 Intake air temperature (IAT) sensor mounted in an air cleaner housing. Some air temperature sensors are located in the intake manifold.

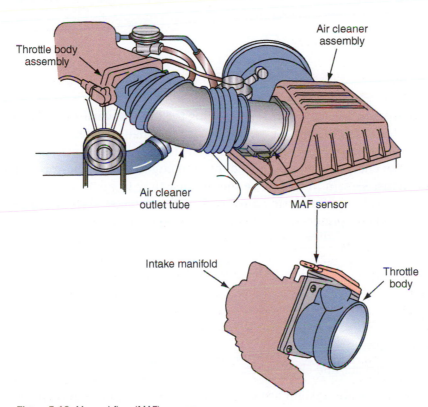

Throttle body
assembly

Air cleaner
assembly

Air cleaner
outlet tube

MAF sensor

Intake manifold

Throttle
body

Figure 5-13 Mass airflow (MAF) sensors.

Vacuum in any four-stroke engine is created by the downward movement of the piston during the intake stroke. With the intake valve open and the piston moving down, a partial vacuum is created because of the throttle plate is blocking the free-flow of air into the intake, there-by causing pressure lower than atmospheric. This partial vacuum is continuous in a multi-cylinder engine, since at least one cylinder is always at some stage of its intake stroke.

The amount of vacuum created is partially related to the positioning of the throttle plates. The throttle plate not only admits air or air-fuel mixture into the intake manifold but also helps control the amount of vacuum available during engine operation. At closed-throttle idle, the vacuum available is usually between 15 and 22 inches. At a wide-open throttle acceleration, the vacuum can drop to zero. Vacuum is highest during a closed-throttle deceleration, since the RPM is higher than it is at idle, but the throttle is closed.

While some **emission control system** output devices are solenoid or linkage controlled, many operate on vacuum. This vacuum is usually controlled by solenoids that are opened or closed, depending on electrical signals received from the **powertrain control module (PCM)**.

The **powertrain control module (PCM)** is the electronic device comprising microprocessors that controls powertrain management.

A BIT OF HISTORY

From the 1920s to the 1960s, intake manifold vacuum was supplied to two components: the distributor vacuum advance and the windshield wipers. From the 1970s to the 1990s, a wide variety of vacuum and electric/vacuum emission and computer system components had been added to the average automobile. Intake manifold vacuum was used for such items as cruise control, air conditioning, computer output control devices, and emission components. However, most systems are now controlled electronically. One important example of a vacuum-operated device that is still in use is the brake booster.

EXHAUST SYSTEM COMPONENTS

The exhaust system is responsible for collecting the exhaust gas from each cylinder and discharging this gas at the rear of the vehicle. If the exhaust system cannot remove the exhaust gas, then engine performance suffers. If the exhaust is still in the cylinders, then the intake charge cannot be pulled into the engine. While performing this function, the exhaust system must silence the exhaust flow to an acceptable level outside and inside the vehicle. Catalytic converters in the exhaust system reduce emission levels. The main components in a typical exhaust system are as follows:

1. Exhaust manifolds
2. Exhaust pipe and seal
3. Catalytic converter
4. Muffler
5. Resonator
6. Tailpipe
7. Heat shields
8. Hangers, brackets, and clamps

All the parts of the system are designed to conform to the available space of the vehicle's undercarriage and yet be a safe distance above the road.

Exhaust Manifolds

Exhaust manifolds are typically made from cast-iron, stainless steel, or heavy-gauge steel. The exhaust manifold contains an exhaust port for each exhaust port in the cylinder head, and a flat machined surface on this manifold fits against a matching surface on the exhaust port area in the cylinder head. Some exhaust manifolds have a gasket between the manifold and the cylinder head (**Figure 5-14**). In other applications, the machined surface fits directly against the matching surface on the cylinder head. The exhaust passages from

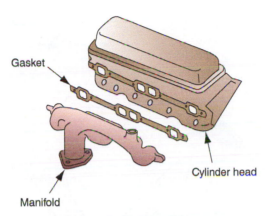

Gasket

Cylinder head

Manifold

Figure 5-14 Exhaust manifold and gasket.

Cylinder Head

Exhaust Outlet

Figure 5-15 A cylinder head with the exhaust manifold incorporated into the casting.

each port in the manifold join into a common single passage before they reach the manifold flange. An exhaust pipe is connected to the exhaust manifold flange.

On a V-type engine, an exhaust manifold is bolted to each cylinder head. Some late-model engines are being equipped with exhaust manifolds that are integral with the cylinder head (**Figure 5-15**). Integrating the exhaust and cylinder head helps reduce the weight and width of the engine. Because the cylinder heads have water jackets, the exhaust can warm the engine up to operating temperature faster, reducing cold emissions levels. This design also lowers engine noise levels and increases the amount of under-hood room available.

Exhaust system components are designed for a specific engine. The pipe diameter, component length, **catalytic converter** size, muffler size, and exhaust manifold design are engineered to provide proper exhaust flow, silencing, and emission levels on a particular engine. Exhaust headers are used in place of exhaust manifolds on some engines (**Figure 5-16**). Each time a power stroke occurs and an exhaust valve opens, a positive pressure occurs in the exhaust manifold. A **negative pressure** occurs in the exhaust manifold between the positive pressure pulses, especially at lower engine speeds.

Some **exhaust headers** are tuned so the exhaust pulses enter the exhaust manifold between the exhaust pulses from other cylinders, preventing interference between the exhaust pulses. If the exhaust pressure pulses interfere with each other, the exhaust flow

Shop Manual
Chapter 5, page 224

A positive pressure can be defined as a pressure higher than atmospheric pressure.

A **catalytic converter**, located in the exhaust system, chemically reacts to exhaust gases and heat to reduce postcombustion emissions.

A **negative pressure** is a pressure less than atmospheric pressure.

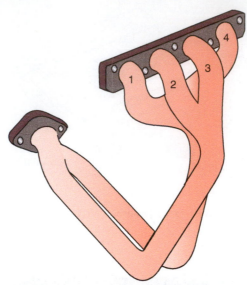

Figure 5-16 Efficiency can be improved with tuned exhaust headers.

is slowed, causing a decrease in volumetric efficiency. Proper exhaust manifold tuning actually creates a vacuum, which helps draw exhaust out of the cylinders and improve volumetric efficiency. The interference between exhaust pulses is most noticeable on a V8 engine compared with a V6 or four-cylinder engine.

Exhaust Pipe and Seal

The exhaust pipe is connected from the exhaust manifold to the catalytic converter. On inline engines, the exhaust pipe is a single pipe, but on V-type engines, the exhaust pipe is connected to each manifold flange. These two pipes are connected into a single pipe under the rear of the engine. This single pipe is then attached to the catalytic converter. Exhaust pipes can be made from stainless steel or zinc-plated steel, and some exhaust pipes are double walled. In some exhaust systems, an **intermediate pipe** is connected between the exhaust pipe and the catalytic converter. Some exhaust systems have a heavy tapered steel or steel composition sealing washer positioned between the exhaust pipe flange and the exhaust manifold flange. Other exhaust pipes have a tapered end that fits against a ball-shaped surface on the exhaust manifold flange. Bolts or studs and nuts retain the exhaust pipe to the exhaust manifold (**Figure 5-17**). Some V-type engines have dual exhaust systems with separate exhaust pipes and exhaust systems connected to each exhaust manifold.

Two or more oxygen sensors are threaded into the exhaust manifolds or pipes (**Figure 5-18**). These sensors inform the computer about the oxygen content in the exhaust.

Shop Manual
Chapter 5, page 217

Automotive Pollutants and Catalytic Converters

Three major automotive pollutants are carbon monoxide (CO), unburned hydrocarbons (HC), and oxides of nitrogen (NO_x). Gasoline is a hydrocarbon fuel containing hydrogen and carbon. Since the combustion process in the cylinders is never 100 percent complete, some unburned HC is left over in the exhaust.

Carbon monoxide (CO) is partially burned fuel. CO is formed by a lack of oxygen. If there were enough oxygen, then there would only be CO_2 in the exhaust.

Oxides of nitrogen (NO_x) are caused by high cylinder temperature. Nitrogen and oxygen are both present in air. If the combustion chamber temperatures are

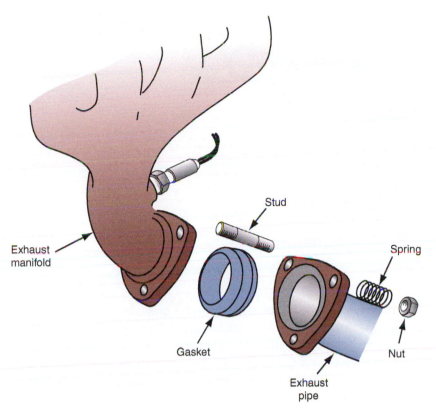

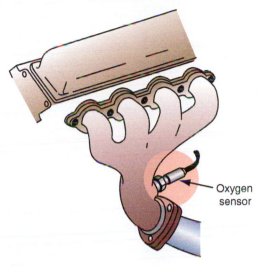

Oxygen sensor

Figure 5-18 Oxygen sensor location in exhaust manifold.

Stud

Spring

Exhaust manifold

Gasket

Nut

Exhaust pipe

Figure 5-17 Exhaust pipe retaining hardware.

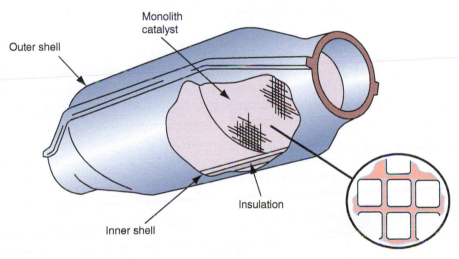

Monolith catalyst

Outer shell

Insulation

Inner shell

Figure 5-19 Monolithic-type catalytic converter.

above 2,500°F (1,371°C), some of the oxygen and nitrogen combine to form NO_x. In the presence of sunlight, HC and NO_x combine to form smog.

Catalytic converters are usually monolithic type. In a **monolithic-type catalytic converter**, the exhaust gas passes through a honeycomb ceramic block (**Figure 5-19**). The ceramic block has a thin coating of platinum, palladium, and rhodium and is mounted in a stainless-steel container.

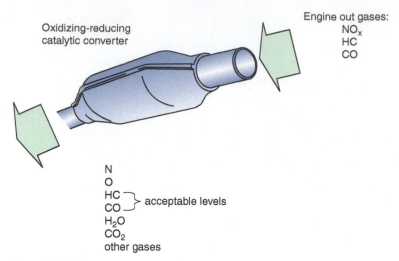

Oxidizing-reducing catalytic converter

Engine out gases:
NO_x
HC
CO

N
O
HC ⎤
CO ⎦ } acceptable levels
H_2O
CO_2
other gases

Figure 5-20 A three-way catalyst.

In a three-way catalytic converter, the **reducing catalyst** is positioned in front of the **oxidation catalyst**. A three-way catalytic converter lowers NO_x emissions as well as CO and HC. The three-way catalyst reduces NO_x into nitrogen and oxygen (**Figure 5-20**) and oxidizes HC and CO into H_2O and CO_2. All modern vehicles use the three-way catalyst. An engine that is improperly tuned causes severe overheating of the catalytic converter. Examples of improper tuning would be a rich air-fuel mixture or cylinder misfiring. The converter will try to burn all of the fuel in the exhaust and cause overheating.

Some vehicles are equipped with a mini-catalytic converter, which provides a close coupled converter that is either built in the exhaust manifold or is located next to it (**Figure 5-21**). It is primarily used to clean the exhaust during engine warm-up. These converters are commonly called warm-up converters or preheaters. On some engines, a mini-catalytic converter is built into the exhaust manifold or is bolted to the manifold flange.

Maniverters

Maniverters (**Figure 5-22**) are catalytic converters that are very close to the exhaust manifold. The maniverter can use the proximity to the manifold to take advantage of the heat generated in the combustion chamber to help maintain the heat needed to keep the converter functioning at peak performance.

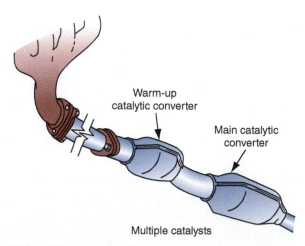

Warm-up
catalytic converter

Main catalytic
converter

Multiple catalysts

Figure 5-21 Some vehicles use a pre-catalytic converter along with the main converter.

Figure 5-22 The maniverter is a converter that is part of the exhaust manifold.

Sulfur in the Converter

During oxidation catalyst operation, small amounts of sulfur in the gasoline combine with oxygen in the air to form oxides of sulfur (SO_x). Sulphur dioxide (SO_2) gas is also formed during the oxidation converter operation. This SO_2 gas is the same gas produced by rotting eggs, and when mixed with rainwater, this SO_2 gas can turn into acid rain. This prompted the EPA to limit the amount of sulfur that is in gasoline as well as in diesel fuels. Excessive amounts of sulfur can poison the converter.

OBD II regulations mandate the manufacturers to provide a way to warn the driver of a problem with the catalytic converter. This is accomplished by placing an oxygen sensor upstream and downstream of the converter. The computer will compare both sensor voltages. If the readings are the same, there is a problem with the converter or with the engine. This will light the malfunction indicator light (MIL).

Diesel Exhaust Treatment

Tailpipe emissions on a diesel engine include HC, CO, and NO_x emissions, similar to a gasoline engine. Diesels are using several strategies to reduce emissions, such as a diesel oxidation converter (DOC), selective catalyst reduction (SCR), and a diesel particulate filter (DPF).

Diesel oxidation converter. The diesel oxidation converter works similarly to the gasoline catalytic converter and oxidizes HC and CO and converts them to H_2O and CO_2. The engine control module (ECM) uses temperature sensors in front of and behind the DOC to monitor the operation of the converter.

Selective catalyst reduction. The selective catalyst reduction system uses a urea solution (diesel exhaust fluid or DEF) to reduce NO_x into N_2 and CO_2. The urea solution is injected into the exhaust stream and collects on the SCR surface. The DEF reacts with the NO_x in the exhaust and reduces NO_x into N_2 and O_2. The DEF is injected into the exhaust stream as needed by NO_x sensors installed in the exhaust stream before and after the SCR.

Diesel particulate filter (**Figure 5-23**). Diesel engines also produce **particulate** emissions, which are small carbon particles, often called soot. Emissions of particulates have come under scrutiny because they have been listed as possibly contributing to lung cancer. A ceramic monolith inside the diesel particulate converter reduces particulate emissions (**Figure 5-24**).

Particulate emissions on a diesel engine may be called soot.

Figure 5-23 Diesel particulate filter.

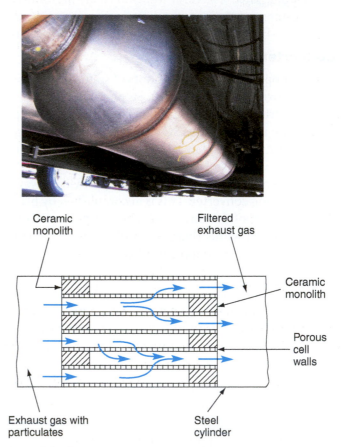

Ceramic monolith

Filtered exhaust gas

Ceramic monolith

Porous cell walls

Exhaust gas with particulates

Steel cylinder

Figure 5-24 The diesel particulate filter and selective catalyst reduction elements are contained in a single housing.

Some late-model diesels are equipped with particulate traps that have the ability to be cleaned, called regeneration. These particulate traps are regenerated (cleaned) when the PCM adds an excess amount of fuel at certain times, causing the soot to burn off. Pressure sensors can determine the amount of exhaust backpressure. Backpressure builds as the DPF becomes full of soot. When the backpressure gets to a predetermined level, the regeneration process starts and burns the soot from the filter, allowing it to be reused. Eventually, the DPF may have to be replaced, if it fills with ash, which does not burn and cannot be removed.

Mufflers

The **muffler** is an oval-shaped, or cylindrical, component made from coated and aluminized steel or stainless steel. Inlet and outlet pipes extend from the ends of the muffler.

Inside the muffler, the exhaust gas flows through a series of perforated tubes and a tuning chamber to silence the exhaust. The perforated tubes inside the muffler cancel out and silence the pressure pulsations in the exhaust each time an exhaust valve opens. The muffler is located behind the catalytic converter in the exhaust system. On many vehicles, the muffler is positioned just behind the center of the vehicle, but space requirements on some vehicles demand muffler installation near the rear of the car. When the muffler is positioned near the rear of the vehicle, it runs cooler and may experience more internal condensation. Mufflers rust on the inside if excessive internal condensation occurs.

The most common type of muffler is the **reverse-flow muffler** (**Figure 5-25**), which changes the direction of exhaust flow inside the muffler. Some mufflers are a straight-through design in which the exhaust passes through a single perforated tube.

There have been several important changes in recent years in the design of mufflers. Most of these changes have been centered at reducing weight and emissions, improving fuel economy and simplifying assembly. These changes include the following:

New materials. More and more mufflers are being made of aluminized and stainless steel. Using these materials reduces the weight of the units as well as extends their life.

Double-wall design. Retarded engine ignition timing that is used on many small cars tends to make the exhaust pulses sharper. Many small cars now use a double-wall exhaust pipe to better contain the sound and reduce pipe ring.

Rear-mounted mufflers. More and more often, the only space left under the car for the muffler is at the very rear. This means that the muffler runs cooler than before and is more easily damaged by condensation in the exhaust system. This moisture, combined with nitrogen and sulfur oxides in the exhaust gas, forms acids that rot the muffler from the inside out. Many mufflers are being produced with drain holes drilled into them.

Backpressure. Even a well-designed muffler will produce some **backpressure** in the system. Backpressure reduces an engine's volumetric efficiency or ability to "breathe." Excessive backpressure caused by defects in a muffler or other exhaust system part can slow or stop the engine. However, a small amount of backpressure can be used intentionally to allow a slower passage of exhaust gases through the catalytic converter. This slower passage results in more complete conversion to less harmful gases.

A **muffler** is a component installed in the exhaust system to reduce or eliminate the engine's postcombustion noise.

Shop Manual
Chapter 5, page 226

Reverse-flow mufflers cause the exhaust to change direction internally.

Backpressure is due to a resistance in the exhaust system that causes an opposing force against the outgoing gases.

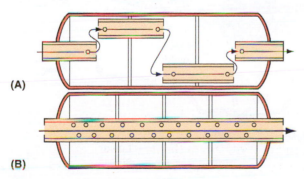

(A)

(B)

Figure 5-25 (A) Reverse-flow muffler and (B) straight-through muffler.

Resonator

On some vehicles, there is an additional muffler known as a **resonator** or silencer. This unit is designed to further reduce the sound level of the exhaust. It is located toward the end of the system and generally looks like a smaller, rounder version of a muffler. The resonator is constructed like a straight-through muffler and is connected to the muffler by an intermediate pipe. The resonator on some cars is an integral part of the tailpipe, forming a one-piece unit. Resonators have been eliminated on nearly all late-model cars as manufacturers attempt to reduce production costs and weight.

Tailpipe

The **tailpipe** carries the flow of exhaust from the muffler to the rear of the vehicle. Some vehicles have an integral resonator in the tailpipe. This resonator is similar to a small muffler and provides additional exhaust silencing. In some exhaust systems, the resonator is clamped into the tailpipe. Tailpipes have many different bends to fit around the chassis and driveline components. All exhaust system components must be positioned away from the chassis and driveline to prevent rattling. The tailpipe usually extends under the rear bumper, and the end of this pipe is cut at an angle to deflect the exhaust downward. Chrome tailpipe extensions are available in auto parts stores. These extensions are attached to the tailpipe with lock screws.

Heat Shields, Clamps, Gaskets, and Hangers

Since exhaust system components become extremely hot, many vehicles have heat shields between some of these components and the chassis (**Figure 5-26**). Without the heat shields, some chassis components such as the floor pan, may become hot enough to burn the padding under the floor mat.

Hangers are used to secure the exhaust system components to the chassis without transferring engine vibration to the chassis. The hanger is clamped to an exhaust system

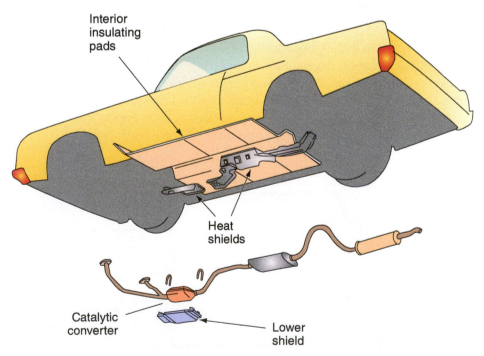

Figure 5-26 Location of heat shields.

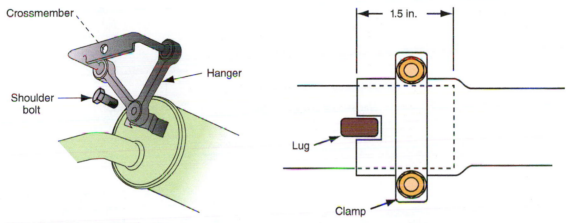

Figure 5-27 Exhaust system hanger.

Figure 5-28 Exhaust pipe clamp installation.

component, and a hanger bracket is bolted to the chassis (**Figure 5-27**). A piece of fabric-reinforced rubber between the hanger clamp and bracket prevents vibration transfer from the exhaust system to the chassis. Exhaust system clamps are bolts curved into a U-shape with two nuts that retain a bar across the U-shaped bolt.

Exhaust system components, such as the muffler and exhaust pipe, are designed to slide over each other. These exhaust system components should overlap 1½ to 3 inches and the clamp should be installed between the center and end of this overlap (**Figure 5-28**). When the clamp is tightened, the two pipes are squeezed together to prevent component movement and provide a seal between the pipes.

FORCED INDUCTION

The power generated by the internal combustion engine is directly related to the amount of air that can be pulled into the cylinders. Adding more fuel is relatively easy, especially with fuel injection. As an engine increases speed, it is even harder to get in the necessary air to burn the extra fuel needed. If air is forced into the cylinders, we can add more fuel. The more fuel and air we can burn, the more power we can produce (within practical limits). We are effectively increasing engine compression.

Two approaches can be used to increase engine compression. One is to modify the engine to increase the compression ratio. This has been done in many ways, including the use of domes or high-top pistons, altered crankshaft strokes, or changes in the shape and structure of the combustion chamber.

Another, less expensive way to increase compression (and engine power) without physically changing the shape of the combustion chamber is to simply increase the intake air charge. By pressurizing the intake mixture before it enters the cylinder, more air and more fuel to go with the air can be packed into the combustion chamber. The two processes of artificially increasing the amount of airflow into the engine are known as turbocharging and supercharging.

TURBOCHARGERS

A **turbocharger** is used to increase engine power by compressing the air that goes into the engine's combustion chambers. It does not require a mechanical connection between the engine and the pressurizing pump to compress the intake gases. Instead, it relies on

A **turbocharger** is an exhaust-driven device that increases the air intake volume, thus creating more horsepower.

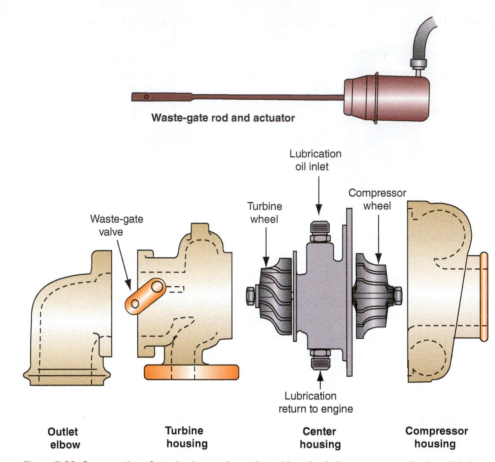

Waste-gate rod and actuator

Lubrication
oil inlet

Compressor
wheel

Turbine
wheel

Waste-gate
valve

Lubrication
return to engine

**Outlet
elbow**

**Turbine
housing**

**Center
housing**

**Compressor
housing**

Figure 5-29 Cross section of a turbocharger shows the turbine wheel, the compressor wheel, and their connecting shaft.

the rapid expansion of hot exhaust gases exiting the cylinders. These gases spin the turbine blades (hence the name turbocharger) of the pump. Because exhaust gas is a waste product, the energy developed by the turbine is said to be free since it theoretically does not use any of the engine power it helps produce.

A typical turbocharger, usually called a turbo, consists of the following components (**Figure 5-29**):

- Turbine or hot wheel
- Shaft
- Compressor or cold wheel

- Waste-gate valve
- Actuator
- Center housing and rotating assembly (CHRA). This component contains the bearings, shaft, turbine seal assembly, and compressor seal assembly.

The turbocharger is normally located close to the exhaust manifold. Since exhaust gas expansion also helps spin the impeller faster, placing the turbocharger close to the exhaust manifold helps its efficiency. An exhaust pipe runs between the exhaust manifold and the turbine housing to carry the exhaust flow to the turbine wheel. Another pipe connects the compressor housing intake to an injector throttle plate assembly.

A turbocharger is often called a turbo.

Ten psi of turbo boost means that air is being fed into the engine at 24.7 psi when the engine is operating at sea level.

Shop Manual
Chapter 5, page 227

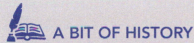

A BIT OF HISTORY

Turbochargers have been common in heavy-duty applications for many years but were not widely used in the automotive industry until the 1980s. Turbochargers had two traditional problems that prevented their wide acceptance in automotive applications. Older turbochargers had a lag, or hesitation, on low-speed acceleration, and there was the problem of bearing cooling. Engineers greatly reduced the low-speed lag by designing lighter turbine and compressor wheels with improved blade design. Water cooling combined with oil cooling provided improved bearing life. These changes made the turbocharger more suitable for automotive applications.

Basic Operation

A turbocharger contains a turbine wheel and a compressor wheel mounted on a common shaft. This shaft is supported on bearings in the turbocharger housing, and both wheels contain blades. The exhaust gas from the cylinders is directed past the turbine wheel, and the force of the exhaust gas against the turbine wheel blades causes the turbine wheel and the shaft to rotate. Since the compressor wheel is positioned on the opposite end of this shaft, the compressor wheel must rotate with the shaft (**Figure 5-30**).

The compressor wheel is mounted in the air intake, and as the compressor wheel rotates, it forces air into the intake manifold. Since most turbocharged engines are port injected, the fuel is injected into the intake ports. The rotation of the compressor wheel compresses the air and the fuel in the intake manifold, creating a denser air-fuel mixture. This increased intake manifold pressure forces more air-fuel mixture into the cylinders to provide increased engine power.

Turbocharger wheels rotate at very high speeds in excess of 100,000 rpm. Therefore, turbocharger wheel balance and bearing lubrication are very important. The turbocharger shaft must reach a certain rpm before it begins to pressurize the intake manifold. Some turbochargers begin to pressurize the intake manifold at 1,250 engine rpm and reach full boost pressure in the intake manifold at 2,250 rpm.

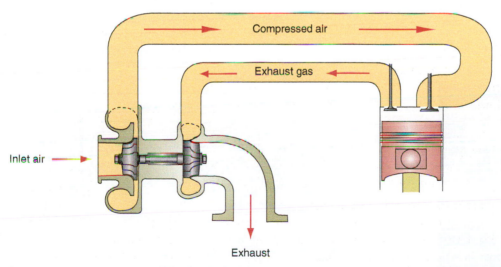

Figure 5-30 Basic turbocharger with turbine and compressor wheel.

> A **waste-gate** diaphragm may be referred to as a bypass valve controller.

Air is typically drawn into the cylinders by the difference in pressure between the atmosphere and engine vacuum. However, a turbocharger is capable of pressurizing the intake charge above normal atmospheric pressure. *Turbo boost* is the term used to describe the positive pressure increase created by a turbocharger. For example, 10 psi of boost means the air is being fed into the engine at 24.7 psi (14.7 psi atmospheric plus 10 pounds of boost).

Boost Pressure Control

> The **waste-gate** is used to control turbo boost pressure.

If the turbocharger boost pressure is not limited, excessive intake manifold and combustion pressure may damage engine components. Many turbochargers have a **waste-gate** diaphragm mounted on the turbocharger. A linkage is connected from this diaphragm to a waste-gate valve in the turbine wheel housing (**Figure 5-31**).

Older and even some later model systems used a vacuum-controlled waste-gate. The diaphragm spring holds the waste-gate valve closed. Boost pressure from the intake manifold is supplied to the waste-gate diaphragm (**Figure 5-32**). When the boost pressure in the intake manifold reaches the maximum safe limit, the boost pressure pushes the waste-gate diaphragm and opens the waste-gate valve. This action allows some exhaust to bypass the turbine wheel, which limits turbocharger shaft rpm and boost pressure.

On many late-model engines, the boost pressure supplied to the waste-gate diaphragm is controlled by a computer-operated solenoid. In many systems, the PCM pulses the waste-gate solenoid on and off to control boost pressure. The PCM can be programmed to momentarily allow a higher boost pressure on sudden acceleration to improve engine performance.

Turbocharger Cooling

Exhaust flow past the turbine wheel creates very high turbocharger temperature, especially under high engine load conditions. Many turbochargers have coolant lines connected from the turbocharger housing to the cooling system in addition to oil cooling. (**Figure 5-33**). All turbochargers have oil supplied to the bearings when the engine is running. Full oil pressure is supplied from the main oil gallery to the turbocharger bearings

Figure 5-31 Waste-gate diaphragm mounted to a turbocharger.

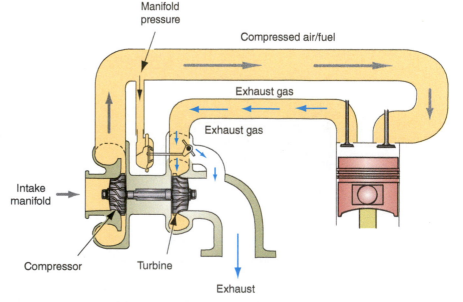

Figure 5-32 A boost pressure hose is connected from the intake manifold to the waste-gate diaphragm.

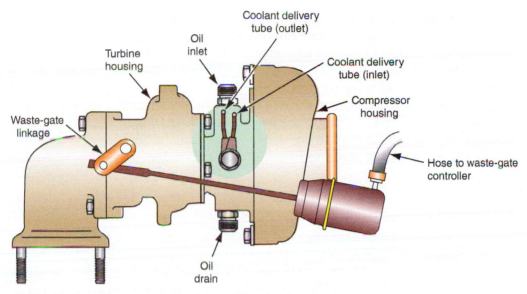

Figure 5-33 Coolant lines connected to the turbocharger housing.

and shaft to lubricate and cool the bearings. Seals at the turbocharger bearings prevent oil from entering the intake manifold or exhaust. This oil is drained from the turbocharger housing back into the crankcase.

Some heat is also dissipated from the turbocharger to the surrounding air. If the turbocharger does not utilize the cooling system to help cool the turbocharger, whenever the engine is shut off immediately after heavy-load or high-speed operation, the oil may burn to some extent in the turbocharger bearings. When this action occurs, hard carbon particles, which destroy the turbocharger bearings, are created. The coolant circulation through the turbocharger housing lowers the bearing temperature to help prevent this problem. When turbochargers that depend on oil and air cooling have been operating at heavy load or high speed, idle the engine for at least 1 minute before shutting it off. This action will help prevent turbocharger bearing failure.

The PCM limits the amount of boost to prevent detonation and engine damage by controlling the waste-gate.

Turbo Lag

Increases in horsepower are normally evidenced by an engine's response to a quick opening of the throttle. The lack of throttle response is felt with some turbocharged systems. This delay, or turbo lag, occurs because exhaust gas requires a little time to build enough energy to spin the blower up to speed.

Turbo lag occurs when the turbocharger is unable to meet the immediate demands of the engine. This causes the power from the engine to temporarily lag behind the need.

Lack of throttle response is referred to as **turbo lag**.

Scheduled Maintenance

Three main turbocharger killers are:

1. Lack of oil.
2. Contaminants in the oil.
3. Ingestion of foreign material through the air intake.

To prevent these turbocharger killers from causing premature turbocharger failure, engine oil and filters should be changed at the vehicle manufacturer's recommended intervals. The engine oil level must be maintained at the specified level on the dipstick.

Shop Manual
Chapter 5, page 235

A non-turbocharged engine can be called a naturally aspirated engine.

The air cleaner element and the air intake system must be maintained in satisfactory condition. Dirt entering the engine through an air cleaner will damage the compressor wheel blades. When coolant lines are connected to the turbocharger housing, the cooling system must be maintained according to the vehicle manufacturer's maintenance schedule to provide normal turbocharger life.

Turbocharged engines have a lower compression ratio than a naturally aspirated engine, and many parts are strengthened in a turbocharged engine because of the higher cylinder pressure. Therefore, many components in a turbocharged engine are not interchangeable with the parts in a naturally aspirated engine.

Shop Manual
Chapter 5, page 236

SUPERCHARGERS

A **supercharger** is an engine-driven device that forces air into the intake. A turbocharger is an exhaust-driven device.

Supercharging fascinated auto engineers even before they decided to steer with a wheel instead of a tiller. The 1906 American Chadwick had a **supercharger**. Since then many manufacturers have equipped engines with superchargers. Supercharged Duesenbergs, Hispano-Suizas, and Mercedes-Benzes were giants among luxury-car marques as well as winners on the race tracks in the 1920s and 1930s. Then after World War II, supercharging started to fade, although both Ford and American Motors sold supercharged passenger cars into the late 1950s. However, after being displaced first by larger V8 engines and then by turbochargers, the supercharger started to make a comeback with 1989 models. Some automobile manufacturers are offering superchargers on performance cars as an alternative to the turbo (**Figure 5-34**).

Basic Design

The supercharger is belt driven from the crankshaft by a ribbed V-belt (**Figure 5-35**). A shaft is connected from the pulley to one of the drive gears in the front supercharger housing, and the driven gear is meshed with the drive gear. The rotors inside the supercharger are attached to the two drive gears (**Figure 5-36**). The supercharger forces air into the intake manifold like the turbocharger, but engine power, instead of exhaust flow, is used to turn the supercharger.

The drive gear design prevents the rotors from touching. However, there is a very small clearance between the drive gears. In some superchargers, the rotor shafts are supported by roller bearings on the front and needle bearings on the back. During the manufacturing process, the needle bearings are permanently lubricated. The ball bearings are lubricated by synthetic base high-speed gear oil. A plug is provided for periodic checks of the front bearing lubricant. Front bearing seals prevent lubricant loss into the supercharger housing.

A supercharger may be called a blower.

Figure 5-34 An intake manifold with supercharger from a late-model V-8.

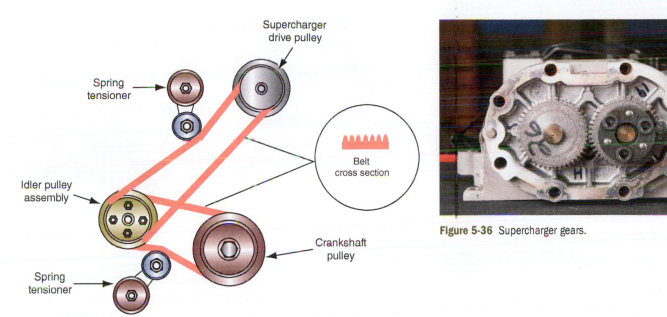

Spring tensioner

Supercharger drive pulley

Belt cross section

Idler pulley assembly

Crankshaft pulley

Spring tensioner

Figure 5-35 Supercharger pulleys and drive belts.

Figure 5-36 Supercharger gears.

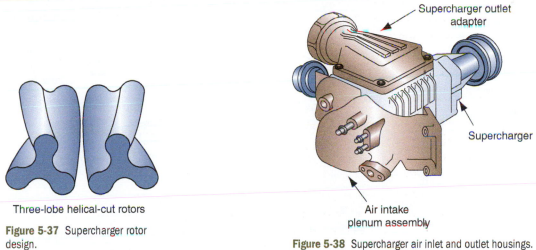

Three-lobe helical-cut rotors

Figure 5-37 Supercharger rotor design.

Supercharger outlet adapter

Supercharger

Air intake plenum assembly

Figure 5-38 Supercharger air inlet and outlet housings.

Supercharger Operation

Many superchargers have three-lobe helical-cut rotors for quieter operation and improved performance (**Figure 5-37**). Intake air enters the inlet plenum at the back of the supercharger, and the rotating blades pick up the air and force it out the top of the supercharger (**Figure 5-38**).

The blades rotate in opposite directions and act like a pump as they rotate. This pumping action pulls air through the supercharger inlet and forces the air from the outlet. There is a very small clearance between the meshed rotor lobes and between the rotor lobes and the housing (**Figure 5-39**).

Air flows through the supercharger system components in the following order:

1. Air flows through the air cleaner and mass airflow sensor into the throttle body. (The air cleaner and mass airflow sensor is not shown.)
2. Airflow enters the supercharger intake plenum.
3. From the intake plenum, the air flows into the rear of the supercharger housing (**Figure 5-40**).

Housing

Lobe straight-cut rotors

Lobe helical-cut rotors

Figure 5-39 Straight- and helical-cut rotors.

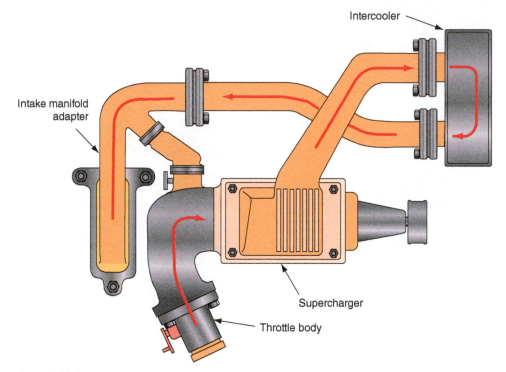

Intercooler

Intake manifold
adapter

Supercharger

Throttle body

Figure 5-40 Supercharger airflow.

4. The compressed air flows from the supercharger to the intercooler inlet.
5. Air leaves the intercooler and flows into the intercooler outlet tube.
6. Air flows from the intercooler outlet tube into the intake manifold adapter.
7. Compressed, cooled air flows through the intake manifold into the engine cylinders.
8. If the engine is operating at idle or very low speeds, the supercharger is not required. Under this condition, airflow is bypassed from the supercharger through a butterfly valve to the intake manifold adapter (**Figure 5-41**).

The bypass butterfly valve (**Figure 5-42**) is operated by an air bypass actuator diaphragm as follows:

1. When manifold vacuum is 7 in. Hg or higher, the bypass butterfly valve is completely open, and a high percentage of the supercharger air is bypassed to the supercharger inlet.

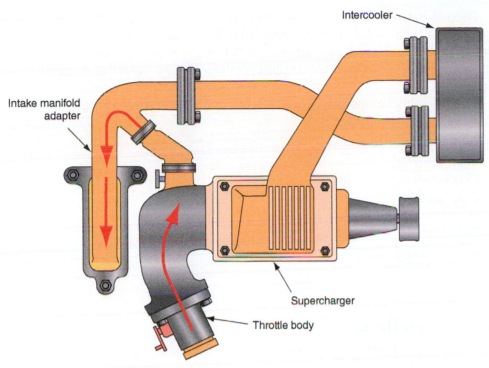

Figure 5-41 Supercharger air bypass actuator.

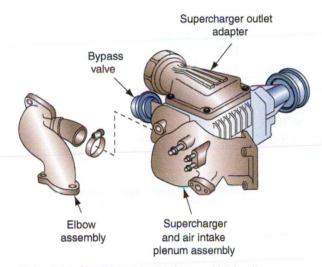

Figure 5-42 Supercharger bypass hose and intake elbow.

2. If the manifold vacuum is 3 in. Hg to 7 in. Hg, the bypass butterfly valve is partially open and some supercharger air is bypassed to the supercharger inlet, while the remaining airflow is forced into the engine cylinders.

3. When the vacuum is less than 3 in. Hg, the bypass butterfly valve is closed and all the supercharger airflow is forced into the engine cylinders.

On some superchargers, the pulley size causes the rotors to turn at 2.6 times the engine speed. Since supercharger speed is limited by engine speed, a supercharger wastegate is not required.

Belt-driven superchargers provide instant low-speed action compared to exhaust-driven turbochargers, which may have a low-speed lag because of the brief time interval required to accelerate the turbocharger shaft. Compared to a turbocharger, a supercharger turns at much lower speed.

Friction between the air and the rotors heats the air as it flows through the supercharger. The intercooler dissipates heat from the air in the supercharger system to the atmosphere, creating a denser air charge. When the supercharger and the intercooler supply cooled, compressed air to the cylinders, engine power and performance are improved.

The following components are reinforced in the supercharged engine because of the higher cylinder pressure:

Engine block
Main bearings
Crankshaft bearing caps
Crankshaft
Steel crankshaft sprocket
Timing chain
Cylinder head
Head bolts
Rocker arms

Supercharger Designs

In a **Roots-type** supercharger, a pair of rotor vanes is driven by the crankshaft.

While there have been a number of supercharger designs on the market over the years, the most popular is the **Roots-type**. The pair of three-lobed rotor vanes in the Roots supercharger is driven by the crankshaft. The lobes force air into the intake manifold.

Of course, the key to the supercharger's operation is primarily the design of the rotors. Some Roots-type superchargers use straight-lobe rotors that result in uneven pressure pulses and consequently relatively high noise levels. Therefore, the supercharger used with most of today's engines uses a helical design for the two rotors. The helical design evens out the pressure pulses in the blower and reduces noise. It was found that a 60-degree helical twist works best for equalizing the inlet and outlet volumes.

Another benefit of the helical rotor design is it reduces carryback volumes—air that is carried back to the inlet side of the supercharger because of the unavoidable spaces between the meshing rotors—which represents a loss of efficiency.

The rotors are held in a proper relationship to each other by timing gears. They are supported by ball bearings in the front and needle roller bearings in the back. The needle bearings are greased for life. The ball bearings are generally lubricated by gear oil, and a plug is provided for periodic checks of the oil level at the ball bearings. The oil is a synthetic base specifically produced for high-speed use. When servicing the gear case, it is very important to keep dirt and moisture out.

To handle the higher operating temperatures imposed by supercharging, an engine oil cooler is usually built into the engine system. This water-to-oil cooler is generally mounted between the engine front cover and oil filter.

Superchargers can be enhanced with electrically operated clutches and bypass valves. These allow the same computer that controls fuel and ignition to kick the boost on and off precisely as needed. This results in far greater efficiency than a full-time supercharger.

Another popular supercharger design, especially in Europe, is the G-Lader supercharger, which is based on a 1905 French design. As shown in **Figure 5-43**, spiral ramps in both sides of the rotor intermesh with similar ramps in the housing. Unlike most superchargers, the rotor of the G-Lader does not spin on its axis; rather, it moves around an eccentric shaft. This motion draws air in and squeezes it inward through the spiral, which compresses it and then forces it through ducts in the center into the engine. Airflow is essentially constant, so intake noise is lower than that of a Roots blower. Because there is only a slight wiping motion between the spiral and the housing, wear is minimal.

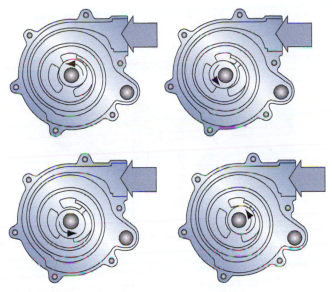

Figure 5-43 Operation of a G-Lader supercharger.

A comparison of supercharger and turbocharger operations is given in **Figure 5-44**.

COMPARISON OF SUPERCHARGER AND TURBOCHARGER

	Supercharger	**Turbocharger**
Efficiency loss	Mechanical friction and power to rotate	Increased exhaust backpressure
Packaging	Belt driven	Major revisions required to package—new exhaust manifold
Lube system	Self-contained	Uses engine oil and coolant
System noise	Pressure pulsation and gear noise	High-frequency whine and waste-gate
Durability	In use by OEMs; seems to be durable	In use by OEMs; improvements have been made and durability has increased
Performance	30% to 40% power increase over naturally aspirated Better low-end torque No lag—dramatic improvement in start-up and passing acceleration, particularly with automatic transmission	30% to 40% power increase over naturally aspirated No loss of horsepower to operate
Temperatures	No change from naturally aspirated	Increased under-hood temperatures

Figure 5-44 Comparison of supercharger and turbocharger.

INTERCOOLERS

The **intercooler** (**Figure 5-45**) cools the air from a supercharger or turbocharger before it reaches the combustion chamber. Cooling the air makes it denser. Therefore, its oxygen content increases, and it increases the engine power from a given boost pressure. It also lowers the temperature produced in the combustion chamber. These factors help reduce engine knock and increase engine output. Intercoolers are like radiators in that heat from the air passing through them is removed and dissipated to the atmosphere. An air-to-air

The intercooler is used on some supercharged and turbocharged vehicles to cool the intake air before entering the engine.

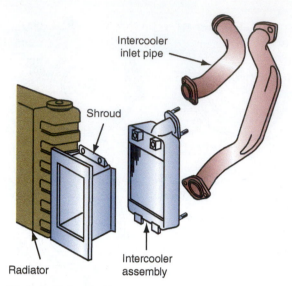

Figure 5-45 A typical intercooler system.

intercooler is normally located next to the conventional engine cooling radiator. Intercoolers can be air or water cooled.

Without an intercooler, the pressurized air is very hot. When this hot air is compressed during the engine's compression stroke, it will become even hotter. The hot air-fuel mixture in the cylinder may ignite on its own and at the wrong time, resulting in detonation.

A turbocharger or supercharger changes the effective compression ratio of an engine simply by packing in air that has a pressure greater than atmospheric pressure. For example, an engine that has a compression ratio of 8:1 and receives 10 pounds of boost will have an effective compression ratio of 10.5:1. This is why these boost devices increase power.

As air pressure increases, the temperature of the air also increases. The idea behind an intercooler is simply to cool the intake air charge, which will allow the turbocharger or supercharger to work to its maximum benefit. Cooler air going in means we have a denser air charge, and we can compress that air charge more than would be possible with ambient air.

SUMMARY

- If the air cleaner allows dust and abrasives to enter the engine, cylinder walls, pistons, and piston rings are scored.
- The heat-resistant plastisol seal on the top and bottom of the air cleaner element must contact the air cleaner body to provide proper sealing.
- Some air cleaner elements contain oil-wetted, resin-impregnated pleated paper for longer life.
- Some heavy-duty air cleaners have an oil-wetted polyurethane cover over the pleated paper element.
- The intake manifold conducts clean air from the throttle body or carburetor to the intake ports.

- In some engines, the exhaust crossover passage in the intake manifold heats the intake to improve fuel vaporization and engine performance.
- Since the fuel is injected at the intake ports in a port fuel injection (PFI) engine, the intake manifold heating requirements are reduced and the manifold may be designed to improve airflow and increase engine performance.
- Direct fuel injection vehicles use the intake manifold to deliver air only; the fuel is injected directly into the combustion chamber.

- Intake manifold vacuum is higher with the throttle closed, and this vacuum decreases as the throttle is opened.
- All domestic and import vehicles must have a vacuum schematic decal in the under-hood area.
- Exhaust manifolds are made from cast iron, nodular iron, stainless steel, or heavy-gauge steel.
- Exhaust manifolds are connected to the exhaust port area on the cylinder head.
- Some exhaust manifolds have a gasket between the manifold and the cylinder head, while other exhaust manifolds have a tight fit between the machined surfaces on the manifold and the cylinder head.
- Tuned exhaust headers are designed so the exhaust pulses from the other cylinders happen at different times, increasing volumetric efficiency.
- The exhaust pipe is connected between the exhaust manifolds and the catalytic converter.
- Automotive pollutants include carbon monoxide (CO), unburned hydrocarbons (HC), and oxides of nitrogen (NO_x).
- A catalytic converter is usually a monolithic type.
- The ceramic honeycomb in a catalytic converter has a thin coating of platinum, palladium, or rhodium.
- A three-way catalyst provides the same function as the two-way catalyst and also reduces NO_x to nitrogen and oxygen.
- Diesel engines produce particulate emissions as well as the other emissions produced by a gasoline engine.
- The diesel oxidation converter (DOC) works similarly to the gasoline catalytic converter and oxidizes HC and CO and converts them to H_2O and CO_2.
- The selective catalyst reduction system uses a urea solution (diesel exhaust fluid, or DEF) to reduce NO_x into N_2 and CO_2.
- Diesel engines also produce particulate emissions, which are small carbon particles, often called soot. Emissions of particulates have come under scrutiny because they have been listed as possibly contributing to lung cancer.
- Mufflers may be reverse-flow or straight-through design.
- In a turbocharger, the exhaust gas from the cylinders flows past the turbine wheel, causing it to rotate at high speed.
- The turbine wheel and the compressor wheel in a turbocharger are mounted on a common shaft.
- The compressor wheel in a turbocharger forces more air into the intake manifold. Since the intake manifold is pressurized, more air flows into the engine cylinders, resulting in increased engine power.
- The waste-gate valve in a turbocharger is controlled by a waste-gate diaphragm, and this diaphragm is operated by boost pressure.
- When the waste-gate diaphragm opens the waste-gate valve, some exhaust bypasses the turbine wheel, which limits boost pressure.
- On many applications, the PCM operates a solenoid connected in the hose between the intake manifold and the waste-gate diaphragm. The PCM operates this solenoid to control boost pressure supplied to the waste-gate diaphragm and limit boost pressure.
- Turbocharger bearings are cooled by oil, coolant, and the surrounding air.
- If turbocharger bearings are not properly cooled, the oil in the bearings may burn to some extent after a hot engine is shut off. This action results in hard carbon formations in the bearings, which cause premature bearing failure.
- Proper maintenance of the lubrication and cooling systems is very important on a turbocharged engine.
- A supercharger is belt driven from the engine.
- Supercharger rotors usually contain three rotors that force air into the intake manifold.
- The supercharger system has a bypass butterfly valve connected to a vacuum diaphragm, and intake manifold vacuum is supplied to this diaphragm.
- When the engine is operating at idle or very low speed, the vacuum diaphragm opens the bypass butterfly valve, which allows a high percentage of the supercharger airflow to be bypassed to the intake manifold.
- When intake manifold vacuum drops below 3 in. Hg, the vacuum diaphragm closes the bypass butterfly valve, which forces all the supercharger air to flow into the intake manifold.
- Since the supercharger speed is limited by the pulley size, a waste-gate for high-speed operation is not required.
- The compression ratio is lower in a supercharged or turbocharged engine compared to a naturally aspirated engine.
- Many components are strengthened in a supercharged or turbocharged engine because of the higher cylinder pressures.
- An intercooler removes heat from the pressurized air from a turbocharger or supercharger, thereby making it denser.

REVIEW QUESTIONS

Short-Answer Essays

1. List at least three purposes of the air cleaner assembly.

2. Describe the effect of CAFE regulations on vehicles produced by manufacturers.

3. Describe where direct fuel injection delivers fuel.

4. Describe intake manifold vacuum and how it is created.

5. Explain the operation of a three-way catalyst.

6. Explain how a turbocharger or supercharger supplies more engine power.

7. Describe basic turbocharger operation.

8. Describe how the PCM controls turbocharger boost pressure.

9. Describe how the turbocharger bearings are cooled.

10. Explain basic supercharger operation.

Fill-in-the-Blanks

1. Ten psi of turbo boost means that air is being fed into the engine at _____ when the engine is operating at sea level.

2. Some exhaust headers are designed so the _____ from each cylinder occur between the _____ from the other cylinders.

3. If any component in the air intake system offers excessive air restriction, _____ _____ is reduced.

4. At a wide-open throttle acceleration, the _____ can drop to zero.

5. The ceramic block has a thin coating of _____, _____, and _____ and is mounted in a stainless-steel container.

6. The exhaust gas from the cylinders is directed past the _____ wheel in a turbocharger.

7. The _____ wheel in a turbocharger forces more air into the intake manifold.

8. When the _____ in the intake manifold reaches the maximum safe limit, the _____ _____ pushes the waste-gate diaphragm and opens the waste-gate valve.

9. A supercharger is _____ driven from the crankshaft.

10. Supercharger rotor speed is limited by _____.

Multiple Choice

1. *Technician A* says that a three-way converter reduces NO_x emissions.
   Technician B says that a three-way converter oxidizes HC and CO.
   Who is correct?
   A. Technician A   C. Both technicians
   B. Technician B   D. Neither technician

2. *Technician A* says that compression can be raised by physically modifying the engine.
   Technician B says that compression can be effectively raised by using a turbocharger to force air into the engine.
   Who is correct?
   A. Technician A   C. Both technicians
   B. Technician B   D. Neither technician

3. While discussing modern intake systems, *Technician A* says that they contain an EBM.
   Technician B says that they contain an ECT.
   Who is correct?
   A. Technician A   C. Both technicians
   B. Technician B   D. Neither technician

4. *Technician A* says that port fuel injection systems intake manifold runners are designed to be a certain length.
   Technician B says that the runners in the manifold are designed to provide a ram air effect.
   Who is correct?
   A. Technician A   C. Both technicians
   B. Technician B   D. Neither technician

5. *Technician A* says that the exhaust is routed past the compressor wheel of the turbocharger.
   Technician B says that the exhaust is routed past the turbine wheel.
   Who is correct?
   A. Technician A   C. Both technicians
   B. Technician B   D. Neither technician

6. All of the following components are part of the turbocharger *except* the:
   A. Turbine or hot wheel
   B. Shaft
   C. Waste-gate valve
   D. Resonator valve

7. *Technician A* says that the supercharger is belt driven.
   Technician B says that a supercharger uses a waste-gate to control boost pressures.
   Who is correct?
   A. Technician A   C. Both technicians
   B. Technician B   D. Neither technician

8. *Technician A* says that the turbocharger can use an intercooler.
   Technician B says that the supercharger can use an intercooler.
   Who is correct?
   A. Technician A   C. Both technicians
   B. Technician B   D. Neither technician

9. *Technician A* says that the diesel oxidation converter (DOC) breaks down N_2 into NO_x.
   Technician B says that the DOC burns CO and HC.
   Who is correct?
   A. Technician A   C. Both technicians
   B. Technician B   D. Neither technician

10. *Technician A* says that the intake air temperature (IAT) sensor can be located in the air cleaner assembly.
    Technician B says that the IAT sensor can be located in the intake manifold.
    Who is correct?
    A. Technician A   C. Both technicians
    B. Technician B   D. Neither technician

CHAPTER 6

INPUT SENSORS

Upon completion and review of this chapter, you should be able to:

- Explain the difference between analog and digital voltage signals.
- Explain binary coding as it relates to computer input signals.
- Describe the design and operation of the oxygen sensor.
- Describe the design and operation of the air-fuel (A/F) ratio sensor.
- Describe the design and operation of the clutch pedal position (CPP) sensor.
- Explain the importance of the engine coolant temperature (ECT) sensor signal in relation to the output control functions of the computer.

- Describe the operation of the intake air temperature (IAT) sensor.
- Explain the wiring connections located on the throttle position sensor.
- Explain the operation of mass airflow (MAF) sensors.
- Describe the operation of a knock sensor (KS).
- Explain the purpose of the exhaust gas recirculation (EGR) valve position sensor.
- Identify the output control functions affected by the vehicle speed sensor (VSS) signal.

Terms To Know

Accelerator pedal position (APP) sensor

Analog/digital conversion

Analog voltage signal

Binary coding

Clutch pedal position (CPP) switch

Digital voltage signal

Engine coolant temperature (ECT) sensor

Exhaust gas recirculation (EGR) valve

Feedback

Fuel tank vapor pressure sensor

Intake air temperature (IAT) sensor

Knock sensor (KS)

Linear Hall-effect TP sensor

Manifold absolute pressure (MAP) sensor

Negative temperature coefficient (NTC)

Oxygen (O_2) sensor

Park neutral position (PNP) switch

Piezoresistive

Positive temperature coefficient (PTC)

Rotary-type throttle position (TP) sensor

Throttle actuator control (TAC)

Vehicle speed sensor (VSS)

INTRODUCTION

Computer input signals must be understood for the technician to effectively diagnose problems as they arise. Sensors can be used to indicate mechanical movement, position, or temperature. Most sensors can be classified as voltage-generating or reference voltage types. In this chapter we will look at analog and digital signals, signal conditioning, and amplification.

SENSORS

The engine control module (ECM) receives inputs that it checks with programmed values. The inputs can come from other computers, from the driver, from the technician, or through a variety of sensors. The ECM has the ability to watch engine operation through sensors. The computer can monitor the operation of actuators with sensors. Depending on the sensor input, the computer will control the actuator(s) until the programmed results are obtained.

Driver input signals are usually provided by momentarily applying a ground through a switch. The computer receives this signal and performs the desired functions. For example, if the driver wishes to reset the trip odometer on a digital instrument panel, a reset button is depressed. This switch provides a momentary ground that the computer receives as an input and sets the trip odometer to zero. Some late-model vehicles use Hall-effect style switches that do not have mechanical contacts to wear over time.

Switches can be used as an input for any operation that requires only a yes-no or on-off condition. Other inputs include those supplied by means of a sensor and those signals returned to the computer in the form of feedback. **Feedback** means that data concerning the effects of the computer's commands are fed back to the computer as an input signal.

If the computer sends a command signal to actuate an output device, a feedback signal may be sent back from the actuator to inform the computer that the task was performed. The feedback signal will confirm both the position of the output device and the operation of the actuator. Another form of feedback is for the computer to monitor voltage as a switch, relay, or other actuator is activated. Changing positions of an actuator should result in predictable changes in the computer's voltage-sensing circuit. The computer may set a diagnostic code if it does not receive the correct feedback signal.

All sensors perform the same basic function. They detect a mechanical condition (movement or position), chemical state, pressure, or temperature condition and change it into an electrical signal that can be used by the computer to make decisions. The PCM makes decisions based on information it receives from sensors. Each sensor used in a particular system has a specific job to do (e.g., monitor throttle position, vehicle speed, and manifold pressure). Together these sensors provide enough information to help the computer form a complete picture of vehicle operation. Even though there are a variety of different sensor designs, they all fall under one of two operating categories: reference voltage sensors or voltage-generating sensors.

> **Feedback** is the information or data received by the computer from various sensors used to verify the effectiveness of the computer's commands.

> The PCM can also monitor how fast the throttle is moved and can alter fuel injection accordingly.

Reference Voltage Sensors

Reference voltage (vRef) sensors provide input to the computer by modifying or controlling a constant, predetermined voltage signal. This signal, which can have a reference value from 5 to 12 volts, is generated and sent out to each sensor by a reference voltage regulator located inside the PCM. Because the computer knows that a certain voltage value has been sent out, it can indirectly interpret things like motion, temperature, and component position based on what comes back.

Potentiometer

An example of a reference voltage sensor is the potentiometer (**Figure 6-1**). It modifies a voltage to or from the computer, indicating a constantly changing status that can be calculated, compensated for, and modified. That is, this sensor simply modifies a voltage signal from the computer. When varying internal resistance of the sensor allows more or less voltage to ground, the computer senses a voltage change on a monitored signal line. The monitored signal line may be the output signal from the computer to the sensor

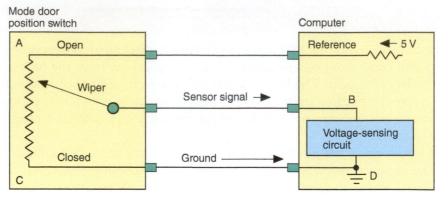

Figure 6-1 A potentiometer sensor circuit measures the amount of voltage drop to determine the position of something, such as the A/C door shown in the illustration.

Since the TP sensor is a very important component in the Throttle Actuator Control (TAC) system, also known as "throttle by wire," in this system there are often at least two TP sensors employed for improved accuracy.

(one- and two-wire sensors), or the computer may use a separate return line from the sensor to monitor voltage changes (three-wire sensors). For example, consider the operation of the throttle position (TP) sensor. During acceleration (from idle to wide-open throttle), the computer monitors throttle plate movement based on the changing reference voltage signal returned by the TP sensor. (The TP sensor is a type of variable resistor known as a rotary potentiometer that changes circuit resistance based on throttle shaft position and speed of movement.) As TP sensor resistance varies, the computer is programmed to respond in a specific manner (e.g., increase fuel delivery or alter spark timing) to each corresponding voltage change.

Thermistors

Another commonly used variable resistor is a thermistor. A thermistor is a solid-state variable resistor made from a semiconductor material that changes resistance in relation to temperature changes. Thermistors are used to monitor engine coolant, intake air, and ambient temperatures. By monitoring the thermistor's resistance value, the PCM is capable of observing small changes in temperature. The PCM sends a reference voltage to the thermistor, normally 5 volts, through a fixed resistor. As the current flows through the thermistor to ground, voltage is dropped by the thermistor. A voltage-sensing circuit compares the voltage sent out by the PCM to the voltage returned to the PCM and determines the voltage drop. Using its programmed values, the computer is able to translate the voltage drop into a temperature reading.

A negative temperature coefficient (NTC) thermistor reduces its resistance as temperature increases. This type is the most commonly used thermistor.

There are two basic types of thermistors: NTC and PTC. A **negative temperature coefficient (NTC)** thermistor reduces its resistance as temperature increases. This type is the most commonly used thermistor. A **positive temperature coefficient (PTC)** thermistor increases its resistance with an increase in temperature. To diagnose thermistors accurately, you must be able to identify the type by looking at a wiring diagram or chart in the service information.

A positive temperature coefficient (PTC) thermistor increases its resistance with an increase in temperature.

Wheatstone Bridges

Wheatstone bridges (**Figure 6-2**) are also used as variable resistance sensors. These are typically constructed of four resistors connected in series-parallel between an input terminal and a ground terminal. Three of the resistors are kept at the same value. The fourth resistor is a sensing resistor. When all four of the resistors have the same value, the bridge is balanced and the voltage sensor will have a value of 0 volts. If the sensing resistor changes value, a change will occur in the circuit's balance. The sensing circuit will receive a voltage reading that is proportional to the amount of resistance change. If the Wheatstone bridge is used to measure temperature, temperature changes will be indicated as a change

The MAP sensor is used to measure changes in engine vacuum, which directly relate to engine load.

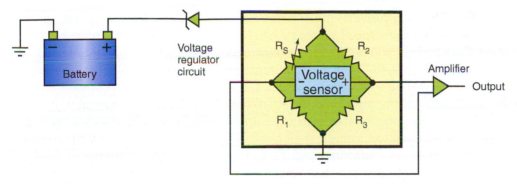

Figure 6-2 Wheatstone bridge.

in voltage by the sensing circuit. **Piezoresistive** devices are used to measure pressure changes when used in conjunction with a Wheatstone bridge.

Wheatstone bridges are one type of the component used in the manifold absolute pressure (MAP) sensor. The MAP sensor changes resistance and therefore output voltage because of a membrane inside the sensor that is printed with piezoelectric strain gauges. The membrane responds to manifold pressure either through a hose or is attached directly to the manifold itself (**Figure 6-3**). A very similar sensor measures fuel tank pressure that is used by the PCM to help determine the presence of leaks in the evaporative control system (**Figure 6-4**). We will be discussing these sensors in more detail later.

In addition to variable resistors, another commonly used reference voltage sensor is a switch. By opening and closing a circuit, switches provide the necessary voltage information to the computer so that vehicles can maintain the proper performance and drivability.

Piezoresistive sensors measure pressure changes by measuring the amount of resistance change in a semiconductor crystal when it is placed under a strain.

Piezoelectric devices generate current when a semiconductor crystal is bent.

Voltage-Generating Sensors

While most sensors are variable resistance/reference voltage, there is another category of sensors—the voltage-generating devices. These sensors include components like the magnetic pulse generators, oxygen sensors (zirconium dioxide), and knock sensors (piezoelectric), which are capable of producing their own input voltage signal. This varying voltage signal, when received by the computer, enables the computer to monitor and adjust for changes in the computerized engine control system.

Magnetic pulse generators use the principle of magnetic induction to produce a voltage signal. They are also called permanent magnet (PM) generators. These sensors are commonly used to send data to the computer about the speed of the monitored

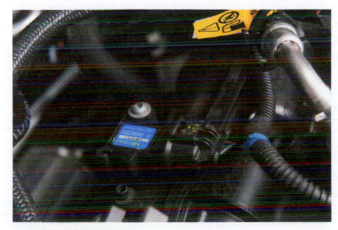

Figure 6-3 A MAP sensor. Note this MAP sensor clips directly onto the intake manifold so the vacuum line is eliminated.

Figure 6-4 A typical fuel tank pressure sensor. This unit is mounted on the fuel sender assembly.

component. This data provides information about vehicle speed, shaft speed, and wheel speed. The signals from speed sensors are used for instrumentation, cruise control systems, antilock brake systems, ignition systems, speed-sensitive steering systems, and automatic ride control systems. A magnetic pulse generator is also used to inform the computer about the position of a monitored device. This is common in engine controls where the PCM needs to know the position of the crankshaft in relation to rotational degrees.

The major components of a pulse generator are a timing disc and a pick-up coil. The timing disc is attached to a rotating shaft or cable. The number of teeth on the timing disc is determined by the manufacturer and the application. If only the number of revolutions is required, the timing disc may have only one tooth, whereas if it is important to track quarter revolutions, the timing disc needs at least four teeth. The teeth will cause a voltage generation that is constant per revolution of the shaft. For example, a vehicle speed sensor may be designed to deliver 4,000 pulses per mile. The number of pulses per mile remains constant regardless of speed. The computer calculates how fast the vehicle is going based on the frequency of the signal. The timing disc is also known as an armature, reluctor, trigger wheel, pulse wheel, or timing core.

The distributor pick-up coil is also known as a stator, sensor, or pole piece. It remains stationary while the timing disc rotates in front of it. The changes of magnetic lines of force generate a small voltage signal in the coil. A pick-up coil consists of a permanent magnet with fine wire wound around it.

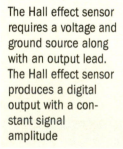

Shop Manual
Chapter 6, page 295

An air gap is maintained between the timing disc and the pick-up coil. As the timing disc rotates in front of the pick-up coil, the generator sends a pulse signal (**Figure 6-5**).

As a tooth on the timing disc aligns with the core of the pick-up coil, it repels the magnetic field. The magnetic field is forced to flow through the coil and pick-up core (**Figure 6-6**). When the tooth passes the core, the magnetic field is able to expand (**Figure 6-7**). This action is repeated every time a tooth passes the core. The moving lines of magnetic force cut across the coil windings and induce a voltage signal.

When a tooth approaches the core, a positive current is produced as the magnetic field begins to concentrate around the coil. When the tooth and core align, there is no more expansion nor contraction of the magnetic field and the voltage drops to zero. When the tooth passes the core, the magnetic field expands and a negative current is produced (**Figure 6-8**). The resulting pulse signal is sent to the PCM.

Hall-Effect Sensors

The Hall effect sensor requires a voltage and ground source along with an output lead. The Hall effect sensor produces a digital output with a constant signal amplitude

The Hall-effect switch operates on the principle that if a current is allowed to flow through thin conducting material that is exposed to a magnetic field, another voltage is produced (**Figure 6-9**).

A Hall-effect switch contains a permanent magnet and a thin semiconductor layer made of gallium arsenate crystal (Hall layer) and a shutter wheel. The Hall layer has a negative and a positive terminal connected to it. Two additional terminals located on either side of the Hall layer are used for the output circuit.

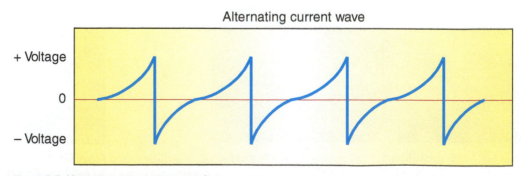

Figure 6-5 Magnetic pulse generator waveform.

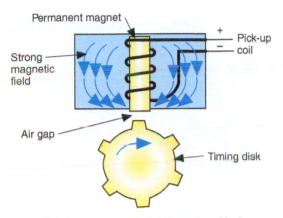

Figure 6-6 A strong magnetic field is produced in the pick-up coil as the teeth align with the core.

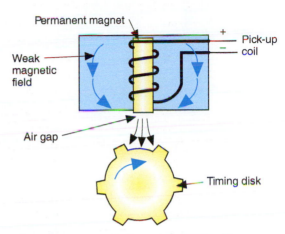

Figure 6-7 The magnetic field expands and weakens as the teeth pass the core.

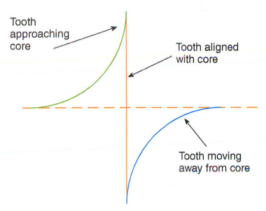

Figure 6-8 Waveform produced by a magnetic pulse generator.

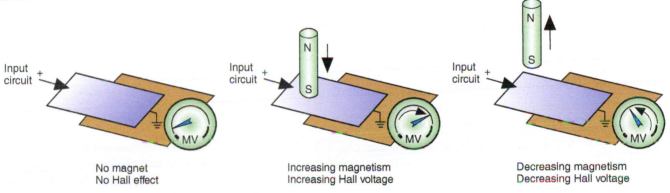

Figure 6-9 Hall-effect principles of voltage induction.

The permanent magnet is located directly across from the Hall layer so that its lines of flux will bisect at right angles to the current flow. The permanent magnet is mounted so that a small air gap is between it and the Hall layer.

A steady current is applied to the crystal of the Hall layer. This produces a signal voltage that is perpendicular to the direction of current flow and magnetic flux. The signal voltage produced is a result of the effect the magnetic field has on the electrons. When the magnetic field bisects the supply current flow, the electrons are deflected toward the

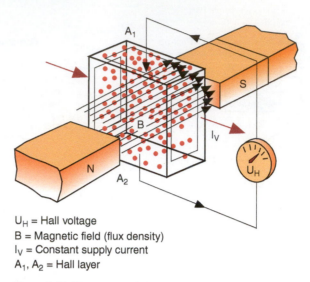

U_H = Hall voltage
B = Magnetic field (flux density)
I_V = Constant supply current
A_1, A_2 = Hall layer

Figure 6-10 The magnetic field causes the electrons from the supply current to gather at the Hall layer's negative terminal. This creates a voltage potential.

Hall layer negative terminal (**Figure 6-10**). This results in a weak voltage potential being produced in the Hall switch.

The shutter wheel consists of a series of alternating windows and vanes. It creates a magnetic shunt that changes the strength of the magnetic field from the permanent magnet. The shutter wheel is attached to a rotational component. As the wheel rotates, the vanes pass through the air gap. When a shutter vane enters the gap, it intercepts the magnetic field and shields the Hall layer from its lines of force. The electrons in the supply current are no longer disrupted, and they return to a normal state. This results in low voltage potential in the signal circuit of the Hall switch.

The signal voltage leaves the Hall layer as a weak analog signal. To be used by the PCM, the signal must be conditioned. It is first amplified because it is too weak to produce a desirable result. The signal is also inverted so that a low input signal is converted into a high output signal. It is then sent through a Schmitt trigger, which is a type of A/D converter, where it is digitized and conditioned into a clean square wave signal. The signal is finally sent to a switching transistor. The computer senses the turning on and off of the switching transistor to determine the frequency of the signals and calculates the speed.

Regardless of the type of sensors used in electronic control systems, the computer is incapable of functioning properly without input signal voltage from sensors.

Magneto-Resistive Sensors

In the family of the Hall-effect sensor, the magneto-resistive sensors are used in measuring rotation, such as in ABS braking, cam sensors, crank sensors, and the like. The magneto-resistive sensor has advantages over the magnetic-reluctance sensor. The magneto-resistive sensor has its own power supply, and, like the Hall-effect switch, the amplitude (strength) of the signal is not affected by the rotational speed. The magnetic-reluctance sensor is not accurate at low speeds (**Figure 6-11**).

VOLTAGE SIGNALS

Voltage does not flow through a conductor. Current flows while voltage is the pressure that pushes the current. However, voltage can be used as a signal. For example, difference in voltage levels, frequency of change, or switching from positive to negative values can be used as a signal.

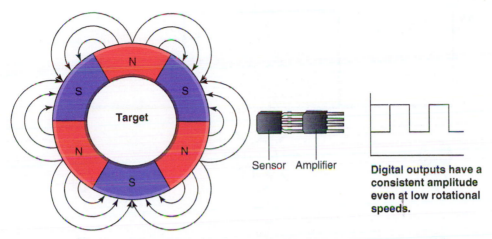

Figure 6-11 The magneto-resistive sensor outputs a digital signal regardless of the rotational speed of the target.

A computer is capable of reading voltage signals. The programs used by the PCM are "burned" into IC chips using a series of numbers. These numbers represent various combinations of voltages that the computer can understand. The voltage signals to the computer can be either analog or digital.

Analog Voltage Signals

An **analog voltage signal** is variable within a certain range. An analog voltage signal is continuously variable within a certain range. When a rheostat is used to control a 5-volt bulb, the rheostat voltage may be anywhere between 0 and 5 volts. If the rheostat voltage is low, a small amount of current flows through the bulb to produce a dim light from the bulb. If the rheostat voltage is 5 volts, higher current flow produces increased light brilliance. As the rheostat voltage is decreased, the light becomes dimmer. This is an example of an analog voltage (**Figure 6-12**). Many of the sensors in an automotive computer system produce analog voltages.

Digital Voltage Signal

Digital signals are switched either fully on or fully off. If an ordinary on/off switch is connected to a 5-volt bulb and the switch is off, 0 volt is available at the bulb. When the switch is turned on, a 5-volt signal is sent to the bulb and the bulb is illuminated to full brilliance.

Digital signals are switched either fully on or fully off. An analog voltage signal is continuously variable within a certain range.

Shop Manual
Chapter 6, page 305

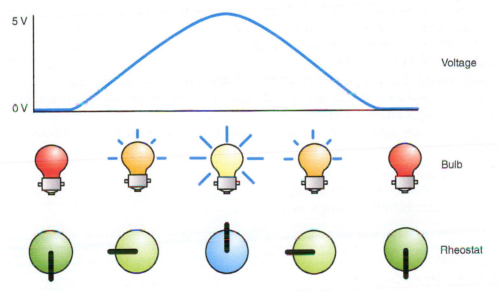

Figure 6-12 Analog voltage signal.

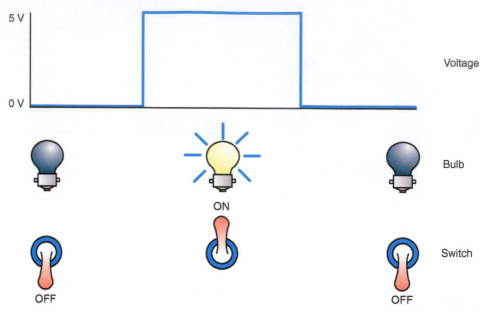

Figure 6-13 Digital square-wave voltage signal.

If the switch is turned off, the voltage at the bulb returns to 0 volt, and the bulb goes out. The voltage signal supplied through the switch is either 0 or 5 volts, or we could say the voltage signal is either high or low. This type of voltage signal is referred to as a digital signal. If the switch is turned on and off rapidly, a square-wave digital voltage signal is applied from the switch to the bulb (**Figure 6-13**).

In an automotive computer, the microprocessor contains a large number of miniature switches that are capable of producing many digital voltage signals per second. A **digital voltage signal**, also called a square-wave signal, is used to control various relays and components in the system. The microprocessor can vary the length of time that the digital signal is high or low for precise control (**Figure 6-14**). Pulse width is the amount of time the signal is turned on. For a computer to produce a varying output—for example, to control a blower motor speed—the computer can turn the output voltage on and off rapidly, with varying amounts of on time. If the signal were off for more time than it was on, the motor would run at a slower speed; if it were on more than off, it would run faster.

> A digital signal may be called a square-wave signal.

Binary Code

We have explained that a digital signal is either high or low. A numeric value may be assigned to digital signals. For example, a low digital signal may be given a value of 0 and a high digital signal may be given a value of 1. This assignment of numeric values to digital signals is called **binary coding**. The word *binary* means two values, and in the binary coding system the two values are 0 and 1 (**Figure 6-15**). In an automotive computer, information is transmitted in binary code form. Conditions, numbers, and letters can be represented by a series of zeros and ones.

> A **binary code** is the application of numeric values to digital signals.

Many of the input sensors operate in the 0 to 5 volts range. The TP sensor may produce these voltages:

    Closed throttle — 0 to 2 volts
    Part-open throttle — 2 to 4 volts
    Wide-open throttle — 4 to 5 volts

A numeric value may be assigned by the computer to each of these voltages:

    0 to 2 volts — 1
    2 to 4 volts — 2
    4 to 5 volts — 3

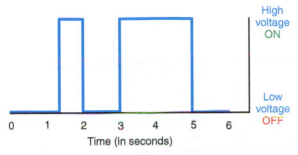

Figure 6-14 Digital voltage signals with variable on time.

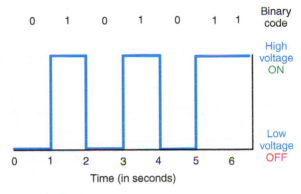

Figure 6-15 Applying numbers to digital voltage signals in a binary code.

INPUT CONDITIONING

Amplification

Some input sensors, such as the oxygen (O_2) sensor produce a very low voltage signal of less than 1 volt. This signal also has an extremely low current flow. Therefore, this type of signal must be amplified, or increased, before it is sent to the microprocessor. This amplification is accomplished by the amplification circuit in the input conditioning chip inside the computer (**Figure 6-16**).

> Input signal amplification means increasing these signals so they are useful to the computer.

Analog to Digital (A/D) Conversion

Because many input sensors produce analog signals and the microprocessor operates on digital signals, something must change the sensor analog signals to digital signals. This job is done by the A/D converter in the computer input conditioning chip (**Figure 6-17**) and is called **analog/digital conversion**.

The A/D converter continually scans the analog input signals at regular intervals. If the A/D converter scans the TP sensor signal and finds this signal at 5 volts, the A/D converter assigns a numeric value of 3 to this specific voltage. The A/D converter then changes this numeric value to a binary code of 1 1 (**Figure 6-18**).

> Digital devices will not operate with analog voltage inputs. The inputs must be a specific "on" and "off" voltage value with no variable value in between. This is an on/off or a digital signal. Analog (variable) voltages must be converted to an on/off signal. **Analog/digital conversion** is accomplished with an integrated circuit component within a processor assembly called an analog-to-digital (A/D) converter. Most computer input sensors send an analog voltage signal that must be converted, amplified, and buffered into a digital signal before the ECM will be able to read the data.

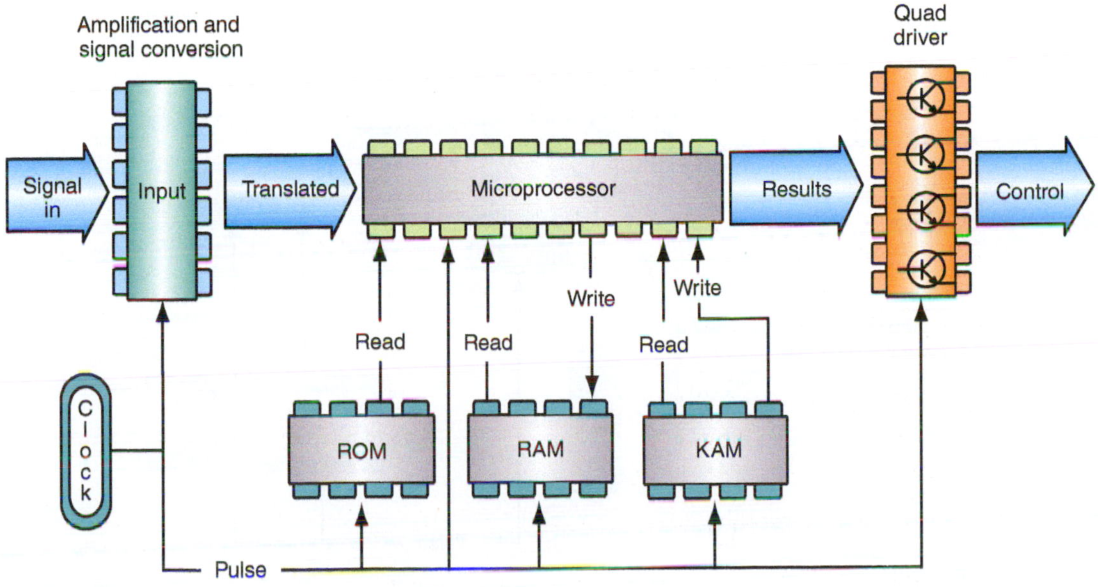

Figure 6-16 Amplification circuit in the computer input conditioning chip.

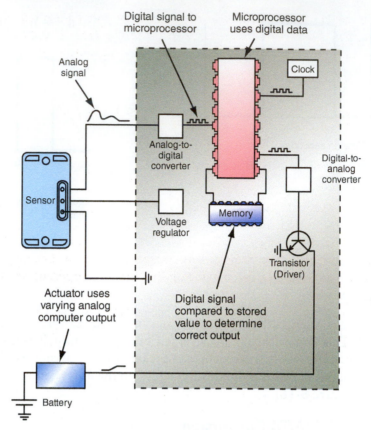

Figure 6-17 Analog/digital converter in the input conditioning chip.

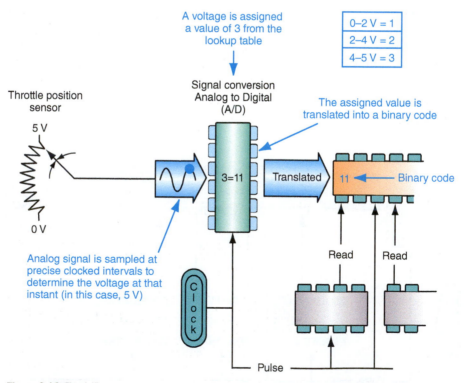

Figure 6-18 The A/D converter in the computer assigns a numerical value to the input voltages and changes the numerical value to binary code.

Therefore, we can understand that the A/D converter continually scans the input sensor signals and assigns numeric values to these voltages. The A/D converter then translates the numeric value to a binary code. In some automotive computers, the input conditioning chip is combined with the microprocessor.

INPUT SENSORS

Oxygen Sensor Design

The **oxygen** (O_2) **sensor** is threaded into the exhaust manifold or into the exhaust pipe near the engine. Some manufacturers refer to this sensor as an exhaust gas oxygen (EGO) sensor or a heated exhaust gas oxygen (HEGO) sensor. An oxygen-sensing element in the center of the O_2 sensor is surrounded by a steel shell. The oxygen-sensing element is made from zirconia in most O_2 sensors. The steel shell has a hex-shaped area on which a socket may be installed for removal and installation purposes. Threads on the lower end of the steel shell match the threads in the exhaust pipe or manifold opening where the sensor is installed (**Figure 6-19**).

A steel cover is installed over the top of the sensor. On many O_2 sensors, the steel cover is loosely installed on the sensor, which allows a constant supply of oxygen from the atmosphere inside the oxygen-sensing element. In some O_2 sensors, the top of the sensor is tightly sealed and oxygen enters the sensor through the signal wire.

A shield covers the bottom end of the sensor that protrudes into the exhaust manifold or pipe. Flutes in this shield help swirl the exhaust gas continually around the sensor element when the engine is running.

Most OBD II O_2 sensors have three or four wires. Two wires are connected to an electric heating element in the sensor. Voltage is supplied from the ignition switch to this heater when the ignition switch is on. Since the O_2 sensor does not produce a satisfactory signal until it reaches about 600°F (315°C), the internal heater provides faster sensor warm-up time and helps keep the sensor hot during prolonged idle operation. The internal O_2 sensor heater maintains higher sensor temperatures, which helps burn deposits off the sensor. When the O_2 sensor has an internal heater, the sensor may be placed farther away from the engine in the exhaust stream, thus giving engineers more flexibility in sensor location. Some O_2 sensors have four wires—a signal wire, heater wire, and two ground wires. In these four-wire sensors, the heater and the sensing element have individual ground wires. A replacement O_2 sensor must have the same number of wires as the original sensor.

The **oxygen (O₂) sensor** sends an analog voltage signal to the computer in relation to the amount of oxygen in the exhaust stream. A rich mixture has little left over oxygen and a lean mixture has a lot of excess oxygen.

Shop Manual
Chapter 6, page 295

Earlier oxygen sensors had one or two wires because they did not use a heater. Most oxygen sensors used since 1996 have been heated oxygen sensors.

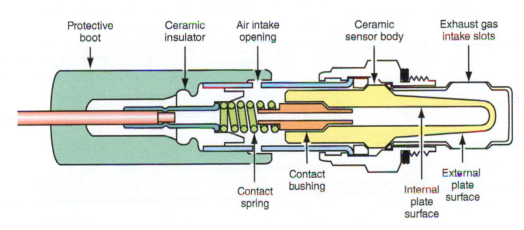

Figure 6-19 Typical oxygen sensor design.

Oxygen Sensor Operation

The zirconia oxygen sensor (O_2S) is used to detect the amount of oxygen in the exhaust. The oxygen sensor produces a voltage between 0 and 1 volt. It works by the interaction of reference oxygen and the oxygen that is in the exhaust. The zirconia is an insulator when cold and becomes a conductor at about 600°F (315°C). The internal plate surface is in contact with atmospheric oxygen, and the external plate is in contact with exhaust oxygen. At the high temperatures that the oxygen sensor operates, the internal plate becomes the positive terminal and the external plate becomes the negative terminal. The greater the difference in the amount of oxygen between the plates, the greater the voltage produced, up to 1 volt. There is a bias voltage applied from the ECM that is usually at around 0.44 volt. This is the dividing line between rich and lean exhaust values. Anything above 0.44 volt is considered a rich signal, and anything below 0.44 volt is considered a lean signal. The activity of the oxygen sensor is monitored by the number of cross counts over a period of time. If the signal has no cross counts, either the sensor is inoperative or cold, or the exhaust is fixed rich or lean.

A lean air-fuel ratio provides excess quantities of oxygen in the exhaust stream because the mixture entering the cylinders has an excessive amount of air in relation to the amount of fuel. Therefore, air containing oxygen is left over after the combustion process. When the exhaust stream has high oxygen content, oxygen from the atmosphere is also present inside the O_2 sensor element. When oxygen is present on both sides of the sensor element, the sensor produces a low voltage.

A rich air-fuel ratio contains excessive fuel in relation to the amount of air entering the cylinders. A rich air-fuel mixture produces very little oxygen in the exhaust stream because the oxygen in the air is nearly completely consumed. When the exhaust stream with very low oxygen ion content strikes the O_2 sensor, there is high oxygen ion content from the atmosphere inside the sensor element. With different oxygen levels inside and outside of the sensor element, the sensor produces from around 0.2 volt lean up to around 1 volt when rich. As the air-fuel ratio cycles from lean to rich, the O_2 sensor voltage changes in a few milliseconds.

In a gasoline fuel system, the stoichiometric, or ideal, air-fuel mixture is 14.7:1. This indicates that for every 14.7 pounds of air entering the air intake, the fuel injectors supply 1 pound of fuel. At the stoichiometric air-fuel ratio, combustion is most efficient and nearly all the oxygen in the air is mixed with fuel and burned in the combustion chambers. Fuel injection systems always maintain the air-fuel ratio near stoichiometric under most operating conditions. As the air-fuel ratio cycles slightly rich and lean from the stoichiometric ratio, the O_2 sensor voltage cycles from high to low (**Figure 6-20**).

Besides the oxygen sensor's role in air-fuel ratio control, it is also important in determining the condition of the catalytic converter in OBD II vehicles. A heated oxygen sensor placed before the converter is called the pre-catalyst O_2 sensor, and another behind the converter is called the post-catalyst O_2 sensor (**Figure 6-21**). The PCM looks at the post-catalyst O_2 compared to the pre-catalyst to make a determination on the condition of the catalyst. The post-converter O_2 sensor should show very little activity compared to the pre-converter O_2 sensor, since most of the oxygen would have been used to convert HC and CO inside the converter. If both O_2 sensors have the same amount of activity, then the PCM assumes that the catalyst has failed and the appropriate code is set. Many OBD II vehicles with V-type engines have a pre-catalyst O_2 sensor in each bank.

Air-Fuel Ratio Sensor

The O_2 sensor has done a reliable job of indicating air-fuel ratio, but it would be advantageous to have a sensor that could indicate air-fuel ratio over a broader range. The traditional O_2 sensor has a narrow range of operation around 14.7:1.

It is important to note that some manufacturers such as Chrysler, use a higher bias voltage on some of their oxygen sensors, typically between 2 and 4 volts. The basic operation is the same. An oxygen sensor with a bias voltage of 2 volts would operate between 1 and 3 volts, while a sensor with a bias voltage of 4 volts would operate between 3 and 5 volts.

The zirconia/platinum is the most commonly used oxygen sensor.

In an O_2 sensor, lean conditions produce low voltage (around 0.2 volt), and rich conditions produce high voltage (around 1 volt).

Certain types of room temperature vulcanizing (RTV) sealant or antifreeze entering the combustion chambers will contaminate the O_2 sensor. Always use the RTV sealant recommended by the vehicle manufacturer.

Always use a digital voltmeter or oscilloscope to check the O_2 sensor.

The sensor threads should be coated with an anti-seize compound prior to sensor installation, or sensor removal may be difficult the next time when it is necessary.

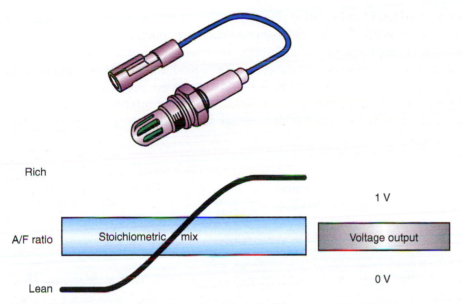

Figure 6-20 Oxygen sensor voltage in relation to air-fuel ratio.

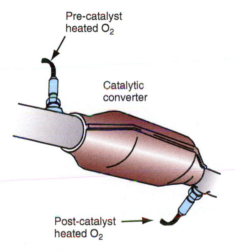

Figure 6-21 Pre-catalyst and post-catalyst heated O_2 sensors.

AUTHOR'S NOTE The conventional O_2 sensor can detect lean and rich conditions, but not precisely how rich or lean. If the ECM cannot determine the mixture with accuracy, the fuel system has to go into open loop for fuel control. Open loop basically means that the computer looks at programming to arrive at a best guess on fuel delivery based on rpm and MAF readings. This is not as accurate as an actual reading. If the actual reading is available, the computer can better deliver the exact amount of fuel needed. Better fuel delivery means lower emissions and improved performance and fuel economy.

There are many different styles of A/F ratio sensors. Most work in a similar fashion, but some can have different current flow directions and voltages from those described here.

The air-fuel (A/F) ratio sensor (**Figure 6-22**) can indicate from about 12:1 to 19:1, as shown in **Figure 6-23**. The A/F ratio sensor looks very much like the conventional oxygen sensor, but there are fundamental differences in construction and operation (**Figure 6-24**).

First, the A/F ratio sensor operates at $1,200°F$ ($650°C$), about twice as hot as the conventional oxygen sensor, so a heater is mandatory. Also, the A/F ratio sensor changes amperage and voltage to indicate A/F ratio. When the mixture is lean, the current is generated in a positive direction with a voltage above 3.3 volts. When the mixture is rich,

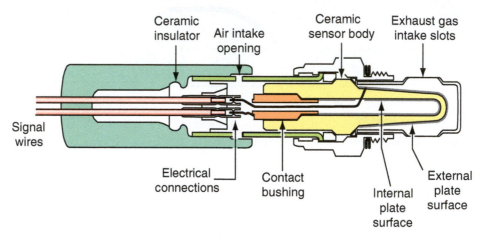

Figure 6-22 The air-fuel ratio sensor is very similar to the oxygen sensor in appearance but varies in operation.

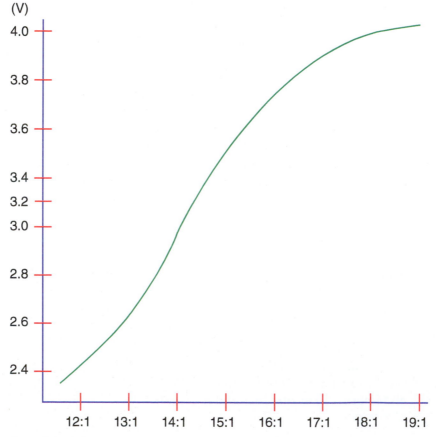

Figure 6-23 A/F ratio sensor voltage and corresponding air-fuel ratio.

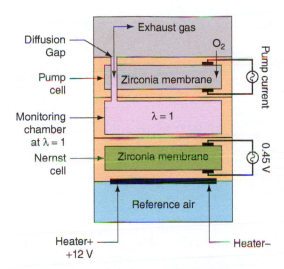

Figure 6-24 The wide range oxygen sensor internal elements.

the current is generated in a negative direction with a voltage below 3.3 volts. At 14.7:1, there is no current produced. The current flow originates inside the ECM and is in the range of 1.5 mA or less. It is used to pump oxygen ions into and out of a special chamber that houses a device called a Nernst cell. Think of the Nernst cell as a tiny conventional oxygen sensor. The ECM is trying to maintain the Nernst cell at a constant 0.44 V, which equates to 14.7:1. By measuring the amount of oxygen pumping current required to keep the Nernst cell chamber balanced, the ECM can determine the air-fuel ratio. If the mixture is lean, the oxygen pump pulls oxygen out of the chamber; if rich, the pump pulls oxygen into the chamber. The A/F sensor does not toggle rich to lean in operation like the oxygen sensor. The A/F sensor can indicate not only rich and lean like the oxygen sensor, but also just how rich or lean the mixture is by varying current direction and voltage. The computer can make better decisions on fuel management since it has more accurate information. When looking at a scan tool reading for the A/F sensor, remember that early OBD II regulations require oxygen sensors to read out between 0 and 1 volt, so the scan tool may convert the signal to correspond to the conventional sensor, or the reading may be translated into a lambda reading. A lambda reading less than 1 indicates a rich condition; a lambda reading greater than 1 indicates a lean condition. The PCM runs a fixed injection amount to check the operation of the A/F sensor. This leads to an accurate diagnostic on the sensor itself that is performed by the ECM during the monitoring for the A/F sensor.

Engine Coolant Temperature sensor

The **engine coolant temperature (ECT) sensor** is often threaded into the intake manifold, and the lower end of the sensor is immersed in engine coolant. A typical ECT sensor contains a thermistor that provides high resistance when cold and much lower resistance at higher temperatures. A typical ECT sensor may have 269,000 ohms at −40°F (−40°C) and 1,200 ohms at 248°F (120°C). The ECT sensor has two wires connected between the sensor and the computer. One of these wires is a voltage signal wire, and the other wire provides a ground. The computer senses the voltage drop across the ECT sensor. This voltage changes in relation to the coolant temperature and sensor resistance. For example, at low coolant temperature and high sensor resistance, the voltage drop across the sensor might be 4.5 volts, whereas at high coolant temperature and low sensor resistance, this voltage drop would be approximately 0.3 volt (**Figure 6-25**).

Shop Manual
Chapter 6, page 299

If the coolant temperature increases, the voltage drop across the engine coolant temperature (ECT) sensor decreases.

The **engine coolant temperature (ECT) sensor** sends an analog voltage signal to the computer in relation to coolant temperature. This signal is high at low coolant temperature and low at high coolant temperature.

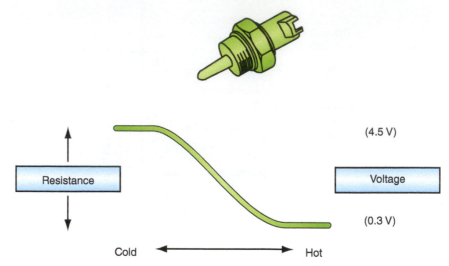

Figure 6-25 Engine coolant temperature sensor resistance and voltage drop.

The computer must know the coolant temperature to complete many of the necessary decisions regarding output control functions. For example, the computer must provide a richer air-fuel ratio when the engine is cold and a leaner stoichiometric air-fuel ratio once the engine is at normal operating temperature. Therefore, the computer must know the engine coolant temperature from the ECT signal to provide the correct air-fuel ratio.

A cold engine requires more spark advance for improved performance, whereas a hot engine needs less spark advance to prevent detonation. The computer must know engine coolant temperature and other inputs to provide the correct spark advance.

Some emission devices, such as the **exhaust gas recirculation (EGR) valve**, are not required on a cold engine, but this valve must be operational on a hot engine under certain operating conditions. The computer uses the ECT signal and other inputs to control the EGR valve in addition to electric cooling fan operation and idle speed. The computer also uses the ECT to help monitor the thermostat. If the coolant temperature does not rise fast enough, the computer can set a thermostat diagnostic trouble code (DTC). This is because an engine that does not reach operating temperature or is delayed in reaching operating temperature has higher exhaust emission levels.

Intake Air Temperature Sensor

Some **intake air temperature (IAT) sensors** can be threaded into the intake manifold, and the lower end of the sensor protrudes into one of the air passages in the intake manifold. On some applications, the IAT sensor is mounted in the air cleaner and senses air intake temperature at this location. Some IAT sensors are included in the MAF housing. The IAT sensor contains a thermistor that has ohm and voltage drop readings that are similar to the ECT sensor (**Figure 6-26**). The IAT sensor is a voltage reference sensor because the voltage drop of a 5-volt signal applied across the sensor is correlated by the ECM to calculate the air temperature.

A voltage signal wire and a ground wire are connected between the IAT sensor and the computer. Cold intake air is denser than warm air; therefore, a richer air-fuel ratio is required. When the IAT sensor signal indicates colder intake air temperature, the computer provides a richer air-fuel ratio. When the IAT indicates hotter air flowing into the intake, the ECM lowers ignition timing to help prevent spark knock. The IAT signal is analog, and like the ECT, the IAT sensor temperature is equated to a voltage drop inside

Shop Manual
Chapter 6, page 285

The **air intake temperature sensor (IAT)** signal is very similar to the engine coolant temperature sensor signal.

The IAT sensor is incorporated into the MAF sensor in some applications.

the computer. Some supercharged engines use a second IAT sensor in the intake manifold to compare the intake manifold temperature to the air entering the air cleaner for better fuel control. Many engine systems that use an MAF sensor incorporate the IAT into the MAF sensor housing (**Figure 6-27**).

Throttle Position Sensor

A **rotary-type throttle position (TP) sensor** contains a potentiometer that varies resistance as it is rotated by the throttle shaft. In computer-controlled fuel injection systems, this sensor is positioned on the end of the throttle shaft in the throttle body. TP sensors on vehicles with throttle actuator control TAC are located in the TAC housing.

> The throttle position sensor analog voltage signal increases in relation to throttle opening.

> A **rotary-type throttle position (TP) sensor** has a potentiometer that changes resistance based on the throttle position.

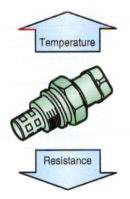

Thermistor temperature to resistance values		
°F	°C	Ohms
210	100	185
190	90	240
160	70	450
100	38	1,800
40	4	7,500
32	0	10,000
0	−18	25,000
−40	−40	100,700

Figure 6-26 Intake air temperature sensor resistance and voltage drop.

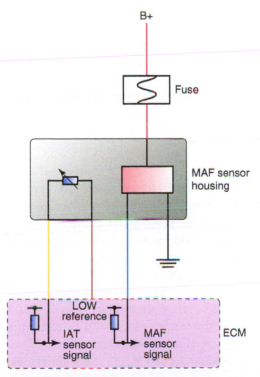

Figure 6-27 Many MAF sensors include a built-in IAT sensor.

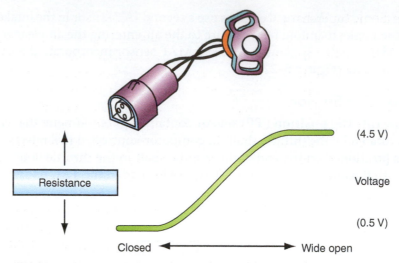

Figure 6-28 Throttle position sensor resistance and voltage drop.

Three wires are connected from the TP sensor to the computer. When the ignition switch is on, the computer supplies a constant 5-volt reference voltage to the sensor on one of these wires. One of the other wires is a TP sensor signal wire from the sensor to the computer, and the third wire is a ground wire between these two components. Sensor ground wires are usually either black or black with a colored tracer.

The voltage signal from a typical TP sensor is 0.5 to 1 volt at idle and 4.5 volts at wide-open throttle. This signal informs the computer regarding the exact throttle position (**Figure 6-28**).

Vehicles with **throttle actuator control (TAC)**, commonly known as "throttle by wire," generally have at least two potentiometer-based TP sensors located on the throttle body (although they are in the same housing) to enhance accuracy and reliability.

The computer also knows how fast the throttle is opened from the TP sensor signal. When the engine is suddenly accelerated, a richer air-fuel ratio is required with the additional airflow into the engine. If the computer receives a TP sensor signal indicating sudden acceleration, the computer supplies the necessary richer air-fuel ratio. The computer also uses the TP sensor signal to control other outputs, such as transmission shifting and stability control operation.

Shop Manual
Chapter 6, page 292

Throttle actuator control systems use an electric motor controlled by the ECM to operate the throttle plates instead of a throttle cable.

Hall-Effect TPS

A new style of throttle position sensor does not depend on a potentiometer. This sensor is also used on the TAC systems. The Hall-effect TP sensor is a bit different from the shutter style used on most ignition system and crank position sensors. The Hall-effect TP sensor uses two magnets of opposite polarity and two linear Hall-effect sensor-integrated circuits: Circuit A is the actual throttle position voltage, and B is used to check the integrity of circuit A. The scan tool will show a throttle valve between 10 and 24 percent at idle (depending on the desired idle speed) and a fully open throttle anywhere between 64 and 96 percent. If a failure occurs, the default value is about 16 percent throttle opening. Electrically, the output of the sensor looks much like the action of two potentiometers used previously. This **linear Hall-effect TP sensor** has the advantage of having no contacts to wear or corrode and is even referred to as a noncontact TP sensor.

Shop Manual
Chapter 6, page 293

On most TP sensors, there is no provision for sensor adjustment. The lowest reading seen by the PCM is recognized as 0 percent throttle.

Manifold Absolute Pressure Sensor

The **manifold absolute pressure sensor** is usually mounted in the engine compartment. A hose can be connected from the intake manifold to a vacuum inlet on the MAP sensor, or more commonly, the sensor is plugged directly into the intake manifold and the vacuum line is eliminated. Three wires are connected from the sensor to the computer. The computer supplies a constant 5-volt reference voltage to the sensor on one of these wires, and the other wires are a signal wire and a ground wire. These wiring connections are the same for most MAP sensors. When the ignition switch is turned on before the engine is started, many MAP sensors act as a barometric pressure sensor. Because the engine is not running, the barometric pressure and the intake manifold pressure are the same. This key-on-engine-not-running MAP sensor signal informs the computer regarding atmospheric pressure that varies in relation to altitude and atmospheric conditions. Some applications have a separate barometric pressure (BARO) sensor. The BARO sensor can be mounted directly to the ECM circuit board. The MAP sensor and BARO sensor are very similar in operation, with the exception that the BARO sensor input is always open to the atmosphere.

The MAP sensor senses the difference between absolute vacuum and manifold pressure. When the engine is idling, high intake manifold vacuum is about 18 in. Hg. If the engine is operating at or near wide-open throttle, the manifold vacuum may be 2 in. Hg. This condition may be called high MAP because the manifold pressure is closer to atmospheric pressure.

The computer uses the MAP sensor signal to determine the engine load. When the MAP sensor signal indicates wide-open throttle, heavy load conditions, the computer provides a richer air-fuel ratio. The computer supplies a leaner air-fuel ratio if the MAP sensor signal indicates light load, moderate cruising speed conditions. The MAP sensor can also confirm EGR valve operation. When the EGR is activated, the manifold pressure increases and the ECM is able to confirm proper operation.

Most types of MAP sensors have a silicon piezoresistive diaphragm that reacts to pressure changes by varying resistance. This varying resistance modifies the voltage signal monitored by the PCM. These sensors are voltage reference sensors (**Figure 6-29**).

When the manifold vacuum stresses the silicon diaphragm, a voltage signal is sent from the sensor to the computer in relation to the amount of vacuum. On a typical MAP sensor, the signal voltage changes from 1 volt to 1.5 volts at idle to 4.5 volts at wide-open throttle. The signal voltage may be checked with a digital voltmeter, oscilloscope, or scan tool.

> MAP sensors produce an analog voltage signal in relation to intake manifold vacuum. This signal is low when the intake manifold vacuum is high, and the signal increases as the vacuum decreases in relation to engine load.

> **Shop Manual**
> Chapter 6, page 305,
> Photo Sequence 12

Combination MAP and Manifold Air Temperature Sensors

Some manufacturers are using technological advancements to produce two-in-one sensors to reduce manufacturing costs. In addition, some space savings is gained with one fewer component, wire, and connector. One example of this is the manifold absolute pressure/manifold air temperature (MAP/MAT) sensor (**Figure 6-30**).

This sensor shares a common PCM power feed and ground and has two signal return wires for each sensor function. The MAP/MAT sensor provides two separate outputs critical to air-fuel optimization. These outputs include one voltage output proportional to engine intake manifold pressure and one voltage output proportional to manifold air temperature.

Fuel Tank Vapor Pressure Sensor

The **fuel tank vapor pressure sensor** (**Figure 6-31**) is another reference voltage (vRef) sensor closely related to the MAP sensor in operation but designed to measure very small changes between the in-tank pressure and atmospheric pressure outside the tank. The fuel tank vapor pressure sensor allows the PCM to determine whether the evaporative emission control system (EVAP) is properly sealed. The PCM can command a vacuum or allow vapor pressure to build inside the tank during testing. If these pressure conditions

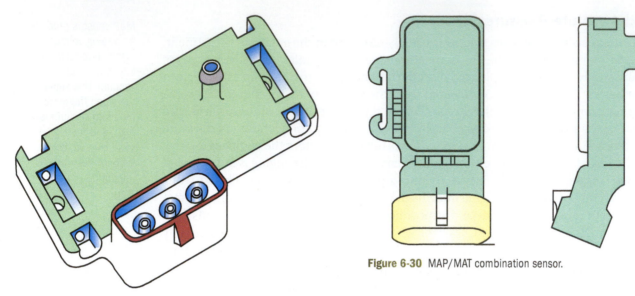

Figure 6-29 Manifold absolute pressure sensor with a silicon diaphragm.

Figure 6-30 MAP/MAT combination sensor.

Figure 6-31 A fuel tank pressure sensor.

Many technicians are confused by the term *absolute pressure*. Remember that a vacuum is a pressure less than atmospheric, and is not a perfect vacuum. A perfect vacuum is the absence of any pressure that occurs about 29.9 in. Hg. So there is always some pressure in the intake manifold, even on a deceleration. It may help to think about this in the metric system: around 100 kPa is normal atmospheric pressure near sea level and 0 kPa is the absence of pressure.

cannot be met, the PCM can set the appropriate trouble code to aid in diagnosis. A typical voltage versus pressure chart is shown in **Figure 6-32**.

Barometric Pressure Sensors

The BARO reading is generally taken by the computer when the key is turned on (before the engine starts cranking) by taking a quick reading of the MAP sensor. With the engine off, MAP and BARO readings will be the same. But some manufacturers have started using a separate barometric pressure sensor (**Figure 6-33**) on turbocharged engines to give the ECM a better idea of boost pressures or to be able to update the BARO reading while the vehicle is being driven.

Fuel Pressure Sensor

The fuel pressure sensor is used by the fuel pressure control module to help determine the necessary output of the fuel pump to match engine conditions. Having a pump run at full output wastes electrical energy. At idle, the pump current can be drastically reduced, and at full throttle, it can be increased to meet the needs of the engine. The fuel pressure sensor

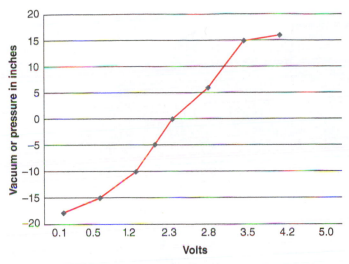

Figure 6-33 A separate BARO sensor is used in some applications.

Figure 6-32 A typical fuel tank pressure sensor chart showing voltage at different pressures.

Figure 6-34 A clutch pedal position switch.

is a typical vRef sensor with a 5-volt input, a ground, and a signal wire. Some Direct Fuel Injection (DFI) systems have a high-pressure sensor on the fuel rail that measures the output pressure of the mechanical high-pressure fuel pump.

Clutch Pedal Position Switch

The **clutch pedal position (CPP) switch** (**Figure 6-34**) informs the computer that there is no load on the engine, as well as prevents the engine from starting unless the clutch is depressed. There are two main types: some are simple switches, and some are three-wire potentiometers that operate as vRef sensors much like the TP sensor. GM vehicles use three-wire sensors that are designed to read around 1 volt with the clutch pedal applied and 4 volts with the clutch pedal released. A percentage of clutch pedal released may be available on the scan tool, with 0 percent equaling a released clutch to 100 percent on an applied clutch.

Shop Manual
Chapter 6, page 283

Mass Airflow Sensors

Vehicle manufacturers commonly use a mass airflow (MAF) sensor in electronic fuel injection (EFI) systems. The MAF sensor determines the mass of air entering the engine. By measuring the mass, the computer knows the temperature and humidity of the intake air. The computer can determine the best timing and injector pulse for a complete burn

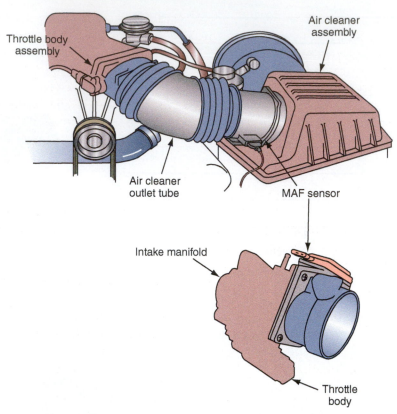

Figure 6-35 The MAF sensor can be attached near the air cleaner or directly on the throttle body.

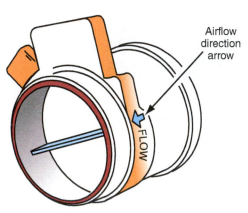

Figure 6-36 Some MAF sensors are marked in the direction of airflow.

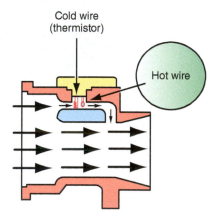

Figure 6-37 Hot-wire-type MAF sensor.

of the fuel. The MAF sensor can be mounted in the hose between the air cleaner and the throttle body or mounted directly on the intake manifold itself so that all the intake air must flow through the sensor (**Figure 6-35** and **Figure 6-36**). A hot wire is positioned in the airstream through the sensor, and an ambient temperature sensor wire is located beside the hot wire. This ambient temperature sensor wire senses intake air temperature, and this wire may be referred to as a cold wire (**Figure 6-37**). The MAF sensor takes a 12-volt power signal and produces a digital output signal. The MAF sensor is not a voltage reference sensor.

When the ignition switch is turned on, the module in the MAF sensor sends enough current through the hot wire to maintain the temperature of this wire 392°F (200°C) above

The modern MAF sensor outputs a frequency based on the amount of airflow through the sensor. It does not alter a reference voltage, but uses ignition voltage for power.

the ambient temperature sensed by the cold wire. If the engine is accelerated suddenly, the rush of cold air tries to cool the hot wire. The module immediately sends more current through the wire to maintain the wire temperature at 392°F (200°C) above the cold wire temperature. The module sends a varying frequency signal to the PCM. This signal is directly proportional to the intake airflow. When this signal is received, the PCM supplies more fuel to go with the increased intake airflow, and this action maintains the exact air-fuel ratio required by the engine. The MAF does not use a reference voltage, but does use ignition voltage for power. The MAF outputs a varying frequency signal that the ECM can use to determine the mass of air entering the engine.

Shop Manual
Chapter 6, page 308

Knock Sensor

The **knock sensor (KS)** is threaded into the engine block, intake manifold, or cylinder head. In some applications, the knock sensor is called a detonation sensor. Many engines now contain two knock sensors for improved detonation control. The knock sensor contains a piezoelectric sensing element that changes a vibration to a voltage signal (**Figure 6-38**). An internal resistor is connected in parallel with the piezoelectric sensing element.

The **knock sensor (KS)** changes a vibration to an analog voltage signal and sends this signal to the computer.

When the engine detonates, a vibration is present in the engine block and cylinder head castings. The knock sensor changes this vibration to a voltage signal and sends the signal to the computer. When this signal is received, the computer reduces spark advance to eliminate the detonation. A typical knock sensor signal would be 300 millivolts (mV) to 500 mV depending on the severity of detonation (**Figure 6-39**).

Shop Manual
Chapter 6, page 304

Vehicle Speed Sensor

The **vehicle speed sensor (VSS)** is connected in the speedometer cable or mounted in the transaxle opening where the speedometer drive was located previously or in four-wheel drive vehicles, in the transfer case (**Figure 6-40**).

The **vehicle speed sensor (VSS)** sends a voltage signal to the computer in relation to vehicle speed.

In some VSSs, the speedometer drive rotates a magnet inside a coil of wire. This type of sensor produces an AC voltage signal proportional to the vehicle speed. In some other VSSs, a magnet with eight poles rotates past a set of reed points, and these rotating poles open the points eight times per sensor revolution. Each time the points open, a signal is sent to the computer. The computer uses the VSS signal for converter clutch lockup, cruise control operation, transmission shifting, and speedometer information.

Shop Manual
Chapter 6, page 304

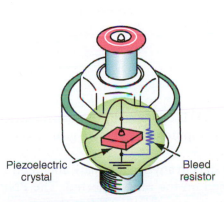

Figure 6-38 Knock sensor with piezoelectric sensing device and parallel resistor.

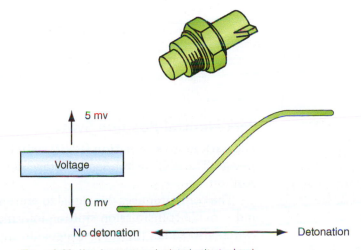

Figure 6-39 Knock sensor and related voltage signal.

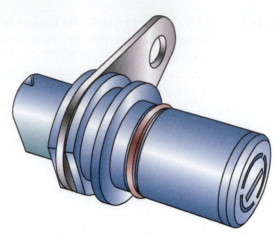

Figure 6-40 A VSS is a voltage-generating sensor.

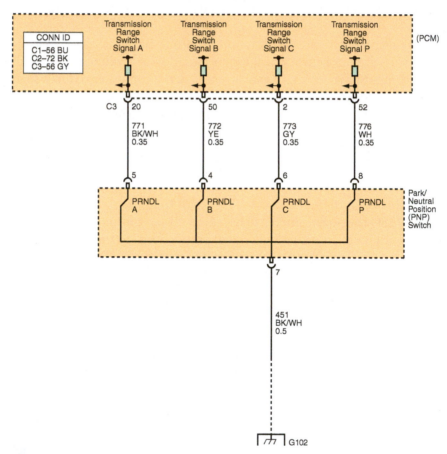

Figure 6-41 The park neutral position switch schematic.

Park Neutral Position Switch

The **park neutral position (PNP) switch** (**Figure 6-41**) is operated by the transmission linkage. The PNP switch sends voltage signals to the computer in relation to gear selector position.

The computer uses this signal to prevent engine start in any gear but park and neutral and provide transmission shifting information based on the range selected. The PNP switch also sends the reverse signal to the PCM or BCM, which operates the reverse lamps. The PNP switch may be referred to as a neutral-park switch or a transmission range switch. The PNP switch relays the range information by using different on/off combinations of the four internal switches, which are fed directly into the PCM.

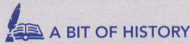

A BIT OF HISTORY

Some of us look back to the days before electronic sensors or computers and wish vehicles were still the same now. Many fail to remember that they had very poor drivability when cold, poor fuel mileage, and high emissions. We still have problems, but vehicles are much better than they were 20 years ago.

Brake Switch

The brake switch changes the voltage state of the computer when the brake is applied and again when it is released. This signal can be used as an input to control idle speed, EGR operation, torque converter clutch, or cruise control disengagement, or to modify ignition timing. The brake switch is also a critical component of the antilock braking system (ABS).

Air-Conditioning Refrigerant Pressure Switch

Air-conditioning systems are integrated into the PCM and engine management systems to signal the A/C refrigerant pressure. Although many variations are in use, many are three-wire pressure switches much like the MAP sensor, but adapted for higher pressures. If the refrigerant pressure is too low or too high, the PCM will prevent compressor clutch engagement. The PCM can also use this switch to control the condenser fan to control high side pressures.

Shop Manual
Chapter 6, page 284

Engine Position Sensors

The crankshaft position (CKP) sensor indicates the rpm of the engine and the crankshaft position. This signal is used for timing control, fuel injection pulse, and precise idle speed control. Although there are several types of sensors, all will provide a voltage signal to the PCM and change proportionately in relation to specific engine position and speed. This may be referred to as a crankshaft position sensor (**Figure 6-42**). The camshaft sensor (CMP) informs the computer of the camshaft position (**Figure 6-43**) and can be either a Hall-effect design or a variable reluctance/magnetic pulse generator type. The camshaft position sensor makes sequential fuel injection possible, because the crankshaft position sensor cannot tell if cylinder number one is on the exhaust or compression stroke. Remember that the crankshaft makes two revolutions to the camshafts, one revolution during the four-stroke cycle.

Power Steering Switch

The power steering (PS) switch simply tells the PCM when there is high pressure in the PS system due to high pump demands such as turning the steering wheel when parking. The PCM then compensates for the engine load by appropriately increasing the idle speed.

System Voltage Monitor

Without proper voltage supply or adequate charging system performance, the engine management system will not operate as designed. The PCM monitors system voltage and is constantly adjusting to compensate for varying load conditions. For example, during high electrical load demands, the PCM can modify the charging system rate, fuel injector pulse width, and ignition timing as needed.

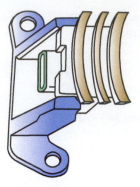

Figure 6-42 A Hall-effect crankshaft position sensor.

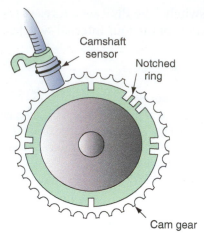

Figure 6-43 A camshaft position sensor.

THROTTLE ACTUATOR CONTROL SYSTEM

Throttle actuator control (TAC) system is used to control the throttle opening by the PCM (**Figure 6-44**). The throttle is actually moved by a reversible DC electric motor that is mounted on the throttle body. **Figure 6-45** shows a block diagram of a typical TAC system. The advantages of a TAC, or which some call "throttle by wire," are many. One of the best is the fact that the throttle becomes a computer output based on input from the driver. This way, the PCM can calculate the best throttle opening for the desired operation, instead of reacting to the actions of the driver directly. This also allows for increased control for traction control and stability control systems. The PCM can back off the throttle to help prevent slippage due to a driver over-applying the accelerator. Idle control systems are also eliminated, because the PCM controls the idle by throttle opening. Cruise control systems are also easily integrated into the design. Any problems that are detected result in the shutdown of the system, sometimes completely, sometimes with severely limited functionality because of safety concerns.

The TAC system has often been called "throttle by wire."

Figure 6-44 Throttle actuator control system.

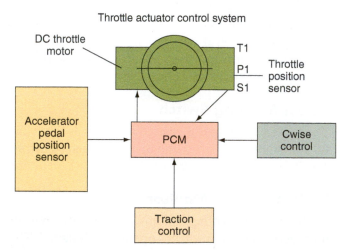

Figure 6-45 Block diagram of a throttle actuator control system.

The **accelerator pedal position (APP) sensor** is usually mounted on the accelerator pedal itself, as shown in **Figure 6-46**. The APP sensor has three potentiometers to make sure that the signal is reliable. **Figure 6-47** shows a wiring diagram for both the APP and TP sensors.

The TP sensor used with this system is also backed up by using two sensors instead of one. One sensor checks the integrity of the other; if there is a discrepancy, the system is shut down. Most of these sensors are potentiometer type; however, some are Hall-effect type.

> The APP sensor tells the exact location of the accelerator pedal.

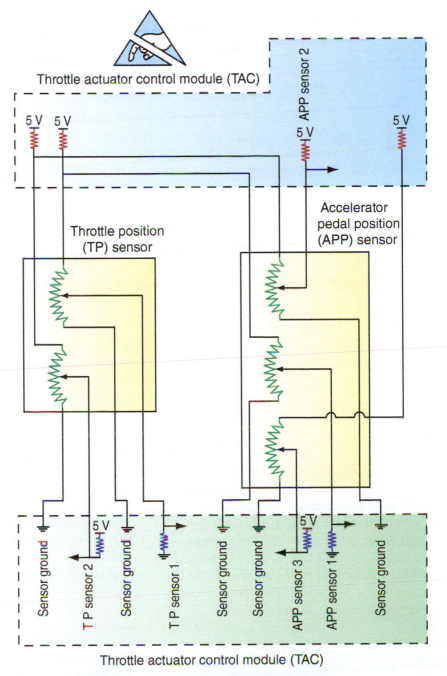

Figure 6-46 The accelerator pedal position sensor is usually located on the accelerator pedal.

Figure 6-47 Schematic for throttle actuator control and related hardware.

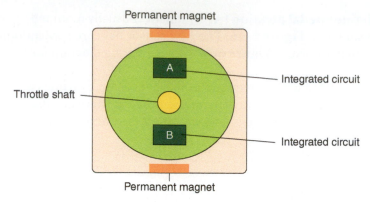

Permanent magnet

Throttle shaft

Integrated circuit

Integrated circuit

Permanent magnet

Figure 6-48 A Hall-effect TP sensor. The magnets rotate with the throttle shaft, and the integrated circuits measure movement by measuring the effect of the magnetic field.

Accelerator Pedal Position Sensor

Vehicles that use "fly by wire" throttle control have an accelerator position sensor that sends the accelerator position information to the TAC module and the PCM. The throttle is actuated by an electric motor on the throttle shaft. There are three individual sensors in the accelerator position sensor, all of which have their own power, ground, and sensor return wiring, along with two TP sensors located on the throttle shaft. The TP sensor is used by the PCM as feedback of the correct position and operation. The obvious redundancy of sensors is to ensure that this signal is correct (**Figure 6-48**).

SUMMARY

- Reference voltage (vRef) sensors provide input to the computer by modifying or controlling a constant, predetermined voltage signal.

- Most sensors presently in use are variable resistors or potentiometers. They modify a voltage to or from the computer indicating a constantly changing status that can be calculated, compensated for, and modified.

- A thermistor is a solid-state variable resistor made from a semiconductor material that changes resistance in relation to temperature changes.

- There are two basic types of thermistors: NTC and PTC. A negative temperature coefficient thermistor reduces its resistance as temperature increases. This type is the most commonly used thermistor. A positive temperature coefficient thermistor increases its resistance with an increase in temperature.

- Voltage-generating sensors include components such as the magnetic pulse generators, oxygen sensors (zirconium dioxide), and knock sensors (piezoelectric), which are capable of producing their own input voltage signal.

- An analog voltage signal is continuously variable within a certain range.

- A digital voltage signal is either high or low.

- The process of assigning numeric values to digital voltage signals is called binary coding. Some input sensor signals must be amplified by the amplification circuit in the computer input conditioning chip.

- Since the input sensors produce analog voltage signals and the computer operates on digital signals, an analog/digital (A/D) signal conversion must be completed by the A/D converter in the input conditioning chip.

- Oxygen (O_2) sensors may be made from zirconia or Titanium.

- The voltage signal from the O_2 sensor increases as the air-fuel ratio becomes richer and the computer adjusts the air-fuel ratio in response to this signal.

- O_2 sensors may be contaminated by RTV sealants or antifreeze entering the combustion chambers.

- (O_2) sensor damage occurs if the sensor is tested with an analog voltmeter.

- The air-fuel (A/F) ratio sensor measures a broader range of mixtures than the oxygen sensor.
- The engine coolant temperature (ECT) sensor resistance increases as the coolant temperature decreases.
- The computer senses the voltage drop across the ECT sensor.
- The intake air temperature (IAT) sensor has ohm and voltage drop values that are similar to those of the engine coolant temperature sensor.
- The throttle position (TP) sensor sends a voltage signal to the computer in relation to the amount and speed of throttle opening.
- The manifold absolute pressure (MAP) sensor sends a voltage signal to the computer that compares manifold pressure (vacuum) in relation to absolute (or perfect) vacuum.
- MAP sensors provide an ordinary analog voltage signal to the computer.

- Mass airflow (MAF) sensors send a varying frequency signal to the computer in relation to the total mass of air entering the engine.
- The knock sensor changes the vibration caused by engine detonation to a voltage signal.
- The exhaust gas recirculation valve position (EVP) sensor informs the computer regarding the EGR valve position.
- The vehicle speed sensor (VSS) sends a voltage signal to the computer in relation to the vehicle speed.
- The park neutral position (PNP) switch informs the computer regarding gear selector position.
- The accelerator position sensor informs the PCM of the requested throttle position.
- The TAC system controls the throttle angle and is a "fly by wire" system.
- The fuel tank vapor pressure sensor measures pressure inside the fuel tank for EVAP system testing.

REVIEW QUESTIONS

Short-Answer Essays

1. Compare a digital voltage signal with an analog signal.

2. Describe the operation of the knock sensor.

3. Describe the operation of the Wheatstone bridge.

4. Describe what is meant by feedback.

5. Describe the differences between an oxygen sensor and an air-fuel ratio sensor.

6. Explain the conditions that cause an oxygen (O_2) sensor to produce high and low voltage.

7. Briefly describe the operation of the clutch pedal position switch.

8. What is meant by the term *pulse width*?

9. Explain the operation of a hot-wire-type mass airflow (MAF) sensor.

10. Describe the purpose of the voltage signals from the park neutral position switch to the computer.

Fill-in-the-Blanks

1. Digital devices will not operate with _____ voltage inputs.

2. A digital voltage signal may be called a _____ _____ signal.

3. The term *binary* means _____ values.

4. A reference voltage sensor provides input by _____ or _____ a constant, predetermined voltage signal.

5. There are two basic types of thermistors: _____ and _____.

6. In an O_2 sensor, lean conditions produce _____ voltage.

7. If the coolant temperature increases, the voltage drop across the engine coolant temperature (ECT) sensor _____.

8. The three wires connected to a manifold absolute pressure (MAP) sensor are the ground wire, signal wire, and _____ wire.

9. The crankshaft position (CKP) sensor indicates the _____ of the engine and the crankshaft _____.

10. The air-fuel (A/F) ratio sensor operates at _____, about twice as hot as the conventional oxygen sensor.

Multiple Choice

1. *Technician A* says that an analog signal is variable within a given voltage range.
Technician B says that a digital signal is either on or off.
Who is correct?
A. Technician A   C. Both technicians
B. Technician B   D. Neither technician

2. *Technician A* says that the TP sensor is a reference voltage sensor.
Technician B says the oxygen sensor is a voltage-generating sensor.
Who is correct?
A. Technician A   C. Both technicians
B. Technician B   D. Neither technician

3. The signal from most sensors must be amplified and converted from:
A. Analog to digital
B. Binary to digital
C. Digital to analog
D. Voltage reference to voltage generating

4. *Technician A* says that computer feedback is information sent back to the computer to confirm that a command was carried out by an actuator.
Technician B says that feedback can also be in the form of a monitored voltage signal to a switch or relay.
Who is correct?
A. Technician A   C. Both technicians
B. Technician B   D. Neither technician

5. *Technician A* says that the air intake temperature (IAT) sensor signal is very similar to the engine coolant temperature sensor signal.
Technician B says that cold intake air is denser than warm air.
Who is correct?
A. Technician A   C. Both technicians
B. Technician B   D. Neither technician

6. *Technician A* says that the Hall-effect switch contains an electromagnet.
Technician B says that the Hall-effect switch also uses a shutter wheel connected to a rotational component.
Who is correct?
A. Technician A   C. Both technicians
B. Technician B   D. Neither technician

7. *Technician A* says that the oxygen sensor produces a high-voltage signal when the exhaust is lean.
Technician B says that the oxygen sensor produces a low-voltage signal when the exhaust is rich.
Who is correct?
A. Technician A   C. Both technicians
B. Technician B   D. Neither technician

8. *Technician A* says that the air-fuel ratio sensor produces below 3.3 volts in a negative direction when the exhaust is rich.
Technician B says that the air-fuel ratio sensor operates at a lower temperature than the conventional oxygen sensor.
Who is correct?
A. Technician A   C. Both technicians
B. Technician B   D. Neither technician

9. *Technician A* says that the engine coolant temperature (ECT) sensor signal is used by the PCM to provide a lean mixture when the engine is cold.
Technician B says that the signal is used by the PCM to provide less spark advance on a hot engine.
Who is correct?
A. Technician A   C. Both technicians
B. Technician B   D. Neither technician

10. *Technician A* says that the intake air temperature (IAT) sensor is a voltage-generating sensor.
Technician B says that the MAF sensor is a reference voltage sensor.
Who is correct?
A. Technician A   C. Both technicians
B. Technician B   D. Neither technician

CHAPTER 7
FUEL SYSTEMS

Upon completion and review of this chapter, you should be able to:

- Describe the four performance characteristics of gasoline.
- Describe the various types of gasoline and additives.
- Describe the different alternative fuels, including diesel.
- Explain the differences between direct fuel injection (DFI) and port fuel injection (PFI) systems.
- Describe fuel tank design and mounting on the vehicle.
- Describe a fuel tank filler and filler cap design.
- Describe the three different types of fuel lines.
- Explain the five different types of fuel line fittings.

- Describe the different fuel filter designs and mountings.
- Describe the operation of an electric fuel pump, including the operation of the relief valve and check valve.
- Describe pulse width modulated fuel pump systems.
- Describe the purpose of the inertia switch in a fuel pump circuit.
- Explain the purpose of the fuel pressure sensor in direct fuel injection systems.
- Explain the operation of the direct fuel injection system.
- Explain the differences in the E-85 fuel systems.

Terms To Know

Antiknock

Cetane

Charcoal canister

Compressed natural gas (CNG)

Detonation

Ethanol

Flame front

Fuel pressure regulator

Fuel rail

Fuel rail pressure sensor (FRPS)

Inertia switch

Lean mixture

LP-gas

Octane number

Onboard refueling vapor recovery (ORVR)

Oxygenate

Paraffin

Pre-ignition

Purge solenoid

Restraints control module (RCM)

Rich mixture

Fuel rail pressure and temperature sensor (FRPT)

Vapor lock

Vaporization

Viscosity

 A BIT OF HISTORY

The first use of fuel injection on a production American engine was a mechanical system installed by General Motors (GM) on some 1957 Corvettes and Pontiac Bonnevilles.

INTRODUCTION

This chapter focuses on an engine system that has been and is being modified year after year to meet the fuel economy and emissions demands of the government in addition to the engine performance demands of the public. For an engine to be efficient, it must receive the correct amount of fuel mixed with the correct amount of air. Providing this is the purpose of the fuel injection and fuel delivery systems. The basics of these systems is covered in this chapter. Later chapters of this book provide more detailed coverage of fuel injection systems.

Before any discussion on fuel systems can take place, the properties of the most commonly used fuel in automobiles, gasoline, must be studied. This chapter takes a look at the composition and qualities of gasoline as well as the other alternate fuels being used by some engines.

Fuel injection involves spraying or injecting fuel directly into the engine's combustion chamber or intake manifold (**Figure 7-1**).

The automobile's fuel system is both simple and complicated. It is simple in the systems that transfer fuel to the engine and complex in the fuel injection system that mixes fuel with air in the correct amounts and proportion to meet all the needs of the engine. The basic parts of the fuel delivery system include the fuel tank, fuel lines, filters, and pumps (**Figure 7-2**).

GASOLINE COMPOSITION

Crude oil, as removed from the earth, is a mixture of hydrocarbon compounds ranging from gases to heavy tars and waxes. The crude oil can be refined into products such as lubricating oils, greases, asphalts, kerosene, diesel fuel, gasoline, and natural gas. Before its widespread use in the internal combustion engine (ICE) of automobiles, gasoline was an unwanted by-product of refining for oils and kerosene.

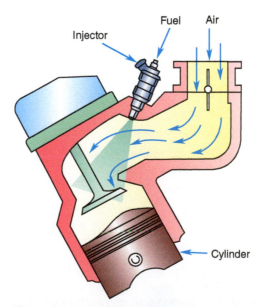

Figure 7-1 In a port fuel injection system, air and fuel are mixed in the intake manifold runners very close to the intake valve(s) of the combustion chamber.

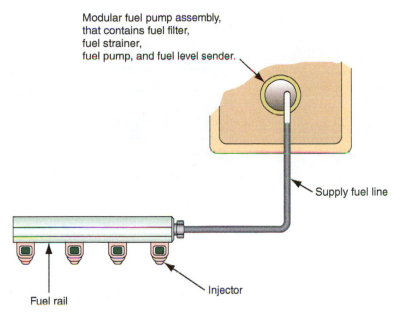

Modular fuel pump assembly,
that contains fuel filter,
fuel strainer,
fuel pump, and fuel level sender.

Supply fuel line

Injector

Fuel rail

Figure 7-2 A basic fuel injection system.

Gasoline contains hydrogen and carbon molecules. The chemical symbol for this liquid is C_8H_{15}, which indicates that each molecule of gasoline contains eight carbon atoms and fifteen hydrogen atoms. Gasoline is a colorless liquid with excellent **vaporization** capabilities.

> Although the chemical symbol for gasoline is C_8H_{15}, gasoline is actually a blend of many different chemical compounds.

GASOLINE QUALITIES

Many of the performance characteristics of gasoline can be controlled in refining and blending to provide proper engine function and vehicle drivability. The major factors affecting fuel performance are antiknock quality, volatility, sulfur content, and deposit control.

Antiknock Quality

Two important factors affect the power and efficiency of a gasoline engine—compression ratio and **detonation**. An engine's output and efficiency increase with a higher compression ratio. The better the efficiency, the less fuel is consumed to produce a given power output. To have a high compression ratio requires an engine of greater structural integrity to withstand higher internal pressures. Also, due to higher temperatures in the combustion chamber, the octane level of the fuel in a high compression engine must be able to resist burning too early, or detonation and spark knock can result in engine damage. Compression ratios now generally range from 8:1 to 10:1. High-performance cars may have higher compression ratios that require premium fuel. Turbocharging and supercharging also require higher octane levels.

Normal combustion occurs gradually in each cylinder. The **flame front** advances smoothly across the combustion chamber until all the air-fuel mixture has been burned (**Figure 7-3**). Detonation occurs after spark ignition when the flame front fails to reach a pocket of air-fuel mixture before the temperature in that area reaches the point of self-ignition. Normal burning at the start of the combustion cycle raises the temperature and pressure of everything inside the cylinder. The last part of the air-fuel mixture is both heated and pressurized, and the combination of those two factors can raise it to the self-ignition point. At that moment, the remaining mixture burns almost instantaneously.

> **Detonation** is best described as abnormal combustion.

> Contributing factors to detonation include overheating, low-octane fuel, inoperative **exhaust gas recirculation (EGR)** valve, carbon deposits raising compression, and lean air-fuel mixtures.

> The **flame front** is the edge of the burning area inside the combustion chamber.

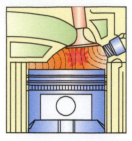

| Spark occurs just before the end of the compression stroke. | Flame front begins to form; piston is just before TDC. | In detonation, a second flame front starts. The heat of this unintended combustion rapidly raises the pressure inside the cylinder. | With increased pressure, the remaining fuel burns more rapidly than normal, causing a detonation in the cylinder instead of a controlled burn. |

Figure 7-3 The stages of combustion that lead to detonation.

The two flame fronts create a pressure wave between them that can destroy cylinder head gaskets, break piston rings, and burn pistons and exhaust valves. When detonation occurs, a hammering, pinging, or knocking sound may be heard, or at higher speeds, engine noise may mask the sound. Detonation can be very harmful to an engine because the shock wave it produces can damage pistons, rings, cylinder heads, and valves.

Pre-ignition is another term used to describe abnormal combustion; it is used to describe the igniting of fuel prematurely by a hot spot in the combustion chamber, such as a glowing piece of carbon. One cause of pre-ignition is a spark plug with an improper heat range (**Figure 7-4**). With pre-ignition, the fuel is ignited before the spark. The piston is still too early in the compression stroke, which further raises the combustion chamber temperature. The motor is working against the fuel burn, resulting in power loss and

Pre-ignition is defined as ignition started by a source other than the spark plug.

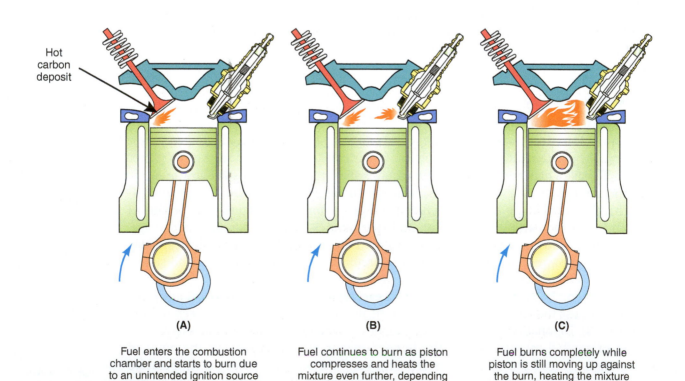

(A)
Fuel enters the combustion chamber and starts to burn due to an unintended ignition source inside the combustion chamber.

(B)
Fuel continues to burn as piston compresses and heats the mixture even further, depending on timing spark plug firing may start another flame front.

(C)
Fuel burns completely while piston is still moving up against the burn, heating the mixture even more, possibly to the melting point of the piston.

Figure 7-4 The process of pre-ignition.

possible engine damage. Pre-ignition can start very soon in the compression process depending on the ignition source. One difference between pre-ignition and detonation is that pre-ignition is usually confined to one or two cylinders, while detonation usually involves most of the cylinders of the engine.

An **octane number** or rating was developed by the petroleum industry so the **antiknock** quality of a gasoline could be rated. The octane number is a measure of the fuel's tendency not to produce knock in an engine. The higher the octane number, the lower the tendency to knock. By itself, the antiknock rating has nothing to do with fuel economy or engine efficiency.

> An octane number is a measure of a fuel's resistance to spark knock, also known as its antiknock quality.

Two commonly used methods for determining the octane number of motor gasoline are the motor octane number (MON) method and the research octane number (RON) method. Both use a laboratory single-cylinder engine equipped with a variable head and knock meter to indicate knock intensity. The test sample is used as fuel, and the engine compression ratio and air-fuel mixture are adjusted to develop a specified knock intensity. There are two primary standard reference fuels, isooctane and heptane. Isooctane does not knock in an engine, but it is not used in gasoline, because of its expense. Heptane knocks severely in an engine. Isooctane has an octane number of 100. Heptane has an octane number of 0.

A fuel of unknown octane value is run in the special test engine, and the severity of knock is measured. Various proportions of heptane and isooctane are run in the test engine to duplicate the severity of the knock of the fuel being tested. When the knock caused by the heptane-isooctane mixture is identical to the test fuel, the octane number is established by the percentage of isooctane in the mixture. For example, if 85 percent isooctane and 15 percent heptane produce the same severity of knock as the fuel in question, the fuel is assigned an octane number of 85.

The octane rating required by law and the one displayed on gasoline pumps is the antiknock index (AKI) and is actually an average of the motor octane and research octane numbers. This formula is commonly found on the gasoline pumps at service stations and indicates the method used to establish a particular octane rating. This method actually uses an average of the two numbers. The formula is R + M/2 or RM/2, meaning research plus motor divided by two.

Factors that affect knock include the following:

- *Lean fuel mixture.* A **lean mixture** burns slower than a **rich mixture**. The heat of combustion is higher, which promotes the tendency for unburned fuel in front of the spark-ignition flame to detonate.
- *Overadvanced ignition timing.* Advancing the ignition timing induces knock. Slowing ignition timing suppresses knock.
- *Compression ratio.* Compression ratio affects knock because cylinder pressures are increased with the increase in compression ratio.
- *Valve timing.* Valve timing that fills the cylinder with more air-fuel mixture promotes higher cylinder pressures, increasing the chances for detonation.
- *Turbocharging.* Turbocharging or supercharging forces additional fuel and air into the cylinder. This induces higher cylinder pressures and promotes knock.
- *Coolant temperature.* Hot spots in the cylinder or combustion chamber due to inefficient cooling or a damaged cooling system raise combustion chamber temperatures and promote knock.
- *Cylinder-to-cylinder distribution.* If an engine has poor distribution of the air-fuel mixture from cylinder to cylinder, the leaner cylinders could promote knock.
- *Excessive carbon deposits.* The accumulation of carbon deposits on the piston, valves, and combustion chamber causes poor heat transfer from the combustion chamber. Carbon accumulation also artificially increases the compression ratio. Both conditions cause knock.

> **Lean mixture** refers to having too little fuel or too much air.

> **Rich mixture** refers to having an excessive amount of fuel or not enough air.

- *Air inlet temperature.* The higher the air temperature when it enters the cylinder, the greater is the tendency to knock.
- *Combustion chamber shape.* The optimum combustion chamber shape for reduced knocking is hemispherical with a spark plug located in the center (**Figure 7-5**). This hemi-head allows for faster combustion, allowing less time for detonation to occur ahead of the flame front.
- *Octane number.* Only when an engine is designed and adjusted to take advantage of the higher octane gasoline can the value of the fuel be obtained. Most modern engines are designed to operate efficiently with regular grade gasoline and do not require a high-octane premium grade.

Volatility

Liquid gasoline will not burn. The fuel must vaporize prior to entering the combustion chamber. Various types and grades of fuel have different rates of volatility. Volatility is the fuel's ability to change from a liquid to a vapor. Typically, gasoline is blended for a certain region or climate several times a year. Volatility that is too high can cause hot weather drivability and emission problems as well as **vapor lock**. Volatility that is too low can cause cold weather starting problems, including slow warm-ups, as well as a buildup of deposits in the combustion chamber, in the crankcase, and on the spark plugs. To ensure complete combustion, complete vaporization must occur. The volatility of gasoline affects the following performance characteristics or driving conditions:

> The formation of gasoline vapors in the fuel system is called vapor lock.

- *Cold starting and warm-up.* A fuel can cause hard starting, hesitation, and stumbling during warm-up if it does not vaporize readily. A fuel that vaporizes too easily in hot weather can form vapor bubbles in the fuel line, resulting in vapor lock or loss of performance.
- *Temperature.* Because a highly volatile fuel vaporizes at a lower temperature than a less volatile fuel, winter grade gasoline is more volatile than summer grade gasoline.
- *Altitude.* Gasoline vaporizes more easily at high altitudes. Therefore, volatility is controlled in blending according to the elevation of the place where fuel is sold.
- *Crankcase oil dilution.* A fuel must vaporize well to prevent diluting the crankcase oil with liquid fuel. If parts of the gasoline do not vaporize, droplets of liquid break down the oil film on the cylinder wall, causing scuffing or scoring. The liquid eventually enters the crankcase oil and results in the formation of sludge, gum, and varnish accumulation, as well as decreases the lubrication properties of the oil.
- *Drivability.* Poor vaporization can also affect combustion efficiency and power output.

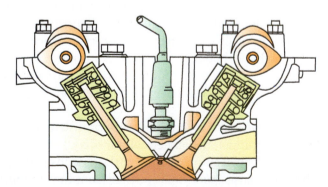

Figure 7-5 Cross section of a hemispherical combustion chamber.

Sulfur Content

Sulfur has been limited to 10 parts per million by the latest Environmental Protection Agency (EPA) regulations, which is much less than the previous 80 parts per million. When the sulfur in the fuel is burned, it combines with the oxygen to form sulfur dioxide. This sulfur dioxide can then combine with water to form highly corrosive sulfuric acid as acid rain. Corrosion caused by this acid is the leading cause of exhaust valve pitting and exhaust system deterioration.

Deposit Control

Several additives are put in gasoline to control harmful deposits, including gum or oxidation inhibitors, detergents, metal deactivators, and rust inhibitors.

BASIC FUEL ADDITIVES

At one time, all a gasoline-producing company had to do to produce their product was pump the crude oil from the ground, run it through the refinery to separate it, dump in a couple of grams of lead per gallon, and deliver the finished product to the service station. Of course, automobiles were much simpler then and what they burned was not very critical. As long as gasoline vaporized easily and did not cause the low compression engines to knock, everything was fine.

Times have changed. Today, refiners are under constant pressure to ensure that their product passes a series of rigorous tests for seasonal volatility, minimum octane, existent gum, and oxidation stability as well as add the correct deicers and detergents to make the product competitive in a price-sensitive marketplace. Gasoline additives, primarily used in unleaded gasolines, have different properties and a variety of uses.

- *Anti-icing or deicer.* Isopropyl alcohol is added seasonally to gasoline as an anti-icing agent to prevent gas line freeze-up in cold climates.
- *Metal deactivators and rust inhibitors.* These additives are used to inhibit reactions between the fuel and the metals in the fuel system that can form abrasive and filter plugging substances.
- *Gum or oxidation inhibitors.* Some gasoline contains aromatic amines and phenols to prevent formation of gum and varnish. During storage, harmful gum deposits can form due to the reaction of some gasoline components with each other and with oxygen. Oxidation inhibitors are added to promote gasoline stability. They help control gum, deposit formation, and staleness. Stale gasoline becomes cloudy and smells like paint thinner.
- *Detergents.* The use of detergent additives in gasoline has been the subject of some public confusion. Detergent additives are designed to do only what their name implies—clean certain critical components inside the engine. They do not affect octane.
- *Ethanol.* By far the most widely used gasoline additive today is **ethanol**, or grain alcohol. Blending 10 percent ethanol into gasoline is seen as a comparatively inexpensive octane booster that results in an increase of 2.5 to 3 road octane points. Fuel with 10 percent ethanol is called E10. In addition to octane enhancement, ethanol blending keeps the fuel injectors clean due to detergent additive packages found in most of the ethanol marketed. It also inhibits fuel system and injection corrosion due to additive packages and decreases carbon monoxide emissions at the tailpipe due to the higher oxygen content of blended fuel.
- *E-85.* Ethanol as a motor fuel has been explored for some time. The problem with unblended ethanol is that it burns slower than gasoline and is not as volatile. One solution has been found in the use of a fuel called E-85, which is being produced for

Gasoline formulations change between summer and winter. Winter fuel is more volatile than summer fuel.

Gum content is influenced by the age of the gasoline and its exposure to oxygen and certain metals such as copper.

If gasoline is allowed to evaporate, the residue left can form gum and varnish.

Shop Manual
Chapter 7, page 340

Gasoline has about 112,000–120,000 BTUs per gallon depending on blend and grade.

E-85 has about 85,000 BTUs per gallon.

Diesel fuel has about 130,000 BTUs per gallon.

Ethanol can be produced from a renewable resource such as corn.

use in specially equipped vehicles. E-85 is a mixture of 15 percent gasoline and 85 percent ethanol. The vehicles have a sensor built into the fuel system to detect the amount of ethanol in the fuel tank, or the amount of ethanol is calculated by the powertrain control module (PCM). The PCM is programmed to change the operating characteristics of the engine to accept the fuel. The gasoline is added to aid volatility of the fuel, needed especially for good cold starting. Vehicles that are equipped to burn E85 are called flex fuel vehicles, or FFVs. E85 fuel is generally cheaper than regular gasoline, but FFVs also tend to get fewer miles per gallon than regular production vehicles. Most experts say that E85 needs to be about 20 percent cheaper than regular gasoline for the owner to break even on cost over regular gasoline. Studies have been done that indicate that FFVs have slightly better emission levels than regular gasoline. The ethanol in the fuel works as an **oxygenate**, which means it adds oxygen to the fuel as it burns. This additional oxygen tends to lower carbon monoxide emissions.

- *Methanol.* Methanol is the lightest and simplest of the alcohols and is also known as wood alcohol. It can be distilled from coal, but most of what is used today is derived from natural gas. Many automakers continue to warn motorists about using a fuel that contains more than 10 percent methanol and cosolvents by volume. Methanol is recognized as being far more corrosive to fuel system components than ethanol, and it is this corrosion that has automakers concerned.

ALTERNATIVE FUELS

Tighter federal, state, and local emissions regulations have led to a search for an alternative fuel. Many factors are considered when determining the viability of an alternative fuel, including emissions, cost, fuel availability, fuel consumption, safety, engine life, fueling facilities, weight and space requirements of fuel tanks, and the range of a fully fueled vehicle. Currently, the major competing alternative fuels include ethanol, methanol, propane, and natural gas. Diesel fuel has been in use for many years, and its properties are included in this section.

Ethanol and methanol were presented earlier under other gasoline additives. Propane is a petroleum-based pressurized fuel used as a liquid. It is a constituent of natural gas. Natural gas comes in two forms: **compressed natural gas (CNG)** and liquefied natural gas (LNG).

Compressed natural gas is a petroleum-based pressurized fuel.

LP-Gas

The *LP* in **LP-gas** stands for *liquefied petroleum*.

LP-gas is a by-product of crude oil refining, and it is also found in natural gas wells. Fuel-grade LP-gas is almost pure propane with a little butane and propylene usually present. Because of its high propane content, many people simply refer to LP-gas as propane. LP-gas has a higher octane than regular gasoline, measuring about 105, but LP-gas engines get slightly less fuel mileage compared to gasoline.

Propane burns clean in the engine and can be precisely controlled. Because it vaporizes at atmospheric temperatures and pressures, it does not puddle in the intake manifold. This means it emits less hydrocarbons and carbon monoxide. Emission controls on the engine can be simpler. Cold starting is easy, down to much below zero. At normal cold temperatures, the propane engine fires easily and produces power without surge or stumble. Propane has a higher octane rating than gasoline and can be used in an engine with a higher compression ratio. However, power loss is about 5 percent due to lower volumetric efficiency.

One of the most noticeable differences between propane and gasoline is that propane is a dry fuel. It enters the engine as vapor. Gasoline, on the other hand, enters the engine as tiny droplets of liquid sprayed in through a fuel injector.

The propane fuel system is a completely closed system that contains a supply of pressurized LP-gas. Because the fuel is already under pressure, no fuel pump is needed.

From the pressurized fuel tanks, the fuel flows to a vacuum filter fuel lock. This serves as a filter and a control, allowing fuel to flow to the engine. Fuel flows to a converter or heat exchanger where it changes from a liquid to a gas. When the propane flows through the converter, it expands as it changes into a gas.

Natural Gas

Vehicles have been designed with gasoline/CNG, diesel/CNG, and dedicated CNG engine applications. Compressed natural gas (**Figure 7-6**) vehicles offer several advantages over gasoline:

The designation "Designated Vehicle" means the vehicle is designed to use one type of fuel.

- The fuel costs less.
- It is the cleanest alternative fuel, generating up to 99 percent less carbon monoxide than gasoline, no particulates, almost no sulfur dioxide, and 85 percent less reactive hydrocarbons than gasoline.
- Natural gas vehicles are safer. The fuel tanks used for CNG are aluminum or steel cylinders with walls that are ½- to ¾-inch thick. They can withstand severe crash tests, direct gunfire, dynamite explosions, and burning beyond any standard sheet metal gasoline tank. Because it is lighter than air, natural gas dissipates quickly. It also has a higher ignition temperature.

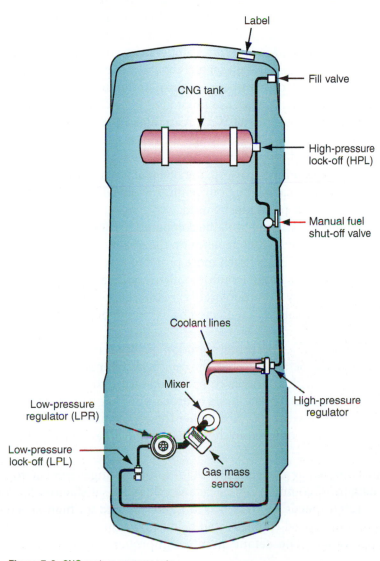

Figure 7-6 CNG system components.

- It generally reduces vehicle maintenance since it burns cleanly. Oil changes may not be needed before 12,000 miles, and spark plugs could last longer.
- Natural gas is abundant and readily available in the United States.

The chief disadvantage of CNG at present is its lack of availability to most users. Fuel facilities are needed in greater numbers than are currently in existence due to the relatively shorter range of CNG vehicles. The space taken by the CNG cylinders and their weight, about 300 pounds, would also be considered disadvantages in most applications.

Diesel Fuel

Diesel fuel is also made from petroleum. At the refinery, the petroleum is separated into three major components: gasoline, middle distillates, and all remaining substances. Diesel fuel comes from the middle distillate group, which has properties and characteristics different from gasoline.

The shape of the fuel spray, turbulence in the combustion chamber, beginning and duration of injection, and the chemical properties of the diesel fuel all affect the power output of the diesel engine. The significant chemical properties of diesel fuel are described briefly in the following.

Viscosity is a measure of a fluid's resistance to flow.

- *Viscosity*. The **viscosity** of diesel fuel varies with temperature. The viscosity of diesel fuel directly affects the spray pattern of the fuel into the combustion chamber. Fuel with a high viscosity produces large droplets that are hard to burn. Fuel with a low viscosity sprays in a fine, easily burned mist. However, if the viscosity is too low it does not adequately lubricate and cool the injection pump and nozzles.

Paraffin is a wax substance present in the middle distillates of crude oil.

- *Wax appearance point and pour point*. Temperature affects diesel fuel more than it affects gasoline. This is because diesel fuels contain **paraffin**. As temperatures drop past a certain point, wax crystals begin to form in the fuel. The point where the wax crystals appear is the wax appearance point (WAP) or cloud point. The better the quality of the fuel, the lower is the WAP. As temperatures drop, the wax crystals grow larger and restrict the flow of fuel through the filters and lines. Eventually, the fuel, which may still be liquid, stops flowing because the wax crystals plug a filter or line. As the temperature continues to drop, the fuel reaches a point where it solidifies and no longer flows. This is called the pour point. In cold climates it is recommended that a low-temperature pour point fuel be used.

- *Volatility*. Gasoline is extremely volatile compared to diesel fuel. The amount of carbon residue left by diesel fuel depends on the quality and the volatility of that fuel. Fuel that has a low volatility is much more prone to leaving carbon residue. The small, high-speed diesels found in automobiles require a high-quality and highly volatile fuel because they cannot tolerate excessive carbon deposits. Large, low-speed industrial diesels are relatively unaffected by carbon deposits and can run on low-quality fuel.

Cetane is the measurement of the fuel's ignition quality.

- *Cetane number or rating*. Diesel fuel's ignition quality is measured by the **cetane** rating. Much like the octane number, the cetane number is measured in a single-cylinder test engine with a variable compression ratio. The diesel fuel to be tested is compared to cetane, a colorless, liquid hydrocarbon that has excellent ignition qualities. Cetane is rated at 100. The higher the cetane number, the shorter is the ignition lag time (delay time) from the point the fuel enters the combustion chamber until it ignites.

In fuels that are readily available, the cetane number ranges from 40 to 55, with values of 40 to 50 being most common. These cetane values are satisfactory for medium-speed engines whose rated speeds are from 500 to 1,200 rpm and for high-speed engines rated over 1,200 rpm. Low-speed engines rated below 500 rpm can use fuels in the above 30 cetane number range. The cetane number improves with the addition of certain compounds such as ethyl nitrate, acetone peroxide, and amyl nitrate. Amyl nitrate is commercially available for this purpose.

■ *Diesel fuel grades.* Minimum quality standards for diesel fuel grades have been set by the American Society for Testing Materials. Two grades of diesel fuels, Number 1 and Number 2, are available. Number 2 diesel fuel is the most popular and is widely distributed. Number 1 diesel fuel is less dense than Number 2 with lower heat content. Number 1 diesel fuel is blended with Number 2 to improve starting in cold weather. In the winter, diesel fuel is likely to be a mixture of Number 1 and Number 2 fuels. In moderately cold climates, the blend may be 90 percent Number 2 to 10 percent Number 1. In very cold climates, the ratio may be as high as 50:50. Diesel fuel economy can be expected to drop off during the winter months due to the use of Number 1 diesel in the fuel blend.

Electric Vehicles

Almost all automobile manufacturers globally have developed an electric-powered vehicle. Electric vehicles, or EVs, are powered by electric motors with energy stored in battery packs (**Figure 7-7**). Recently EVs have finally entered the mass market. The advances in battery technology have made the EV a viable option. Tesla Motors has been building high-end all-electric vehicles for some time now. Nissan has released the Leaf, Mitsubishi has released the "*i*", and Ford is releasing an electric version of the Focus. Also, several small companies are developing, or have already developed or released electric vehicles. The primary batteries used are Li-ion and nickel metal hydride. The battery packs are generally made by wiring many small batteries in series. Tesla advertises a range of 300 miles on their 80-kilowatt battery, while the Nissan Leaf is expected to go about 100 miles, and the Mitsubishi "*i*" is touted to go 62 miles on a single charge, not using climate control. GM advertises a range of 238 miles on the Chevrolet Bolt. Manufacturers of EVs are recommending preheating or precooling the vehicle in the garage while it is plugged in to its charger. Driving conditions, temperature, and load can affect how long the vehicle can be driven. Recharging the batteries (**Figure 7-8**) can take a significant amount of time (around 4 to 5 hours at 220 V), but recharging at night, off-peak hours when the cost is significantly less, is also an option. Most EVs can be charged from a 110-volt receptacle, but a faster charging 240-volt charger is suggested. One of the biggest disadvantages to the EV is the driving range and lack of recharging stations, but EVs could be driven by the vast majority of commuters at the present time. All the listed EVs are using regenerative braking to help recharge the batteries. High-efficiency accessories in addition to permanent magnet electric drive motors also extend range. Lower fuel and

Diesel particulate emissions have become a concern. These particles can penetrate the tissues in the lungs.

 Caution

Never attempt to perform any repair on an electric vehicle or hybrid vehicle without being trained. The high voltages can KILL YOU if you do not follow very specific procedures.

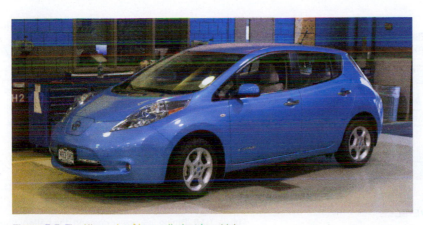

Figure 7-7 The Nissan Leaf is an all-electric vehicle.

Figure 7-8 An EV battery pack is made up of many cells.

maintenance costs are the main advantages to owning an EV. Disadvantages include the high cost of replacement batteries and the dangers of high operating voltage.

Hybrid Electric Vehicles

Hybrid vehicles use two sources of power, a small internal combustion engine that provides the main source of vehicle power plus a high-voltage electric motor that serves to assist the internal combustion engine during acceleration (**Figure 7-9**). The high-voltage components such as the high-voltage battery pack are located either under the vehicle or in the trunk area. A control unit is also associated with the high-voltage battery to control its output. The internal combustion engine, the electric motor, and the transmission are located in the front of the vehicle. The electric motor is situated between the engine and transmission, or in some cases is part of the transmission itself (**Figure 7-10**).

The high-voltage DC motor also serves as a starter for the internal combustion engine and as a generator to recharge the high-voltage battery through the internal combustion engine and through regenerative braking (**Figure 7-11**). In stop-and-go city traffic, when demand is low, the electric motor can be used alone to power the wheels. Conventional 12-volt electrical circuits and a 12-volt battery are supplied with current through the main battery with a voltage reduction DC/DC converter.

The hybrid system also uses a battery condition module and a motor control module working together to prevent under- or overcharging of the main battery by turning off either the electric assist when the battery is undercharged or the regenerative braking if

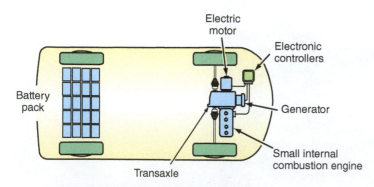

COMPONENTS OF A HYBRID VEHICLE

Figure 7-9 Main components of a hybrid vehicle.

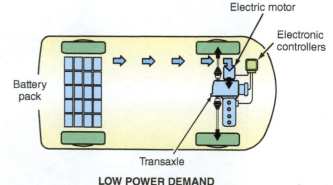

LOW POWER DEMAND

Figure 7-10 During low power demand, the battery pack can provide power to allow the electric motor to move the vehicle.

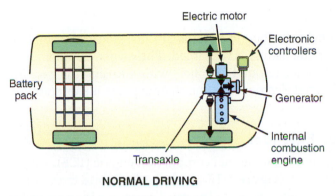

NORMAL DRIVING

Figure 7-11 During normal driving, the internal combustion engine provides power and recharges the battery.

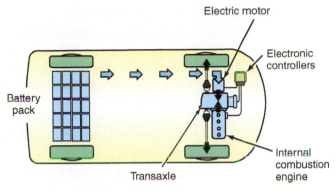

FULL ACCELERATION

Figure 7-12 Both electric motor and gas engine power the vehicle on acceleration.

the battery is in danger of overcharge. If there is a low battery state of charge or a problem in the electric motor, a conventional starter is used to start the internal combustion engine (**Figure 7-12** and **Figure 7-13**).

Parallel Hybrids

In the parallel hybrid, two power sources are used to power the drive wheels, the electric motor, and the internal combustion engine (ICE). During low-speed driving, the electric motor powers the drive wheels. When higher speeds are needed, the ICE will automatically restart to provide additional power. Also, if the battery is low or the electrical demand is high, the ICE will also restart. As the ICE is used to charge the battery, it may restart even if the vehicle is stopped (if safe to do so as determined by transmission selection and brake pedal position). Because the electric motor is connected directly to the ICE, the main engine start is quick and quiet compared to the usual gasoline vehicle.

The constantly variable transmission (CVT) can use power from either the gasoline engine or the electric motor or both at the same time. As with any hybrid, when the vehicle comes to a stop, the main engine can be shut down to improve fuel economy if the engine is warm (**Figure 7-14**), the electrical demands are low, and the outside temperature is above about 40°F (4.4°C).

The Honda Insight and the Toyota Prius were among the first widely produced parallel hybrid vehicles. Since that time, several others are in production using similar technologies. Do not attempt to service these vehicles without specific service information for the vehicle you are working on, since hybrid vehicles have high-voltage circuits that may cause severe injuries or property damage.

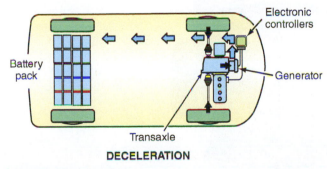

DECELERATION

Figure 7-13 During deceleration, the car slowing down is converted to electrical energy to charge the batteries.

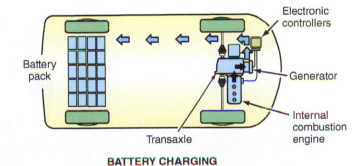

BATTERY CHARGING

Figure 7-14 The gasoline engine will run with the vehicle stopped to charge the batteries if necessary.

Plug-In Hybrids

Some hybrid vehicles are plug-in hybrids. Plug-in hybrids are just that: the vehicle is plugged in to household current and the battery is recharged. The owner can then drive on full electric power for a short distance (around 15 miles) without using the gasoline engine. Although they have a very limited electric range, the charging time is short (around 3 hours).

Series Hybrid/EV with Range Extender

The Chevrolet Volt uses an electric motor, an internal combustion engine (ICE), a high-voltage battery, and a conventional battery (**Figure 7-15**). In the series hybrid the gasoline engine drives a generator to power the electric motors. The gasoline engine is not intended to recharge the battery or power the drive wheels. The owner will use a charging station at home to recharge the battery. As long as the commute is less than 40 miles, the electrical energy is provided off the home electrical grid. Of course, the series hybrid also has the advantage of being able to use no power while sitting in traffic (as do all hybrids). An advantage of the series hybrid design over the EV is the fact that the owner does not have to worry about recharging to prevent the vehicle from going dead. The gasoline motor can extend the range of the vehicle.

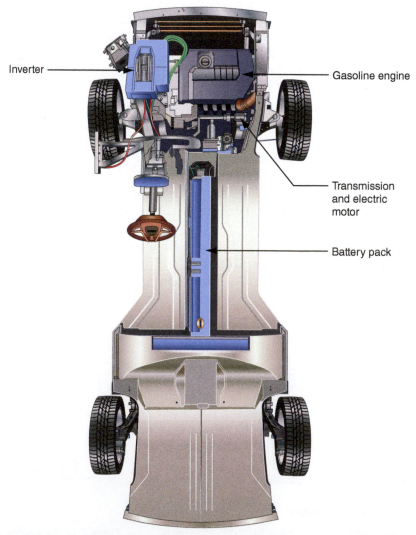

Inverter

Gasoline engine

Transmission and electric motor

Battery pack

Figure 7-15 A cut-away view of a series hybrid vehicle.

Using the gasoline engine strictly as a generator has the benefit of not having to provide torque at varying operating conditions like a normal vehicle or even a parallel hybrid. The engine can be run at optimum rpm and fuel mileage potential. Advantages of the series hybrid with range extender are that the gasoline engine does not have to run to recharge the batteries. So if the vehicle is driven less than 40 miles daily, as the majority of workers drive, it would use no gasoline at all.

Fuel Cells

A fuel cell is a device that converts fuel into electricity. Fuel cell technology has been used in several nonautomotive applications such as aeronautics, spacecraft, and submarines. Fuel cells are being used to drive the entire vehicle or to power only certain accessories or vehicle subsystems. The fuel cells do not store electricity. When necessary, the electricity is converted to drive electric motors that propel the vehicle. A fuel cell system consists of a fuel processor, air supply subsystem, cooling subsystem, and various controls. Combining these subsystems with the fuel cell creates a complete power-generating system. A single cell is sometimes referred to as a proton exchange membrane, or PEM. A PEM fuel cell (**Figure 7-16**) is made of two plates sandwiched together with a plastic membrane coated with a catalyst in the center.

Hydrogen from methanol, natural gas, or petroleum along with oxygen from the air is fed through channels in the plates—hydrogen on one side and oxygen on the other. Chemically speaking, the hydrogen and oxygen want to be together. The shortest way to join is through the membrane, but only part of the hydrogen atom, the proton, can pass through the membrane. The other part, the electron, has to take the long way around through an external circuit, or load, creating useful electricity. The oxygen side of the PEM attracts the protons and the electrons that have traveled through the external circuit, generating the by-products of water and heat. As long as hydrogen is supplied, a fuel cell will operate continuously. By combining single cells, a fuel cell stack is formed to produce the required amount of power. When the vehicle's power demands exceed the amount the fuel cell can generate, additional current is drawn from storage batteries. One advantage of the fuel cell is that it charges the batteries directly and there is no need for plug-in charging as with other electric vehicles. Fuel cell technology continues to evolve and may be found in production vehicles within the next decade. **Figure 7-17** shows the layout of a typical fuel cell vehicle. The Toyota Mirai is being released as a hydrogen fuel–only vehicle. Honda also has a new version of a previous hydrogen-fueled vehicle named the FCX.

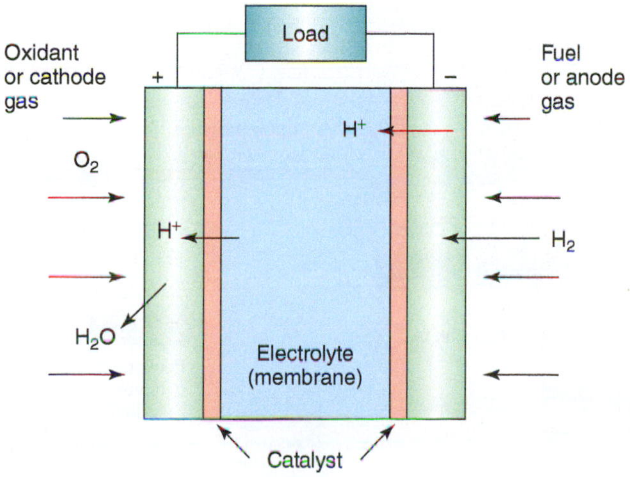

Unlike the classical definition of a series hybrid, the Volt ICE does not charge the battery pack.

The Chevrolet Volt is described as an EV with extended range. The gasoline engine has been called a "range extender."

Oxidant or cathode gas

Fuel or anode gas

O_2

H^+

H^+

H_2

H_2O

Load

Electrolyte (membrane)

Catalyst

Figure 7-16 Basic operation of a fuel cell.

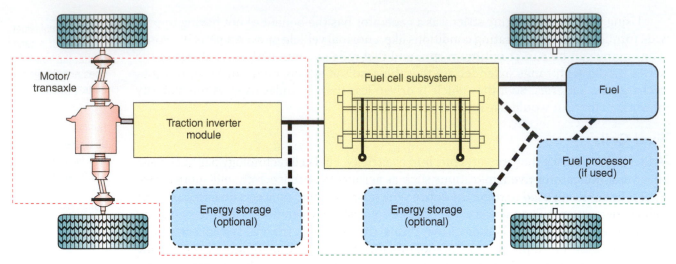

Figure 7-17 Layout of a typical fuel cell vehicle.

FUEL DELIVERY SYSTEM

The components of a typical gasoline delivery system are fuel tanks, fuel lines, fuel filters, and fuel pump control modules. **Figure 7-18** shows a typical DFI fuel delivery system. The direct fuel injection system has low-pressure and high-pressure components of the fuel system. The low-pressure fuel system consists of a conventional in-tank electric fuel pump that delivers fuel to a high-pressure mechanical fuel pump that is usually driven by the camshaft. The high-pressure fuel pump sends fuel to the injectors via high-pressure fuel lines. The high-pressure system contains a **fuel rail pressure sensor (FRPS)** that the PCM uses to determine fuel pressure at the **fuel rail**. The high-pressure pump pressure is regulated by an electrical **fuel pressure regulator** mounted directly to the pump itself. There is much more information on these specific fuel systems in Chapter 8.

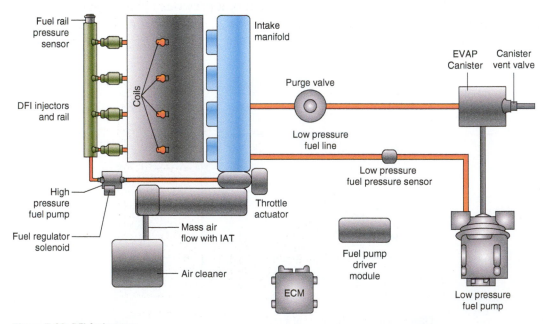

Figure 7-18 DFI fuel system

Fuel System with Return Line

Earlier fuel delivery systems used a fuel return line. The return line returned fuel that was not used by the fuel injectors by way of the fuel pressure regulator that was mounted on the fuel injector rail. Most of these systems had a replaceable fuel filter that was located in the fuel supply line.

Returnless Fuel Systems

Late-model vehicles have eliminated the fuel return line and started placing the regulators in the fuel tank as part of the fuel pump module (**Figure 7-19**). Elimination of the fuel return line prevents the return of warm fuel to the fuel tank. Warm fuel will agitate the fuel in the tank and cause an excess of fuel evaporation, which places a heavier load on the evaporative emissions system. The charcoal canister has to retain these fuel vapors because they are a hydrocarbon emission if released to the atmosphere. Also, because the fuel system does not have to pump unused fuel from the fuel rail and then back to the tank, the fuel pump is not pumping the quantity of fuel that it did with the return line. This allows for the fuel filter to become a "lifetime" filter located in the fuel pump assembly that is only serviced when the fuel pump is serviced. Additionally, the cost of a fuel line on each vehicle manufactured is quite significant, so it also makes economic sense.

Evaporative Emissions and Fuel Tanks

Evaporative emission controls were introduced in the late 1960s after it was discovered that about 20 percent of the vehicle hydrocarbon (HC) emissions were due to gasoline evaporation. Modern fuel systems must be able to prevent unburned fuel release in the form of vapor. A method had to be found to allow the fuel to expand and contract without the release of HC. The fuel tank had to be vented in some fashion because when fuel was consumed, the tank would develop an internal vacuum, making fuel difficult (if not impossible) to pump from the tank. Also, if the tank were full of fuel and the temperature increased, the fuel tank would be placed under pressure by the volatility of the fuel.

The answer was to store the fuel vapors that were being released in a canister filled with activated charcoal, usually called a **charcoal canister** (**Figure 7-20**). When the vehicle was at operating temperature, the engine control unit (ECM) opened a **purge solenoid** that allowed the fuel vapors to be drawn into the intake manifold and burned. Fresh air is pulled into the charcoal canister through the normally open canister vent. This action frees the charcoal to store more vapors when needed. The evaporative emission systems are closely monitored by the OBD II system for leaks and will be covered in more detail in a later chapter.

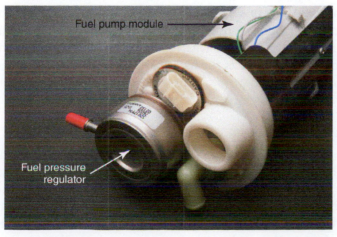

Figure 7-19 A fuel pressure regulator for a returnless fuel system is part of the fuel sender module.

Figure 7-20 The charcoal canister is located near the fuel tank on most late-model vehicles.

The fuel tank vent allows for normal expansion and contraction of the fuel vapor inside the tank through a normally open solenoid. (The solenoid can be closed by the ECM or through a scan tool by the technician to check for leaks in the evaporative emission system.) The charcoal canister removes the fuel vapor from the expanding gases inside. Fuel tank expansion has also been addressed by adding an internal expansion tank to the fuel tank or making the fuel tank impossible to completely fill, leaving space at the top of the tank for expansion (**Figure 7-21**). Without this extra space for expansion, liquid fuel might run out of the fuel tank vent if a completely full tank was exposed to high temperatures.

The gas cap usually contains an overpressure and vacuum release valve in the case of failure of the evaporative emission system to adequately vent the system.

Added beginning in the 1998 model year evaporative emission controls is that of **onboard refueling vapor recovery (ORVR)**. It has been discovered that more HC emissions were released when refueling a vehicle than were released by burning a tank of fuel! This led to a redesign of some components in the fuel filler neck and tank assembly to help prevent these emissions. ORVR is covered in more depth in Chapter 10.

Plug-In Hybrid Fuel Tank Differences

The nature of the plug-in hybrid vehicle may use fuel at a very slow rate, especially if the owner usually drives within the battery-powered-only range of the vehicle. Because fuel can tend to spoil after 90 days or so, some of these vehicles place the fuel tank under pressure to prevent any fuel from evaporating and leaving gum and varnish in the fuel tank. The ECM has to remove the pressure when the owner presses the switch to unlatch the gas tank cover. The canister collects the hydrocarbons present and vents the pressure so that the owner does not have to worry about the gas cap (and possible fuel) will not blow out when the cap is opened.

Fuel Tank Construction

Fuel tanks can be constructed of pressed corrosion-resistant steel, aluminum, or molded reinforced polyethylene plastic. Aluminum and molded plastic fuel tanks are becoming more common as manufacturers attempt to reduce the overall weight of the vehicle. Metal tanks are usually ribbed to provide added strength. Welded seams and heavier gauge steel are often used on exposed sections for added strength. Fuel tanks are fitted with insulators between the top of the tank and the chassis to protect the tank and prevent noise from transferring into the passenger compartment (**Figure 7-22**).

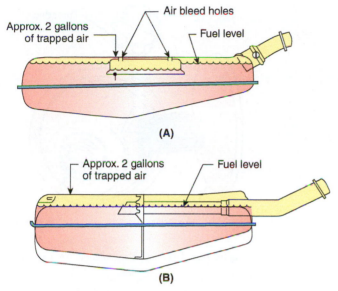

Air bleed holes

Approx. 2 gallons of trapped air

Fuel level

(A)

Approx. 2 gallons of trapped air

Fuel level

(B)

Figure 7-21 Two types of gas tank overflow protection.

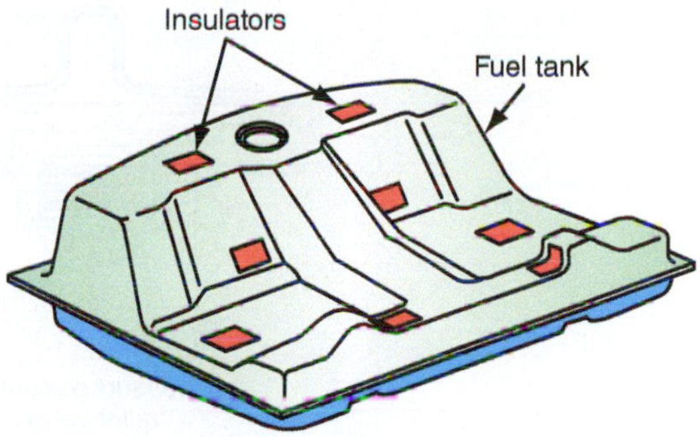

Insulators

Fuel tank

Figure 7-22 Insulators positioned between the top of the fuel tank and the chassis.

Most tanks have slosh baffles or surge plates to prevent the fuel from splashing around inside the unit. In addition to slowing down fuel movement, the plates tend to keep the fuel pickup or sending assembly immersed in the fuel during hard braking and acceleration. The plates or baffles also have holes or slots in them to permit the fuel to move from one end of the tank to the other.

The fuel tank is provided with an inlet filler tube and cap. The filler pipe is usually supported by a bracket that is bolted to the chassis. The location of the fuel inlet filler tube depends on the tank design and tube placement. It is usually positioned behind the filler cap or a hinged door in the center of the rear panel or in the outer side of either rear fender panel. All vehicles manufactured today have a restrictor in the filler tube that prevents the entry of an incorrect fuel delivery nozzle at the gas pumps (**Figure 7-23**). The filler pipe can be a rigid one-piece tube soldered to the tank or a three-piece unit. The three-piece unit has a lower neck soldered to the tank and an upper neck fastened to the inside of the body sheet metal panel.

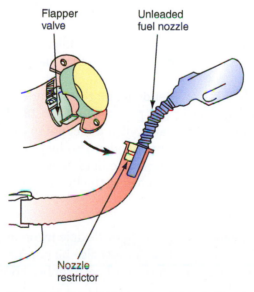

Flapper valve

Unleaded fuel nozzle

Nozzle restrictor

Figure 7-23 Unleaded fuel nozzle restrictor in filler tube.

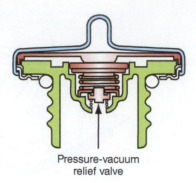

Figure 7-24 Pressure-vacuum gasoline filler cap.

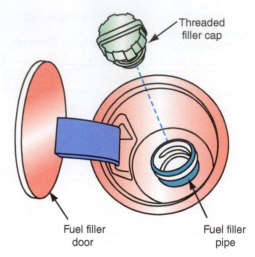

Figure 7-25 Filler door and threaded filler cap.

Fuel Tank Filler Caps

Gasoline filler caps usually have some type of pressure-vacuum relief valve arrangement (**Figure 7-24**). Under normal operating conditions, the valve is closed, but whenever pressure or vacuum is more than the calibration of the cap, the valve opens. Once the pressure or vacuum has been relieved, the valve closes. This prevents dangerous buildup of pressure in the event that the fuel tank vent should happen to fail.

The gasoline filler cap may be threaded into the upper end of the filler pipe. These caps require several turns counterclockwise to remove (**Figure 7-25**). The long threaded area on the cap is designed to allow any remaining fuel tank pressure to escape during cap removal. The cap and filler neck have a ratchet-type, torque-limiting design to prevent overtightening. Some fuel caps have a cap that is pushed down and turned one-quarter. These are called a bayonet mount. These caps were developed to prevent the OBD II system from setting a code for an evaporative emissions control leak. Some owners had difficulty making the screw-on caps tight enough to prevent these leaks. Many late-model vehicles do not use a conventional cap. They use a fuel filler portal that allows the pump nozzle to be installed in the filler neck without removing a cap (**Figure 7-26**).

Starting with the 1976 model year, a Federal Motor Vehicle Safety Standard (FMVSS 301) required a control on gasoline leakage from passenger cars, certain light trucks, and buses after they were subjected to barrier impacts and rolled over. A rollover check valve located in the fuel sender is pictured in **Figure 7-27**.

A check valve might also be fitted in the fuel tank filler cap, and most caps' pressure-vacuum relief valve settings have been increased so that fuel pressure cannot open them in a rollover.

Inside the fuel tank there is also a sending unit that includes a pick-up tube and float-operated fuel gauge, and an electric fuel pump (**Figure 7-28**). Electric fuel pumps are combined with the sending unit, and are called a fuel pump assembly. There is a mesh screen filter at the bottom of the fuel pump sender. This helps prevent rust, dirt, and sediment from being drawn into the fuel pump and lines, which can cause it to fail and clog the fuel filter. A ground wire is often attached to the fuel tank unit.

Fuel Lines and Fittings

Shop Manual
Chapter 7, page 344

Fuel lines can be made of either metal tubing or flexible nylon or synthetic rubber hose. The latter must be able to resist gasoline. It must also be non-permeable, so gas and gas vapors cannot evaporate through the hose. Ordinary rubber hose, such as that used for vacuum lines, deteriorates when exposed to gasoline. Only hoses made for fuel systems should be used for replacement. Similarly, vapor vent lines must be made of material that

Figure 7-26 A cap-less fuel filler port.

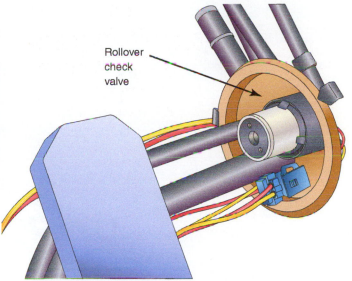

Figure 7-27 Rollover leakage protection on a fuel-injected vehicle.

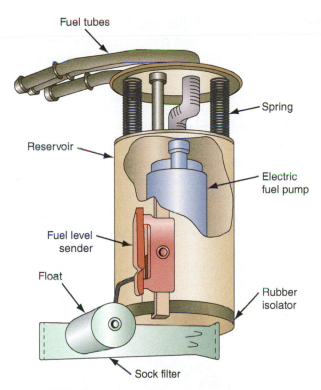

Figure 7-28 A fuel pump module.

resists attack by fuel vapors. Replacement vent hoses are usually marked with the designation *EVAP* to indicate their intended use. The inside diameter of a fuel delivery hose is generally larger ($5/16$ to $3/8$ inch) than that of a fuel return hose (¼ inch).

The fuel lines carry fuel from the fuel tank to the fuel injection assembly. These lines are usually made of rigid metal, although some sections are constructed of rubber hose to allow for car vibrations. Clips retain the fuel lines to the chassis to prevent line movement and damage (**Figure 7-29**). This fuel line, unlike filler neck or vent hoses, must work under pressure or vacuum. Because of this, the flexible synthetic hoses must be stronger. This is especially true for the hoses on fuel injection systems where pressures reach 64 psi or more. For this reason, flexible fuel line hose must also have special resistance properties.

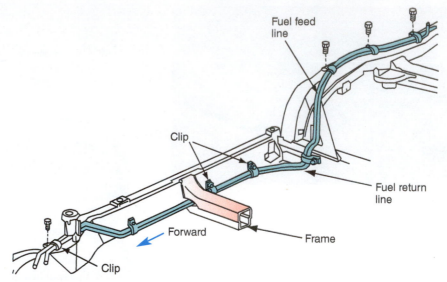

Figure 7-29 Gaps between the frame and the tank are joined with flexible hose, and the tubing is held to the chassis by clips.

Many auto manufacturers recommend that flexible hose only be used as a delivery hose to the fuel metering unit in a fuel injection system. It should not be used on the pressure side of the injector systems. This application requires a special high-pressure hose. On direct fuel injection systems, the high-pressure pump and fuel rail and lines have to be able to withstand pressure in excess of 2,000 psi. The high-pressure fuel line from the high-pressure pump to the fuel rail, which holds the injectors, must be replaced after it is removed. It should not be reused to prevent an extremely high-pressure leak.

Fuel supply lines from the tank to the injectors are routed to follow the frame along the under-chassis of vehicles. Generally, rigid lines are used extending from near the tank to a point near the fuel pump. To absorb engine vibrations, the gaps between the frame and the tank or fuel pump are joined by short lengths of flexible hose.

Sections of fuel line are assembled together by fittings. Some of these fittings are a threaded-type fitting, while others are a quick-release design. Quick-disconnect fuel line fittings may be hand releasable or a special tool may be required to release these fittings. Many fuel lines have quick-disconnect fittings with a unique female socket and a compatible male connector (**Figure 7-30**). These quick-disconnect fittings are sealed by an O-ring

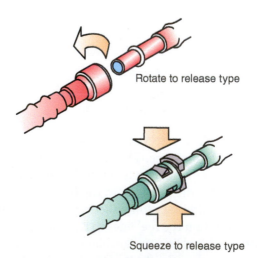

Rotate to release type

Squeeze to release type

Figure 7-30 Quick-disconnect hand-releasable fuel line fittings.

inside the female connector. Some of these quick-disconnect fittings have hand-releasable locking tabs, while others require a special tool to release the fitting (**Figure 7-31**).

One type of threaded-type tube fittings is the *double-flare* (**Figure 7-32**). The double-flare is made with a special tool that has an anvil and a cone. Another common fuel line connection on many Japanese vehicles is the "banjo" bolt style of attachment. It uses a special bolt with a drilled passage and a set of sealing washers (**Figure 7-33**).

Some fuel lines have threaded fittings with an O-ring seal to prevent fuel leaks (**Figure 7-34**). These O-ring seals are usually made of Viton, which resists deterioration from gasoline. On some other fuel lines, the fuel hose is clamped to the steel line, and the hose and clamp must be properly positioned on the steel line (**Figure 7-35**).

Shop Manual
Chapter 7, page 345

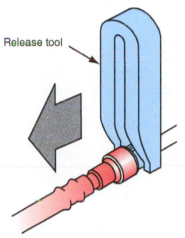

Figure 7-31 Fuel line quick-disconnect separator tool.

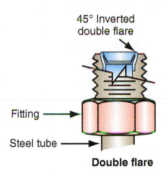

Figure 7-32 A double-flare fitting.

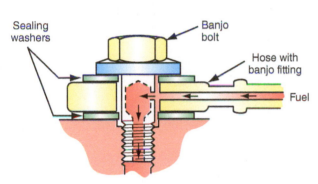

Figure 7-33 A banjo-style fuel line fitting.

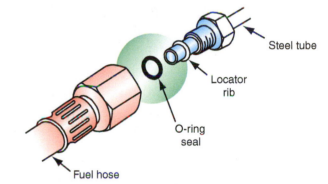

Figure 7-34 Threaded fuel line fitting with an O-ring seal.

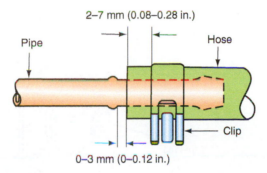

Figure 7-35 Fuel hose clamped to the steel tubing.

DFI high-pressure pump to fuel rail lines are made of stainless steel and cannot be reused.

To control the rate of vapor flow from the fuel tank to the vapor storage tank, a plastic or metal restrictor may be placed in either the end of the vent pipe or the vapor-vent hose itself.

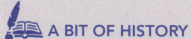

A BIT OF HISTORY

The first successful mechanical fuel pump was introduced in 1927 by the AC Spark Plug Company.

Fuel Filters

Automobiles and light trucks usually have an in-tank strainer and a gasoline filter (**Figure 7-36**). The strainer, located in the gasoline tank, is made of a finely woven fabric. The purpose of this strainer is to prevent large contaminant particles from entering the fuel system where they could cause excessive fuel pump wear or plug fuel-metering devices.

On fuel-injected vehicles with a return line, the fuel filter is connected in the fuel line between the fuel tank and the engine. Many of these filters are mounted under the vehicle (**Figure 7-37**), and others are mounted in the engine compartment. Most fuel filters contain a pleated paper element mounted in the filter housing, which may be made from metal or plastic. Fuel filters on fuel-injected vehicles usually have a metal case. On many fuel filters, the inlet and outlet fittings are identified and the filter must be installed properly. An arrow on some filter housings indicates the direction of fuel flow through the filter.

Since the advent of returnless fuel systems, many new vehicles have been produced without an accessible fuel filter. The filter is located in the fuel sender and listed as a "lifetime" filter. The reasoning behind this is that the returnless system does not pump as much fuel through the filter, since the return line is not constantly feeding fuel into the tank. If the filter does clog on these vehicles, then the tank must be dropped and the modular fuel pump removed to replace the filter.

Shop Manual
Chapter 7, page 349

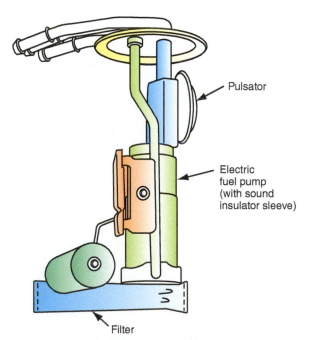

Figure 7-36 An in-tank pump with a fuel filter at the pump inlet.

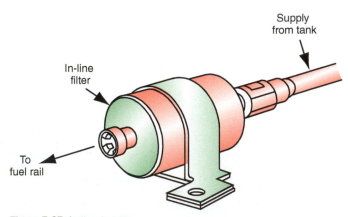

Figure 7-37 In-line fuel filter.

Electric Fuel Pumps

Electric fuel pumps have been used exclusively with fuel injection systems for many years. The electric fuel pump is usually combined with the fuel gauge's sending unit, a pressure regulator and a lifetime fuel filter called a fuel pump module (**Figure 7-38**). A schematic diagram for an electric fuel pump is shown in **Figure 7-39**. These combined units are often called *modular fuel pump assemblies.*

Vehicles equipped with electronic fuel injection (EFI) have electric fuel pumps mounted in the fuel tank.

The electric fuel pump in the fuel tank contains a small direct current (DC) electric motor with an impeller mounted on the end of the motor shaft. A pump cover is mounted over the impeller, and this cover contains inlet and discharge ports. When the armature and impeller rotate, fuel is moved from the tank to the inlet port, and the impeller grooves pick up the fuel and force it around the impeller cover and out the discharge port (**Figure 7-40**).

Fuel moves from the discharge port through the inside of the motor and out the check valve and outlet connection, which is connected via the fuel line to the fuel filter and under-hood fuel system components. A pressure relief valve near the check valve opens if the fuel supply line is restricted and pump pressure becomes very high. When the relief valve opens, fuel is returned through this valve to the pump inlet. This action protects fuel system components from high fuel pressure. Each time the engine is shut off, the check valve prevents fuel from draining out of the under-hood fuel system components into the fuel tank. A fuel filter is attached to the pump inlet. This filter prevents dirt or water from entering the pump.

Although it is dangerous to have a spark near gasoline, the in-tank fuel pump is safe because there is no oxygen to support combustion in the tank.

⚠ **Caution**

Never introduce shop air into a fuel tank for testing, because of the risk of explosion.

Figure 7-38 Combination of fuel pump and fuel gauge sending unit. This is sometimes referred to as a modular reservoir assembly (MRA).

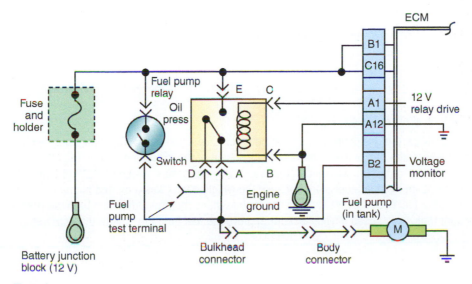

Figure 7-39 Typical wiring diagram for an electric fuel pump.

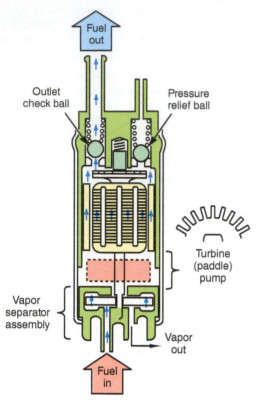

Outlet
check ball

Pressure
relief ball

Fuel
out

Turbine
(paddle)
pump

Vapor
separator
assembly

Vapor
out

Fuel
in

Figure 7-40 Electric fuel pump.

Figure 7-41 A DFI high-pressure fuel pump.

Direct Fuel Injection High-Pressure Pump

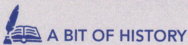

The pump in the tank is often called a lift pump.

As we discussed earlier, the direct fuel injection system uses a conventional electric fuel pump along with a high-pressure fuel pump (**Figure 7-41**). The high-pressure fuel pump is mechanical and is driven from the camshaft. The PCM controls the volume, and thereby the pressure that the high-pressure pump produces. The direct-injection high-pressure system also has a fuel pressure sensor that the ECM uses to monitor fuel pressure and react to the needs of the engine.

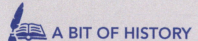 **A BIT OF HISTORY**

General Motors introduced the in-tank electric fuel pump on the 1969 Buick Riviera and then again in 1971 on the Chevrolet Vega.

A BIT OF HISTORY

In the early part of this century, many cars did not have a fuel pump. On some cars, like the Model A Ford, the fuel tank was mounted in the front of the instrument panel. The filler cap was positioned in front of the windshield. This was a very dangerous position for a fuel tank if the vehicle was involved in a collision. Because the gasoline tank was mounted higher than the carburetor, the fuel flowed by gravity from the tank to the carburetor. Other cars in those years had a vacuum tank, which used engine vacuum to move the fuel from the tank to the carburetor.

The main reasons for the fuel pump being located in the fuel tank are (1) to keep the fuel pump cool while it is operating and (2) to keep the entire fuel line pressurized to prevent premature fuel evaporation.

■ *Inertia switch*. On Ford products, an **inertia switch** is connected in series in the fuel pump circuit. A steel ball inside the inertia switch is held in place by a permanent magnet. If the vehicle is involved in a collision, this ball pulls away from the magnet and strikes a target plate to open the points in the switch (**Figure 7-42**). This inertia switch action opens the fuel pump circuit and stops the fuel pump.

A reset button on top of the inertia switch must be pressed to close the switch and restore fuel pump operation. The inertia switch is located in the trunk area on most Ford cars. A decal in the trunk of late-model cars indicates the inertia switch location. On trucks, the inertia switch is located under the dash.

Late-model Fords use the **restraints control module (RCM)** to send a signal to shut down the fuel pump in case an accident has replaced the inertia switch. (See "FPDM" later in the chapter.)

■ *Oil pressure switch*. In a late-model General Motors fuel pump circuit, the PCM supplies voltage to the winding of the fuel pump relay when the ignition switch is turned on. This action closes the relay points, and voltage is supplied through the points to the in-tank fuel pump (**Figure 7-43**). The fuel pump remains on while the engine is cranking or running. If the ignition switch is on for 2 seconds and the engine is not cranked, the PCM shuts off the voltage to the fuel pump relay and the relay points open to stop the pump.

If the ignition switch is on and the fuel line is broken during an accident, PCM and fuel pump relay action is a safety feature that prevents the fuel pump from pumping gasoline from the ruptured fuel line. An oil pressure switch is connected parallel to the fuel pump relay points. If the fuel line ruptures, the engine will quit running and the pump will shut down from the lack of an rpm signal to the PCM. This arrangement also provides for a backup of the fuel pump relay. If the relay becomes defective, voltage is supplied through the oil pressure switch points to the fuel pump. This action keeps the fuel pump operating and the engine running even though the fuel pump relay is defective.

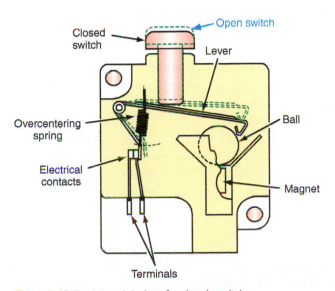

Figure 7-42 The internal design of an inertia switch.

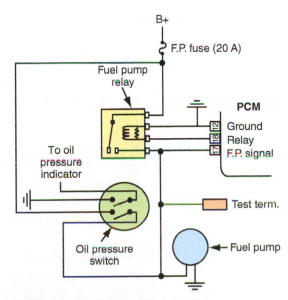

Figure 7-43 Typical GM fuel pump circuit.

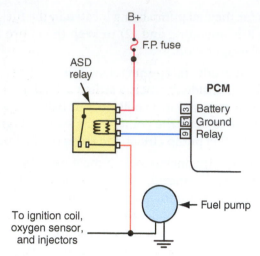

Figure 7-44 Chrysler fuel pump circuit with an ASD relay.

When the engine is cold, oil pressure is not immediately available and the engine may be slow to start if the fuel pump relay is defective.

■ *ASD relay.* The fuel pump relay in some Chrysler EFI systems is referred to as an automatic shutdown (ASD) relay. If the ignition switch is turned on, the PCM grounds the windings of the fuel pump relay and the relay points close. The ASD relay points supply voltage to the fuel pump, positive primary coil terminal, oxygen sensor heater, and the fuel injectors in some systems (**Figure 7-44**).

On Chrysler products the PCM grounds the ASD relay winding when the ignition switch is turned on, and the relay remains closed while the engine is cranking or running. If the ignition switch is on for a half second and the engine is not cranked, the PCM opens the circuit from the ASD relay winding to ground. Under this condition, the ASD relay points open and voltage is no longer supplied to the fuel pump, positive primary coil terminal, injectors, and oxygen sensor heater.

Late-model Chrysler fuel pump circuits have a separate ASD relay and a fuel pump relay. In these circuits, the fuel pump relay supplies voltage to the fuel pump, and the ASD relay powers the positive primary coil terminal, injectors, and oxygen sensor heater. The ASD relay and the fuel pump relay operate the same as the previous ASD relay. The PCM grounds both relay windings through the same wire.

■ *Circuit opening relay.* On many Toyota vehicles, the fuel pump relay is called a circuit opening relay. The wiring diagram (**Figure 7-45**) shows the main relay is powered by the 25-amp EFI fuse. The relay coil is controlled by power supplied by the ECM. The ECM uses this output to control the main relay. The main relay supplied voltage to the contacts of the circuit opening relay. The ECM also controls the circuit opening relay. The ECM turns on the circuit opening relay when it sees an rpm signal from the crankshaft position sensor by grounding the coil of the circuit opening relay. This, in turn, causes the contacts to close in the circuit opening relay, supplying current to operate the fuel pump.

Fuel Pump Driver Module

Some manufacturers are using a multiple-speed fuel pump to help tailor the demand of the vehicle to the fuel pump output. The PCM determines the fuel pump speed needed and sends this information to the fuel pump driver module (FPDM). Ford's fuel pump

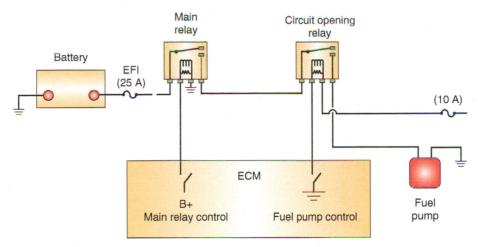

Figure 7-45 Wiring diagram for the circuit opening relay.

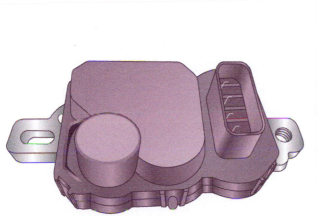

Figure 7-46 Ford fuel pump control module.

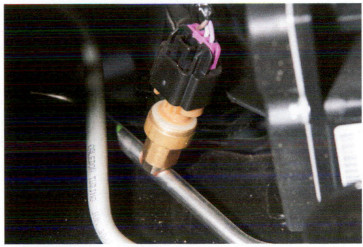

Figure 7-47 A fuel rail pressure and temperature (FRPT) sensor.

driver module (**Figure 7-46**) is one example. With the 3.7-liter port fuel-injected engine, the fuel pump driver module sends either 10 or 12 volts to the fuel pump depending on the need for fuel. The 2.0-liter and 3.5-liter engines with direct fuel injection system use a closed-loop pressure control fuel system. A **fuel rail pressure and temperature (FRPT) sensor** is used to provide feedback to the PCM, so fuel pressure can be modified according to load (**Figure 7-47**). The temperature sensor can also tell the PCM that the fuel is in danger of vaporizing and the PCM can increase fuel pressure to prevent a vapor lock. The fuel pump driver module switches to a higher pressure in this instance. Remember that a returnless system pumps much less fuel than a return type system. The return system was always pumping fuel through the system, which tended to cool the rail. A schematic drawing of Ford's FPDM circuit is shown in **Figure 7-48**. Note that the FPDM on late-model Fords is also connected to the restraints control module, which, in turn, can turn the fuel pump off in case of a crash, taking the place of the inertia switch that was used earlier.

The restraints control module is the module that controls the air bag deployment.

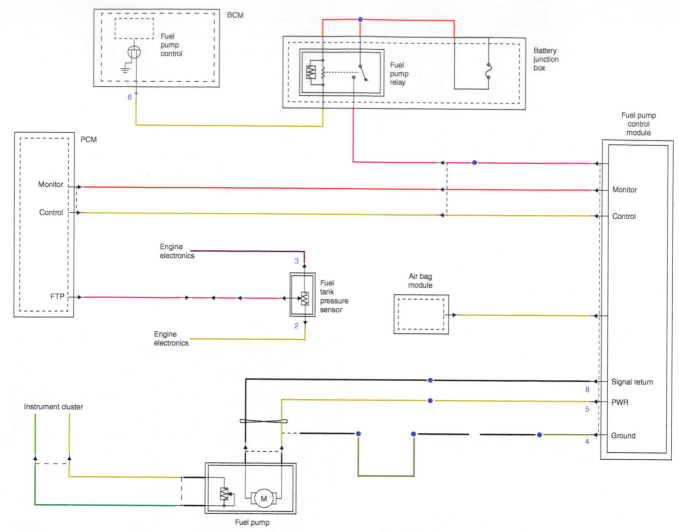

Figure 7-48 DFI fuel pump driver module and associated components.

SUMMARY

- Gasoline is made from crude oil.
- The major factors affecting fuel performance are antiknock quality, volatility, and deposit control.
- An octane number is a measure of gasoline's antiknock quality.
- The significant chemical properties of diesel fuel are its wax appearance point, pour point, viscosity volatility, and cetane number.
- Because of its high propane content, LP-gas is often referred to as propane. Propane is a dry fuel, meaning that it enters the engine as a vapor. Gasoline, on the other hand, enters as tiny droplets of liquid.
- Natural gas costs less than gasoline. It is the cleanest alternative fuel. Therefore, it generally reduces vehicle maintenance. It is safe and abundant and readily available in the United States.
- Returnless fuel systems have the fuel pressure regulator located in the fuel tank.
- The evaporative emissions system helps prevent evaporation of fuel into the atmosphere creating hydrocarbon emissions.
- DFI, or direct fuel injection, systems inject fuel directly into the combustion chamber.
- There are two types of electronic fuel injection systems in use today: port fuel injection and direct fuel injection.

- Fuel supply systems include the evaporative control system that prevents raw fuel vapors from leaving the tank. They have slash baffles or surge plates to prevent the fuel from splashing around. Each tank has an inlet filler tube and a nonvented cap. Rollover protection devices prevent gasoline from leaking out of the fuel tank in an accident. A liquid vapor separator stops liquid fuel or bubbles from reaching the vapor storage canister or the engine crankcase.
- The fuel lines carry fuel from the tank to the fuel pump, fuel filter, and fuel injection metering pump. They are made of either metal tubing or flexible nylon or synthetic rubber hose.
- An inertia switch or a signal from the restraints control module to the fuel pump driver module opens the fuel pump circuit immediately if the vehicle is involved in a collision.
- The oil pressure switch connected in parallel with the fuel pump relay operates the fuel pump if the fuel pump relay is defective.
- Returnless fuel systems may have a fuel pump module that controls fuel pump output.
- Direct fuel injection systems have a high-pressure mechanical fuel pump as well as an electric in-tank fuel pump.

REVIEW QUESTIONS

Short-Answer Essays

1. Describe the fuel system for a direct fuel-injected vehicle, including the high- and low-pressure systems.

2. Describe an advantage of using compressed natural gas to fuel a motor vehicle.

3. Name the two forms of compressed natural gas commonly used as an automotive fuel.

4. Name the components of a typical fuel delivery system.

5. Name an advantage of a returnless fuel system.

6. Briefly describe the operation of a parallel hybrid electric vehicle.

7. Name the differences in number one and number two diesel fuels.

8. The 2.0-liter and 3.5-liter Ford engines with direct fuel injection use a fuel rail pressure and temperature (FRPT) sensor. Describe the purpose of the fuel rail pressure and temperature sensor.

9. Describe the purpose of the relief valve and one-way check valve in an electric fuel pump.

10. Describe the reasoning behind a "lifetime" fuel filter.

Fill-in-the-Blanks

1. Gasoline _____ are nonventing and usually have some type of pressure-vacuum relief valve arrangement.

2. _____ _____ were introduced in the late 1960s after it was discovered that about 20 percent of the vehicle hydrocarbon (HC) emissions were due to gasoline evaporation.

3. If the fuel is ignited in the cylinder before the spark, _____ has occurred.

4. In an electric fuel pump, the relief valve opens if the fuel supply line becomes _____.

5. E-85 is a mixture of _____ percent gasoline and _____ percent ethanol.

6. On direct fuel injection systems, the high-pressure pump and lines have to be able to withstand pressure in excess of _____ psi.

7. A _____ _____ is a device that converts fuel into electricity.

8. An octane number is a measure of gasoline's _____ quality.

9. A lean mixture burns _____ than a rich mixture.

10. In a gasoline engine, _____ occurs when the flame front fails to reach a pocket of air-fuel mixture before the temperature in that area reaches the point of self-ignition.

Multiple Choice

1. *Technician A* says that a high-compression engine has greater efficiency than a low-compression engine of the same displacement.

   Technician B says that a high-compression engine has to have greater structural integrity.

   Who is correct?

   A. Technician A   C. Both technicians
   B. Technician B   D. Neither technician

2. *Technician A* says that the octane number of gasoline refers to the power it contains.

   Technician B says that there are two ways to determine the octane of fuel, the MON method and the RON method.

   Who is correct?

   A. Technician A   C. Both technicians
   B. Technician B   D. Neither technician

3. *Technician A* says that the ability of fuel to go from a vapor to a liquid is called volatility.

   Technician B says that summer fuel is more volatile than winter fuel.

   Who is correct?

   A. Technician A   C. Both technicians
   B. Technician B   D. Neither technician

4. *Technician A* says that the addition of 10% ethanol raises the octane of gasoline by 2.5 to 3 points.

   Technician B says that ethanol increases CO emissions when burned with gasoline.

   Who is correct?

   A. Technician A   C. Both technicians
   B. Technician B   D. Neither technician

5. *Technician A* says diesel fuel is more volatile than gasoline.

   Technician B says that the viscosity of diesel fuel changes with temperature.

   Who is correct?

   A. Technician A   C. Both technicians
   B. Technician B   D. Neither technician

6. While discussing DFI fuel injection systems,
   Technician A says that DFI systems have a low-pressure fuel system.
   Technician B says that DFI systems have a high-pressure fuel system.
   Who is correct?

   A. Technician A   C. Both technicians
   B. Technician B   D. Neither technician

7. *Technician A* says that fuel line can be joined by threaded-style of fittings.

   Technician B says that some fittings are made to be quickly released either by hand or with a special tool.

   Who is correct?

   A. Technician A   C. Both technicians
   B. Technician B   D. Neither technician

8. *Technician A* says that a parallel hybrid vehicle uses both the electric motor and the gasoline engine to directly power the drive wheels.

   Technician B says that the series hybrid uses just the gasoline engine to power the drive wheels, and the battery is used to power accessories.

   Who is correct?

   A. Technician A   C. Both technicians
   B. Technician B   D. Neither technician

9. *Technician A* says that many Chrysler fuel control systems have an ASD relay.

   Technician B says that Toyotas use a circuit closing relay.

   Who is correct?

   A. Technician A   C. Both technicians
   B. Technician B   D. Neither technician

10. *Technician A* says that some consumers have difficulty installing a gas cap properly.

    Technician B says that some vehicles are equipped with a fuel filler portal that does not require a gas cap.

    Who is correct?

    A. Technician A   C. Both technicians
    B. Technician B   D. Neither technician

CHAPTER 8
ELECTRONIC FUEL INJECTION

Upon completion and review of this chapter, you should be able to:

- Explain the principles of operation of a fuel injection system.
- Explain direct fuel injection (DFI) fuel systems.
- Explain DFI advantages.
- Explain DFI fuel injection strategy.
- Explain variable valve timing (VVT) and its purpose.
- Explain the design and function of some major manufacturers' electronic fuel injection (EFI) systems.
- Describe how the computer supplies the correct air-fuel ratio on an EFI system.
- Explain how the computer provides air-fuel ratio enrichment while starting a cold engine equipped with EFI.
- Describe the operation of the pressure regulator in a returnless EFI system.
- Explain the purpose of the intake manifold tuning valve (IMTV).
- Explain how an idle air control (IAC) bypass air motor controls idle speed.
- Explain the operation of the throttle actuator control (TAC) system.

Terms To Know

Direct fuel injection (DFI)	Mass airflow (MAF) sensor	Speed density
Idle air control (IAC)	Pintle	Throttle actuator control (TAC)
Intake manifold tuning valve (IMTV)	Poppet nozzle	Variable valve timing (VVT)
	Port fuel injection (PFI)	

INTRODUCTION

Fuel injection has been around for many years and has been refined many times over the decades. Today we normally see either **port fuel injection (PFI)** systems or **direct fuel injection (DFI)** systems. PFI (**Figure 8-1**) system sprays fuel just behind the intake valve, which prevents fuel from condensing in the manifold and has eliminated the need for a heated intake. Earlier port fuel systems were multiport fuel injection (MFI), and injected fuel was bank fired, first firing half the injectors, and then the other half in 180 degrees of crank rotation regardless of intake valve position. Sequential fuel injection was a port fuel injection system that delivered fuel when the intake valve was opening for a more accurate delivery, and some fuel mileage and idle performance gains were also realized. DFI (**Figure 8-2**) followed PFI. DFI allows for lean burns as well as increased engine performance, fuel economy, and lower emissions. DFI injects fuel directly into the combustion chamber at very high pressures. DFI allows the flexibility

A few years back it was fairly common to call sequential fuel injection "SFI," to differentiate between sequential and non-sequential (multiport or MFI). Now most in the field seem to be calling these systems simply "port fuel" or "PFI."

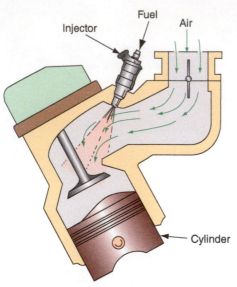

Figure 8-1 In a port fuel injection (PFI) system, air and fuel are mixed right outside the combustion chamber.

Figure 8-2 A direct fuel injection (DFI) system. Note that fuel is injected directly into the cylinder.

DFI or direct fuel injection is also called gasoline direct injection (GDI) or simply direct injection.

"EFI" is a term that can be used to describe any type of electronic fuel injection system.

of injecting fuel whenever conditions are the very best. DFI can inject fuel regardless of the position of the intake valve.

The objective of this chapter is to review the common components and differences between most EFI systems and the principles of various types of EFI. Some typical manufacturers' EFI systems are included, along with an in-depth look at some of the EFI systems used on domestic and imported vehicles. If you are familiar with EFI components and principles, you will be able to understand the many different EFI systems when you encounter them.

The chapter begins with a review of sensor operation and basic fuel injection principles and then covers some manufacturer-specific systems in detail.

 A BIT OF HISTORY

Although fuel injection has been around for decades, it was not until the 1980s that most major manufacturers started changing fuel systems from carburetors or computer-controlled carburetors to EFI systems. This action was taken to improve fuel economy, performance, and emissions levels in response to government regulations for increased fuel economy and the public demand for good engine performance across all modes of operation. A major obstacle with carburetors was cold engine operation and emissions levels.

INPUT SENSORS

The ability of the fuel injection system to control the air-fuel ratio depends on its ability to properly time the injector pulses with the intake stroke of each cylinder and its ability to vary the injector "on" time according to changing engine demands. Both tasks require the use of electronic sensors that monitor the operating conditions and positions of the engine camshaft and crankshaft.

AIRFLOW SENSORS

Injector "on" time is also called pulse width.

To control the proportion of fuel to air in the air-fuel charge, the fuel system must be able to measure the amount of air entering the engine. Several sensors have been developed to do just that. There are two different ways to determine airflow into the engine: with a mass airflow (MAF) sensor or a speed density system that uses a complex calculation to determine airflow into the engine. Both systems use most of the same sensors, with the exception of the MAF sensor in the case of the speed density systems. Speed density systems always have a manifold absolute pressure (MAP) sensor to measure engine load and in some cases barometric pressure.

Shop Manual
Chapter 8, page 392

Intake Air Temperature Sensor

The denser the air becomes, the more the air molecules are packed into a volume of air and the more that volume of air weighs.

Cold air is denser than warm air. Cold, dense air can burn more fuel than the same volume of warm air because it contains more oxygen.

Most systems do this by using an air temperature sensor (IAT) mounted in the air cleaner or intake manifold, or incorporated into the MAF sensor in the induction system. The air sensor measures air temperature and sends an electronic signal to the control computer. The computer uses this input along with the air mass input in determining the amount of oxygen entering the engine. Additionally, the IAT plays a role in ignition timing. If the air is warm going into the intake, it will be much warmer on the compression stroke. If the air is cold, then timing can be advanced further without the risk of spark knock. The PCM can take this into account and adjust the timing accordingly.

Mass Airflow Sensor

A **mass airflow (MAF) sensor** is used to measure the mass of the incoming air (**Figure 8-3**). As explained previously, the denser the air, the more oxygen it contains. From a measurement of mass, the electronic control unit adjusts the fuel delivery for the oxygen content in a given mass of air. The accuracy of air-fuel ratios is greatly enhanced when matching fuel to air mass. If the computer knows the amount of air entering the engine and the operating conditions, the powertrain control module (PCM) can calculate the exact amount of fuel it needs to deliver.

A mass airflow (MAF) sensor measures intake air speed, volume, temperature, and pressure.

The MAF sensor converts airflowing past a heated sensing element into an electronic signal. The principal type of MAF sensor in common use today is the hot wire type. A very thin wire (about 0.2 mm thick) is used as the heated element. The element temperature is set at 176°F (80°C) to 392°F (200°C) above the incoming air temperature.

As the volume and density (mass) of airflow across the heated element changes, the temperature of the element is affected and the current flow to the element is adjusted to maintain the desired temperature of the heating element. The varying current flow

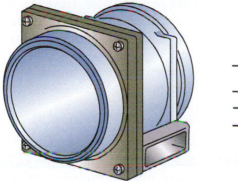
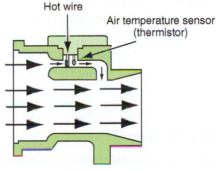

Figure 8-3 An MAF sensor.

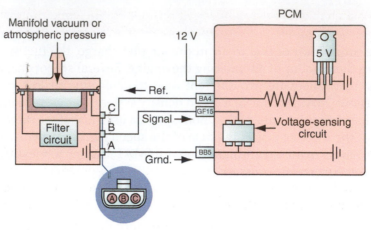

Figure 8-4 A typical MAP sensor and wiring to PCM.

Shop Manual
Chapter 8, page 392

Some fuel injection systems use a speed density calculation based on information from the TP sensor, RPM, and MAP sensor instead of the MAF sensor.

Barometric pressure is the air around us. Barometric pressure is most affected by altitude, and can also be affected to a lesser degree by the weather around us.

Manifold pressure is the pressure inside the intake manifold. Because it is generally less than atmospheric pressure, it is termed a "vacuum." The pressure inside the manifold will be at or near atmospheric pressure during wide-open throttle operation.

When a MAP has a low-voltage signal, vacuum is high. When vacuum is low, the voltage signal is high.

parallels the particular characteristics of the incoming air (hot, dry, cold, humid, high/low pressure). The electronic control unit monitors the changes in current to determine air mass and to calculate precise fuel requirements.

Manifold Absolute Pressure Sensor

The MAP sensor (**Figure 8-4**) measures changes in the intake manifold pressure that result from changes in engine load and speed. The pressure measured by the MAP sensor is the difference between an absolute vacuum and manifold pressure. At closed throttle, the engine produces a low MAP value. A wide-open throttle produces a high value. This high value is produced when the pressure inside the manifold is the same as pressure outside the manifold and 100 percent of the outside air is being measured. This MAP output is the opposite of what is measured on a vacuum gauge. The use of this sensor also allows the control computer to adjust automatically for different altitudes. Some systems use a separate barometric pressure (BARO) sensor to measure barometric pressure. The MAP sensor reading can be taken at key on, engine off, and used as a BARO reading.

The PCM sends a voltage reference signal to the MAP sensor. As the MAP changes, the electrical resistance of the sensor also changes. The PCM uses the amount of voltage drop across the sensor. The PCM can determine the manifold pressure by monitoring the sensor output voltage. A high-pressure, low-vacuum condition, such as wide-open throttle, requires more fuel. A low-pressure, high-vacuum condition requires less fuel. Like an airflow sensor, a MAP sensor relies on an air temperature sensor to adjust its base pulse signal to match incoming air density.

Some EFI systems with MAF sensors do not have MAP sensors. However, there are engines with both of these sensors. In those applications, the engine design characteristics require a fuel delivery strategy known as idle mode fueling to enhance idle quality and maintain low rpm performance. Such designs will use the MAP input for idle range and the MAF for off-idle operations. In addition, the MAP can be used as a backup in the event of a MAF-related malfunction. When the EFI system has a MAF sensor, the computer calculates the intake airflow from the MAF and rpm inputs.

Other EFI Systems Sensors

In addition to airflow, air mass, or manifold absolute pressure readings, the ECM relies on input from a number of other system sensors. These inputs further adjust the injector pulse width to match engine operating conditions. Operating conditions are communicated to the control computer by the following types of sensors:

Coolant Temperature. The coolant temperature sensor signals the PCM when the engine needs cold enrichment, as it does during warm-up. This adds to the base pulse but decreases to zero as the engine warms up.

Throttle Position (TP). The TP sensor(s) on the throttle body signals the PCM of how far the throttle is opened and how fast it was opened.

Engine Speed and Position. The crankshaft position (CKP) sensor sends the PCM a tachometer signal reference pulse corresponding to engine speed. The CKP sensor (**Figure 8-5**) signal is one of the most critical signals that the PCM will see. Without the crankshaft position (CKP) signal, the PCM does not know that the engine is running, how fast it is running, or the position of the crankshaft.

Camshaft(s) Position. The camshaft position (CMP) sensor is very necessary for sequential fuel injection systems and coil over plug ignition systems that comprise the majority of vehicles on the road today. The PCM needs to know the position of the camshaft to be able to time a very precise injection, such as PFI and DFI. Remember that the crankshaft turns two revolutions to one revolution of the camshaft. Two cylinders will be at top dead center (TDC) at the same time, one on the exhaust stroke, one on the compression. The crankshaft position sensor (CKP) tells the PCM which cylinder is on the intake stroke. Additionally, many DFI systems have variable cam timing on both the intake and exhaust camshafts. The PCM needs to know exactly where the camshafts are located at all times.

> The "absolute" in "manifold absolute pressure" refers to psia, or pounds per square inch absolute. Psia adds atmospheric pressure to the gauge reading. MAP sensors read any pressure above a perfect vacuum "0" absolute or 29.9 in. Hg vacuum.

Oxygen and A/F Sensor. To review, the oxygen sensor is a very important player in any EFI system (**Figure 8-6**). The oxygen or A/F sensor has the responsibility of telling the fuel injection system how the mixture is being burned. The PCM can then add fuel or subtract fuel to correct the mixture as needed.

The A/F sensor has been instrumental in allowing the PCM to know more precisely what the air-fuel ratio measurement is during operation away from the normal 14.7:1. The PCM can more accurately deliver the needed fuel over a broader range of operating conditions. This precise control of fuel under many conditions has made the A/F sensor attractive to manufacturers. Now instead of knowing whether the engine was rich or lean of 14.7:1, the PCM can determine how rich or lean the mixture is precisely. For example, if a situation calls for a mixture of 9:1, then the A/F sensor can determine if the mixture is correct for the condition by measuring the mixture.

> The A/F ratio sensor has several names, one of the most common of which is the universal exhaust gas oxygen sensor (UEGO). It is also called the wide-band oxygen sensor by other manufacturers.

Figure 8-5 A crankshaft position (CKP) sensor.

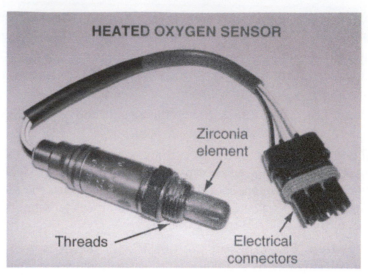

Figure 8-6 An oxygen sensor.

Stoichiometric ratio refers to the amount of air to fuel. Ideal combustion is supported at a ratio of 14.7 parts of air to 1 part of fuel, or 14.7:1.

FEEDBACK FUEL CONTROL

The feedback system is easy to understand. The prime objective of this system is to maintain an air-fuel ratio as close to 14.7 to 1 as possible during most driving, or to deliver a richer or leaner fuel mixture when conditions warrant. A richer fuel mixture is desired during high-load conditions, such as acceleration. A very lean mixture is required during deceleration. Fuel delivery is electronically controlled by the powertrain control module (PCM). This is accomplished by increasing or decreasing the amount of fuel to be delivered to the engine based on driving demands and operating conditions. Outputs or commands from the PCM correct the current mixture condition it is sensing. A number of sensors give input to the PCM to indicate the current engine demand and operating environment such as load, air, and coolant temperature, barometric pressure, and vehicle and engine speed, to name a few. The oxygen sensor or A/F sensor in the exhaust manifold responds to the presence or absence of oxygen in the post-combustion exhaust. The PCM is programmed to add fuel to the combustion chamber (enrich the mixture) when it "sees" a lean condition (low oxygen sensor voltage input). The PCM increases the amount of fuel the injector delivers by lengthening the pulse width. A comparison of injector pulse widths shows a normal waveform at stoichiometry in **Figure 8-7**. **Figure 8-8** shows the waveform when the injector pulse width has been lengthened. It will also subtract fuel

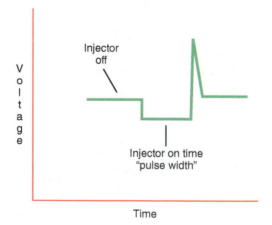

Figure 8-7 Typical injector waveform at normal operation.

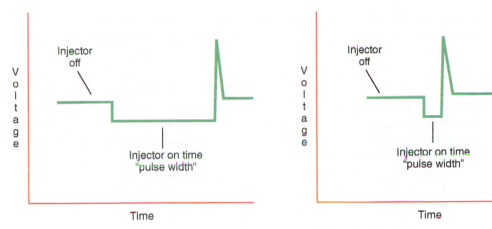

Figure 8-8 On acceleration, the injector pulse width becomes longer to deliver more fuel. The same would be seen if the PCM were correcting for a lean condition.

Figure 8-9 During deceleration, or trying to compensate for a rich mixture. Notice how short the pulse width becomes.

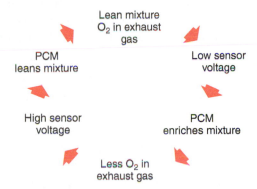

Figure 8-10 Closed-loop operation.

(lean the mixture) when the exhaust condition is rich (high oxygen sensor voltage). (See **Figure 8-9**.) This process is part of an ongoing cycle of fuel management known as closed loop (**Figure 8-10**). During the warm-up period or under certain driving conditions, the engine may demand a high fuel requirement. In this situation, the PCM will ignore the oxygen sensor and deliver a fixed rate of fuel. This condition is referred to as open loop. Depending on the engine system, the rich-lean cycle can occur several times per second, precisely controlling fuel and emissions.

Pulse Width

The length of time that the computer grounds the injector is referred to as pulse width. Under most operating conditions, the computer provides the correct injector pulse width to maintain the stoichiometric air-fuel ratio. For example, the computer might ground a PFI injector for 2 milliseconds at idle speed and 7 milliseconds at wide-open throttle to provide the stoichiometric air-fuel ratio.

> Pulse width is the amount of time an injector is open and injecting fuel.

Fuel Trims

As we discussed earlier, the ECM controls fuel delivery based on inputs from the various sensors and the feedback from the oxygen or A/F ratio sensors. Fuel control includes the adaptive fuel control system called fuel trim. There are two types of fuel trim: short term and long term. Short-term fuel trim normally changes rapidly, as in times of acceleration and deceleration, and provides for immediate feedback regarding the needs of fuel control. Long-term fuel trims move slower and take care of long-term conditions of the fuel management. For example, say a small vacuum leak has developed in the intake manifold. First, short-term fuel trim would keep adding fuel to correct for the lean mixture. Long-term fuel trim would then begin to move to add fuel to the base calculation and bring short-term trim back to near zero. Once the short-term fuel trim is back near zero, the long-term fuel trim would stop adding fuel. In contrast to a long-term lean mixture, if a rich mixture was occurring, the short-term fuel trim would subtract fuel initially, then the long-term fuel trim would take over and subtract fuel from the long-term calculation. Long-term fuel trim is retained in computer memory after the key has been turned off, but short-term fuel trims are reset every time the key is turned off. Long-term fuel trim only updates when the vehicle is at operating temperature. Long-term fuel trim adaptation is used even in open loop fuel calculations.

Basic Modes of Operation

Cranking Enrichment. The intake air temperature (IAT) sensor and the coolant temperature sensor (ECT) monitor the temperature of the engine and incoming air to determine the best fuel injection amount to make the engine run properly in cold or hot weather. This initial injection must be made without being able to see the oxygen sensor (or A/F sensor), because they do not produce a good signal when cold. This cold operation is based on the ECM/PCM's programming and is therefore called **open loop** operation.

Altitude Compensation. As the car operates at higher altitudes, the thinner air needs less fuel. Altitude compensation in a fuel injection system is accomplished by monitoring the MAP sensor when the key is turned on and before the engine is started. There are also strategies to measure barometric pressure during deceleration, closed throttle. And some systems actually have the barometric pressure sensor (BARO) mounted directly to the ECM's circuit board.

Coasting Shutoff. Coasting shutoff can be found on a number of control systems. It can improve fuel economy as well as reduce emissions of hydrocarbons and carbon monoxide. Fuel shutoff is controlled in different ways depending on the type of transmission (manual or automatic). The PCM makes a coasting shutoff decision based on a closed throttle as indicated by the throttle position or idle switch or on engine speed as indicated by the signal from the ignition coil. When the PCM detects that power is not needed to maintain vehicle speed, the injectors are turned off until the need for power exists again.

Additional Input Information Sensors

- Knock sensor
- Accessory operation
- Gearshift lever position
- Battery voltage
- Vehicle speed sensor (VSS)
- Exhaust gas recirculation (EGR) valve position

Figure 8-11 shows the typical inputs and outputs of the PCM responsible for controlling the EFI system.

SPEED DENSITY EFI

In an electronic fuel injection (EFI) system, the computer must know the amount of air entering the engine so it can supply the stoichiometric air-fuel ratio. In EFI systems with a MAP sensor, the computer program is designed to calculate the amount of air entering

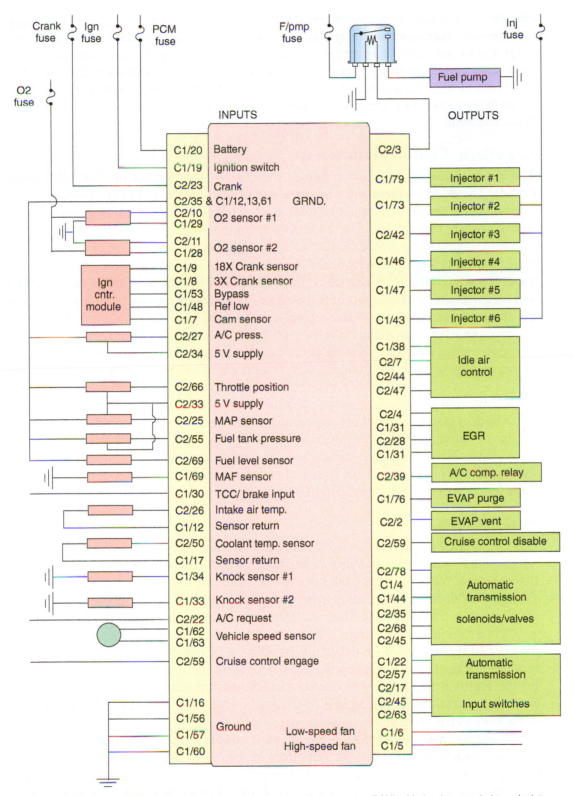

Figure 8-11 Input sensors from all engine systems supply the control computer (PCM) with the data needed to calculate the correct injector pulse for driving conditions and driver demands.

A **speed density** EFI system is one in which the computer calculates the air entering the engine from the MAP and engine speed inputs.

the engine from the MAP and rpm input signals. The crankshaft position sensor supplies an RPM signal to the ECM. This type of EFI system is referred to as a **speed density** system because the computer calculates the air intake flow from the engine rpm, or speed, altitude, intake air temperature, throttle position, coolant temperature, and volumetric efficiency. Therefore, the computer must have accurate signals from these inputs to maintain the stoichiometric air-fuel ratio. EGR flow must also be calculated and subtracted since it does not burn. Other inputs are used by the computer to fine-tune the air-fuel ratio. For example, if the TP sensor input indicates sudden acceleration, the computer momentarily supplies a richer air-fuel ratio. Some manufactures have basically used the speed density system instead of the MAF system. Chrysler traditionally has used the speed density system.

Powertrain Control Module

The heart of the fuel injection system is the computer or powertrain control module (PCM). The PCM receives and processes signals from all the system sensors and then transmits programmed electrical pulses to the fuel injectors. Both incoming and outgoing signals are sent through a wiring harness and a multiple-pin connector. The PCM is also called an engine control module (ECM), depending on the application or the manufacturer. The ECM is part of a network of computers that share information with each other about the entire vehicle. For instance, the ECM shares information about vehicle speed and load with the transmission control module (TCM), body control module (BCM), and the antilock braking system (ABS) control module (EBCM).

Electronic feedback in the PCM means the unit is self-regulating and is controlling the injectors on the basis of operating performance or parameters rather than on preprogrammed instructions. For example, a PCM with a feedback loop reads signals from the oxygen sensor or air/fuel sensor, varies the pulse width of the injectors, and again reads the signals from the oxygen sensor or air/fuel sensor. This is repeated until the injectors are pulsed for just the amount of time needed to achieve the proper amount of oxygen into the exhaust stream. While this interaction is occurring, the system is operating in closed loop. When conditions such as starting or wide-open throttle demand that the signals from the oxygen sensor be ignored, the system operates in open loop. During open loop, injector pulse length is controlled by set parameters contained in the PCM's memory.

Shop Manual
Chapter 8, page 400

Fuel Injectors

Fuel injectors are electromechanical devices that meter and atomize fuel so it can be sprayed into the intake manifold. Fuel injectors are mounted in the fuel rail. O-rings are used to seal the injector at the intake manifold, throttle body, and fuel rail mounting positions. These O-rings provide thermal insulation to prevent the formation of vapor bubbles and promote good hot-start characteristics. They also dampen potentially damaging vibration.

The **pintle** is the component in the injector that opens and closes against the injector seat, thus controlling fuel output.

When the injector is electrically energized, a solenoid-operated valve opens, and a fine mist of fuel sprays from the injector tip. Two different valve designs are commonly used.

The first consists of a valve body and a nozzle or needle valve that has a special ground **pintle** (**Figure 8-12**). A movable armature is attached to the nozzle valve, which is pressed against the nozzle body sealing seat by a helical spring. The solenoid winding is located at the back of the valve body.

When the solenoid winding is energized, it creates a magnetic field that draws the armature back and pulls the nozzle valve from its seat. When the solenoid is de-energized, the magnetic field collapses and the helical spring forces the nozzle valve back on its seat.

The second popular valve design uses a ball valve and valve seat. In this case, the magnetic field created by the solenoid coil pulls a plunger upward, lifting the ball valve from its seat. Once again, a spring is used to return the valve to its seated or closed position.

Port fuel injectors can be either top fuel feeding or bottom fuel feeding (**Figure 8-13**). Top feed injectors are primarily used in port injection systems that operate using high fuel system pressures. Bottom feed injectors were used in some Subaru port fuel systems. Bottom feed injectors are able to use fuel pressures as low as 10 psi. A typical wiring diagram for an injector circuit is shown in **Figure 8-14**.

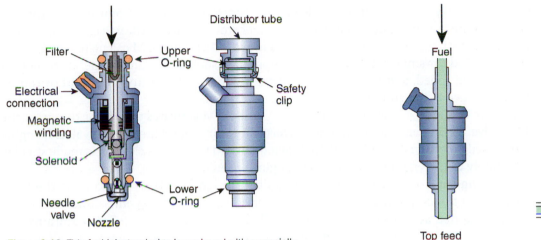

Figure 8-12 This fuel injector design is equipped with a specially ground pintle for precise fuel control.

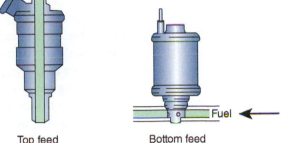

Figure 8-13 Examples of bottom and top feed injectors.

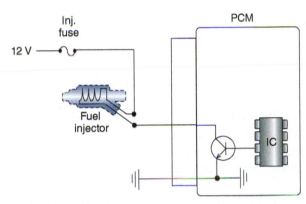

Figure 8-14 Injector diagram.

AUTHOR'S NOTE Fuel injector technology is constantly improving, but the injector must still operate even in the most adverse conditions. Consider harsh temperature extremes, various fuels or additives that may be used, and the manner in which the vehicle is driven. Eventually, the injector can begin to wear and result in a range of drivability symptoms. For example, a restricted injector can cause either a lean or rich mixture depending on the restriction. Suppose the pintle area has a deposit that distorts the spray pattern. The distorted spray pattern may limit fuel atomization, causing poor combustion. (Fuel must be atomized to burn properly.) If the restriction prevents the pintle from closing, fuel would likely continue out of the injector, causing a rich mixture. Any of these problems can result in an engine misfire, poor performance, poor fuel economy, or increased emissions. Many manufacturers only recognize specific types of injector cleaners and methods for fuel injector cleaning. Refer to the manufacturer's recommendations to ensure warranty compliance before attempting injector cleaning or service.

Idle Speed Control

Although most new vehicles are using throttle actuator control, there are still many vehicles on the road using **idle air control (IAC)** systems. In older port EFI systems, engine idle speed is controlled by bypassing a certain amount of airflow past the throttle valve in the throttle body housing.

The **idle air control (IAC)** valve controls the amount of intake air entering the engine to maintain proper idle speeds on port fuel engines.

The IAC system consists of an electrically controlled stepper motor or actuator that positions the IAC valve in the air bypass channel around the throttle valve. The air bypass channel is part of the throttle body casting (**Figure 8-15**). The PCM calculates the amount of air needed for smooth idling based on input data such as coolant temperature, engine load, engine speed, and battery voltage. It then signals the actuator to extend or retract the IAC valve in the air bypass channel.

If the engine speed is lower than desired, the PCM activates the motor to retract the IAC valve. This opens the channel and diverts more air around the throttle valve. If engine speed is higher than desired, the valve is extended and the bypass channel is made smaller. Air supply to the engine is reduced and engine speed falls.

During the cold starts, idle speed can be as high as 2,100 rpm to raise the temperature of the catalytic converter for proper control of exhaust emissions. Idle speed that is attained after a cold start is controlled by the PCM. The PCM maintains idle speed for approximately 40 to 50 seconds even if the driver attempts to alter it by kicking the accelerator. After this preprogrammed time interval, depressing the accelerator pedal rotates the throttle position sensor (TPS) and signals the PCM to reduce idle speed.

Vehicles with throttle actuator control, also called "throttle by wire," do not need IAC; the throttle position is controlled by the PCM.

PORT FUEL INJECTION

Port fuel injection (PFI) is a term that may be applied to any fuel injection system that has its injectors located in the intake ports.

Port fuel injection (PFI) systems use one injector at each cylinder. They are mounted in the intake manifold near the cylinder head where they can inject a fine, atomized fuel mist as close as possible to the intake valve (**Figure 8-16**). Fuel lines run to each cylinder from a fuel manifold, usually referred to as a fuel rail. The fuel rail assembly on a PFI system of V6 and V8 engines usually consists of a left- and right-hand rail assembly. The two rails can be connected either by crossover and return fuel tubes or by a mechanical bracket arrangement. Since each cylinder has its own injector, fuel distribution is exactly equal. With little or no fuel to wet the manifold walls, there is no need for manifold heat to prevent fuel condensation on the manifold walls. Fuel does not collect in puddles at the base of the manifold.

Port fuel vehicles that are manufactured today have the pressure regulator in the fuel tank.

The throttle body in a PFI system controls the amount of air that enters the engine, as well as the amount of vacuum in the manifold. It also houses and controls the IAC motor and the throttle position sensor. The TP sensor enables the PCM to know where the throttle is positioned at all times.

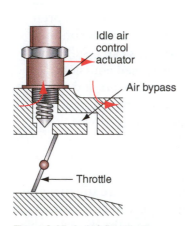

Figure 8-15 An IAC fitted in the casting of a throttle body.

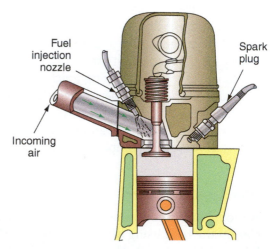

Figure 8-16 Port fuel injection systems use an injector at each cylinder.

The throttle body is a single-cast aluminum housing with a single throttle blade attached to the throttle shaft. The TP sensor and the IAC valve/motor are also attached to the housing. The throttle shaft is controlled by the accelerator pedal or the throttle actuator control. The throttle shaft extends the full length of the housing. The throttle bore controls the amount of incoming air that enters the air induction system. Some systems have a small amount of engine coolant routed through a passage in the throttle body to prevent icing during cold weather.

Older Port fuel injection systems with a fuel return-line are equipped with fuel pressure regulators on the fuel rail (**Figure 8-17**) that sense manifold vacuum and continually adjust the fuel pressure to maintain a constant pressure drop across the injector tips at all times. Returnless systems have the pressure regulator in the tank that holds fuel pressure steady by sensing fuel pressure alone.

In current PFI systems the injectors deliver fuel just as the intake valve opens for a particular cylinder. This results in a smoother idle and a slight increase in fuel mileage. The PCM controls each injector individually by switching the ground side of the injector under normal operation.

AUTHOR'S NOTE The earlier port fuel injection systems from the 1980s were called multiport fuel injection (MFI or MPFI). Most of these systems were bank fired, firing half the injectors, then the other half in 180 degrees of crank rotation, regardless of the position of the intake valve. Modern port fuel injectors spray at the optimum time, during the intake stroke, which is called sequential fuel injection requires the use of a camshaft position sensor. The older multiport fuel injection systems were not as efficient as the sequential port fuel systems. Today, both types are generally referred to as port fuel or PFI systems.

Port Fuel Injection System Design

As you saw in Chapter 7, an in-tank modular fuel pump assembly contains the pressure regulator and a lifetime fuel filter. The fuel pump may also use a fuel pump driver module to regulate fuel pump speed according to load.

The lower end of each port injector is sealed in the intake manifold with an O-ring seal, and a similar seal near the top of the injector seals the injector to the fuel rail. Most port fuel injector rails have a Schrader valve. A pressure gauge may be connected to this

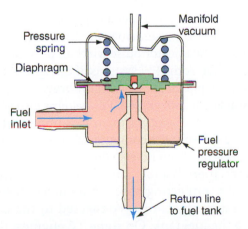

Figure 8-17 Operation of a vacuum-operated fuel pressure regulator. Low vacuum increases fuel pressure.

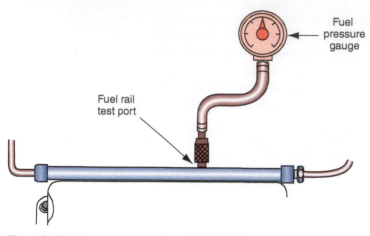

Figure 8-18 Fuel gauge connected to a Schrader valve.

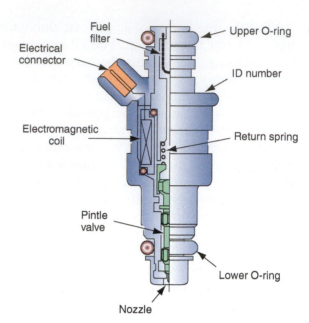

Figure 8-19 A typical fuel injector used in port fuel injection systems.

valve when testing fuel pressure (**Figure 8-18**). A dust cap on the Schrader valve must be removed prior to gauge installation.

Each injector has a movable armature in the center of the injector, and a pintle with a tapered tip is positioned at the lower end of the armature. A spring pushes the armature and pintle downward so the pintle tip seats in the discharge orifice. The injector coil surrounds the armature, and the two ends of the winding are connected to the terminals on the side of the injector. An integral filter is located inside the top of the injector. When the ignition switch is turned on, voltage is supplied to one injector terminal and the other terminal is connected through the computer. Each time the computer completes the circuit from the injector winding to ground, current flows through the injector coil, and the coil magnetism moves the plunger and pintle upward. Under this condition, the pintle tip is unseated from the injector orifice, and fuel sprays out this orifice into the intake port (**Figure 8-19**).

The computer is programmed to ground the injectors well ahead of the actual intake valve openings so the intake ports are filled with fuel vapor before the intake valves open (**Figure 8-20**). In PFI systems, the computer supplies the correct injector pulse width to provide the stoichiometric air-fuel ratio under all operating conditions. Under heavy load, even a sequential port fuel injection system may fire the injectors regardless of valve position. The computer increases the injector pulse width to provide air-fuel ratio enrichment while starting a cold engine. On some PFI systems, if the ignition system is not firing, the computer stops operating the injectors. This action prevents severe flooding from long cranking periods while starting a cold engine. PFI systems also severely decrease injector pulse width while the engine is decelerating, to provide improved emission levels and fuel economy. On some of these systems, the computer stops operating the injectors while the engine is decelerating in a certain rpm range, which is called deceleration enleanment.

On V6 and V8 engines, there is a fuel rail on each side of the intake manifold. Fuel is supplied to one rail, and a connecting hose carries fuel to the second rail. On return-line style systems, a pressure regulator is connected to the end of the fuel rail, and excess fuel is returned to the fuel tank. On some V6 engines, the fuel rail is U-shaped and each injector is mounted in this one-piece rail (**Figure 8-21**). Many fuel rails are made from steel or aluminum alloy. Plastic-type fuel rails have been installed recently

A Schrader valve is best described as being like an air valve on a tire.

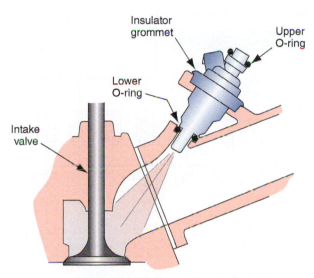

Figure 8-20 A port fuel injector sprays fuel into the intake port and fills the port with fuel vapors before the intake valve opens.

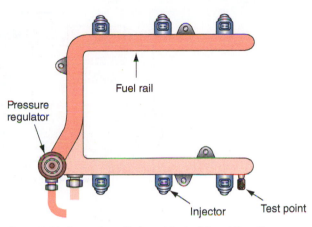

Figure 8-21 A one-piece, U-shaped fuel rail for a V6 engine.

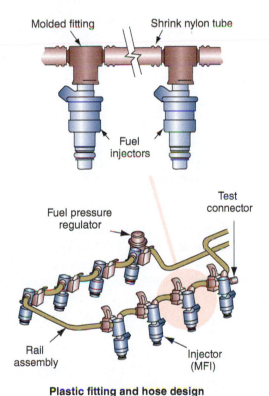

Plastic fitting and hose design

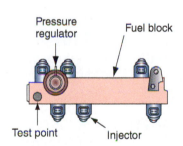

Machined aluminum block design

Figure 8-22 Plastic and one-piece fuel rails.

on some cars. These fuel rails transfer less heat to the fuel and reduce the possibility of fuel boiling in the rail (**Figure 8-22**).

Shop Manual
Chapter 8, page 399

Toyota Camry Fuel System

An example of a PFI Toyota Camry model is shown in **Figure 8-23**. Note the flow of power to the PCM and injectors. Look at the wiring diagram to see that each injector has its own connections to the PCM. The PCM grounds injectors in sequence as the respective cylinder is on the intake stroke, making this system sequential injection. Also in this

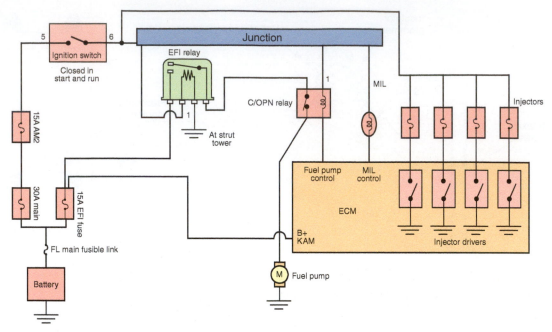

Figure 8-23 Toyota sequential fuel injection ECM, fuel pump, and injector power wiring diagram.

figure, take note of the power supply to the system. Power flows through the FL Main fusible link from the battery, which powers the 30A main fuse, and the 15A EFI fuse.

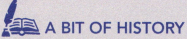 **A BIT OF HISTORY**

Since the 1970s, we have seen quite a shift in fuel delivery systems—from carburetors to throttle body injection to port fuel injection and now direct fuel injection.

- The 15A EFI fuse has two connections, one feeds terminal 5 of the EFI relay and the other feeds the BATT terminal of the PCM. Because the 15A EFI fuse is powered with the key off, the BATT terminal is used to preserve the keep alive memory or KAM.
- The 30A main fuse feeds the 15A AM2 fuse that feeds pin 5 of the ignition switch. When the ignition switch is turned to run, pin 6 of the ignition switch feeds both the fuel injector's power supply and a junction shown near the circuit opening relay (relay shown on the diagram as C/OPN).
- The junction has three outputs; one provides power to the MIL lamp. (The PCM will provide a ground to the LED to illuminate it as required.) The second output goes to the circuit opening relay terminal 1 and the EFI relay terminal 1.
- When the EFI terminal 1 is powered, the relay turns on because it has a ground at the strut tower. The main relay supplies power to many of the output devices of the PCM, as well as the circuit opening relay. (The circuit opening relay is what most manufacturers would call a fuel pump relay.) The circuit-opening relay is turned on when the PCM grounds terminal FC, completing the current path for the relay coil and activating the fuel pump.

Toyota ECM Inputs and Outputs

The Toyota PFI diagram has many inputs and outputs. Although this Toyota system (**Figure 8-24**) has some differences from the domestic versions, many of input sensors and output actuators are the same. The illustration and list are not all inclusive; only the most important ones are noted here.

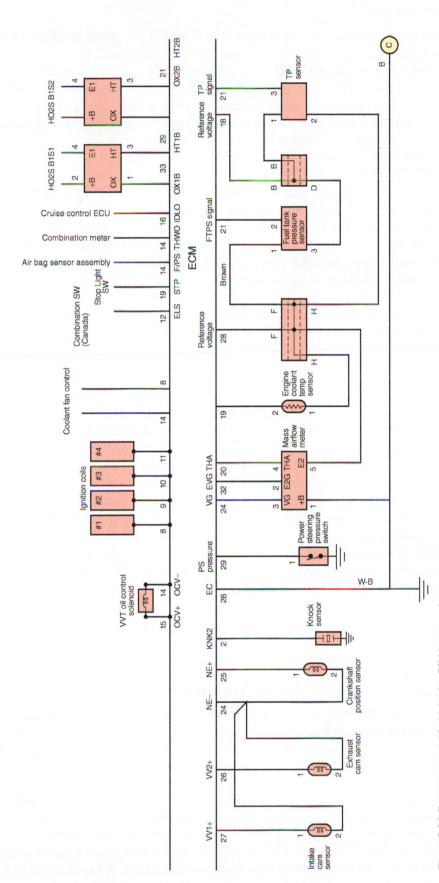

Figure 8-24 Toyota sequential fuel injection ECM inputs and outputs.

Computer Inputs

- *Power steering oil pressure switch.* Informs the PCM of high power steering pressure so the idle speed and fuel delivery can be altered to prevent stalling.
- *Engine coolant temperature (ECT).*
- *Fuel tank pressure (FTP) sensor.* Provides input to the PCM concerning pressure in the fuel tank. The evaporative emission control monitor looks at vapor pressure in the fuel tank to evaluate the condition of the evaporative control system, which is covered in more detail in Chapter 11.
- *Mass airflow (MAF) sensor.*
- *Throttle position (TP) sensor.*
- *Heated oxygen sensor bank 1 sensor 1.* Informs the PCM of the oxygen content of the exhaust gas. Lean mixtures have more oxygen than rich mixtures.
- *Heated oxygen sensor bank 1 sensor 2.* Also informs the PCM of the exhaust oxygen content, but is located behind the catalytic converter. This sensor allows the PCM to monitor the effectiveness of the converter. If the converter is working properly, most of the oxygen should be used up in the converter, so very little oxygen should pass through the converter.
- *Intake camshaft and exhaust camshaft sensors.*
- *Crankshaft sensor.*
- *Knock sensor.*

Computer Outputs

> Vapor pressure is the pressure that builds in the tank due to the volatility of gasoline. More volatile gasoline has more vapor pressure than less volatile gasoline.

- *Radiator fan control.* The PCM monitors and controls the coolant fan motor operation.
- *Camshaft timing oil control valve.* This vehicle has camshaft timing control. The camshaft timing can be varied by a special sprocket on the camshaft. The PCM can control cam timing on one (or both depending on the model and system used) of the camshaft(s) by varying oil pressure to a camshaft actuator.
- *Combination meter.* Toyota calls their instrument cluster a combination meter. Some of the outputs to the combination meter are speedometer, MIL lamp, temperature lamp, and charging voltage. The combination meter is also called an instrument panel cluster (IPC). The instrument cluster receives most of its information from the vehicle's computer network by means of serial data—the information that computers use to send information to one another.
- *Cruise control module.* The computer network, which includes the PCM, can also be used to share data with the cruise control module, such as engine speed, brake pedal position, and so on.
- *Injectors.* The injectors are obviously another important PCM output.

Returnless Fuel System Pressure Regulators

> Returnless fuel systems do not circulate excess fuel all the way to the fuel rail and back to the tank, lowering the strain on fuel pumps, fuel filters, and the EVAP system.

Manufacturers are presently using a returnless fuel delivery system. In these systems, the fuel pressure regulator and filter are mounted in the top of the modular fuel pump assembly that also contains the fuel gauge sending unit in the fuel tank (**Figure 8-25**). The fuel line from the fuel rail under the hood is connected to the filter with a quick-disconnect fitting.

Fuel enters the filter through the fuel supply tube in the center of the regulator and filter assembly. Fuel pressure is applied against the regulator seat washer, which is seated by the seat control spring (**Figure 8-26**). At the specified regulator pressure, the seat is forced downward against the spring and fuel flows past the seat into the cavity around the seat control spring. Fuel returns from this cavity to the fuel tank. When the pressure drops slightly, the seat closes again. With the returnless fuel system, only the fuel needed by the engine is filtered, thus allowing the use of a smaller fuel filter. The advantage of the returnless system is that no warm fuel is returned to the tank, making the EVAP system work

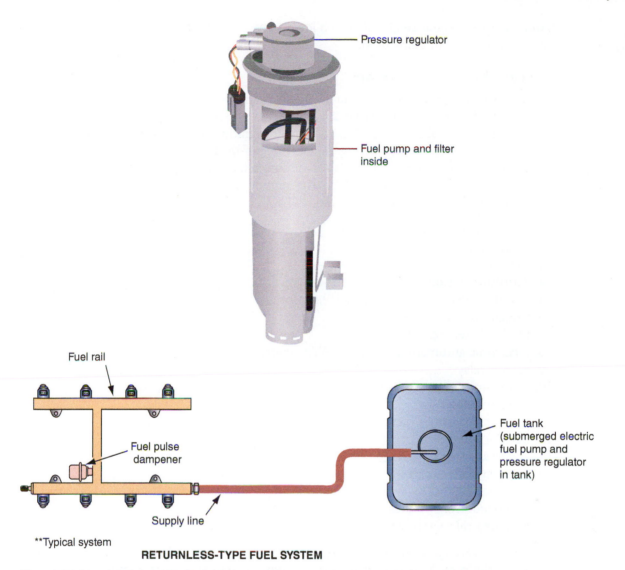

RETURNLESS-TYPE FUEL SYSTEM

Figure 8-25 Returnless fuel system with the pressure regulator mounted in the fuel tank with the fuel pump and fuel gauge sending unit.

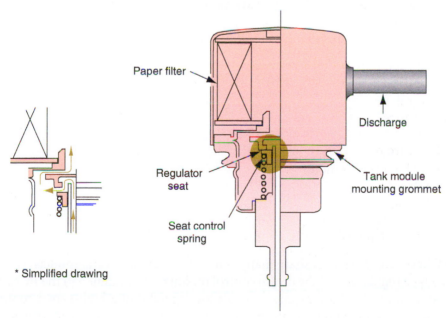

Figure 8-26 Pressure regulator and fuel filter from a returnless fuel system. Operation of a vacuum-operated fuel pressure regulator. Low pressure increases fuel pressure.

Shop Manual
Chapter 8, page 411

harder; a return line is removed, which lowers cost; and the fuel pump does not have to pump fuel that is not being used.

Typical Chrysler PFI System

In a Chrysler PFI system on a 3.5-liter engine, each injector has a separate ground wire connected into the PCM (**Figure 8-27**). Voltage is supplied through the automatic shutdown (ASD) relay points to the injectors when the ignition switch is turned on, and a separate fuel pump relay supplies voltage to the fuel pump. This engine is equipped with an electronic ignition system, and the crank and cam sensors are inputs for this system. Since these inputs are connected to the PCM, the ignition module is contained in the PCM.

As an example, the Chrysler 3.5-liter PFI system inputs are as follows:

1. Dual heated oxygen (O_2) sensor
2. Crank sensor
3. Cam sensor
4. Throttle position (TP) sensor
5. Manifold absolute pressure (MAP) sensor
6. Coolant temperature sensor
7. Intake air temperature sensor
8. Electronic automatic transaxle (EATX) computer
9. Starter relay
10. Generator field
11. Dual knock sensors
12. Data links to other computers such as the EATX
13. Battery voltage
14. Power steering switch
15. Brake switch
16. Ignition switch
17. Cruise control switches
18. Park neutral switch, starter relay

V6 and V8 engines have dual oxygen sensors that provide improved control of the air-fuel ratio in each bank of cylinders. An oxygen sensor is usually mounted in each exhaust manifold. Dual knock sensors are located in many V6 and V8 engines to improve spark knock or detonation control. The knock sensors are positioned in each side of the block or cylinder heads.

The outputs on the Chrysler 3.5-liter PFI system are as follows:

1. Automatic shutdown (ASD) relay
2. Fuel pump relay
3. Ignition coil
4. Spark advance
5. Injectors
6. Cruise control
7. Automatic idle speed motor
8. Purge solenoid
9. EGR solenoid
10. Manifold solenoid
11. Dual radiator fan relays

The Chrysler PFI systems share many similarities across several models. For example, the voltage regulator and the cruise control module are contained in the PCM board. The PFI system on the 3.5-liter engine has a low-speed and a high-speed cooling fan relay. At a specific coolant temperature, the PCM grounds the low-speed relay winding,

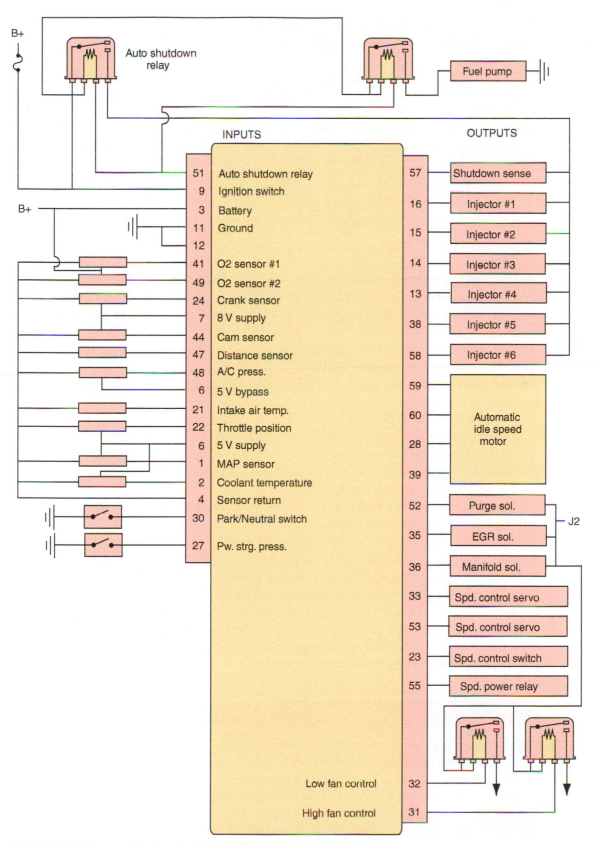

Figure 8-27 PCM inputs and outputs for Chrysler's PFI system used on 3.5-liter engine.

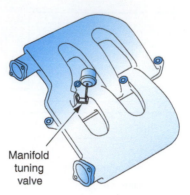

Figure 8-28 Intake manifold tuning valve for a Chrysler 3.5-liter engine.

which closes the relay points and supplies voltage to the fan motor. If the engine coolant temperature continues to increase, the PCM grounds the high-speed cooling fan relay winding, which closes the fan relay points and supplies voltage to the high-speed fan motor.

The manifold solenoid controls the vacuum supplied to the intake manifold tuning valve. This solenoid is mounted on the right shock tower, and the manifold tuning valve is positioned near the center of the intake manifold (**Figure 8-28**). The manifold contains a pivoted butterfly valve that opens and closes to change the length of the intake manifold air passages. This butterfly valve is mounted on a shaft, and the outer end of the shaft is connected through a linkage to a diaphragm in a sealed vacuum chamber. A vacuum hose is connected from the outlet fitting on the manifold solenoid to the vacuum chamber in the manifold tuning valve. Another vacuum hose is connected from the inlet fitting on the manifold solenoid to the intake manifold.

One terminal on the manifold solenoid winding is connected to the ignition switch, and the other terminal on this winding is connected to the PCM. While the engine is running at lower rpm, the PCM opens the manifold solenoid circuit. Under this condition, the solenoid shuts off the manifold vacuum to the intake manifold tuning valve, and the butterfly valve closes some of the air passages inside the intake manifold.

At higher engine rpm, the PCM grounds the manifold solenoid winding and energizes the solenoid. This action opens the vacuum passage through the solenoid and supplies vacuum to the intake manifold tuning valve. Under this condition, the butterfly valve is moved so it opens additional air passages inside the intake manifold to improve airflow and increase engine horsepower and torque.

DIRECT FUEL INJECTION

With the development of fuel injection systems, technological advancements continue to evolve. The most recent fuel delivery system to be released is the DFI system. The typical PFI system delivers a fuel spray before the intake valve. Similar to the diesel injection system, the DFI system delivers the fuel charge directly into the combustion chamber (**Figure 8-29**). The fuel pressure is much higher than the typical fuel injections systems and can reach 2,200 psi. Under this amount of pressure, the fuel is vaporized as it is injected into the cylinder. With DFI, mixtures can be substantially leaner (up to 35:1). This can improve fuel economy by as much as 30 percent and dramatically reduces hydrocarbon emissions on a cold engine. The introduction of the DFI systems are phasing out traditional port fuel systems and becoming commonplace as system application expands to more engine systems.

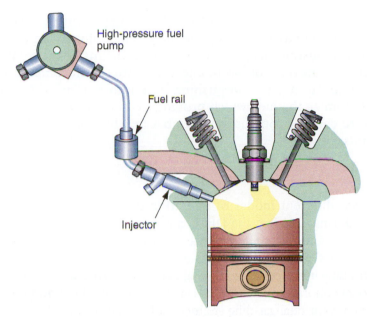

Figure 8-29 A direct fuel injection fuel pump and injector.

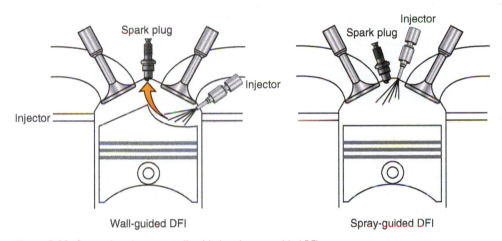

Figure 8-30 Comparison between wall-guided and spray-guided DFI.

Types of DFI

There are two different types of direct injection: wall guided and spray guided (**Figure 8-30**). Wall-guided injection sprays the fuel onto a specially designed piston that concentrates fuel in the vicinity of the spark plug. The richer mixture in the vicinity of the spark plug will ignite relatively easily, allowing for lean mixtures. The spray-guided process may not use a special piston, but the injector sprays fuel directly at the spark plug.

The DFI engine uses a special piston design, high-pressure swirl injectors, 12:1 compression, and intake manifold design to help burn a leaner fuel mixture than possible with conventional fuel injection. Since the injection of fuel is not dependent on the opening of the intake valve, the fuel can be injected during the compression stroke or the intake stroke, depending on the mode of operation. Because the intake carries only air, the intake can be further enhanced for optimal performance. The fuel pump on the DFI engine is increased by a camshaft-driven mechanical pump producing fuel pressures of around 2,200 psi. This high pressure is necessary for a few reasons—the extra pressure helps keep the fuel from boiling in the injector from the high heat of the combustion chamber. Also, the pressure inside the combustion chamber is very high, and the injector has to inject fuel into the chamber against this pressure to keep combustion gas from pushing back into the injector. Lastly, the high pressure helps atomize the fuel, which promotes complete combustion, and the fuel injection helps to cool the combustion chamber that has the side effect of helping to reduce NO_x emissions.

DFI Modes

The DFI has two different modes of operation called stratified and homogeneous (**Figure 8-31**). During cruising, the DFI engine is operating in the stratified or direct injection lean mode, and fuel is injected late in the compression cycle. Piston design and injector spray pattern produce this stratified air-fuel charge by concentrating the fuel around the spark plug in the cylinder and make the fuel burn at a very lean 35:1 compared to 14.7:1 in a conventional engine. DFI is capable of saving 35 to 40 percent of the fuel consumed by a conventional gasoline engine at idle and cruising.

During acceleration, high speed, or heavy load, the fuel spray occurs in the intake stroke, like that of a conventional fuel-injected engine. This mode is called homogeneous, and since the fuel in this mode is injected during the intake stroke, it is much like regular PFI.

The fuel injector can also be pulsed twice: once during a stratified charge mode and once during the homogeneous mode as required for a smooth transition.

Cold start emissions reduction. DFI systems have the ability to fire the injector twice during cold start operation. This rich mixture helps bring the catalytic converter to operating temperature faster.

Emissions

One early concern with DFI was NO_x emissions developed under high compression and high temperatures produced by the lean burning fuel. Engineers now say that the cooling effect of the highly atomized fuel, along with advances in variable valve timing have reduced these problems.

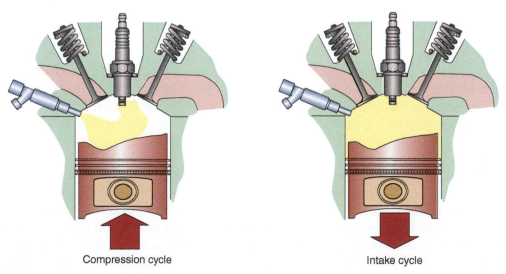

Compression cycle   Intake cycle

Figure 8-31 Two modes of DFI. Note piston design.

Typical Direct Fuel Injection System

DFI systems share many, if not most, components from the PFI systems, with some important differences. DFI systems will be returnless. Many PFI engines did take advantage of returnless fuel systems, but with DFI, the high fuel pressures needed to atomize the fuel (up to almost 2,200 psi) would be impractical to maintain from the tank to the injectors. DFI fuel systems will consist of a traditional in-tank pump plus a camshaft-driven PCM-regulated mechanical high-pressure fuel pump. The PCM can control the output pressure of the high-pressure fuel pump by pulse width modulation of an inlet valve on the high-pressure pump from a low of about 500 psi at idle to around 2,200 psi maximum. Different models will, of course, have different fuel pressures. As was discussed in Chapter 7, there are low-pressure and high-pressure fuel systems on DFI-equipped vehicles, low pressure from the pump in the fuel tank to the fuel rail, and a high-pressure camshaft-driven mechanical pump located on the fuel rail. The high-pressure pump has an inlet solenoid that is controlled by the ECM. This solenoid limits the amount of fuel admitted to the fuel rail and controls the high pressure at the fuel rail. Fuel pressure sensors keep the ECM informed of the fuel pressure in the system so the ECM can adjust the inlet solenoid as needed. The extreme high pressure of DFI requires an opening voltage of 65 volts. The schematic diagram for a GM DFI system is shown in **Figure 8-32**. This higher voltage is produced by a capacitor that is charged between injector firings (the circuitry is not shown in the schematic diagrams). The PCM supplies both power and ground to the injectors in most vehicles equipped with DFI.

Ford Direct Fuel Injection System

Ford DFI systems control both the power and ground sides for each injector independently from the PCM as shown in **Figure 8-33**. The power feed is controlled to provide a capacitive boost to the injector supply voltage by circuitry inside the PCM to help it

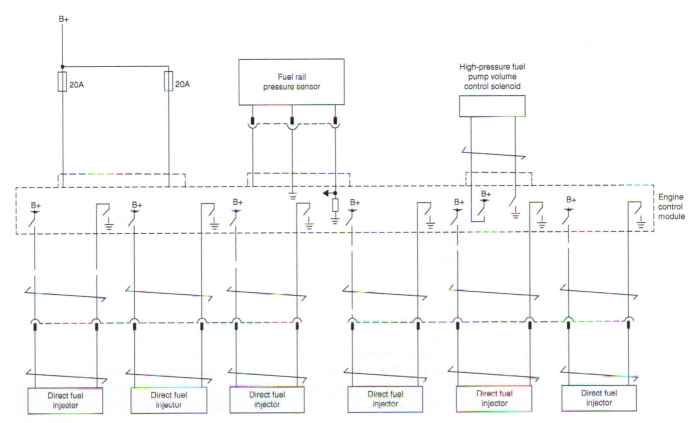

Figure 8-32 Direct fuel injection injector and pressure control.

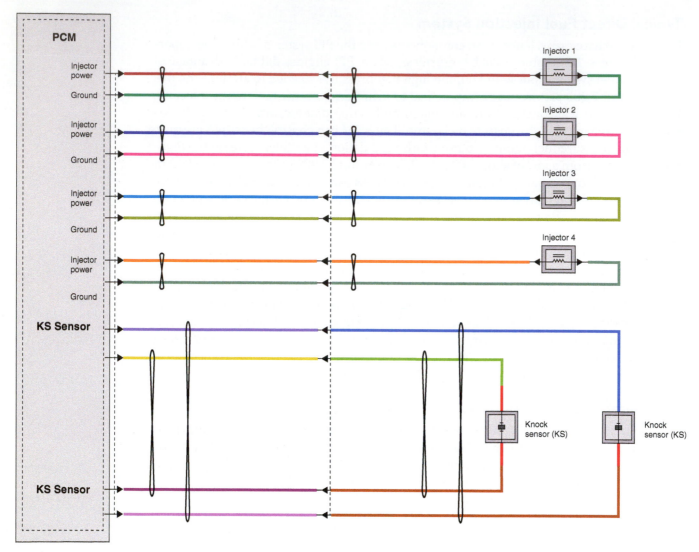

Figure 8-33 Ford style DFI injector controls.

open under high pressures in the combustion chamber. Ground is supplied as well to trigger the injector firing itself. Port fuel injection systems usually have a common power feed and then the grounds only for each injector are completed by the PCM. Fuel injectors are considered actuators.

In **Figure 8-34**, you can see the PCM with some actuators, such as:

1. Canister vent valve
2. Canister purge valve
3. Two variable camshaft position solenoids that work to control camshaft timing

Also in the diagram are some sensors, notably the oxygen sensor and the A/F ratio sensor, which is called the universal exhaust gas oxygen sensor (UEGO) by Ford, as well as from some other manufacturers. The A/F ratio sensor is located before the catalytic converter and the HO2S sensor is located behind the converter.

The DFI injectors (**Figure 8-35**) are of a special design to accommodate high fuel pressures and harsh conditions. Remember that the tips of the injectors are located in the combustion chamber itself, with both high temperatures and pressures during their lifetime. DFI injectors have to be able to seal against combustion pressures, both at the seal on the outside of the injector and at the injector pintle to prevent combustion pressures

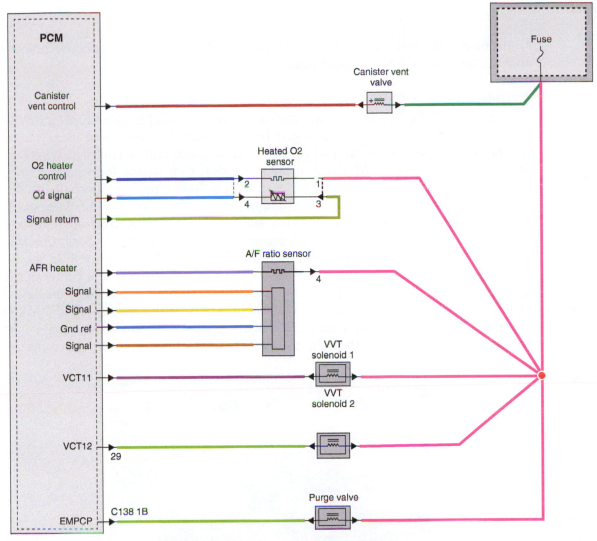

Figure 8-34 Ford GDI schematic showing camshaft sensors, UEGO sensor, and heated oxygen sensor.

Figure 8-35 A direct fuel injector. Notice how long the injector tip extends and the seal at the bottom.

from entering the fuel system. DFI injectors will require some special service techniques that will be covered in the Shop Manual.

Variable Valve Timing

The majority of the DFI vehicles will have some sort of **variable valve timing (VVT)** control, turbocharging or supercharging, or both. Traditional camshaft and lifter arrangements are a compromise of idle quality, engine performance, and fuel mileage. With variable valve timing, the PCM can determine the ideal cam timing for a given engine operation. Variable valve timing allows engineers to have a perfect camshaft for idle, midrange, and full power. Variable valve timing often has the added benefit of being able to eliminate the EGR valve, the AIR system, and their associated controls and hardware. There are several types of valve timing mechanisms, but most are hydraulically operated by solenoids in the cylinder head (**Figure 8-36**). **Figure 8-37** and **Figure 8-38** show a view of a valve timing cam sprocket with the front cover removed. The center of the sprocket is attached to the camshaft. The outside of the sprocket is turned by the timing chain. The PCM controls pressurized oil through a pulse width modulated (PWM) oil control valve. If oil is directed to the left side of the oil cavity, the cam is advanced relative to the sprocket. If oil goes to the right, the cam is retarded. In most models, the cams can

Figure 8-36 Variable valve timing actuators from a Pontiac Solstice.

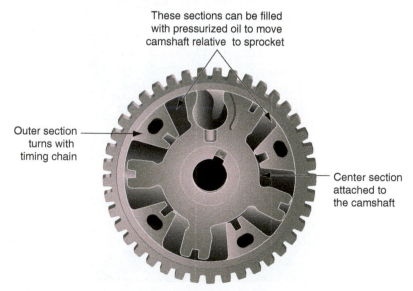

Figure 8-37 Variable valve timing is accomplished through a special cam sprocket.

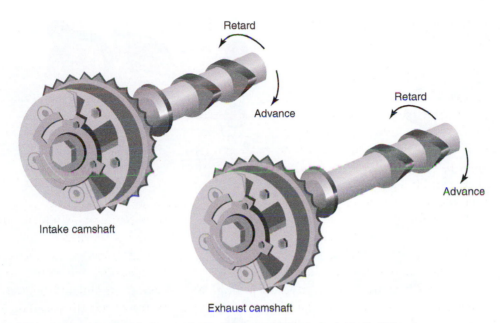

Figure 8-38 Intake and exhaust camshafts can be advanced and retarded by the VVT system.

be moved constantly within a 24-degree window. Both intake and exhaust cams can be fitted with variable valve timing, and some DOHC setups have variable valve timing on all four camshafts. More valve overlap is desirable at high speeds to increase scavenging of the combustion chamber, lowering the resistance to the flow of intake air and exhaust. Less overlap is desired at idle and for low-end torque.

VVT has several capabilities; not all of these applications are used on all vehicles.

1. *Late intake valve closing.* The intake valve stays open past bottom dead center (BDC). This actually pushes some of the air-fuel mixture back out of the combustion chamber. This partially pressurizes the intake manifold charge on the following intake strokes. Late intake valve closing has been shown to reduce pumping losses by around 40 percent during partial load conditions, and to decrease (NOx) emissions by 24 percent. The late closing reduces the amount of intake charge that has to be compressed, lowering compression heat and reducing NOx emissions.

2. *Early intake valve closing.* Reduces pumping losses at idle, resulting in around 6 percent less fuel consumption at idle.

3. *Early intake valve opening.* Opening the intake valve early—while still in the exhaust stroke—allows some of the exhaust gas to go into the intake manifold and cool. It is then taken back in with the intake charge on the following intake stroke, which is in effect EGR. This allows the elimination of the EGR valve.

4. *Valve overlap.* Valve overlap has been around for many years. Leaving the intake and exhaust valves open for a small amount of time during the exhaust stroke is called scavenging, the effect being that the leaving exhaust gases help pull the intake charge into the cylinder. More overlap is helpful at high engine speeds, but less or no overlap is desirable at idle speed. VVT can control the time and the amount precisely.

A Unique but Popular Fuel Injection System: Central Sequential Fuel Injection

Central sequential fuel injection (CSFI) was implemented on trucks beginning in 1996. The system has an electrical injector for each cylinder that feeds a **poppet nozzle** in the fuel meter assembly. The fuel meter assembly also includes the fuel pressure regulator

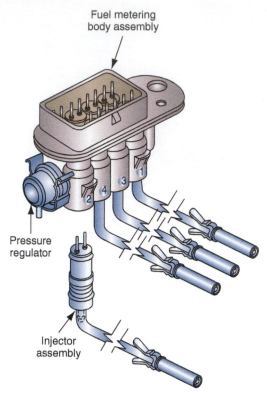

Fuel metering
body assembly

Pressure
regulator

Injector
assembly

Figure 8-39 Central sequential fuel injection.

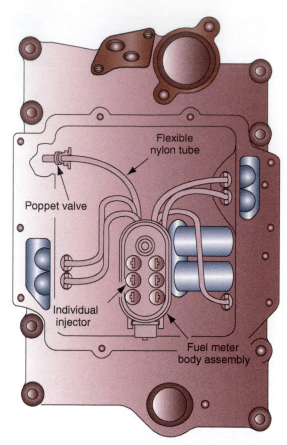

Flexible
nylon tube

Poppet valve

Individual
injector

Fuel meter
body assembly

Figure 8-40 A CPFI fuel meter and poppet nozzles from the lower half of the intake manifold.

(**Figure 8-39**). The electrical injector is fired from the fuel meter assembly, which forces the poppet nozzle check ball off its seat, spraying fuel into the cylinder. The separate electrical injectors allow the injection to be sequential instead of multiport. **Figure 8-40** shows the CSFI assembly located inside the lower intake manifold.

Intake Manifold Tuning Valve

The two-piece aluminum intake manifold has an integral throttle body. An **intake manifold tuning valve (IMTV)** assembly is mounted in the top of the intake manifold. The IMTV assembly contains an electric solenoid that operates a rectangular-shaped valve inside the manifold. Two zip tubes are connected from the throttle body to dual plenums in the upper half of the intake manifold (**Figure 8-41**). Each plenum feeds three tuned intake runners.

The IMTV rectangular valve is mounted in an opening between the two halves of the upper intake manifold. While the engine is operating at lower rpm, the IMTV remains closed, and airflow in the two halves of the upper intake is separated.

When the engine rpm reaches 3,025 to 4,650 and the throttle opening exceeds 36 percent, the PCM energizes the IMTV solenoid. This action opens the rectangular valve, and the two halves of the intake are now joined through this valve opening. Airflow in the intake increases to improve horsepower and torque.

One end of the IMTV relay winding is connected to the ignition switch and the other end to the PCM. When the engine operating conditions require IMTV operation, the PCM grounds the IMTV relay winding. This action closes the relay points. Current flows through the IMTV relay points to the IMTV solenoid winding in the intake manifold to open the rectangular valve (**Figure 8-42**).

The **intake manifold tuning valve (IMTV)** makes the intake manifold an "active" manifold; that is, it responds to changing driver demands.

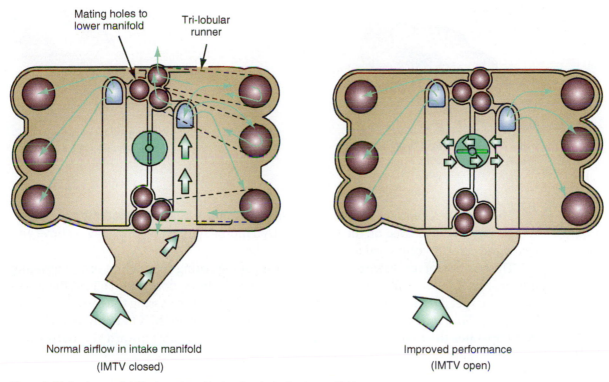

Normal airflow in intake manifold
(IMTV closed)

Improved performance
(IMTV open)

Figure 8-41 Intake manifold tuning valve with zip tubes in the intake manifold.

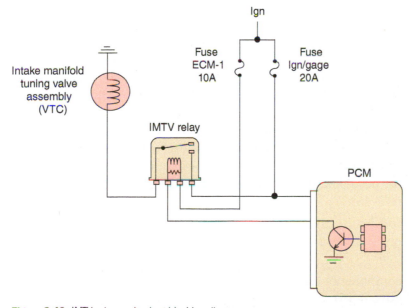

Figure 8-42 IMTV relay and solenoid wiring diagram.

IDLE SPEED CONTROL SYSTEMS

Idle speed control is a function of the PCM. Based on operating conditions and inputs from various sensors, the PCM regulates the idle speed to control emissions.

Idle Air Control Bypass Air (IAC BPA) Motors

Older PFI systems have an idle air control bypass (IAC BPA) motor in the throttle body to control idle speed. In any of these systems, the IAC BPA motor is threaded or bolted into the throttle body assembly (**Figure 8-43** and **Figure 8-44**).

The IAC is used on many port fuel EFI systems.

Figure 8-43 IAC BPA motor that threads into a throttle body.

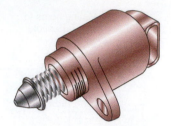

Figure 8-44 IAC BPA valve that bolts onto a throttle body.

The IAC BPA motor contains a reversible DC motor that is operated by the PCM. A tapered valve is located on the end of the IAC motor stem with air passages above and below the throttle connected to this tapered valve.

The position of this valve controls idle speed by regulating the amount of air bypassing the throttle. If the motor opens the tapered valve and allows more air to bypass the throttle, engine speed is increased. Decreasing the amount of air bypassing the throttle reduces idle speed. Four wires from the motor are connected to the PCM. On many systems, the PCM operates the motor to move the plunger and tapered valve inward and outward in steps or counts.

Throttle Actuator Control System

As technology advances, vehicle systems will begin to use electronics rather than mechanical devices to control outputs. This is commonly referred to as drive by wire. In other words, various systems are controlled electronically rather than mechanically. Emerging systems include brake-by-wire and steer-by-wire. One example of this is the cable less accelerator or the **throttle actuator control (TAC)** system. The driver depresses the accelerator pedal, and a DC motor on the throttle body instantly moves the throttle plate, causing acceleration. Depending on the driver's command, the DC motor will move or maintain the throttle plate in any position. The TAC system uses various engine and other vehicle inputs to determine and control the position of the throttle opening by regulating the angle of the throttle blade. This system does not use a mechanical cable attached to the accelerator pedal and throttle body. A typical TAC system has the following components (**Figure 8-45**):

Shop Manual
Chapter 8, page 397

- Throttle body with a throttle position sensor and DC motor unit
- Accelerator pedal position (APP) sensor
- TAC module

All the components are integrated as one system to accurately control the throttle position (**Figure 8-46**).

Accelerator Pedal Position (APP) Sensor. The APP is mounted on the accelerator pedal assembly and relays the driver's inputs for throttle control. Multiple signals generated from the single sensor must coordinate to determine exact pedal movement. This prevents one signal from causing an accelerator system malfunction, such as unintended acceleration. Three signals are produced from a single sensor. The APP sensor number one signal voltage should increase at the same time the accelerator is depressed from under 1 volt at closed throttle (0 percent pedal travel) to over 2 volts at wide-open throttle (100 percent pedal travel). The APP sensor number two signal voltage should decrease from over 4 volts at 0 percent pedal travel to less than 2.9 volts at 100 percent pedal travel. APP sensor number three signal voltage should decrease from over 3.8 volts at 0 percent pedal travel to less than 3.1 volts at 100 percent pedal travel.

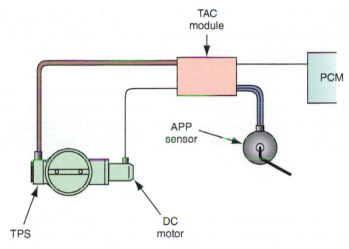

Figure 8-45 Typical throttle actuator control (TAC) system arrangement.

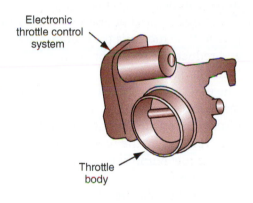

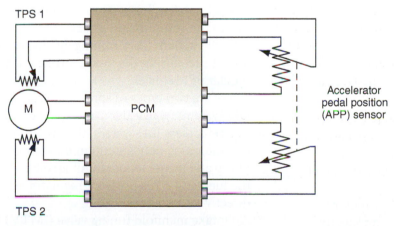

Figure 8-46 The PCM monitors several signals simultaneously.

Throttle Body Assembly. The throttle body unit used in the TAC system is similar to other types of throttle bodies. However, a DC motor is mounted on the throttle body because there is no mechanical throttle cable to the accelerator pedal to move the throttle plate. Another difference is the TP sensor. Instead of a traditional TP sensor, the TAC system uses a unique dual signal sensor. The TP sensor actually consists of two sensors in one housing. Separate 5-volt and low-reference circuits are used and are connected to the TAC control module. Like the APP sensor multiple signals, they must agree. The TP sensor number one signal voltage increases at the same time the throttle opens. The voltage increases from about 1 volt at 0 percent throttle to over 3.5 volts at 100 percent throttle. TP sensor number two signal voltage decreases from about 3.8 volts at 0 percent to less than 1 volt at 100 percent throttle at the same time the throttle opens.

TAC Control Module. The TAC module is the main control unit for the electronic throttle system. The PCM and TAC module communicate on a dedicated circuit. They monitor the commanded throttle position and compare it to the actual throttle position. This is accomplished by monitoring the APP and TPS. These two values must agree and be within a calibrated value of each other.

SUMMARY

- Two types of EFI systems are in common use today: port fuel injection and direct fuel injection. In the port fuel injection system, there is one injector at each cylinder. The DFI system injects the fuel directly into the combustion chamber.
- The EFI system includes a fuel delivery system, system sensors, electronic control unit, and fuel injectors.
- The MAF sensor determines the mass of air entering the engine.
- The MAP sensor measures changes in the intake manifold pressure that results from changes in engine load and speed.
- The heart of the fuel injection system is the electronic control unit. The PCM or ECM receives signals from all the system sensors, processes them, and transmits programmed electrical pulses to the fuel injectors.
- Two types of port fuel injectors are currently in use: top feed and bottom feed. Both injectors are used in port fuel injection systems.
- Two methods are used to control idle speed on engines equipped with fuel injection: an auxiliary air valve and throttle actuator control (TAC).
- Many of the same input sensors are used on DFI systems and PFI systems.
- In a speed density EFI system, the computer uses the MAP, IAT, ECT, and engine rpm inputs to calculate the amount of air entering the engine. The computer then calculates the required amount of fuel to go with the air entering the engine.
- In any EFI system, the fuel pressure must be high enough to prevent fuel boiling.
- In an EFI system, the computer supplies the proper air-fuel ratio by controlling injector pulse width.
- In an EFI system, the computer increases injector pulse width to provide air-fuel ratio enrichment while starting a cold engine.

- In a PFI system with sequential injection, each injector has an individual ground wire connected to the computer.
- Compared to steel or aluminum alloy fuel rails, plastic-type fuel rails transfer less heat to the fuel and reduce the possibility of fuel boiling in the fuel rail.
- The pressure regulator on a return style system maintains the specified fuel system pressure and returns excess fuel to the fuel tank.
- In a DFI system, the fuel pressure from the electric fuel pump is increased by a mechanical fuel pump.
- Fuel pressure may be as high as 2,200 psi in a DFI system.
- The DFI system has a stratified mode that injects fuel in the compression stroke.
- The DFI system has a homogeneous mode that injects fuel during the intake stroke much like a regular PFI vehicle.
- There are two types of spray techniques used on DFI systems: wall guided and spray guided.
- In a returnless fuel system, the pressure regulator and filter assembly are mounted with the fuel pump and gauge sending unit assembly on top of the fuel tank. This pressure regulator returns fuel directly into the fuel tank.
- An intake manifold tuning valve (IMTV) controls the air passages inside the intake manifold to provide improved airflow in the manifold.
- An IAC BPA motor controls idle speed by controlling the amount of air bypassing the throttle.
- An IAC BPA valve opens and closes the air bypass passage around the throttle to control idle speed.
- The TAC system is used to control throttle position electronically. TAC uses an accelerator pedal position (APP) sensor to relay the driver's inputs to the PCM.

REVIEW QUESTIONS

Short-Answer Essays

1. Explain the advantages of a *returnless* style fuel system.

2. Explain the DFI stratified mode of operation.

3. Describe how variable valve timing works to improve engine performance.

4. Explain the difference between port fuel injection (PFI) and direct fuel injection (DFI).

5. Describe the meaning of pulse width.

6. Explain the purpose of a coolant temperature sensor on an EFI system.

7. Explain the purpose of the APP sensor.

8. What is meant by sequential firing of fuel injectors?

9. Describe the purpose of a manifold absolute pressure (MAP) sensor.

10. Explain the speed density method of intake air measurement.

Fill-in-the-Blanks

1. _____ is the amount of time an injector is open and injecting fuel.

2. When the IAC BPA motor increases the amount of air bypassing the throttle, engine rpm is _____.

3. In a(n) _____ EFI system, the computer uses the MAP and engine rpm inputs to calculate the amount of air entering the engine. The computer then calculates the required amount of fuel to go with the air entering the engine.

4. Port fuel injection systems use one injector for each _____.

5. To control the proportion of fuel to air in the air-fuel charge, the fuel system must be able to measure the amount of _____ entering the engine.

6. The MAF (sensor) determines the _____ of air entering the engine.

7. The _____ receives and processes signals from all the system sensors and then transmits programmed electrical pulses to the fuel injectors.

8. On some EFI systems, if the _____ is not firing, the computer stops operating the injectors.

9. With the _____ fuel system, only the fuel needed by the engine is filtered, thus allowing the use of a smaller fuel filter.

10. On a direct fuel injection (DFI) vehicle, fuel is injected into the _____ _____.

Multiple Choice

1. *Technician A* says that PFI sprays fuel just behind the intake valve.
   Technician B says that DFI systems spray fuel behind the exhaust valve.
   Who is correct?
   A. Technician A
   B. Technician B
   C. Both technicians
   D. Neither technician

2. *Technician A* says that DFI allows for lean air-fuel mixtures that produce lower emissions.
   Technician B says that DFI has good performance but sacrifices fuel economy.
   Who is correct?
   A. Technician A
   B. Technician B
   C. Both technicians
   D. Neither technician

3. *Technician A* says that on a Chrysler PFI system on a 3.5-liter engine each injector has its own separate ground wire from the PCM.
   Technician B says that the 3.5-liter Chrysler engine uses an ASD relay to supply power to the injectors.
   Who is correct?
   A. Technician A
   B. Technician B
   C. Both technicians
   D. Neither technician

4. *Technician A* says that DFI systems have higher fuel pressures than a typical PFI system.
   Technician B says that DFI systems have a maximum of 1,000 psi.
   Who is correct?
   A. Technician A
   B. Technician B
   C. Both technicians
   D. Neither technician

5. All of the following are true of MAP sensors *except*:
   A. Measure changes in manifold pressure that result from changes in engine speed and load
   B. High MAP value at closed throttle
   C. MAP value that is opposite of a vacuum gauge reading
   D. MAP sensor used to adjust for different altitudes

6. *Technician A* says that when conditions such as starting or wide-open throttle demand that the signals from the oxygen sensor be ignored, the system operates in open loop.

   Technician B says that when the PCM is using feedback from the oxygen sensor to control fuel injection, it is in closed loop.

   Who is correct?

   A. Technician A
   B. Technician B
   C. Both technicians
   D. Neither technician

7. *Technician A* says that most port fuel injection systems supply the injectors a full battery voltage.

   Technician B says that the port fuel injector ground wire is routed to the PCM.

   Who is correct?

   A. Technician A
   B. Technician B
   C. Both technicians
   D. Neither technician

8. While discussing TAC systems,

   Technician A says that an A/C electric motor is used to move the throttle instead of a cable.

   Technician B says that the TP sensor used in TAC systems is actually four sensors built into one housing.

   Who is correct?

   A. Technician A
   B. Technician B
   C. Both technicians
   D. Neither technician

9. *Technician A* says that variable valve timing engines use more valve overlap at idle speeds.

   Technician B says that all manufacturers only use an actuator on the exhaust camshaft for variable valve timing.

   Who is correct?

   A. Technician A
   B. Technician B
   C. Both technicians
   D. Neither technician

10. *Technician A* says that one type of DFI is called wall guided.

    Technician B says another type of DFI is called spray guided.

    Who is correct?

    A. Technician A
    B. Technician B
    C. Both technicians
    D. Neither technician

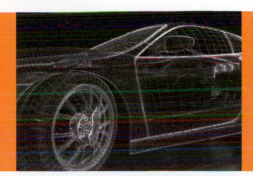

CHAPTER 9

IGNITION SYSTEMS

Upon completion and review of this chapter, you should be able to:

- Explain the purpose of the electronic control unit.
- Describe the various types of spark timing systems, including electronic switching systems and their related engine position sensors.
- Explain how the fuel injection system may rely on components of the ignition system.
- Describe the operation of a distributor ignition (DI) system.
- Describe the advantages and operation of distributorless electronic ignition (EI) systems.
- Describe the secondary ignition wiring connections on an EI system, including an explanation of how the spark plugs fire.
- Explain the coil and injector sequencing on a Chrysler system while cranking the engine.
- Describe the operating principles of a coil-on-plug (COP) system.
- Explain the basic operation of ion-sensing ignition.
- Explain the basic operation of compression sense ignition.

Terms To Know

Coil-near-plug (CNP)	Compression sense ignition	ion-sensing ignition
Coil-on-plug (COP)	Distributorless ignition system	Waste spark
Coil pack		

INTRODUCTION

The purpose of the ignition system is to supply the spark production, spark timing control, and spark distribution. The ignition system functions similarly in all engines because a spark will arc across the gap of the spark plug at the right moment, igniting the mixture in the combustion chamber. The delivery of the spark is the function of the secondary ignition system, while the production and timing of the spark is the function of the primary ignition circuit. These principles have not changed in decades. What *has* changed are the components that make these events occur and how they operate. Ignition systems have progressed from fully mechanical breaker point systems with vacuum and mechanically controlled timing advance units to systems that have no moving parts and are purely electronically operated. In this chapter, we will explore the operation and design of distributor ignition (DI) and distributorless electronic ignition (EI) systems.

EI and DI systems do, of course, share many elements. The primary circuit, secondary circuit, and the ignition coil work basically the same way they do in a distributor ignition, with some variations. The module still needs a trigger to tell it when to fire, the module still controls the primary circuit, and the secondary circuit still provides the voltage to fire the spark plug. If you have a good understanding of the fundamentals of the ignition system, DI and EI systems should not be difficult.

The powertrain control module (PCM) controls timing advance using its array of sensors such as the throttle position (TP) sensor, mass air flow (MAF) sensor, manifold absolute pressure (MAP) sensor, engine coolant temperature (ECT) sensor, intake air temperature (IAT) sensor, and others that result in a system that is efficient and provides the most power possible.

Most ignition systems function the same when taken to a very basic level. They will all use components of a similar type, but specific to the manufacturer or application. There are two electrical sides to the ignition coil that are referred to as the primary and the secondary. The primary or low side of the ignition circuit becomes saturated and builds a strong magnetic field around both the primary and secondary windings of the ignition coil. Spark production occurs as a function of the secondary or high-voltage side of the ignition, which is induced into the many turns of the secondary windings by the collapse of the primary windings when the ignition primary circuit is opened. This opening of the primary coil circuit causes a surge of up to 400 volts in the primary windings. Most secondary coils have a voltage potential ranging from 25,000 to 60,000 volts. The typical system operating voltage required to fire the spark plugs at idle will usually range from 8,000 to 12,000 volts. The remaining coil capacity is used as a reserve for when certain operating conditions occur, such as climbing a hill, pulling a trailer, passing another vehicle, and cold engine starts. Some of these conditions may demand the coil's entire potential.

The 12 volts in the primary can surge as high as 400 volts when the primary circuit is turned off, due to the inductive kick.

AUTHOR'S NOTE The ignition system is one of the most critical engine systems because of what it must accomplish to complete the combustion process. It must be able to transform 12 volts into a potential of 60,000 volts, hundreds of times per second, in all kinds of operating conditions and temperatures. Regardless of the type of system, the end result is to fire the correct cylinder at precisely the right time and for the correct duration. Distributor systems, electronic or distributorless systems, and coil-on-plug systems all accomplish the same goal. Get to know the system being worked on. Refer to the service information, and do not be intimidated by a new or unfamiliar system, because they all work using the same basic principles. And when in doubt, don't guess—test it!

PURPOSE OF THE IGNITION SYSTEM

Although today's ignition systems look different compared to the breaker-point-controlled systems used for over 60 years, their purpose is the same. The ignition system has three main jobs:

1. *Spark production.* The ignition system must be able to quickly build high voltage sufficient to satisfy the requirements to ignite the air-fuel mixture and maintain adequate burn time for complete combustion.
2. *Spark timing control.* The ignition system must be able to alter the delivery time of the spark to account for rpm, varying load, and demand conditions.
3. *Spark distribution.* The ignition system must be able to deliver a spark to the correct cylinder at the right time during the compression stroke to begin the combustion process.

When the combustion process is completed, a very high pressure is exerted against the top of the piston. This pressure pushes the piston down on its power stroke. This pressure is the force that gives the engine power. For an engine to produce the maximum

Basically, the ignition system needs to be able to deliver a spark that lasts long enough, that has enough voltage, and is delivered at the right time.

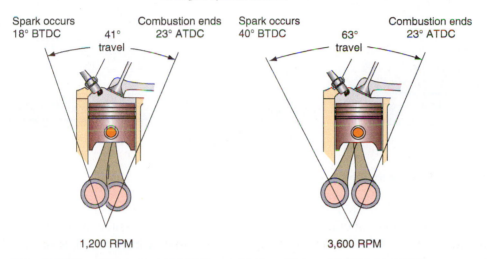

Figure 9-1 With an increase in speed, ignition must begin earlier to end by 23 degrees ATDC.

amount of power that it can, the maximum pressure from combustion should be present when the piston is at 10 to 20 degrees after top dead center (ATDC). Because combustion of the air-fuel mixture within a cylinder takes a short period of time, usually measured in thousandths of a second (milliseconds), the combustion process must begin before the piston is on its power stroke. Therefore, the delivery of the spark must be timed to arrive at some point before the piston reaches top dead center (TDC) in the compression stroke.

Determining how much before top dead center (BTDC) the spark should begin is complicated by the fact that the speed of the piston, as it moves from its compression stroke to its power stroke, increases, and the time needed for combustion stays about the same. This means that the spark should be delivered earlier as the engine's speed increases (**Figure 9-1**). However, as the engine has to provide more power to do more work, the load on the crankshaft tends to slow down the acceleration of the piston and the spark needs to be somewhat delayed.

Figuring out when the spark should begin becomes more complicated with the fact that the rate of combustion varies according to certain factors. Higher compression pressures tend to speed up combustion. Higher octane gasoline ignites less easily and requires more burning time. Increased vaporization and turbulence tend to decrease combustion times. Other factors, including intake air temperature, humidity, and barometric pressure, also affect combustion. Because of all of these complications, delivering the spark at the right time is a difficult task.

> Top dead center (TDC) is the precise position of the piston at the top of its travel in the cylinder as it transitions from an upward to a downward movement. The piston is stationary, or dead, during that transition. Any measurement after TDC is referred to as after top dead center (ATDC).

IGNITION TIMING

Ignition timing refers to the precise time that spark occurs. Ignition timing is specified by referring to the position of the number one piston in relation to crankshaft position (CKP) in degrees relative to TDC. On DI systems, ignition timing reference marks may be located on engine parts and on a pulley or flywheel to indicate the position of TDC number one cylinder (**Figure 9-2**). These reference marks were used by technicians to fine-tune the position of the distributor, and hence the timing, as the distributor was a mechanical correlation between the camshaft and crankshaft positions.

On older DI systems, that had adjustable timing, when the timing marks are aligned at TDC, or 0, the piston in cylinder one is at TDC. Additional numbers on a scale indicate

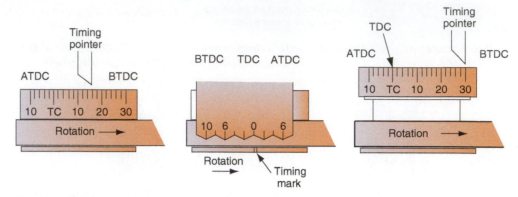

Figure 9-2 Reference timing marks used on some vehicles.

the number of degrees of crankshaft rotation before top dead center (BTDC) or after top dead center (ATDC). In a majority of engines, the initial or base timing is specified at a point between TDC and 20 degrees BTDC on the compression stroke. The ignition system will add timing to the initial or base timing to set it where it can take advantage of the current engine situation, from hard acceleration, to heavy loads, to idle. Remember that the crankshaft must turn twice to every turn of the camshaft, so 0 degrees is TDC, but it can be either on the compression or exhaust stroke. The position of the valves, or checking for compression itself by turning the engine over slowly, might be necessary to determine which stroke the cylinder is on before installing the distributor in the engine. Information on setting base timing can be found in the Shop Manual of this text.

If optimum engine performance is to be maintained, the ignition timing of the engine must change as the operating conditions of the engine change. All of the different operating conditions affect the speed of the engine and the load on the engine. All ignition timing changes are made in response to these primary factors, based on inputs from the engine sensors, such as the MAF, MAP, IAT, ECT, and RPM inputs.

> Any measurement, usually in degrees, prior to the TDC point is referred to as before top dead center (BTDC).

> Ignition systems rely on the inputs from various sensors to control ignition timing.

Engine Speed

At higher engine speeds, the crankshaft turns through more degrees in a given period of time. If combustion is to be completed by 23 degrees ATDC, ignition timing must occur sooner or has to be "advanced." Of course, these figures will vary slightly with different engine designs.

However, air-fuel mixture turbulence increases with rpm. This causes the mixture inside the cylinder to burn faster. Increased turbulence requires the ignition to be tweaked some, depending on the cylinder design.

These two factors must be balanced for best engine performance. Therefore, while the ignition timing must be advanced as engine speed increases, the amount of advance must be decreased to compensate for the increased turbulence.

Engine Load

The load on an engine is related to the work it must do. Driving up hills or pulling extra weight increases engine load. Under load, the pistons accelerate slower and the engine runs less efficiently. A good indication of engine load is the amount of vacuum formed during the intake stroke.

Under light loads and with the throttle plate(s) partially opened, such as during steady driving, a high vacuum exists in the intake manifold. The amount of air-fuel mixture drawn into the manifold and cylinders is small. On compression, this thin mixture produces less combustion pressure and combustion time is slow. To complete combustion by

23 degrees ATDC, ignition timing must be advanced. This advance also improves fuel economy.

Under heavy loads, when the throttle is opened fully, a larger mass of air-fuel mixture can be drawn in and the vacuum in the manifold is low. High combustion pressure and rapid burning result. In such a case, the ignition timing must be slowed to prevent complete burning from occurring too soon in the combustion cycle.

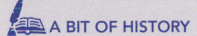

A BIT OF HISTORY

The man credited with the development of the ignition system as we know it today was Charles Kettering. He is also credited with the development of electric starters for Cadillacs in 1911, quick-drying paint, and ethyl gasoline. Kettering was one of the founders of Dayton Engineering Laboratories Company in Dayton, Ohio. This later became known as Delco Electronics, a division of Delphi.

> Engine vacuum is formed during the intake stroke of the cylinder. Engine vacuum is any pressure lower than atmospheric pressure.

> Basic ignition timing is typically advanced with increases in engine speed and slowed with heavy engine load.

Firing Order

Up to this point, the primary focus of discussion has been ignition timing as it relates to any one cylinder. However, the function of the ignition system extends beyond timing the arrival of a spark to a single cylinder. It must perform this task for each cylinder of the engine in a specific sequence.

Each cylinder of an engine produces power once every 720 degrees of crankshaft rotation. Each cylinder must have a power stroke at its own appropriate time during the rotation. To make this possible, the pistons and rods are arranged in a precise fashion. The firing order is arranged to reduce rocking and imbalance problems. Because the potential for this rocking is determined by the design and construction of the engine, the firing order varies from engine to engine. Vehicle manufacturers simplify identifying each cylinder identification by numbering them (**Figure 9-3**). Regardless of the particular firing order used, the number one cylinder always starts the firing order, with the rest of the cylinders following in a fixed sequence.

The ignition system must be able to monitor the rotation of the crankshaft and the relative position of each piston to determine which piston is on its compression stroke. It must also be able to deliver a high-voltage surge to each cylinder at the proper time during its compression stroke. How the ignition system accomplishes this depends on the design of the system. In distributor ignitions, the relationship is mechanical. The distributor is indexed and is driven by the camshaft. The distributor turns and delivers spark to the correct cylinder at the correct time. The electronic ignition systems deliver spark based on signals from the camshaft and crankshaft sensors.

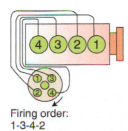

Firing order:
1-3-4-2

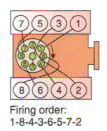

Firing order:
1-8-4-3-6-5-7-2

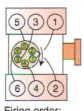

Firing order:
1-4-3-6-5-2

Figure 9-3 Cylinder numbering and firing orders.

BASIC IGNITION SYSTEM

Manufacturers use the electrical principles of induction to accomplish spark production. All ignition systems consist of two interconnected electrical circuits: a primary (low-voltage) circuit and a secondary (high-voltage) circuit (**Figure 9-4**). Depending on the exact type of ignition system, components in the primary circuit include the following:

- Battery
- Ignition switch
- Ignition coil primary
- Triggering device
- Control module

The ignition coil is actually the point where the primary and secondary circuits meet. They are brought together and then separated. The ignition coil is both a primary and a secondary ignition component.

The secondary circuit includes the following components:

- Ignition coil secondary winding
- Distributor cap and rotor (some systems)
- Ignition (spark plug) cables (some systems)
- Spark plugs

Primary Circuit Operation

The ignition coil works on the principle of magnetic induction to change 12 volts in the primary to 60,000 volts or more in the secondary. The primary windings in the coil are relatively large and few in number compared to the secondary windings. The secondary windings are smaller, and there are many more of them. When the primary circuit is switched on, the primary windings build a magnetic field. When the primary circuit is opened, the magnetic field that has been surrounding both sets of windings collapses. This collapse of the magnetic field creates a high-voltage surge in the secondary, which

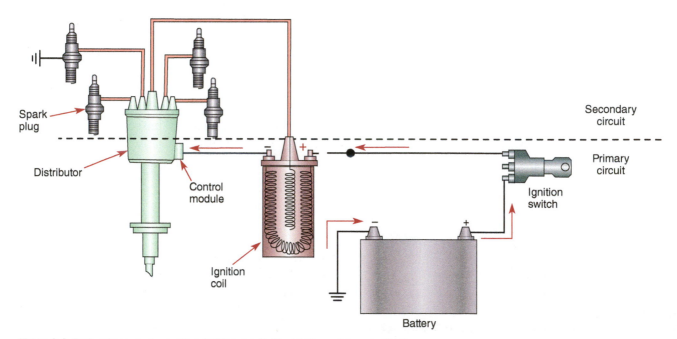

Figure 9-4 Basic primary and secondary ignition circuits. The ignition coil is part of both.

provides the spark. Controlling the precise time the coil switches on or off is the function of the ignition module and/or ECM depending on the design of the system, because it electronically controls the coil ground connection. The positive lead is connected from the ignition switch. The ignition module or ECM is informed when to switch by a triggering device signal or the PCM depending on operating conditions. The amount of time the coil is turned on is called "dwell time," and the time that the coil is switched off is determined by the ignition timing. The triggering device for a DI system is generally a Hall-effect sensor, or a magnetic reluctance sensor, and is generally indexed on the distributor shaft itself.

Dwell

Dwell time is the time that the ignition coil is turned on. Dwell time is important to the operation of the ignition system. If the dwell time is too short, the coil does not have time to saturate. If it is too long, the coil and module can overheat.

Shop Manual
Chapter 9, page 490

Secondary Circuit Operation

The secondary circuit carries high voltage to the spark plugs. The exact manner in which the secondary circuit delivers these high-voltage surges depends on the system design.

In a distributor ignition (DI) system, high voltage from the secondary winding passes through an ignition cable running from the coil to the distributor. The distributor then distributes the high voltage to the individual spark plugs through a set of ignition cables. The cables are arranged in the distributor cap according to the firing order of the engine. A rotor driven by the distributor shaft rotates and completes the electrical path from the secondary winding of the coil to the individual spark plugs. The distributor delivers the spark to match the compression stroke of the piston. The distributor assembly may also have the capability of advancing or slowing ignition timing.

In the SAE J1930 terminology, the term *distributor ignition* replaces all previous terms for electronically controlled distributor-type ignition systems.

The distributor cap is mounted on top of the distributor assembly, and an alignment notch in the cap fits over a matching lug on the housing (**Figure 9-5**). Therefore, the cap can be installed in only one position, thereby ensuring the correct firing sequence.

The rotor is positioned on top of the distributor shaft, and a projection or other keyway inside the rotor fits into a slot in the shaft. This allows the rotor to be installed in only one position. A metal strip on the top of the rotor makes contact with the center distributor cap terminal, and the outer end of the strip passes by the cap terminals as it rotates. This action completes the circuit between the ignition coil and the individual spark plugs according to the firing order.

Shop Manual
Chapter 9, page 499

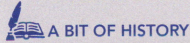

A BIT OF HISTORY

A turning point in the automotive industry came in 1902 when the Humber, an automobile from England, used a magneto electric ignition system.

IGNITION COIL ACTION

All ignition systems share a number of common components. Some, such as the battery and ignition switch, perform simple functions. The battery supplies battery voltage to the ignition primary circuit. Current flows when the ignition switch is in the "start" or the "run" position.

To generate a spark to begin combustion, the ignition system must deliver high voltage to the spark plugs. Because the amount of voltage required to bridge the gap of the spark

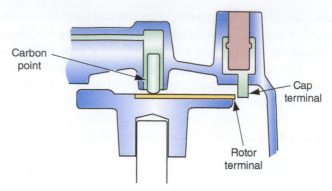

Figure 9-5 Relationship of a rotor and distributor cap.

plug varies with the operating conditions, most late-model vehicles can easily supply 25,000 to 60,000 volts to force a spark across the air gap. Because the battery delivers 12 volts, a method of stepping up the voltage must be used. Multiplying battery voltage is the job of a coil.

The ignition coil is a pulse transformer. It transforms battery voltage into short bursts of high voltage. As explained previously, when a wire is moved through a magnetic field, voltage is induced in the wire. The inverse of this principle is also true—when a magnetic field moves across a wire, voltage is induced in the wire. If a wire is bent into loops forming a coil and a magnetic field is passed through the coil, an equal amount of voltage is generated in each loop of wire. The more loops of wire in the coil, the greater the total voltage induced.

Also, the faster the magnetic field moves through the coil, the higher the voltage induced in the coil. If the speed of the magnetic field is doubled, the voltage output doubles.

An ignition coil uses these principles and has two coils of wire wrapped around an iron core. An iron or steel core is used because it has low inductive reluctance. The primary coil winding is normally composed of 100 to 200 turns of 20-gauge wire. This coil of wire conducts battery current. When a current is passing through the primary coil, it magnetizes the iron core. The strength of the magnet depends directly on the number of wire loops and the amount of current flowing through those loops. The secondary coil of wires may consist of 15,000 to 25,000 or more turns of very fine copper wire.

Because of the effects of counter-electromotive force (EMF) on the current flowing through the primary winding, it takes some time for the coil to become fully magnetized or saturated. Therefore, current flows in the primary winding for some time between firings of the spark plugs. The period of time that there is primary current flow is called "dwell." The length of the dwell period is important.

When current flows through a conductor, it will immediately reach its maximum value as allowed by the resistance in the circuit. If a conductor is wound into a coil, maximum current will not be immediately present. As the magnetic field begins to form, the magnetic lines of force cross over each other. This tends to cause an opposition to current flow, an occurrence called "reactance." Reactance causes a temporary resistance to current flow and delays the flow of current from reaching its maximum value. When maximum current flow is present in a winding, the winding is said to be saturated and the strength of its magnetic field will also be at a maximum.

Saturation Time

The wires in the coil are wound around an iron core because the iron is a better conductor of the magnetic field than air. The better the magnetic field is conducted, the stronger the field can become using the same amount of power.

Saturation takes time, because as the magnetic field builds around each wire in the coil, the magnetic fields oppose each other, called counter-electromotive force. When the field is as strong as possible for the applied current, the field stops growing, or is saturated; the counter-electromotive force equals the current applied to the coil.

Saturation can occur only if the dwell period is long enough to allow for maximum current flow through the primary windings. A less-than-saturated coil will not be able to produce the voltage it was designed to produce. If the energy from the coil is too low, the spark plugs may not fire long enough or not fire at all.

When the primary coil circuit is suddenly opened, the magnetic field instantly collapses. The sudden collapsing of the magnetic field produces a very high voltage in the primary coil and a much higher voltage in the secondary windings. This high voltage is used to push current across the gap of the spark plug. **Figure 9-6** is a simplistic rendering of the coil's primary and secondary circuits.

> Saturation time is the time it takes for the coil to be saturated. Saturation is the point of maximum current flow in the coil's windings.

> **Shop Manual**
> Chapter 9, page 495

Ignition Coil Construction

A laminated soft iron core is positioned at the center of the ignition coil, and an insulated primary winding is wound around this core. The primary winding has approximately 200 turns of heavy wire, and the two ends of this winding are connected to the primary terminals on top of the coil (**Figure 9-7**). An enamel-type insulation prevents the primary windings from touching each other. Paper insulation is also placed between the layers of windings.

Secondary coil windings are made of very fine wire. These windings are on the inside of the primary winding. A similar insulation method is used on the secondary and primary windings. The ends of the secondary winding are usually connected to one of the primary terminals and to the high-tension terminal in the coil tower. When the windings and core

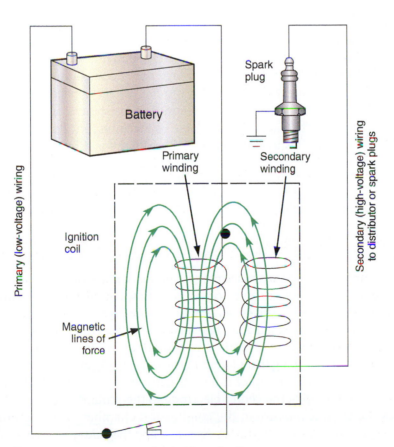

Figure 9-6 Primary and secondary ignition coil action.

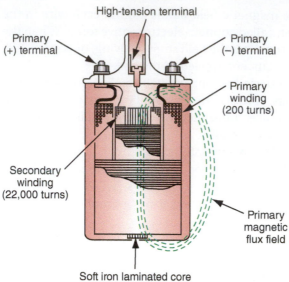

Figure 9-7 Cutaway of a typical ignition coil.

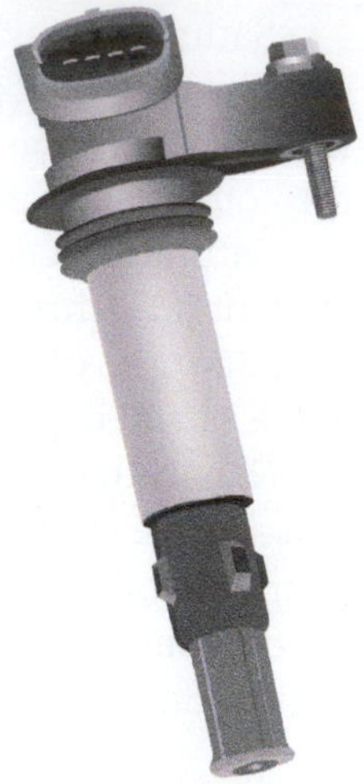

Figure 9-8 A coil over plug coil assembly (with spark plug boot).

The negative terminal of an ignition coil on distributor systems is commonly referred to as the tachometer (Tach) terminal.

Shop Manual
Chapter 9, page 494

assembly are mounted in the coil container, the core rests on a ceramic insulating cup in the bottom of the container. Metal sheathing is placed around the outside of the coil windings in the container. This sheathing concentrates the magnetic field on the outside of the windings. A sealing washer is positioned between the tower and the container to seal the unit. Coils used on COP ignitions are much smaller, as they have to power only one spark plug (**Figure 9-8**).

Secondary Voltage

The typical amount of secondary coil voltage required to jump the spark plug gap is 10,000 to 12,000 volts. Most ignitions have a secondary voltage potential of at least 25,000 to 60,000 volts. The difference between the required voltage and the maximum available voltage is referred to as secondary reserve voltage. This reserve voltage is necessary to compensate for high cylinder pressures and increased secondary resistances as the spark plug gap increases through wear. The maximum available voltage must always exceed the required firing voltage or ignition misfire will occur. If there is an insufficient amount of voltage available to push current across the gap, the spark plug will not fire. It is important to note that a decrease in primary voltage will decrease the output of the secondary. Normally a decrease of 1 volt in the primary will cause a decrease of 5 kV in the secondary.

SPARK PLUGS

The spark plug electrodes provide a gap inside each combustion chamber across which the secondary current flows to ignite the air-fuel mixture in the combustion chambers.

Every type of ignition system uses spark plugs. The spark plugs provide the crucial air gap across which the high-voltage current from the coil flows in the form of an arc. The three main parts of a spark plug are the steel core, the ceramic core or insulator that acts as a heat conductor, and a pair of electrodes, one insulated in the core and the other grounded on the shell. The shell holds the ceramic core and electrodes in a gas-tight assembly and has the threads needed for plug installation in the engine (**Figure 9-9**). An ignition cable connects the secondary to the top of the plug. Current flows through the center of the plug and arcs from the tip of the center electrode to the ground electrode. The resulting spark ignites the air-fuel mixture in the combustion chamber. Most automotive spark plugs also have a resistor between the top terminal and the center electrode. This resistor reduces electromagnetic interference (EMI), which prevents noise on stereo equipment.

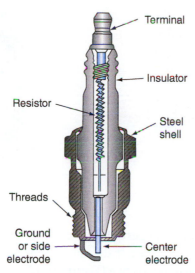

Figure 9-9 Components of a typical spark plug.

Voltage peaks from EMI could also interfere with or damage onboard computers. A resistor is built into the spark plug and, like all other resistances in the secondary, increases the voltage needed to jump the gap of the spark plug. However, it also reduces amperage, extending plug life.

Spark plug reach describes the electrodes, position in the combustion chamber.

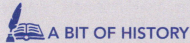 **A BIT OF HISTORY**

Louis S. Clark of the Auto Car Company designed the porcelain spark plug insulation in 1902.

Spark plugs come in many different sizes and designs to accommodate different engines. To fit properly, spark plugs must be of the proper size and reach. Another design factor that determines the usefulness of a spark plug for a specific application is its heat range. The spark plug reach describes how deep the electrode extends into the combustion chamber. The desired heat range depends on the design of the engine and on the type of driving conditions the vehicle is subject to.

The heat range of a spark plug is a statement of how well a spark plug can conduct heat away from its tip.

A terminal post on top of the center electrode is the point of contact for the spark plug cable. The center electrode, commonly made of a copper alloy, is surrounded by a ceramic insulator, and a copper and glass seal is located between the electrode and the insulator. These seals prevent combustion gases from leaking out of the cylinder. Ribs on the insulator increase the distance between the terminal and the shell to help prevent electric arcing on the outside of the insulator. The steel spark plug shell is crimped over the insulation, and a ground electrode, on the lower end of the shell, is positioned directly below the center electrode. There is an air gap between these two electrodes, and the width of this air gap is specified by the auto manufacturer.

The center electrode usually has platinum or iridium tips on modern spark plugs.

A spark plug socket may be placed over a hex-shaped area near the top of the shell for plug removal and installation. Threads on the lower end of the shell allow the plug to be threaded into the engine's cylinder head. The thread length is referred to as reach (**Figure 9-10**). Various thread lengths are designed to match the cylinder head and combustion chamber design.

Shop Manual
Chapter 9, page 464

Some spark plugs have a metal gasket positioned between the seat on the steel shell and the cylinder head, whereas other shells have a tapered seat with a matching seat in the cylinder head that do not use a gasket. Platinum-tipped spark plug electrodes greatly

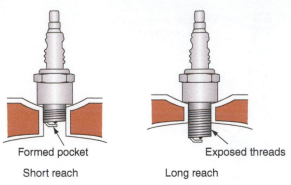

Figure 9-10 Spark plug reach: short versus long.

Figure 9-11 A platinum-tipped spark plug.

extended the life of a spark plug (**Figure 9-11**), and more recently iridium-tipped spark plugs have appeared in several vehicles. Iridium is harder, has a higher melting point than platinum, and has less electrical resistance (**Figure 9-12**).

Spark plugs are designed with different heat ranges to compensate for various operating conditions. When the engine is running, most of the plug's heat is concentrated on the center electrode. Heat is quickly dissipated from the ground electrode because it is threaded into the cylinder head. The spark plug heat path is from the center electrode through the insulator into the shell and to the cylinder head where the heat is absorbed by engine coolant circulating through the cylinder head. Spark plug heat range is determined by the depth of the insulator before it contacts the shell. For example, in a cold-range spark plug, the depth of the insulator is short before it contacts the shell. This cold-type spark plug has a short heat path, which provides cooler electrode operation. A cold-range spark plug has a shorter heat path compared to a hot-range spark plug (**Figure 9-13**).

In a hot spark plug, the insulator depth is increased before it makes contact with the shell. This provides a longer heat path and increases electrode temperature. A spark plug needs to retain enough heat to clean itself between firings but not so hot that it damages itself or causes premature ignition of the air-fuel mixture in the cylinder. If an engine is driven continually at low speeds, the spark plugs may become carbon fouled. Under this condition, a hotter range spark plug may be required. Severe high-speed driving over an extended time period may require a colder range spark plug to prevent electrode burning from excessive combustion chamber heat.

Ignition Cables

Spark plug or ignition cables make up the secondary wiring of a distributor ignition, and on some EI systems. Many EI systems like coil over plug (COP) usually have a very short

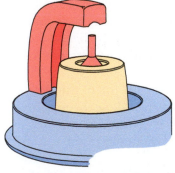

Figure 9-12 Iridium spark plug tips have a higher melting point than platinum.

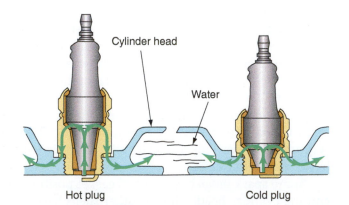

Figure 9-13 Spark plug heat range: hot versus cold.

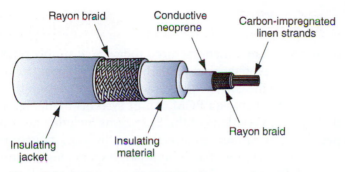

Figure 9-14 Static suppression ignition cable.

secondary path that may be no more than a boot, or in some cases a very short plug wire from the coil to the spark plug. These cables carry the high-voltage current from the distributor to the spark plugs. The cables are not solid wire; rather, they contain fiber cores that act as a resistor in the secondary circuit (**Figure 9-14**). They cut down on radio and television interference, increase firing voltages, and reduce spark plug wear by decreasing current. Metal terminals on each end of the spark plug wires contact the spark plug and the distributor cap terminals. Insulated boots on the ends of the cables strengthen the connections as well as prevent dust and water infiltration and voltage loss.

Shop Manual
Chapter 9, page 499

Triggering Devices/Engine Position Sensors

Primary ignition on/off time (or ignition timing) must coincide with the correct position of the crankshaft, camshaft, and the desired timing advance based on operating conditions (**Figure 9-15**). Several types of crankshaft and camshaft position sensors and a variety of systems are used to trigger the ignition module and provide a signal for timing. Some of the commonly used sensors we will examine include:

- Magnetic reluctance sensor
- Hall-effect sensor
- Magneto-resistive (MR) sensor
- Photoelectric sensor

Shop Manual
Chapter 9, page 456

A BIT OF HISTORY

The first use of automatic spark advance was in 1900. In spite of this development, many cars continued to have driver-operated timing advance controls.

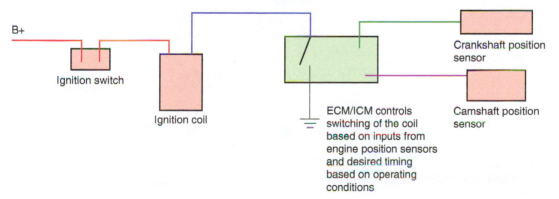

Figure 9-15 Block diagram of interaction between control module, cam and crank signals, and ignition coil control.

ELECTRONIC IGNITION SYSTEMS

Distributorless ignition systems manufactured today, now simply referred to as electronic ignition systems, can be categorized similarly. The ignition control modules (ICMs) can be mounted directly with the coil pack(s), remotely mounted from the coil pack(s), or as an internal function of the vehicle's PCM. All EI systems manufactured today have computer-controlled timing. Variations of the EI systems include one coil for two spark plugs or one coil per plug either mounted directly over the spark plug or connected with a spark plug wire. The single coil mounted directly to the spark plug is commonly referred to as a **coil-on-plug (COP)** system, while the single coil connected to a spark plug with a spark plug wire is referred to as a **coil-near-plug (CNP)**.

Ignition Coils and Primary Circuit Controls

The ignition current control depends on engine speed. At idle, there is more than enough time to saturate the ignition coil between cylinder firings. One of the advantages of EI system is the fact that there is more than one coil and consequently, more time for saturation. The control of timing is the responsibility of the ignition module until the engine starts, at which time the PCM takes over timing based on inputs from the engine sensors such as the TP sensor, MAP, MAF, IAT, ECT, KS, and others.

Triggering Devices (Engine Position Sensors)

The triggering device determines the crankshaft and camshaft position for the PCM. Waste spark systems only need to know crankshaft position (that will be explained later in this chapter). Most EI systems, such as coil-near-plug and coil-on-plug, need to know the position of the camshaft as well as the crankshaft. The Hall-effect and magnetic reluctance sensors have seen heavy use as EI-triggering sensors on EI systems. The triggering device signal has to be amplified to be used by both the ignition control module (ICM) and the PCM. The ICM sends a signal to the PCM to inform it of engine speed. Based on its inputs, the PCM then adds in the proper amount of ignition timing and sends this information back to the ICM, unless the ECM handles the ignition timing duties on its own, which seems to be a trend.

Hall-Effect Sensor

The Hall-effect sensor is a highly accurate triggering device that has found popularity in EI and DI systems (**Figure 9-16**). When a metal vane passes between a magnet and a semiconductor, the magnetic field is temporarily interrupted and the module voltage

Any ignition system has three jobs: provide a spark that is hot enough to light the air-fuel mixture (kV), provide a spark that lasts long enough to properly ignite the air-fuel mixture (duration), and provide a spark that occurs at the precise time needed to ignite the air-fuel mixture.

For the ignition system to be as efficient as possible with low emissions, good performance, and fuel mileage, the ignition timing must be controlled so spark happens at the ideal moment. The ECM uses the cam shaft and crank shaft position sensors, along with temperature, engine load, airflow, oxygen, or AF sensors, to accomplish this task.

Shop Manual
Chapter 9, page 458

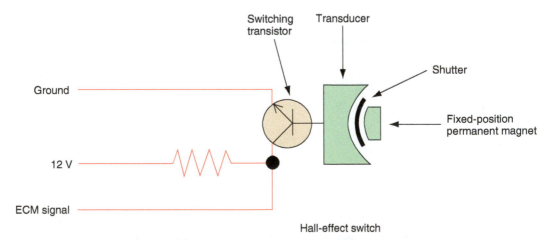

Figure 9-16 The position of the shutter blade in or out of the Hall-effect sensor will produce either an on or an off signal.

supply is pulled high by the sensor. After going through the Schmidt trigger, the waveform is inverted, so a high signal is low and vice versa.

Hall-effect sensors are used as crankshaft and/or camshaft sensors. A camshaft position sensor is shown in **Figure 9-17**. The shutter or interrupter blades can also be part of the engine's lower crankshaft pulley (**Figure 9-18**). The Hall-effect sensor is mounted in a fashion that allows the interrupter blades of the crankshaft pulley to pass through the sensor opening. Some sensors are configured to read two sets of interrupter blades—an inner ring and an outer ring (**Figure 9-19**). An example of a dual output Hall-effect sensor is shown in **Figure 9-20**.

AUTHOR'S NOTE The advantage of a Hall-effect switch over the PM (permanent magnet) generator is the fact that the Hall-effect switch always outputs a signal of the same amplitude, meaning that the peak voltage is always the same, but the frequency changes with rpm. The PM generator produces an analog signal that varies in amplitude and frequency with rpm. If the engine cranks slowly, the signal may not be strong enough to trigger the module.

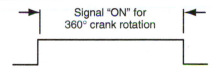

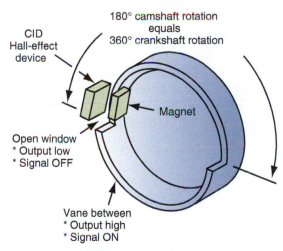

Figure 9-17 A Hall-effect switch used as a cam sensor.

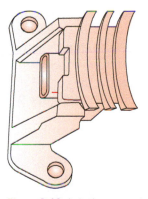

Figure 9-18 A dual vane crankshaft sensor.

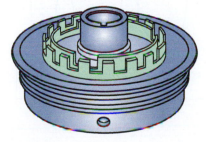

Figure 9-19 Interrupter rings on a crankshaft pulley.

Crankshaft sensor signal

Camshaft sensor signal

Figure 9-20 Relationship between camshaft and crankshaft sensor signals on a Buick 3.8-liter PFI engine.

Pull-Up and Pull-Down Hall-Effect Circuits

The Hall-effect switch can be used to ground or open a circuit with the interrupter blade located within the switch. This is referred to as either a pull-up- or a pull-down-type circuit. The Hall-effect switch may switch a source voltage provided internally by the module to ground or switch an externally sourced supply voltage going to the module. It will be necessary to understand the function of the switch to interpret the expected results when testing.

Permanent Magnet Sensors

The permanent magnet sensor is another very commonly used automotive sensor. **Figure 9-21** shows a commonly used crankshaft sensor. The sensor is basically a magnet and a coil of wires that produce an A/C output voltage as a gear, reluctor ring, or toothed wheel passes nearby to send an analog signal to the PCM and/or the ignition module, which is shown in **Figure 9-22**.

Magneto Resistive Sensor

The magneto-resistive sensor operates on the principle of a magnetic field changing the resistance of a semiconductor material. The MR sensor is triggered by a (in this case) crankshaft reluctor wheel. The MR sensor outputs a digital signal, but the voltage levels go from high to low voltage, not high voltage to ground. A permanent magnet is located between two pickups. The magnetic fields within the sensor react to the passing reluctor wheel. The two pickups read the approaching reluctor tooth. After the tooth passes the first pickup, it immediately passes the second pickup. Both pickups produce identical signals. However, the signal from the second pickup is slightly delayed because of its position in relation to the approaching reluctor tooth. The dual sensors allow the sensor to measure rotational speed to a very low level.

The data received by any of the triggering systems discussed is essential to ignition and fuel injection control. Advances in the triggering systems, electronics, and processors used in many late-model designs make it possible to control ignition timing on a per-cylinder basis and store misfire information in the PCM's memory. In addition, most systems employ some fail-safe strategies in the event of a compound failure. If necessary, the PCM is programmed to default to other incoming signals as a substitute, thereby keeping the engine running until it can be serviced. A magneto-resistive crankshaft sensor is shown in **Figure 9-23**.

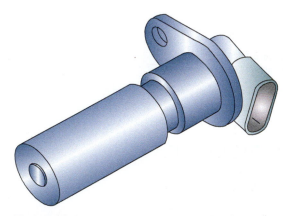

Figure 9-21 A permanent magnet-type crankshaft sensor.

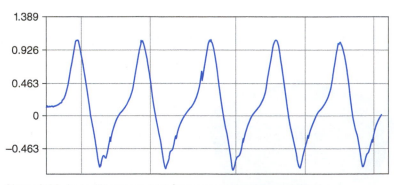

Figure 9-22 An oscilloscope pattern from a magnetic reluctance signal.

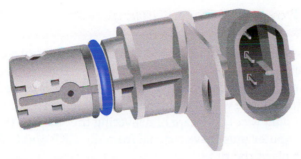

Figure 9-23 A magneto-resistive crankshaft sensor is highly accurate at all speeds.

ELECTRONIC IGNITION SYSTEMS (DISTRIBUTORLESS)

Electronic ignition system spark distribution is controlled by an electronic control unit, the vehicle's computer, or both. Instead of a single ignition coil for all cylinders, each cylinder may have its own ignition coil, or two cylinders may share one coil. The coils are wired directly to the spark plug they control. An ICM, tied into the vehicle's computer control system, controls the firing order and the spark timing and advance. In some cases, the ICM is actually built into the PCM.

In EI systems, a crank sensor located at the front, rear, or center of the crankshaft, depending on the engine design and manufacturer, is used along with a camshaft position sensor to trigger the ignition system based on the position of the engine parts and timing advanced or retarded according to the needs of the engine by the ECM (**Figure 9-24**).

- Fewer moving parts, therefore less friction and wear.
- Flexibility in mounting location, which is important with today's smaller engine compartments.
- Less required maintenance—there is no rotor or distributor cap to service.
- Reduced radio frequency interference because there is no rotor-to-cap gap.
- Elimination of a common cause of ignition misfire—the build up of water and ozone/nitric acid in the distributor cap.
- Elimination of mechanical timing adjustments. Has no mechanical load on the engine to operate.
- Increased available time for coil saturation.
- Increased time between firings, which allows the coil to cool more.

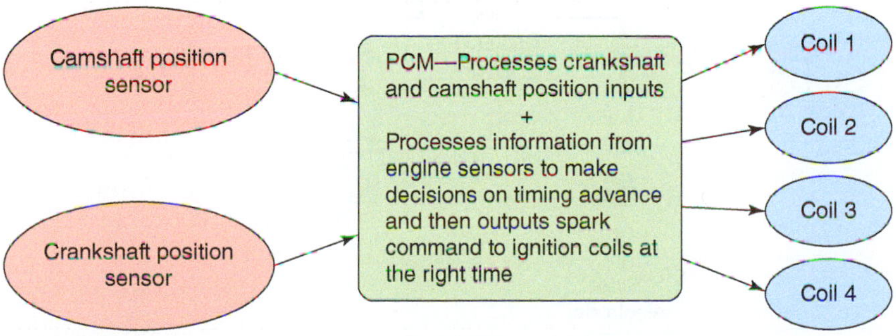

In the SAE J1930 terminology, the term "electronic ignition (EI)" replaces all previous terms for distributorless ignition systems.

Distributorless ignition produces and delivers spark to the spark plugs solely by the use of electronics, thus eliminating the distributor unit and related parts.

```
Camshaft position
sensor
                        PCM—Processes crankshaft        Coil 1
                        and camshaft position inputs
                                   +                    Coil 2
                        Processes information from
                        engine sensors to make
                        decisions on timing advance     Coil 3
                        and then outputs spark
Crankshaft position     command to ignition coils at    Coil 4
sensor                  the right time
```

Figure 9-24 A block diagram of a basic EI system.

BASIC COMPONENTS

Distributorless systems use multiple coils and modules to provide and distribute high secondary voltages directly from the coil to the plug.

Electronic ignition systems electronically perform the functions of a distributor. In place of the distributor are the engine position sensors, crankshaft position and camshaft position sensors. Of course, in a distributor-style ignition, the relationship is mechanical. The distributor is timed to coincide with the number one cylinder at TDC on the compression stroke, the rotor and plug wires take care of delivering the spark to the proper cylinders. In EI systems the computer must have a signal from the CKP and CMP sensors to place the spark at the right place and time.

The computer, ignition module, and position sensors combine to control spark timing and advance. The computer collects and processes information to determine the ideal amount of spark advance for the operating conditions. The ignition module uses crank/cam sensor data to control the timing of the primary circuit in the coils. The ignition module synchronizes the coils' firing sequence in relation to crankshaft position and firing order of the engine. Therefore, the ignition module takes the place of the distributor.

Waste Spark Distributorless Ignition

Waste spark is the spark delivered to the cylinder during the exhaust stroke.

In a distributorless ignition system that uses waste spark, each end of the secondary winding of a coil is connected to a spark plug. Engines with even numbers of cylinders have two pistons at TDC at the same time. One cylinder is under compression, and the other is on the exhaust stroke. Both plugs fire simultaneously, completing the series circuit (**Figure 9-25**). When a coil fires its two spark plugs, the plug that fires into the exhaust stroke is the **waste spark** plug (**Figure 9-26**). Each plug in the set always fires the same way regardless of the stroke of the piston. On the next crankshaft revolution, the previous waste spark plug will be on the compression stroke, and so forth.

It is no mystery how an individual coil pack simultaneously fires two spark plugs. It is nothing more than a standard ignition coil reconfigured to connect to two spark plugs rather than a single ignition coil tower to a coil wire. Once the voltage jumps the gap of the first spark plug, a spark is formed. However, this is not the end of the voltage journey. Once the spark has crossed the gap of the spark plug, it must return to its source to complete the circuit. In other words, it must return to the ignition coil. In a waste spark system, the energy has to jump a second spark plug gap to return to the coil.

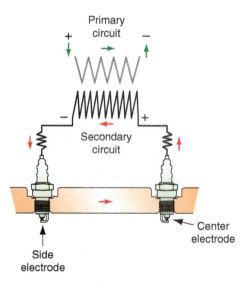

Figure 9-25 Two spark plugs connected to one coil. One spark plug fires from the center electrode to the side electrode; the other spark plug fires from the side electrode to the center electrode.

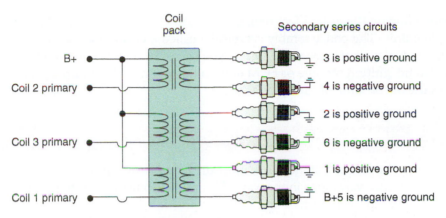

Figure 9-26 Polarity of the spark plugs in an EI system.

Figure 9-27 A waste spark system has one coil for every two cylinders.

Because the polarity of the primary and secondary windings are fixed, one plug always fires in a forward direction and the other in a reverse direction (**Figure 9-27**). The voltage dropped across each plug is determined by its polarity and cylinder pressure. It takes more electrical energy to "jump the gap" of the spark plug under compression than it does with the spark plug in the exhaust stroke (very little pressure).

It takes more energy to fire a plug under compression because air is an electrical insulator, and being compressed makes the "insulation" denser.

Distributorless ignition systems have had many names prior to SAE's J1930 list of acceptable names. Some of the commonly used names were EDIS, DIS, C3I, and IDI.

A BIT OF HISTORY

A very crude but effective ignition system with individual coils for each cylinder was used on the Model T Ford. This system was based on a magneto that rotated with the engine. The magneto sent timed, low-voltage signals to a wooden box that contained four coils. Each of these coils was connected to a spark plug. The pulses from the magneto caused the magnetic field in the coils to build and collapse, firing the spark plugs according to a firing order.

Typically, the voltage to fire the waste spark is about the same as jumping the rotor air gap in the distributor system. Due to the way the secondary coils are wired, when the induced voltage cuts across the primary and secondary windings of the coil, one plug fires in the normal direction—center electrode to side electrode—and the other plug fires just the reverse-side to the center electrode.

Depending on the exact EI system, the ignition coils can be serviced separately or as a complete unit. The coil assembly is typically called a **coil pack** and comprises two or more individual coils (Figure 9-27). On those EI systems that use one coil per spark plug, the electronic ignition module or ECM determines when each spark plug should fire and controls the on/off time of each plug's coil.

Shop Manual
Chapter 9, page 454

EI SYSTEM OPERATION

From a general operating standpoint, most distributorless ignition systems are similar. However, there are variations in the way different distributorless systems obtain a timing reference in regard to crankshaft and camshaft positions.

Most late-model engines have an EI system in which the coils are mounted directly over the spark plugs so that no wiring between the coils and plugs is necessary (**Figure 9-28**). On other systems, the coil packs are mounted remote from the spark plugs (**Figure 9-29**). High-tension secondary wires carry high-voltage current from the coils to the plugs.

Many ignition systems use separate Hall-effect sensors to monitor crankshaft and camshaft positions for the control of ignition and fuel injection firing orders. The crankshaft pulley has interrupter rings that equal in number to half of the cylinders of the engine. The resultant signal informs the PCM as to when to fire the plugs. The camshaft sensor helps the computer determine when the number one piston is at TDC on the compression stroke. When the interrupter ring indicates TDC, at the same time the camshaft sensor indicates number one is on compression, the ignition module knows the cylinder number one is ready for ignition.

Defining the different types of EI systems used by manufacturers focuses on the location and type of sensors used. There are other differences, such as the construction of the coil pack, wherein some are a sealed assembly and others have individually mounted ignition coils. Both the camshaft and crankshaft sensors may be Hall-effect sensors: Some systems may use two magnetic reluctance sensors; some systems use a magneto-resistive sensor for the crankshaft sensor.

Some systems have the camshaft sensor mounted in the front of the timing chain cover (**Figure 9-30**). A magnet on the camshaft gear rotates past the inner end of the camshaft sensor and produces a signal for each camshaft revolution so the ignition module can tell which set of cylinders is coming up on compression during this crankshaft revolution; remember that the crankshaft turns twice for one turn of the camshaft.

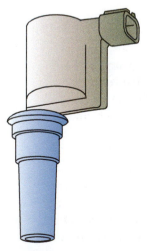

Figure 9-28 A coil-on-plug assembly.

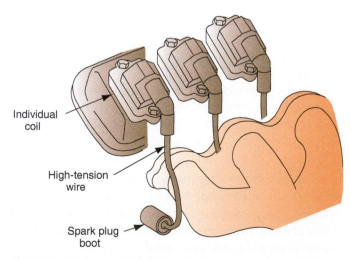

Figure 9-29 Coil-near-plug systems typically have one coil per cylinder.

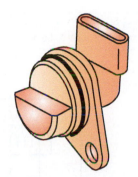

Figure 9-30 Camshaft position (CMP) sensor mounted in the front cover is activated by a magnet on the camshaft pulley.

Engines equipped with VVT have a camshaft sensor for each camshaft used in order for the ECM to not only tell the position of the camshaft, but also check the VVT system for errors.

The development of EI has resulted in reduced emissions, improved fuel economy, and increased component reliability brought about by these systems.

Shop Manual
Chapter 9, page 456

EI SYSTEMS WITH THE CAM SENSOR IN THE TIMING CHAIN COVER

The camshaft sensor in GM's 3.8-liter non-turbocharged SFI V6 engine is positioned in the front of the timing gear (front) cover. A magnet on the camshaft gear rotates past the inner end of the camshaft sensor and produces a signal for each camshaft revolution or every two crankshaft revolutions.

Each coil primary winding is connected into the coil module. When the ignition switch is turned on, 12 volts are supplied through a pair of fuses to ignition module, terminals N and P. Three wires from camshaft and crankshaft sensors are connected to the coil module, and five wires are connected between the PCM and the coil module (**Figure 9-31**).

FAST-START EI SYSTEMS (WASTE SPARK)

A fast-start system is used on GM's 3800 and 3300 engines. Sometimes referred to as a Type-Ill system, this fast-start EI system has a dual crankshaft sensor at the front of the crankshaft (**Figure 9-32**). The cam sensor is mounted in the timing gear cover as on previous 3.8-liter SFI engines. Two Hall-effect switches are located in the dual crankshaft sensor, and two matching interrupter rings are attached to the back of the crankshaft pulley (**Figure 9-33**). The inner ring on the crankshaft pulley has three blades of unequal lengths with unequal spaces between the blades. These blades have 110 degrees, 100 degrees, and 90 degrees of crankshaft rotation each, and the spaces between the blades are 30, 20, and 10 degrees each. On the outer ring, there are 18 blades of equal length with equal spaces between the blades. The signal from the inner Hall-effect switch is referred to as the 3X signal while the outer Hall-effect switch is called the 18X signal. These signals are sent from the dual crankshaft sensor to the coil module. These engines are called fast start because they can start in as little as one-third or 120 degrees of crankshaft rotation. The coil module and coil assembly in this fast-start system are similar to the components found on other systems, but they are not interchangeable.

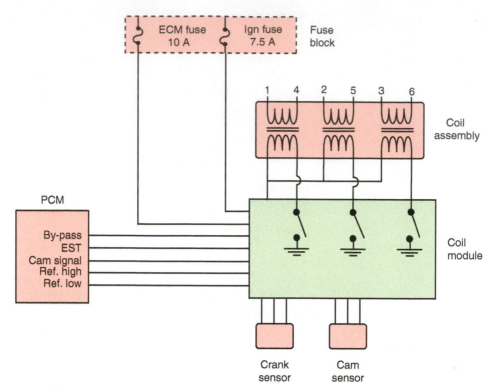

Figure 9-31 Schematic for the EI system on a Buick 3.8-liter PFI engine.

Figure 9-32 A dual Hall-effect crankshaft position sensor.

Figure 9-33 Crankshaft pulley interrupter rings.

The leading-edge 3X signals are spaced 120 degrees apart, and each pair of cylinders that is connected to the same coil reaches TDC on the compression stroke 75 degrees after the 3X leading-edge signal. Cylinders 1-4, 3-6, and 2-5 are paired together by the ignition coils. Because the leading-edge 3X signals are spaced 120 degrees apart, spark plug firings still occur at the correct intervals. An 18X leading-edge signal occurs once every 20 degrees of crankshaft rotation or 18 times during one crankshaft revolution. This sensor also provides a trailing-edge signal the same number of times per crankshaft revolution. Therefore, leading-edge and trailing-edge 18X signals occur every 10 degrees of crankshaft rotation.

The coil module monitors the 18X signal in relation to the 3X signal to provide coil sequencing for the ignition coils. For example, during the 10-degree window on the 3X interrupter blade, one trailing-edge 18X signal is received. While the 20-degree window in the 18X interrupter blade rotates through the Hall-effect switch, one leading-edge and one trailing-edge 18X signal are produced. During the 30-degree window rotation on the 3X interrupter blade, two trailing-edge and one leading-edge 18X signals are sent to the coil module.

The coil module knows which coil to sequence from the number of 18X signals received during each 3X window rotation. For example, when two 18X signals are received, the coil module is programmed to sequence coil 3-6 next in the firing sequence. Within 120 degrees of crankshaft rotation, the coil module can identify which coil to sequence and thus start firing the spark plugs. Therefore, the system fires the spark plugs with less crankshaft rotation during initial starting than previous systems.

Once the engine is running, a 5-volt signal is sent from the PCM through the bypass wire to the coil module. When this signal is received, the system switches to the electronic spark timing (EST) mode, and the PCM uses the 18X signal for CKP and speed information. The 18X signal may be referred to as a high-resolution signal (**Figure 9-34**). A schematic for this system is shown in **Figure 9-35**.

If the 18X signal is not present, the engine will not start. When the 3X signal fails with the engine running, the engine continues to run but will refuse to restart.

In this system, the cam sensor signal is used for injector sequencing, but it is not required for coil sequencing. If the cam sensor signal fails, the PCM logic begins sequencing the injectors after two cranking revolutions. There is a one-in-six chance that the PCM logic will ground the injectors in the normal sequence. When the PCM logic does not ground the injectors in the normal sequence, the engine hesitates on acceleration and will set a diagnostic trouble code (DTC).

> While starting the engine, fast-start EI systems start firing the spark plugs in 120 degrees or less of crankshaft rotation.

EI SYSTEMS WITH A CRANKSHAFT RELUCTOR RING

Electronic ignition systems are used on GM's 2.0-liter, ECOTEC4, and 2.8-, 3.1-, and 3.4-liter V6 engines. The main difference between these EI systems and the previously explained EI systems is in the crankshaft sensor. These other systems have a notched

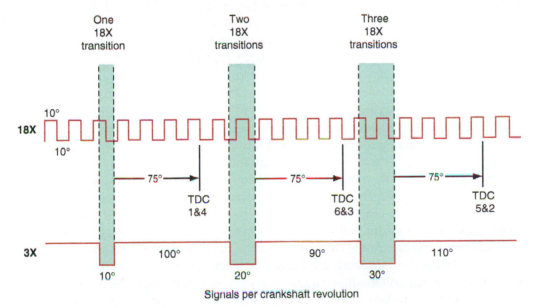

Figure 9-34 3X and 18X crankshaft sensor signals.

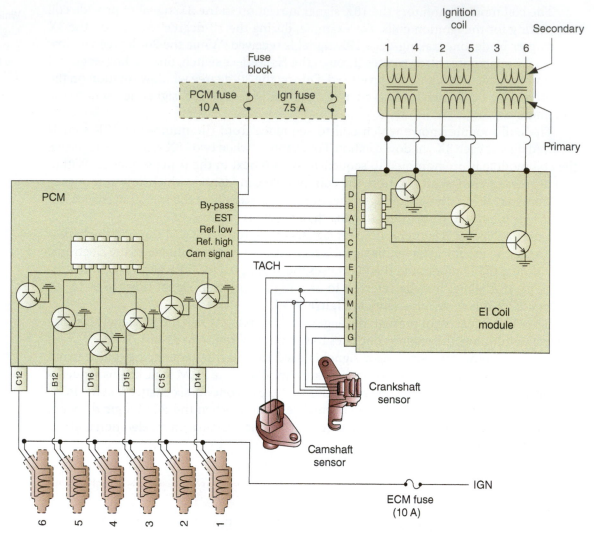

Figure 9-35 Schematic for a GM type 1 EI fast-start engine.

reluctor ring positioned near the center of the crankshaft (**Figure 9-36**). This ring is permanently cast on the crankshaft. A magnetic sensor containing a permanent magnet and a winding is mounted in an opening in the engine block. The tip of this sensor is 0.050 inch. (1.27 mm) from the reluctor ring outer surface. This gap between the magnetic sensor tip and the reluctor ring is not adjustable. The magnetic sensor is retained in the engine block with a bolt and clamp. The coil assembly and coil module are similar to those used on other slow-start and fast-start EI systems.

On the 2.0-liter and 2.2-liter four-cylinder engines and the 2.8-liter and 3.1-liter V6 engines, the coil and module assembly are positioned on one side of the engine block, and the magnetic sensor is located in the opposite side of the block.

The reluctor ring has seven notches on four-cylinder or V6 engines. Six of these notches are spaced 60 degrees apart, and the seventh notch is positioned 10 degrees from the sixth notch. A signal from the seventh notch is referred to as a SYNC signal, which is used by the coil module for coil sequencing. On four-cylinder engines, the coil module is programmed to recognize the SYNC notch and count notch 1. When notch 2 passes the sensor tip, the coil module opens the 2-3 coil primary circuit and fires spark plugs 2 and 3. After this event, the coil module counts notches 3 and 4, and then the module opens the primary circuit of the 1-4 coil when notch 5 rotates past the sensor. This module action fires spark plugs 1 and 4. The coil module counts notches 6 and 7, and the cycle starts repeating (**Figure 9-37**). During engine starting, the 2-3 coil fires before the 1-4 coil.

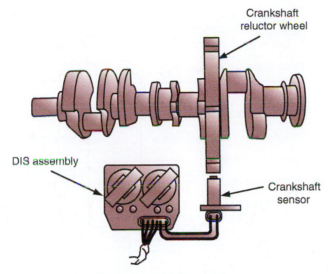

Figure 9-36 EI system on a four-cylinder engine.

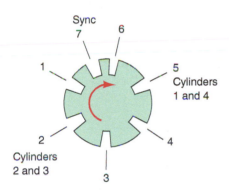

Figure 9-37 Reluctor ring and firing order for a four-cylinder engine.

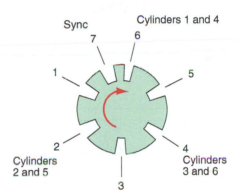

Figure 9-38 Reluctor ring and firing order for a V6 engine.

On the V6 engines, the coil module recognizes the SYNC notch and counts notch 1. When the signal from notch 2 is received, the coil module opens the primary circuit of the 2-5 coil which fires spark plugs 2 and 5 (**Figure 9-38**). Notch 3 is counted next, and the signal from notch 4 is then used to open the primary circuit of the 3-6 coil and fire spark plugs 3 and 6. As the sequence continues, notch 5 is counted and the signal from notch 6 is then used to open the primary circuit of the 1-4 coil and fire spark plugs 1 and 4. The cycle repeats as the crankshaft continues to rotate. During engine starting, the 2-5 coil fires first.

As the engine is running, a low-voltage signal is sent via the reference wire from the coil module to the PCM 60 degrees before the cylinder event notch passes the sensor. When the cylinder event notch passes the sensor, the reference signal changes to high voltage. This reference signal informs the PCM regarding crankshaft position and speed. After a reference signal is received, the PCM scans the inputs and determines the spark advance required by the engine. When this calculation is completed in a few milliseconds, the PCM sends an EST signal to the coil module to open the primary circuit of the next coil at the correct instant to provide the required spark advance (**Figure 9-39**). The PCM can affect the spark advance during engine starting on these EI systems.

Shop Manual
Chapter 9, page 449

Chrysler's EI System with a Crankshaft Reluctor Ring

Again, most systems that use a crankshaft reluctor ring are quite similar in construction and operation. Chrysler's is like the others but uses a different number of teeth on the

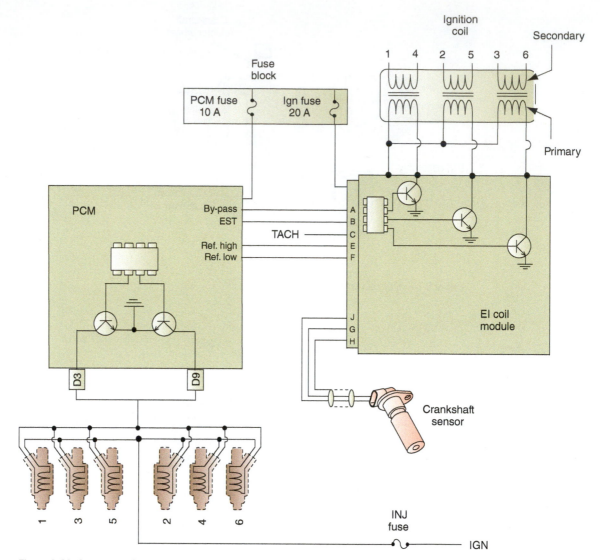

Figure 9-39 Schematic of the EI system for a V6 engine.

reluctor, a camshaft sensor, and a camshaft reluctor. Therefore, the signals received by the control module are also different.

The CKP sensor is mounted in an opening in the transaxle bell housing (**Figure 9-40**). The inner end of this sensor is positioned near a series of notches and slots that are integral with the transaxle drive plate (**Figure 9-41**).

A bolt retains the CKP sensor to the bell housing. A group of four slots is located on the transaxle drive plate for each pair of engine cylinders. Thus a total of 12 slots are positioned around the drive plate. The slots in each group are positioned 20 degrees apart. When the slots on the transaxle drive plate rotate past the CKP sensor, the voltage signal from the sensor changes from 0 to 5 volts. This varying voltage signal informs the PCM regarding crankshaft position and speed. The PCM calculates spark advance from this signal. The PCM also uses the crankshaft timing sensor signal along with other inputs to determine air-fuel ratio. Base timing is determined by the signal from the last slot in each group of slots.

The camshaft reference sensor is mounted in the top of the timing gear cover (**Figure 9-42**). A notched ring on the camshaft gear rotates past the end of the camshaft reference sensor. This ring contains two single slots, two double slots, and a triple slot (**Figure 9-43**).

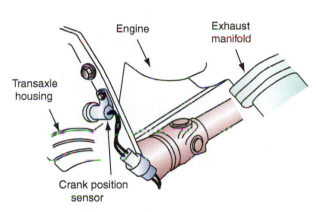

Figure 9-40 Crankshaft position sensor mounted in a transaxle.

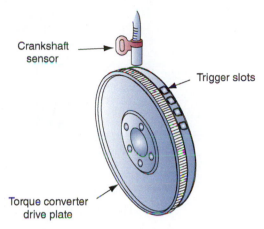

Figure 9-41 Crankshaft position sensor tone ring is attached to the flex plate.

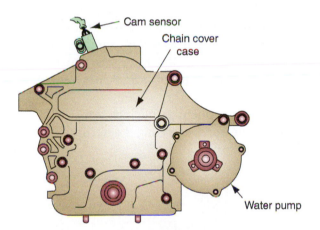

Figure 9-42 Camshaft position sensor mounting in timing cover.

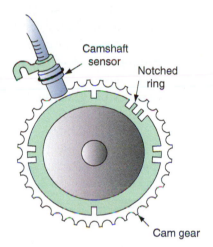

Figure 9-43 Notched ring on camshaft gear.

When a camshaft gear notch rotates past the camshaft reference sensor, the signal from the sensor changes from 0 to 5 volts. The single, double, and triple notches provide different voltage signals from the camshaft reference sensor as they rotate past the sensor. These signals are sent to the PCM. Since the ignition module is integrated into the PCM, an external module is not used on these systems. The PCM determines the exact camshaft and crankshaft position from the camshaft reference sensor signals, and the PCM uses these signals to sequence the coil primary windings and each pair of injectors at the correct instant.

When the engine starts cranking, the spark plugs fire and the injectors discharge fuel within one crankshaft revolution. The PCM determines when to sequence the coils and injectors from the camshaft reference sensor signals. If the camshaft reference sensor or the crankshaft timing sensor is defective, the engine will not start. The spark plug wires from coil number one are connected to cylinders one and four. The spark plug wires from coil number two go to cylinders two and five while the spark plug wires on coil number three are attached to cylinders three and six. The cylinder firing order for the 3.3-liter V6 engine is 1-2-3-4-5-6.

Once the engine is started, the PCM knows the exact crankshaft position and speed from the camshaft reference sensor and crankshaft timing sensor signals. The leading edges of the slots in the transaxle drive plate rotate past the crankshaft timing sensor at 9, 29, 49, and 69 degrees BTDC. When the single camshaft reference sensor slot rotates past the sensor, one high digital signal and one low digital signal is received by the PCM. When these signals are received, the PCM is programmed to fire the number two coil next, which is connected to spark plugs 2 and 5 (**Figure 9-44**).

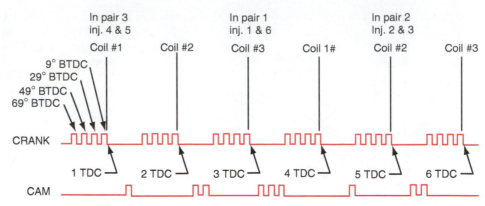

Figure 9-44 EI system coil firing and injector sequencing.

When the engine is cranking, all spark plug firings are at 9 degrees BTDC on the compression stroke. The PCM also sequences injectors 2 and 3 when the single slot rotates past the camshaft reference sensor.

When the double notch rotates past the camshaft reference sensor, the PCM is programmed to fire coil 3 connected to spark plugs 3 and 6, as well as sequence injectors 1 and 6. When the triple notch rotates past the camshaft reference sensor, the PCM is programmed to fire coil 1 connected to spark plugs 1 and 4 and sequence injectors 4 and 5.

Because Chrysler engines are now equipped with sequential fuel injection, the PCM grounds each injector individually, but the proper injector sequencing is determined from the CMP sensor signals.

EI SYSTEMS WITH COILS CONNECTED DIRECTLY TO SPARK PLUGS

These EI systems have the same reluctor ring and magnetic sensor as other EI systems. However, on these EI systems, the spark plug wires are eliminated or dramatically reduced, and the coil secondary terminals are connected directly to the spark plugs. Since the spark plug wires on coil-on-plug have been eliminated, the chance of high-voltage leaks is reduced. **Figure 9-45** shows a typical coil-on-plug system. COP ignitions use a coil mounted over the spark plug for each cylinder. Some of these systems have a module integral with the coil, some have a module that controls all the coil primaries, and some are controlled directly by the PCM.

Shop Manual
Chapter 9, page 451

> **AUTHOR'S NOTE** The name "coil-on-plug" is being used interchangeably with the term "coil-over-plug." This textbook primarily uses "coil-on-plug," but both systems are the same—only the names are different.

Figure 9-46 shows an import COP ignition with an ignition module that controls the individual coils for each cylinder. **Figure 9-47** shows a COP ignition system with the module function taken over by the PCM. Some systems have an ignition module built into each coil assembly.

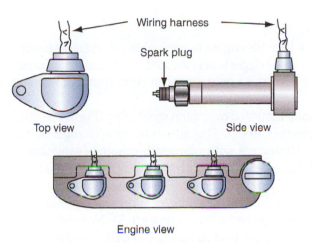

Wiring harness

Spark plug

Top view

Side view

Engine view

Figure 9-45 COP ignition system. Coil is mounted directly over the spark plug.

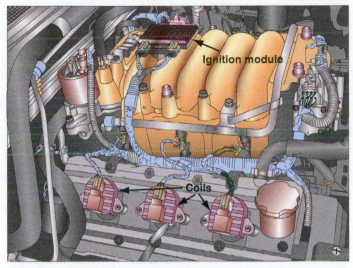

Ignition module

Coils

Figure 9-46 A COP ignition from an Isuzu V6 showing separate module and coils.

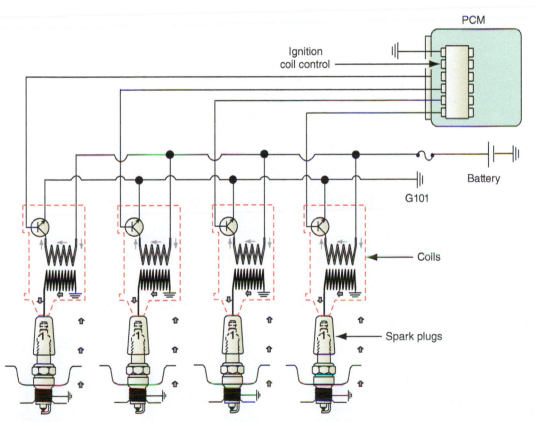

PCM

Ignition coil control

Battery

G101

Coils

Spark plugs

Figure 9-47 A COP ignition controlled entirely by the PCM.

The basic advantage of COP is that there are no plug wires to wear out or be damaged, and these systems are much less likely to be affected by moisture than systems using plug wires.

Shop Manual
Chapter 9, page 453

Ford's 5.4-Liter COP System

Shop Manual
Chapter 9, page 468

Ford's EI system CKP sensor for their 5.4-liter V8 engine is a variable reluctance sensor, which is triggered by a 36 minus 1 (or 35) tooth trigger wheel located inside the front cover of the engine (**Figure 9-48**). The sensor provides two types of information: crankshaft position and engine speed.

The system uses eight ignition coils, one mounted over each spark plug (**Figure 9-49**). Each coil terminal has two cylinder numbers on them. This allows for interchangeability.

Trigger wheel is another name for a sensor reluctor.

The trigger wheel has a tooth every 10 degrees with one tooth missing (sometimes called 36-1). When the part of the wheel that is missing a tooth passes by the sensor, there is a longer-than-normal pause between signals from the sensor. The engine control module (ECM) recognizes this and is able to identify this long pause as the location of piston number one (**Figure 9-50**). The ignition coils schematic can be found in **Figure 9-51**. Note how each coil is controlled through the PCM, which means the module function is built into the PCM.

Honda COP Ignition System

Honda uses a COP design that is very similar to the other systems seen so far (**Figure 9-52**). One difference in this system is the fact that the coil assemblies also have an ignition module built in (**Figure 9-53**). The ICM is a power transistor built into the coil.

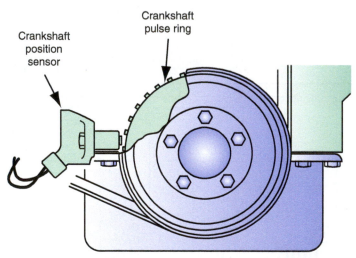

Figure 9-48 Location of the crankshaft position (CKP) sensor on a Ford 5.4-liter V8 engine.

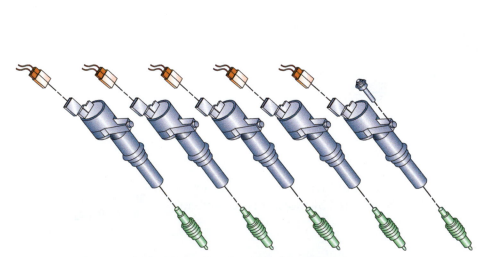

Figure 9-49 The Ford 5.4-liter engine has a coil over each spark plug.

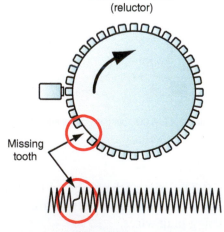

Figure 9-50 Sensor activity to monitor engine speed and crankshaft position (CKP), as well as the location of piston number one.

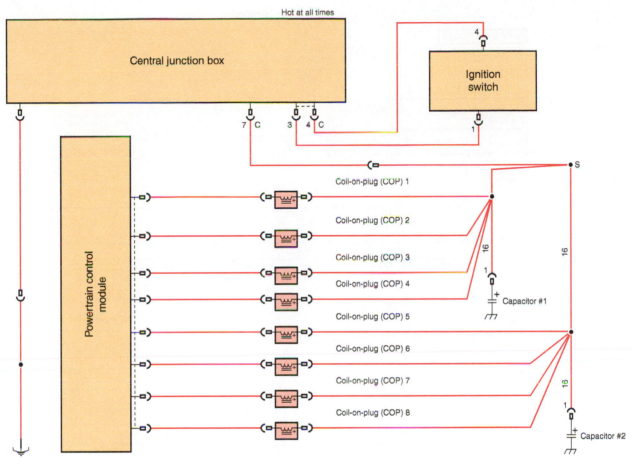

Figure 9-51 The COP wiring diagram for the Ford 5.4-liter V8 engine.

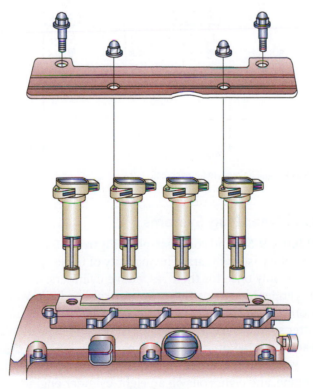

Figure 9-52 Honda COP ignition system.

Ignition System

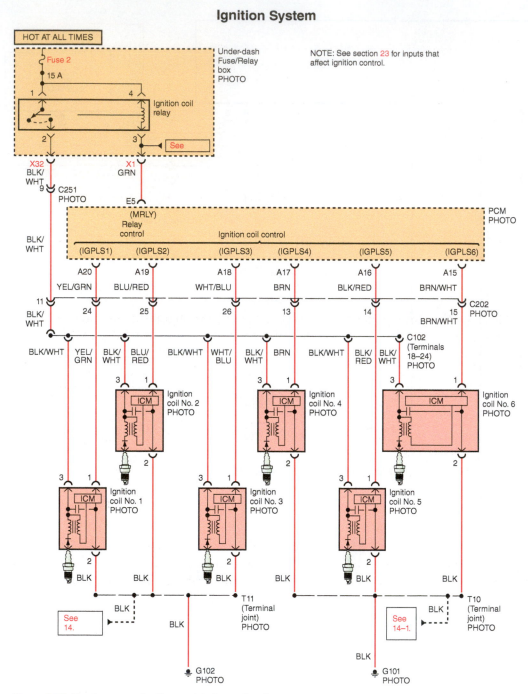

Figure 9-53 Honda uses an ignition module for each coil.

Coil-On-Plug/Coil-Near-Plug Systems

The coil-on-plug (**Figure 9-54**) and coil-near-plug (**Figure 9-55**) systems have emerged on many vehicles during the past few years. The majority of these systems are not waste spark systems. The coils fire only once for their respective cylinders. One coil is fired at a time and only when the cylinder is on its compression stroke. The PCM will use the CMP sensor to determine ignition timing. Having a coil for each cylinder gives the coil even more time to saturate for a hotter spark, which is needed for the extended spark plug service intervals. The ignition coils are mounted on or near the spark plug for each cylinder of the engine. For example, these systems contain four individual coils on a four-cylinder engine, six coils on a six-cylinder engine, and eight coils on an eight-cylinder engine.

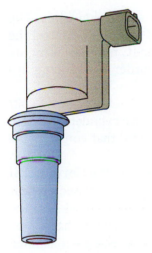

Figure 9-54 A COP assembly.

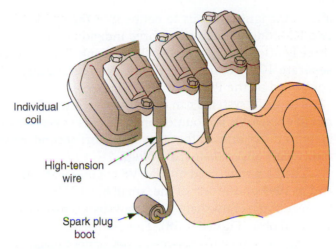

Figure 9-55 A CNP ignition assembly.

Although the ignition control (IC) systems may be configured differently among the various manufacturers, they all generally will operate in similar ways. Many COP systems control the individual ignition coils directly by the PCM. The PCM has individual drivers that control the primary current flow to the ignition coil. When the primary is turned on, battery voltage flows through the primary windings and builds the magnetic field. When the driver in the PCM opens the primary circuit, the coil fires.

GM, as well as many import systems, use individual coils that are connected to their own IC modules (**Figure 9-56**). The modules contain the drivers that control the ignition coil (**Figure 9-57**). The switching of the drivers is a function of the PCM. The ICM is connected to power and ground; the ECM signals the ICM when to fire the spark plug. Some of these systems can give an "ignition confirmation" signal back to the ECM that the ignition did take place.

Some systems, such as GM's 3.5-liter V6 engine, may integrate several drivers in a single housing. In this system, an ignition coil assembly consisting of three coils and one IC module is located in the center of each camshaft cover. There are two of these assemblies in the V6, one on each side. This configuration, like other COP systems,

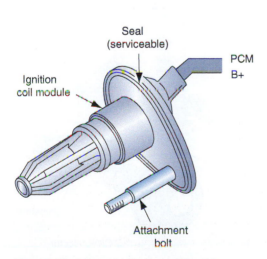

Figure 9-56 Coil-on-plug ignition coil/module assembly.

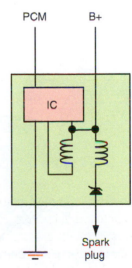

Figure 9-57 Circuit for a COP control module.

allows the ignition coils to connect to the spark plugs without spark plug wires. The two IC modules each have three individual driver circuits that separately control the three ignition coils on each cylinder bank. The PCM controls the IC modules using six separate IC circuits. Individual cylinder timing and firing order sequencing are controlled by the PCM.

The CNP systems can be found on certain V8 GM products. Each individual coil/module assembly is mounted on the rocker covers above its respective cylinder. The CNP systems differ only in that they use short secondary spark plug wire that connects the individual coils to their spark plugs.

The EI system used on GM V8 engines was introduced in 1997 on the Corvette LSI and in 1999 on the rest of the "small block" V8s. Two reluctor rings are pressed on near the rear of the crankshaft. The crankshaft sensor is attached to the engine block near the starter motor (**Figure 9-58**). The crankshaft sensor is somewhat unique in that it uses two magnetic pickups between a permanent magnet. The reluctor wheels are machined into four 90-degree segments. Each machined segment corresponds to a pair of cylinders at TDC. As you recall, a V8 engine has a power stroke every 90 degrees. Additionally, each section is divided into six 15-degree segments. Within each 90-degree segment there are two different sizes of notches, and each 90-degree segment has a different pattern of notches, enabling verification of the exact cylinder pairs within 90 degrees of crankshaft rotation due to the patterns generated by the two reluctor wheels. The crankshaft sensor output is a 24X signal. The PCM will need to "relearn" crankshaft position when the relationship of the crankshaft and crankshaft sensor are disturbed, such as when the engine, crankshaft, or sensor is replaced on this system.

The camshaft sensor (**Figure 9-59**) is also a magneto-resistive sensor that produces a digital on/off pulse once per camshaft revolution. The camshaft sensor is located near the rear center of the block and extends into the engine to read the cam reluctor wheel. The cam sensor reluctor wheel can be part of the cam itself, or it can be pressed onto the cam gear. The configuration of the camshaft sensor is such that it can send a valid signal before the engine starts.

The PCM is in control of the ignition system with input from several sensors such as the throttle position (TP sensor, ECT, MAF, IAT, knock sensor (KS), vehicle speed sensor (VSS), and others.

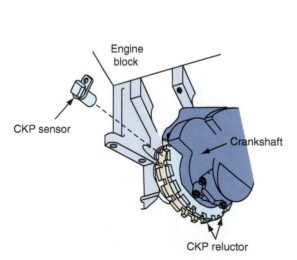

Figure 9-58 A crankshaft reluctor wheel and crankshaft sensor from a late-model Chevrolet engine.

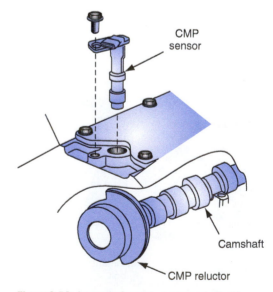

Figure 9-59 A camshaft sensor and camshaft with reluctor wheel from a late-model Chevrolet engine.

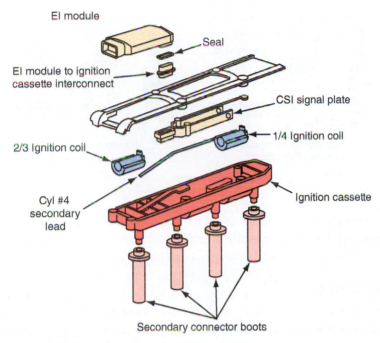

Figure 9-60 The compression sense ignition cassette.

Compression Sense Ignition

One of the latest developments in ignition systems is the compression sense ignition. This ignition system offers the advantage of sequential fuel injection and ignition without a camshaft position sensor on GM's ECOTEC engines (**Figure 9-60**). **Compression sense ignition** uses the basic waste spark ignition system, which, as you remember, fires two spark plugs in series. One cylinder of the firing pair is on compression, and the waste spark is fired in the exhaust stroke. The spark plug in the exhaust stroke takes less energy to fire (3–4 kV) than the spark plug under compression pressure (9–12 kV). The compression sense ignition module is able to determine which plug took the most energy to fire, along with the direction of current flow and uses this information as a substitute for a cam sensor signal, meaning that the cam sensor is not necessary with this ignition type (**Figure 9-61** and **Figure 9-62**). There are times that the substitute cam signal may not be reliable, such as during

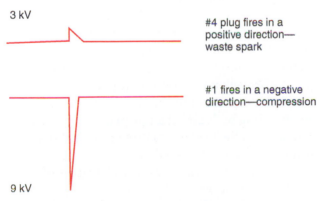

Compression-sensing module "sees" 1 compression and a cam signal is sent to the PCM from the ignition module.

Figure 9-61 The module knows cylinder number one is on compression because the required firing voltage is higher than the companion number four cylinder. The ignition module can then send a cam signal to the PCM.

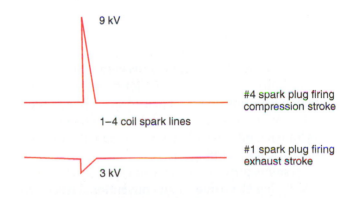

Compression sense ignition "sees" 4 on compression.

Figure 9-62 In this example, the module knows cylinder number four is on compression because the required firing voltage is higher than the companion number one cylinder.

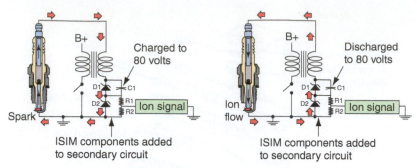

Figure 9-63 Delphi's ion-sensing ignition module.

deceleration. During these times, the PCM can synchronize the injection based on the last good signal received. If there is an ignition problem, the cam signal may not be usable; in this event, the PCM will fire injectors 1-4 and 2-3 as a batch-fire multiport injection system.

Ion-Sensing Ignition

Delphi has developed an ion-sensing ignition that is used by the coil-on-plug style of EI. The spark plug in **ion-sensing ignition** is actually used as a sensor inside the combustion chamber. The system works by applying a voltage across the spark plug gap immediately after the spark plug firing, while the spark gap is still ionized (**Figure 9-63**). The capacitor is charged by the spark plug firing voltage. The ignition module has the ability to analyze the characteristics of the current flow across the gap and inferring information regarding misfire, air-fuel ratio, preignition, and spark knock. The information can be used to eliminate the use of a knock sensor.

SUMMARY

- The electronic control module (ECM) or powertrain control module (PCM) opens and closes the primary circuit either directly or through the ignition control module (ICM).
- Magnetic pulse generators and Hall-effect sensors are the most widely used engine position sensors. At certain times during crankshaft rotation, they generate an electrical signal, which triggers the switching device to control ignition timing.
- Electronic ignition (EI) systems eliminate the distributor. There are two major types: waste spark, which uses one coil for a pair of spark plugs, and COP, or coil-on-plug, which has one coil per cylinder. Coil-near-plug is a variation of COP but does use a short plug wire.
- EI systems provide longer spark duration at the spark plug electrodes than conventional electronic ignition systems, which helps fire leaner air-fuel ratios in today's engines.
- In a waste spark EI system, the coil fires two spark plugs in series, and the current flows down through one spark plug and up through the other spark plug. This is known as waste spark.

- In a waste spark EI system, while a pair of spark plugs is firing, one of the cylinders is on the compression stroke and the other cylinder in the pair is on the exhaust stroke.
- The camshaft sensor signal informs the computer when to sequence the coils and fuel injectors.
- The crankshaft position (CKP) sensor signal provides engine speed and crankshaft position information to the computer.
- Vanes built into the crankshaft pulley, a toothed reluctor wheel on the crankshaft pulley, a toothed wheel cast or built onto the crankshaft, or a notched ring built into the flex plate, may be used with a Hall-effect switch to provide a CKP signal.
- Some EI systems are called fast start because the spark plugs begin firing within 120 degrees of crankshaft rotation.
- In other EI systems, the coils are mounted directly on top of the spark plugs and the spark plug wires are eliminated.
- Compression sense ignition is based on waste spark ignition.
- Ion-sensing ignition is based on COP ignition and uses the spark plug as a sensor.

REVIEW QUESTIONS

Short-Answer Essay

1. Name three different types of primary triggering devices commonly used in ignition systems.

2. Explain the basic operation of a Hall-effect sensor.

3. Give some examples of information that a computer uses to determine the ideal amount of ignition timing advance for a particular operating condition.

4. Explain the advantage of compression sense ignition.

5. What is the purpose of the 18X crankshaft sensor?

6. What does the term *waste spark* mean?

7. Explain the basic operation of an ion-sensing ignition.

8. Describe the pickup and triggering mechanism on an EI system with a reluctor ring mounted near the center of the crankshaft.

9. Explain the necessary requirements of the ignition system.

10. Describe the term "ignition timing'

Fill-in-the Blanks

1. The computer,_____, and _____sensors combine to control spark timing and advance.

2. Ford's EI system crankshaft position (CKP) sensor for its 5.4-liter V8 engine is a(n) _____ _____sensor.

3. The basic advantage of COP is that there are no _____ _____to wear out or be damaged.

4. Engines equipped with _____ have a camshaft sensor for each camshaft.

5. Typically, the voltage to fire the "waste spark" is about the same as jumping the _____ _____ _____in the distributor system.

6. Some Chrysler EI systems use a crankshaft _____mounted in an opening in the transaxle bell housing.

7. Some EI systems are called fast start because the spark plugs begin firing within_____ of crankshaft rotation.

8. Waste spark is the spark delivered to the cylinder on the_____stroke.

9. The magneto-resistive sensor is nothing more than an enhanced _____crankshaft sensor and is able to produce two signals from a single source.

10. GM, as well as many import systems, use individual coils that are connected to their own_____.

Multiple Choice

1. *Technician A* says that the primary circuit of the ignition coil operates basically the same in EI and DI systems.
   Technician B says that the secondary circuit operates basically the same in EI and DI systems.
   Who is correct?
   A. Technician A   C. Both technicians
   B. Technician B   D. Neither technician

2. *Technician A* says that the permanent magnet sensor is basically a magnet and a coil of wire.
   Technician B says that permanent magnet sensors produce a DC signal.
   Who is correct?
   A. Technician A   C. Both technicians
   B. Technician B   D. Neither technician

3. All of the following are examples of advantages of EI *except:*
   A. flexibility in mounting locations.
   B. reduced radio frequency interference.
   C. elimination of distributor cap buildups of water and nitric acid.
   D. reduced spark burn time.

4. While discussing waste spark ignition,
   Technician A says that there are two spark plugs connected in series.
   Technician B says that the two spark plugs are in companion cylinders that are at TDC at the same time.
   Who is correct?
   A. Technician A   C. Both technicians
   B. Technician B   D. Neither technician

5. *Technician A* says that a magnet on the camshaft gear rotates past the inner end of the camshaft sensor.
   Technician B says that Hall-effect sensors are used as crankshaft and/or camshaft sensors.
   Who is correct?
   A. Technician A
   B. Technician B
   C. Both technicians
   D. Neither technician

6. While discussing EI systems that use a crankshaft reluctor ring,
   Technician A says that the tip of the sensor is 0.5 inch from the cast reluctor.
   Technician B says that the gap can be adjusted on these engines.
   Who is correct?
   A. Technician A
   B. Technician B
   C. Both technicians
   D. Neither technician

7. *Technician A* says that some coil on plug systems have an ignition module that is part of the PCM.
   Technician B says that some coil on plug ignition systems have a separate ignition module built into each coil assembly.
   Who is correct?
   A. Technician A
   B. Technician B
   C. Both technicians
   D. Neither technician

8. Two technicians are discussing Chevrolet V8s that have reluctor rings pressed onto the crankshaft.
   Technician A says there are 3 reluctor rings pressed on the crankshaft,
   Technician B says the signal produced from the Chevrolet reluctor ring is a 24x signal.
   Who is correct?
   A. Technician A
   B. Technician B
   C. Both technicians
   D. Neither technician

9. *Technician A* says that the compression sense ignition does not need a camshaft position sensor.
   Technician B says that the compression sense ignition does not need a crankshaft position sensor.
   Who is correct?
   A. Technician A
   B. Technician B
   C. Both technicians
   D. Neither technician

10. *Technician A* says that the ion-sensing ignition system is based on the coil-on-plug system.
    Technician B says that the ion-sensing ignition system uses the spark plug itself as a sensor.
    Who is correct?
    A. Technician A
    B. Technician B
    C. Both technicians
    D. Neither technician

CHAPTER 10
EMISSION CONTROL SYSTEMS

Upon completion and review of this chapter, you should be able to:

- Explain why hydrocarbon (HC) emissions are released from an engine's exhaust.
- Explain how carbon monoxide (CO) emissions are formed in the combustion chamber.
- Describe oxygen (O_2) emissions in relation to air-fuel ratio.
- Describe how carbon dioxide (CO_2) is formed in the combustion chamber.
- Describe how oxides of nitrogen (NO_x) are formed in the combustion chamber.
- Describe the operation of an evaporative control system during the canister purge and non-purge modes.
- Explain the purpose of the positive crankcase ventilation (PCV) system.
- Describe the operation of the PCV system at idle speed, part-throttle, and heavy load conditions.

- Describe the operation of the knock sensor and electronic spark control module.
- Describe the operation of a differential pressure feedback exhaust (DPFE) gas recirculation valve.
- Explain the operation of a digital exhaust gas recirculation (EGR) valve.
- Explain the operation of a linear EGR valve.
- List and describe the various controls commonly used in EGR systems.
- Define the purpose of a catalytic converter.
- Explain how a catalytic converter works.
- Describe the operation of a secondary air injection system.
- Describe Tier 3 emissions.
- Describe LEV II California emissions.
- Explain how an onboard refueling vapor recovery (ORVR) works to reduce HC emission.

Terms To Know

Air diverter (AIRD)
Air injection reactor (AIR)
Blow-by
Carbon monoxide (CO)
Charcoal (carbon) canister
Delta pressure feedback EGR sensor (DPFE) sensor
Electronic vacuum regulator valve (EVRV)

Engine off natural vacuum (EONV)
Enhanced EVAP
Evaporative emission control (EVAP)
Fuel pressure control valve
Fuel tank level sensor
Fuel tank pressure sensor
Hydrocarbon (HC)
Inspection/Maintenance (I/M)

Leak detection pump (LDP)
Oxides of nitrogen (NO_x)
Periodic motor vehicle inspection (PMVI)
Post-combustion control
Pre-combustion control
Purge valve
Tier 3
Vent valve

The name "smog" comes from a combination of the words "smoke" and "fog," which pretty much tells what smog looks like.

The California Air Resources Board is commonly called the CARB.

INTRODUCTION

Emission controls on cars and trucks have one purpose—to reduce the amount of pollutants and environmentally damaging substances released by the vehicles. The consequences of the pollutants are grievous. The air we breathe and the water we drink have become contaminated with chemicals that adversely affect our health. It took many years for the public and the industry to address the problem of these pollutants. Not until smog became an issue did anyone in power really care and do something about these pollutants.

Smog not only appears as dirty air, but also is an irritant to your eyes, nose, and throat. The components necessary to form photochemical smog are HC and NO_x exposed to sunlight in stagnant air. When there is enough HC in the air, it reacts with the NO_x in the air. The energy of sunlight causes these two chemicals to react and form photochemical smog.

LEGISLATIVE HISTORY

The first Clean Air Act prompted Californians to create the California Air Resources Board (CARB). CARB's purpose was to implement strict air standards. These became the standard for federal mandates. One of the CARB's approaches to clean the air was to start **periodic motor vehicle inspection (PMVI)**. The purpose for the PMVI is to inspect a vehicle's emission controls once a year or every other year. This inspection includes a tailpipe emissions test and an under-hood inspection. The tailpipe test certifies that the vehicle's exhaust emissions are within the limits set by law. The under-hood and/or vehicle inspection verifies that the pollution control equipment has not been tampered with or disconnected.

Periodic motor vehicle inspection (PMVI) is basically the same thing as I/M (Inspection/Maintenance).

Many states have established emissions test either annually or biannually. Testing varies among states as to whether vehicles are affected throughout the entire state or just in certain highly populated areas of these states. Most of these testing programs use such protocols as the I/M 240, ASM, OBD II monitor, or similar program. The I/M 240 and ASM tests are performed with the vehicle under a variety of loads at various speeds, depending on the type of test being performed. These types of tests are more accurate than exhaust emission readings taken at idle in obtaining actual emission performance results.

The Acceleration Simulation Mode (ASM) test, which incorporates steady state and transient testing, varies depending on specific state requirements. The ASM tests at a certain percentage of engine load for a defined number of seconds at a specified steady speed and a 90-second transient test.

The OBD II monitor test depends on all the OBD II monitors running to completion. If a fault has been found that prevents all monitors completing, then the vehicle fails the emission test.

The On-Board Diagnostics generation two (OBD II) emission monitor test is performed in conjunction with an exhaust emissions test or as a stand-alone test. As time goes on, more states will likely begin to use the OBD II test in lieu of the I/M 240 or ASM-type tests. The ability of the OBD II system to reveal engine data, performance, and emission control–related history was implemented on all domestic vehicles in 1996. However, a few models had some OBD II capability in 1994 and 1995. The diagnostic trouble codes (DTC) as well as systems monitor data and history is retrieved by connecting a scanner to the under-dash diagnostic connector. If all the OBD II monitors have ran successfully the vehicle passes the emissions test.

The **Inspection/Maintenance (I/M)** programs refer to the periodic vehicle emission testing mandated by each state to conform to EPA regulations.

The **Inspection/Maintenance (I/M)** 240 test requires the use of a chassis (road) dynamometer, commonly called a "dyno." While on the dyno, the vehicle is operated for 240 seconds and under different load conditions. The test drive on the dyno simulates both in-traffic and highway driving and stopping. The emissions tester tracks the exhaust quality through these conditions.

The I/M 240 program also includes a functional test of the evaporative emission control devices and a visual inspection of the total emission control system. If the vehicle fails the test, it must be repaired and certified before it can be registered.

Passenger cars are responsible for 17.8 percent of the total hydrocarbon emissions, 30.9 percent of the total carbon monoxide emissions, and 11.1 percent of the oxides of nitrogen emissions. After 50 years of emission regulations, these figures remain staggering! Imagine what these figures would be if automotive and industrial emissions had remained unregulated during the last 50 years!

There are three main automotive pollutants: hydrocarbons (HC), carbon monoxide (CO), and oxides of nitrogen (NO_x). Particulate emissions are also present in diesel engine exhaust. HC emissions are caused largely by unburned fuel from the combustion chambers. HC emissions can also originate from evaporative sources such as the gasoline tank. CO emissions are a by-product of the combustion process, resulting from excessively rich air-fuel mixtures. NO_x emissions are caused by nitrogen and oxygen uniting at cylinder temperatures above 2,500°F (1,371°C).

Emission standards have been one of the driving forces behind many of the technological changes in the automotive industry. Catalytic converters and other emission systems were installed to meet emission standards. Fuel injection systems are constantly being improved to provide more accurate control of the air-fuel ratio to reduce emission levels and allow the catalytic converter to operate efficiently.

During the last 20 years, emission standards in the United States have become increasingly stringent. New engine systems and technology are being developed to keep pace with the demands of emerging stringent emission standards. Since approximately 90 percent of hydrocarbon emissions occur before the catalytic converter is hot enough to provide proper HC oxidation, direct fuel injection systems are being developed that drastically reduce cold start emissions. A cold start emission strategy that includes retarding the ignition timing and elevated idle is helping meet the demand for cleaner air.

> An engine that runs too cool tends to produce more HC and CO, while an engine that runs too hot produces more NO_x.

NHTSA and EPA Fuel Mileage and Greenhouse Gas Emission Goals— 2017 to 2025

The Energy Policy and Conservation Act (EPCA) has established fuel economy and greenhouse gas (GHG) standards for the years 2017 and beyond. The new CAFÉ standard sets fuel economy standards of 40.3 to 41 mpg for the end of the first phase that is scheduled to run from 2017 to 2021 (**Figure 10-1**). The first phase has already been written into the regulations. The second phase of the fuel economy standards are set to be 48.7 to 49.7 mpg. The EPA has also set GHG standards at 163 grams/mile of CO_2 in the model year 2025 (**Figure 10-2**). The second phase regulations are tentative, based on the progress gained

	2017	2018	2019	2020	2021
Passenger cars	39.6–40.1	41.1–41.6	42.5–44.8	44.2–44.8	46.1–46.8
Light trucks	29.1–29.4	29.6–30.0	30.0–30.6	30.6–31.2	32.6–33.3
Combined	35.1–35.4	36.1–36.5	37.1–37.7	38.3–38.9	40.3–41.0

Figure 10-1 Phase one fuel economy average standards according to the CAFÉ 2017–2025 Fact Sheet.

	2022	2023	2024	2025
Passenger cars	48.2–49.0	50.5–51.2	52.9–53.6	55.3–56.2
Light trucks	34.2–35.8	37.5–39.3	34.9–36.6	38.5–40.3
Combined	42.3–44.3	46.5–48.7	43.0–45.1	47.4–49.7

Figure 10-2 Phase two fuel economy proposals according to the CAFÉ 2017–2025 Fact Sheet.

**Model Year 2025 Fuel Economy and CO₂ Targets for
Various MY 2012 Vehicle Types**

Vehicle Type	Example Models	Example Model Footprint (sq. ft.)	CO₂ Emissions Target (g/mi)	Fuel Economy Target (mpg)
Example Passenger Cars				
Compact car	Honda Fit	40	131	61.1
Mid-size car	Ford Fusion	46	147	54.9
Full-size car	Chrysler 300	53	170	48.0
Example Light Trucks				
Small SUV	4WD Ford Escape	44	170	47.5
Mid-size crossover	Nissan Murano	49	188	43.4
Minivan	Toyota Sienna	55	209	39.2
Large pick-up truck	Chevy Silverado	67	252	33.0

Figure 10-3 NHTSA and EPA Propose to extend the national program to improve fuel economy and greenhouse gases for passenger cars and light trucks.

> According to the California Air Resources Board, hydrocarbon emissions from a vehicle built in 1965 released about 1 ton of hydrocarbons over 100,000 miles of driving. A 1998 vehicle released about 50 pounds and a 2010 vehicle with Level II emissions about 10 pounds. This is quite an accomplishment by the automotive industry.

from the first phase. The fuel mileage ranges are an average of the expected fuel economy of the entire manufacturer's sales. More specific fuel mileage expectations for individual categories are in **Figure 10-3**.

Greenhouse Gas Emission Standards

With the concern over global warming, standards for CO_2 emissions have been developed by the National Highway Traffic Safety Administration and the Environmental Protection Agency (EPA). Many believe that our emissions of greenhouse gases (principally CO_2) are contributing to a warming trend on the earth called global warming, which could lead to continued melting of the polar ice caps and climate change. CO_2 in the atmosphere acts as a blanket to insulate the earth and prevent heat from escaping into space, much like a greenhouse. The standards are being implemented over a time period from 2017 to 2025, to give manufacturers some lead time to develop vehicles that can meet the standard. The standards are designed around a vehicle's footprint. A smaller vehicle will have a lower CO_2 emission standard than a full-size truck (see Figure 10-2). CO_2 is a product of combustion, actually desired as far as efficient burning of fuel is concerned. This means that using less fuel will be a huge factor in meeting these standards, as well as the reason they are accompanied by fuel mileage standards.

DEVELOPMENT OF EMISSION CONTROL DEVICES

In late 1959, California established the first standards for automotive emissions. In 1967, the Federal Clean Air Act was amended to provide for federal standards to apply to motor vehicles.

The first source of emissions to be brought under control was the crankcase. Positive crankcase ventilation (PCV) systems that route crankcase vapors back to the engine's intake manifold were developed and incorporated into 1961 cars and light trucks sold in California. These systems were installed on all cars nationwide beginning with the 1963 models.

Control of unburned hydrocarbons and carbon monoxide in the engine's exhaust was the next major development. An **air injection reactor (AIR)** system was built into cars and light trucks sold in California in 1966. Other systems, including the controlled combustion system, were developed and used nationwide in 1968. Further improvements in the following years improved combustion to reduce hydrocarbon and carbon monoxide emissions.

Fuel vapors from the gasoline tank and fuel system were brought under control with the introduction of evaporation control systems. These systems were first installed in 1970 model cars sold in California and in most domestic-made cars beginning with 1971 models.

Most vehicle manufacturers started to provide emission control systems that reduced oxides of nitrogen as early as 1970. The exhaust gas recirculation system used on some 1972 models was used extensively for 1973 models when federal standards for oxides of nitrogen took effect.

One of the most important developments for lowering emission levels has been the availability and use of unleaded gasoline. Beginning with 1971, engines were designed to operate on unleaded fuels. In 1975, unleaded fuel became mandatory in new cars.

The catalytic converter, a later development, provided a means for oxidizing the carbon monoxide and hydrocarbon emissions in the engine exhaust. Beginning with the 1975 model year, passenger cars and light trucks have been equipped with converters. These early converters oxidized only HC and CO. Converters that have the capability of also chemically reducing NO_x were not developed until later.

Three basic types of emission control systems are used in modern vehicles: evaporative control systems, pre-combustion, and post-combustion.

Most of the pollution control systems used today prevent emissions from being created in the engine either during or before the combustion cycle. The common **pre-combustion control** systems are as follows:

- *Positive crankcase ventilation.* The PCV system removes pollutants that blow by the pistons into the crankcase and recirculates them into the induction system.
- *Engine modification systems.* These systems improve combustion and reduce HC and CO in the exhaust.
- *Exhaust gas recirculation (EGR) systems.* EGR reduces NO_x by diluting the AF mixture with exhaust gas, which is inert, causing a cooler combustion.

Post-combustion control systems clean up the exhaust gases after the fuel has been burned. Catalytic converters are placed into the exhaust to reduce HC and CO to harmless water vapor and carbon dioxide by chemical (thermal) reaction with oxygen in the air. All catalysts now reduce NO_x as well as HC and CO.

> **Air injection reactor (AIR)** systems originally injected air into the exhaust manifold when the vehicle was cold and the exhaust when the vehicle was warm. Modern vehicles with AIR inject air into the exhaust manifold only when the vehicle is cold. Most modern vehicles do not use AIR.

> Original converters controlled only HC and CO emission; later three-way converters also controlled NO_x.

> **Shop Manual**
> Chapter 10, page 555

POLLUTANTS

The gases that are of most concern to environmentalists, engineers, and technicians are HC, CO, NO_x, CO_2, and O_2. The latter two are not really pollutants but are monitored because they are indicators of combustion efficiency.

Hydrocarbons

Hydrocarbon (HC) emissions are unburned fuel and are caused by incomplete combustion. Even an engine in good condition with satisfactory ignition and fuel systems produces some HC. When the flame front in the combustion chamber approaches the cooler cylinder wall, the flame front quenches, leaving some unburned HC.

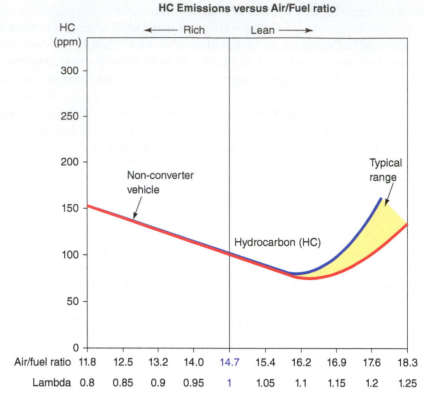

Figure 10-4 Hydrocarbon emissions are high with a lean or a rich air-fuel ratio.

An excessively lean air-fuel ratio also results in cylinder misfiring and high HC emissions. A very rich air-fuel ratio also causes higher-than-normal HC emissions. At the stoichiometric air-fuel ratio, HC emissions are low (**Figure 10-4**). Evaporative emissions from fuel tanks and evaporative systems are also a source of HC emissions.

Because of the catalytic converter, the HC reading at the tailpipe is very low if the engine, ignition, fuel, and emission systems are in normal working condition. When a cylinder misfires, all the unburned gas in the cylinder is delivered to the exhaust system, resulting in a high HC reading. Cylinder misfiring and high HC readings may be caused by ignition defects, such as defective spark plugs, spark plug wires, or coil. Low cylinder compression also causes high HC emissions. Basically, HC emissions can be caused by anything that affects the ability of the engine to burn gasoline effectively. If the mixture is too lean, then there is not enough fuel to support combustion. If the mixture is too rich, then there is too much fuel and not enough air to support combustion. If there is a misfire, then the fuel goes out of the combustion chamber unburned.

Excessive HC emissions may be caused by the following:

- Ignition system misfiring
- Improper ignition timing
- Excessively lean mixture
- Low cylinder compression
- Defective valves, guides, or lifters
- Defective rings, pistons, or cylinders
- Vacuum leaks
- Plugged PCV system
- Excessively rich air-fuel ratio
- Engine oil diluted with gasoline

AUTHOR'S NOTE The evaporative emission control is critical in preventing fuel vapors from escaping into the atmosphere. The source can sometimes be a challenge to locate and can come from such items as the evaporative canister, hose connections, control valves, or gas cap, to name a few. A "smoke machine" can be used to pump smoke into the fuel system under a slight pressure. Looking along the hoses, evaporative controls, fuel tank, gas cap, and so on will assist in locating the leak, which will be evident by a smoky discharge. A liquid fuel leak from the fuel lines or tank can be located with an infrared exhaust analyzer probe. Raw fuel will be evident as a huge increase in HC at the point of the leak.

Carbon Monoxide

Carbon monoxide (CO) is a by-product of combustion. CO is a poisonous chemical compound of carbon and oxygen. It forms in the engine when there is not enough oxygen (or too much fuel) to combine with the hydrocarbons during combustion. When there is enough oxygen in the mixture, carbon dioxide (CO_2) is formed. CO_2 is not a pollutant and is the gas used by plants to manufacture oxygen. CO is primarily found in the exhaust but can also be in the crankcase. CO is odorless and tasteless, but it is toxic in concentrated form.

CO is produced because there is not enough O_2 to completely burn the air-fuel mixture present.

CO emissions are caused by a lack of air or too much fuel in the air-fuel mixture. CO will not occur if combustion does not take place in the cylinders. Therefore, the presence of CO means combustion is taking place. As the air-fuel ratio becomes richer, the CO levels increase (**Figure 10-5**). At the stoichiometric air-fuel ratio, the CO emissions are very low. If the air-fuel ratio is leaner than stoichiometric, the CO emissions remain very low. Therefore, CO emissions are a good indicator of a rich air-fuel ratio, but are not an

CO_2 is a greenhouse gas, even though it is actually an indicator of complete combustion.

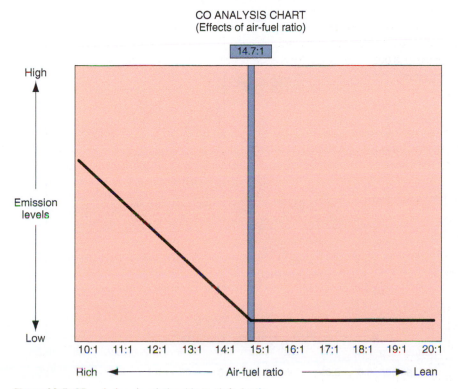

Figure 10-5 CO emissions in relationship to air-fuel ratio.

accurate indication of a lean air-fuel ratio. When cylinder misfiring occurs, there is less total combustion in the engine cylinders. Since CO is a by-product of combustion, cylinder misfiring does not increase CO emissions. If a cylinder is misfiring, CO emissions may decrease slightly.

Excessive CO emissions may be caused by the following:

- Rich air-fuel mixtures
- Dirty air filter
- Faulty injectors
- Higher-than-normal fuel pressures
- Defective system input sensor
- Plugged PCV system
- Excessively rich air-fuel ratio
- Stuck open heat riser valve
- Inoperative or disconnected AIR pump
- Engine oil diluted with gasoline

Oxides of Nitrogen

NO_x is pronounced "knocks."

This pollutant is actually various compounds of nitrogen and oxygen. Both of these gases are present in the air used for combustion. The formation of **oxides of nitrogen (NO_X)** is the result of high combustion temperatures, which can be present in lean air-fuel ratios or high combustion engines (**Figure 10-6**). When combustion temperature reaches more than $2,500°F (1,370°C)$, the N and the O_2 in the air combine to form these oxides of nitrogen. NO_x emissions are a major contributor to the problem of smog. Since outside air is 78 percent N, the gas cannot be prevented from entering

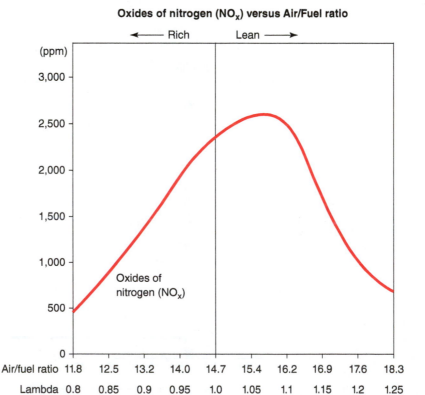

Figure 10-6 Relationship of NO_x emissions to air-fuel ratio.

the combustion chamber. The key to controlling NO_x is to prevent N from joining with O_2 during the combustion process.

One way to combat NO_x emissions is to lower the compression ratio. The problem with this is that fuel economy and emissions are adversely affected when the compression ratio is lowered. Reducing the temperature of the combustion chamber had been the job of the EGR system. Late-model vehicles are using VVT to accomplish the same goal. In either case the reduction in combustion chamber temperatures is a compromise.

The introduction of burned exhaust gases into the combustion chamber lowers the temperature of the combustion chamber and therefore reduces NO_x. The X in NO_x stands for the proportion of oxygen mixed with a nitrogen atom. The X is a variable, which means it could be the number 1, 2, 3, and so on; therefore, the term "NO_x" refers to many different oxides of nitrogen (NO, NO_2, NO_3, and so on). Most of the NO_x emissions from an engine are NO, or nitrous oxide. When NO is released in the air, NO seeks, finds, and combines with an oxygen atom to form nitrous dioxide (NO_2). NO_2 is a very toxic gas and contributes to the formation of smog, ozone, and acid rain.

NO_x is produced by high temperatures and pressure.

Higher-than-normal NO_x emissions may be caused by the following:

- An overheated engine
- Lean air-fuel mixtures
- Vacuum leaks
- Over advanced ignition timing
- Defective EGR system
- Too high compression

Acid rain is NO_2 mixed with water, forming nitric acid.

The first attempt to control NO_x with EGR valves was in the 1970s.

Oxygen

A certain amount of oxygen is required for proper catalytic converter operation. O_2 is not a pollutant; therefore, its presence in the exhaust does not pose any threat to our environment. However, too much oxygen in the exhaust does indicate that an improper mixture was in the cylinders or poor combustion has occurred in the engine. An improper air-fuel mixture will also cause a high reading of oxygen.

If the air-fuel ratio is rich, all the oxygen in the air is mixed with fuel and the O_2 levels in the exhaust are very low. When the air-fuel ratio is lean, there is not enough fuel to mix with all the air entering the engine, and O_2 levels in the exhaust are higher (**Figure 10-7**). Therefore, O_2 levels are a good indicator of a lean air-fuel ratio, and they are affected by catalytic converter operation to some extent because the converter stores oxygen.

Lower-than-normal O_2 emissions may be caused by the following:

- Rich air-fuel mixture
- Dirty air filter
- Faulty injectors
- Higher-than-normal fuel pressures
- Defective system input sensor
- Restricted PCV system
- Charcoal canister purging at idle and low speeds

Higher-than-normal O_2 emissions may be caused by the following:

- An engine misfire
- Lean air-fuel mixtures
- Vacuum leaks
- Lower-than-specified fuel pressures
- Defective fuel injectors
- Defective system input sensor
- Engine mechanical problems

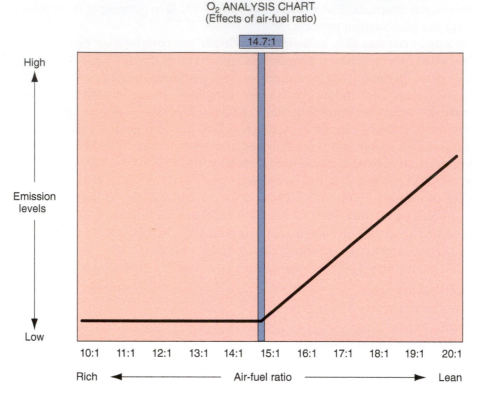

Figure 10-7 O_2 levels in relationship to air-fuel ratio.

Carbon Dioxide

CO_2 also is not a pollutant; however, it has been linked to the greenhouse effect. An ideal by-product of combustion is CO_2. Therefore, large amounts of this gas are desired. If the air-fuel ratio goes from 9:1 to 14.7:1, the CO_2 levels gradually increase from approximately 6 percent to 13.5 percent (**Figure 10-8**). CO_2 levels are highest when the air-fuel ratio is slightly leaner than stoichiometric. At the stoichiometric air-fuel ratio, CO_2 levels begin to decrease.

Lower-than-normal CO_2 emissions may be caused by the following:

- Leaking exhaust system
- Improper air-fuel mixture
- An engine misfire
- Insufficient combustion
- Engine mechanical problems

EVAPORATIVE EMISSION CONTROL SYSTEMS

If gasoline is allowed to evaporate into the air, it becomes a significant source of hydrocarbon emissions. Before the implementation of evaporative controls, it was determined that 20 percent of the vehicle's total emissions were occurring even while the vehicle was sitting in the driveway. Evaporative emissions were implemented in 1968. Prior to evaporative emission controls, the fuel tank was simply vented to the atmosphere. Of course, the fuel tank does have to be vented, the pressure relieved, or both, depending on conditions. Warm fuel will cause pressure to build inside the tank, or as fuel is used, a vacuum would develop inside that the fuel pump would not be able to overcome. To solve this problem, the tank is connected to and vented into a charcoal canister. The evaporative

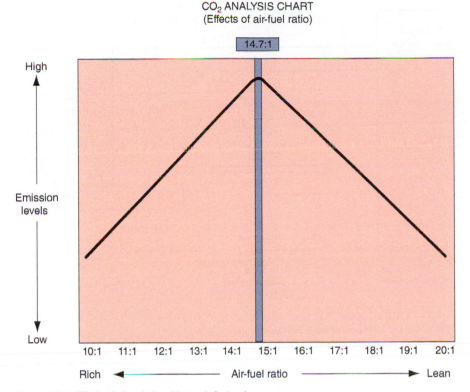

Figure 10-8 CO_2 levels in relationship to air-fuel ratio.

emissions system stores, the vapors until certain vehicle operating conditions are present. The vapors are then drawn out of the canister by vacuum and burned in the engine by the action of the **purge valve**. To help control the expansion of the fuel tank when warm, the tank is designed to never be totally full of fuel. The evaporative emissions system is monitored by the PCM to determine if the fuel system is sealed and HC vapors are not escaping without going through the charcoal canister.

Early evaporative emission systems since the 1980s did have a purge valve that was computer controlled (**Figure 10-9**). The **evaporative emission control (EVAP)** system in use since around 1996 is the **enhanced EVAP**. The **vent valve** and tank pressure sensor

The **evaporative emission control (EVAP)** system is a sealed system. It traps the fuel vapors that would normally escape from the fuel tank into the air. The onboard refueling vapor recovery system catches HC emissions when the vehicle is refueled.

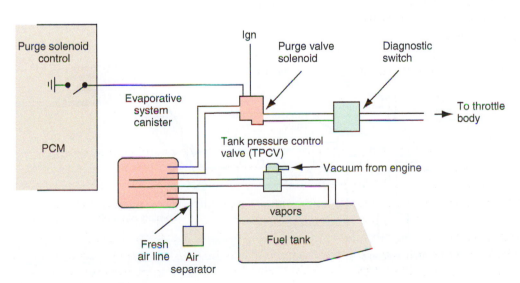

Figure 10-9 Basic EVAP system components.

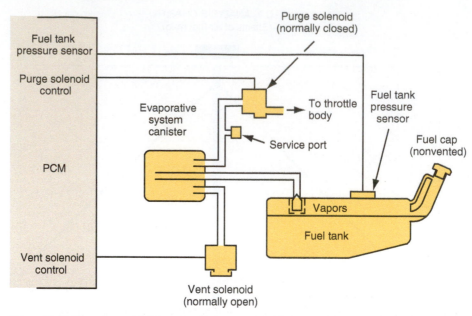

Figure 10-10 The enhanced EVAP system. The purge solenoid, the evaporative system canister, the vent solenoid, the fuel tank pressure sensor, and the fuel cap.

are used for diagnostic purposes, and the fuel level sensor was added to the list of input sensors to the PCM. With enhanced EVAP, the PCM conducts several tests to determine if the system is operational and there are no leaks (**Figure 10-10**). Modern evaporative emission systems must be able to detect a leak as small as 0.020 inch in diameter and use this test to confirm that no leaks are present.

Another part of the evaporative emission system is called onboard refueling vapor recovery (ORVR). It was determined that even though the evaporation of fuel had been drastically reduced, fuel vapors being pushed out of the tank during refueling actually produced greater amounts of HC emissions than burning the tank of fuel. ORVR systems (**Figure 10-11**) store all the vapor produced when refueling directly into the evaporative canister, where they can be trapped until the computer opens the purge valve, at which time the refueling vapors are burned along with the rest of the stored fuel vapors. The

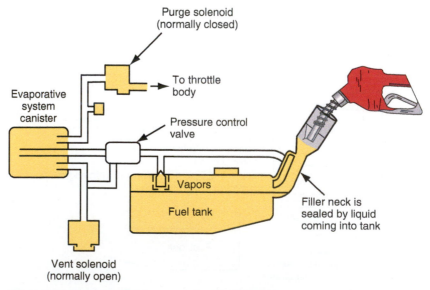

Figure 10-11 Onboard refueling vapor recovery (ORVR) system.

ORVR starts when the driver places the fuel fill nozzle into the tank. The pressure of the tank starts to rise as the fuel goes into the tank. A special filler neck design allows liquid fuel to seal the nozzle end so vapors cannot blow around the filler neck. The pressure control valve opens when pressure higher than atmospheric is present inside the tank, which opens the passage to the charcoal canister, storing the vapors in the canister. ORVR has been mandated since 1998. The filler neck, fuel tank vent, evaporative canister, and rollover check valve were all changed to provide this capability. Major components of the evaporative emission system are as follows for all systems:

> The canister on an ORVR vehicle can become warm while refueling.

- Charcoal canister in which to store the HC vapors until burned.
- Filler cap with vacuum and pressure relief valves (**Figure 10-12**).
- Fuel tank with a domed area to provide for expansion of fuel. The tank cannot be completely filled to help store vapors.
- Purge valve or solenoid to allow for the drawing of the fuel vapor into the engine.
- Fuel tank pressure control valve (TPCV).
- Hoses and tubing to connect the system components.
- Vent valve to control the flow of fresh air into the system.
- Vacuum pressure sensor mounted on the tank.
- Small filler neck that contains a check valve (ORVR-equipped vehicles).
- Check and one-way valves to keep liquid and vapor separated (**Figure 10-13**).
- Fuel level sensor.
- An EVAP pump, used on some vehicles to pressurize the system while the fuel tank pressure sensor monitors for leaks.

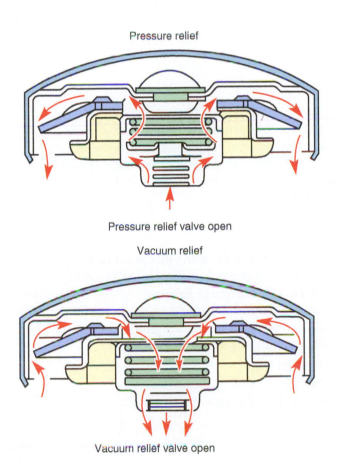

Figure 10-12 Sealed fuel tank cap.

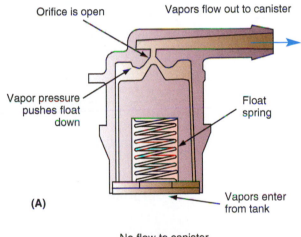

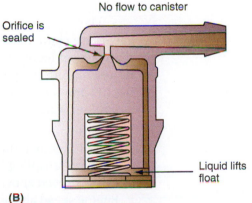

Figure 10-13 (A) Normal operation of vapor separator, (B) with liquid in separator.

Figure 10-14 An onboard refueling vapor recovery (ORVR) charcoal canister.

Note that there are various systems in existence, and always make sure to check specific information on the vehicle before starting work on any system.

A **charcoal (carbon) canister** is used to trap and store HC vapors until they are introduced into the engine where they are burned.

The **charcoal (carbon) canister** is the heart of the EVAP system. The charcoal canister is generally located near the fuel tank. The location near the fuel tank is especially true if the vehicle has the ORVR system. The canister has to be large enough on these vehicles to store the vapors from refueling. Charcoal has the ability to store the vapors and then release them when the PCM determines that the engine can handle the extra fuel load, usually after the engine is warm (**Figure 10-14**). When the PCM decides to purge the canister, fresh air is pulled into the canister with vacuum applied from the engine through the purge valve or solenoid.

AUTHOR'S NOTE A common misconception is that most systems are completely sealed until purged, because of the ability of the system to find leaks as small as 0.020 inch. Actually, on most vehicles, the vent is normally open. Air can flow into and out of the system freely until the vent valve is closed for diagnostic testing. The EVAP system relies on the charcoal canister to catch and hold the HC, while allowing the air to pass.

The purge valve has undergone many changes since the inception of EVAP. Purge valves are computer controlled to allow precise control of the vapors according to the many sensor inputs at the PCM (**Figure 10-15**). The purge solenoid can also be remotely mounted from the purge valve. The canister purge valve is normally closed. It opens the inlet to the purge outlet when the PCM commands purge. Purging is done only when conditions warrant. Typically some of the conditions are as follows:

- 150 seconds since the PCM entered closed loop
- Coolant temperature above 176°F (349°C)
- PCM not enabling injector shutdown, such as during a traction control
- Vehicle speed above 20 mph
- Engine speed above 1,100 rpm
- Temperature sensor not indicating overheating
- Low coolant not indicated

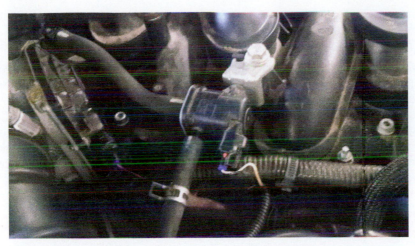

Figure 10-15 The purge valve is activated by the ECM to clean the charcoal vapors from the canister.

The conditions that cause purge to occur will, of course, vary from vehicle to vehicle.

The vent valve is normally open to allow fresh air through the charcoal canister. The vent valve is also PCM controlled. The vent valve can be closed with the purge valve open while the PCM monitors the **fuel tank pressure sensor** for vacuum. Vacuum should build with the vent closed and the purge open. The vent is checked for blockage during normal purging by checking for excessive vacuum with the vent valve open. If there is vacuum drawn on the fuel tank with the vent and purge valves open, then the vent must be blocked. Finally, the purge valve can be checked for leaks by closing both the purge and vent valves and determining if a vacuum begins to build. If it does, the purge valve is leaking.

The most recent use of the vent valve and tank pressure sensor on EVAP systems on some vehicles is the **engine off natural vacuum (EONV)** test. The system runs a pressure and vacuum test based on the natural volatility of the fuel. After driving, the fuel will be warm. If the vent to the tank is closed, the fuel tank vapor pressure will build in a predictable way according to the ambient temperature. After the fuel starts to cool down, again with the vent closed, a natural vacuum will build within the tank that the system can measure. The PCM can use this information to determine if there are any small leaks in the system. If the fuel is too volatile, as determined by the rise in pressure after the engine is turned off, the test is canceled.

The fuel tank pressure sensor is mounted on the fuel tank and can measure vacuum and pressure at the tank. The fuel tank pressure sensor is a PCM input.

The **fuel tank level sensor** is actually the same sensor used for the fuel gauge, but on OBD II vehicles it is used to determine whether the fuel tank pressure or vacuum test should be run. If the tank is not between 15 and 85 percent, the test will not run. The fuel tank sensor can also be checked at start-up and during driving to detect refueling, which could upset the results of the leak test.

The **fuel pressure control valve** is designed to control fuel tank pressure while the vehicle is sitting still. Vapors are stored in the tank with the valve closed. If tank pressure builds too high, the valve opens and lets the vapor into the canister. When the engine is running, the valve has vacuum applied, opening the valve and allowing vapors to be stored in the canister until purging.

Some systems use a **leak detection pump (LDP)** that pressurizes the system and checks for leaks, along with the fuel tank pressure sensor. Most manufacturers have opted to use the EONV test because it does not require as much hardware to operate.

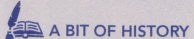

PRE-COMBUSTION SYSTEMS

Systems designed to prevent or limit the amount of pollutants produced by an engine are called pre-combustion emission control devices. Although specific systems and engine designs are classified this way, anything that makes an engine more efficient can be categorized as a pre-combustion emission control. The specific systems discussed in this chapter are as follows:

- Engine design changes
- PCV system
- Ignition control systems
- EGR systems

ENGINE DESIGN CHANGES

In recent years, the basic engine has seen many technological changes. For the most part, the basics of operation have not been changed. Engineers have worked overtime in an attempt to squeeze as much out of small engines as they can. Many of these changes have not only increased the efficiency of the engines but have decreased the pollutants released by the engines. Also, some of the changes were only necessary or brought about to accommodate changes in the engine's fuel and ignition systems.

Better Sealing Piston Rings

One of the first areas of concern for engineers and emission control devices was the crankcase of an engine. PCV systems relieve the crankcase of unwanted gases and pressure. The PCV did not solve the problem of these pressurized gases blowing by the piston rings. Not only do these blow-by gases represent a problem for the engine's lubrication, but they are an indication of wasted energy as well. The blow-by gases start out to be fuel-air mixture. The mixture leaks past the piston rings during the compression and power strokes of the engine. Instead of being used to produce energy, this sampling of mixture is used only to dilute the oil. The dilution of oil with unburned fuel also results in the formation of sludge in the engine, which is very harmful to the lubrication and cooling processes of the oil. Blow-by gases can and are being reduced in some engines through the use of better sealing piston rings and improved cylinder wall surfaces. Many of these better sealing piston rings also have frictional qualities that make them less of a drag. This increases fuel economy and engine power.

> Poor sealing pistons also allow engine oil to enter the combustion chamber while the piston is on its intake stroke.

Combustion Chamber Designs

Combustion chamber design has seen many changes. The aim of these changes is to reduce or eliminate the quench area of the combustion chamber. The quench area (**Figure 10-16**) is any place in the chamber where the flame front of combustion is

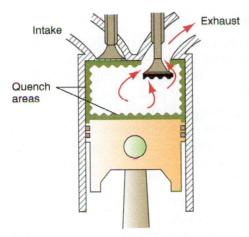

Intake

Exhaust

Quench areas

Figure 10-16 Unburned fuel is left on all surface areas.

cooled as it tries to move into a small area. By removing the quench areas in a combustion chamber, HC and CO emissions can be reduced. Another change in the design of combustion chambers has been the locating of the spark closer to the center of the chamber. The shape of the combustion chamber has also been optimized for the best combustion possible. This provides for a more even burn and allows for leaner air-fuel mixtures. Manufacturers have also worked with designs that cause controlled turbulence in the chamber (**Figure 10-17**). This turbulence improves the mixing of the fuel with the air, which results in improved combustion.

> Manufacturers have begun placing the piston rings closer to the top of the piston to reduce the quench area and the resulting hydrocarbon emissions.

Changes in Compression Ratios

Combustion chamber designs have also affected the compression ratios of engines. By keeping the compression ratio low, combustion temperatures can be kept below the point where NO_x is formed. Initially, when compression ratios dropped, engine performance and fuel economy also dropped. With the application of new technology, engine performance and fuel economy have greatly improved in spite *higher* compression ratios. Direct fuel injection has helped improve emissions and engine performance with high compression ratios. Faster computers, better timing control, direct fuel injection, and variable valve timing strategies are making great strides in automotive technology.

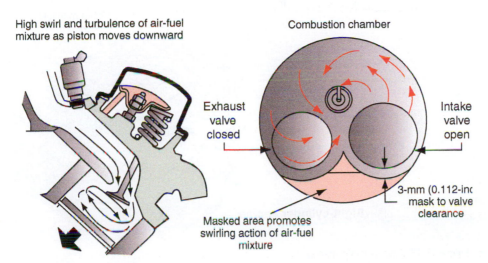

High swirl and turbulence of air-fuel mixture as piston moves downward

Combustion chamber

Exhaust valve closed

Intake valve open

Masked area promotes swirling action of air-fuel mixture

3-mm (0.112-inc mask to valve clearance

Figure 10-17 Combustion chamber designed to control turbulence.

Decreased Friction

Power losses through friction are a major contributor to the overall inefficiency of a gasoline engine. The friction of all of the moving parts of an engine results in a large loss of usable power and energy. By reducing the friction at key points within the engine, engineers have reduced the amount of power lost. By regaining power, fuel economy and engine performance are increased. Emissions are also reduced because less fuel is burned to produce the same amount of usable power. Improved engine oils, new component materials, and weight reductions have had the biggest impact on reducing friction.

Intake Manifold Designs

Intake manifolds have been designed to distribute equal amounts of air to each cylinder. Most of the design changes to intake manifolds have been afforded by the use of port fuel injection (**Figure 10-18**). The use of plastics in the manufacture of intake manifolds have allowed for smoother runners and better heat control of the air. Intake manifolds can be designed to be more efficient at low engine speeds, high engine speeds, or both. Again, increased efficiency results in decreased emissions.

Improved Cooling Systems

Today's engines have a higher normal operating temperature than older engines. The higher engine temperature reduces HC and CO emissions. However, the higher temperature also makes the formation of NO_x harder to control. Most engine cooling systems have been designed to run at high temperatures but are prevented from becoming too hot. This prevention reduces NO_x. The high engine temperature results in increased efficiency because the engine parts around the combustion chamber are hotter. This means less heat

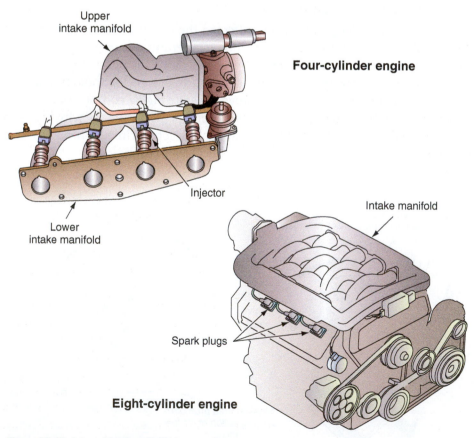

Figure 10-18 A tuned intake manifold.

from combustion is absorbed by the metal engine parts, leaving more heat for power conversion. Today's engine control systems incorporate many features that change air-fuel mixture, ignition timing, and idle speed to control the temperature of the engine.

Shop Manual
Chapter 10, page 539

PCV SYSTEMS

During the last part of the engine's combustion stroke, some unburned fuel and products of combustion—water vapor, for instance—leak past the engine's piston rings into the crankcase. This leakage into the engine crankcase is called **blow-by**. Blow-by gases must be removed from the engine before it condenses in the crankcase and reacts with the oil to form sludge. Sludge, if allowed to circulate with engine oil, corrodes and accelerates the wear of pistons, piston rings, valves, bearings, and other internal working parts of the engine. Blow-by gases must also be removed from the crankcase to prevent premature oil leaks. Because these gases enter the crankcase by the pressure formed during combustion, they pressurize the crankcase. The gases exert pressure on the oil pan gasket and crankshaft seals. If the pressure is not relieved, oil is eventually forced out of these seals.

Blow-by leakage is post-combustion gas that leaks into the crankcase.

Because the air-fuel mixture in an engine never completely burns, blow-by also carries some unburned fuel into the crankcase. If not removed, the unburned fuel dilutes the crankcase oil. When oil is diluted with gasoline, it does not lubricate the engine properly, causing excessive wear.

Combustion gases that enter the crankcase are removed by a positive crankcase ventilation system that uses engine vacuum to draw fresh air through the crankcase. This fresh air dissipates the harmful gases and enters through the air filter or through a separate PCV breather filter located on the inside of the air filter housing.

Because the vacuum supply for the PCV system is from the engine's intake manifold, the airflow through this system must be controlled in such a way that it varies in proportion to the regular air-fuel ratio being drawn into the intake manifold. Otherwise, the additional air that is drawn into the system would cause the air-fuel mixture to become too lean for efficient engine operation. Therefore, a PCV valve is placed in the flow just before the intake manifold to regulate the flow according to vacuum.

The positive crankcase ventilation system has two major functions. It prevents the emission of blow-by gases from the engine crankcase to the atmosphere. These gases were once vented through a road draft tube. Now they are recirculated to the engine intake and burned during combustion. It also scavenges the crankcase of vapors that could dilute the oil and cause it to deteriorate or that could build undesirable pressure in the crankcase. Fresh air from the air cleaner mixes with these vapors and makes them flow to the intake.

The PCV system benefits the vehicle's drivability by eliminating harmful crankcase gases, reducing air pollution, and promoting fuel economy. An inoperative PCV system could shorten the life of the engine by allowing harmful blow-by gases to remain in the engine, causing corrosion and accelerating wear.

The PCV valve is usually mounted in a rubber grommet in one of the valve covers. A hose is connected from the PCV valve to the intake manifold. A clean air hose is connected from the air cleaner to the opposite rocker arm cover. A filter is positioned in the air cleaner end of the clean air hose (**Figure 10-19**). On some systems, the PCV valve is mounted in a vent module and the clean air filter is located in this module (**Figure 10-20**).

When the engine is running, intake manifold vacuum is supplied to the PCV valve. This vacuum moves air through the clean air hose into the rocker arm cover. From this location, air flows through cylinder head openings into the crankcase where it mixes with blow-by gases that escape from the combustion chamber past the piston rings. The mixture of blow-by gases and air flows up through cylinder head openings to the rocker arm cover and PCV valve. Intake manifold vacuum moves the blow-by gas mixture through

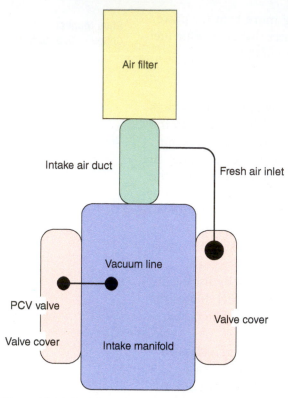

Figure 10-19 The PCV system takes filtered air from the intake air duct to replace air pulled into the intake manifold.

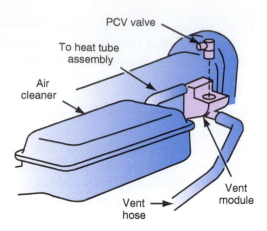

Figure 10-20 PCV valve mounted in a vent module.

the PCV valve into the intake manifold (**Figure 10-21**). The blow-by gases are then moved through the intake valves into the combustion chambers where they are burned.

On many engines, the PCV system delivers blow-by gases to one location in the intake manifold. This type of system may not deliver these gases equally to all the cylinders. This action may result in an air-fuel ratio variation between the cylinders, which results in rougher idle operation. Some engines, such as Ford's 4.6-liter V8, have passages from the PCV valve through the intake manifold to supply blow-by gases equally to each cylinder, resulting in smoother idle operation.

A PCV valve contains a tapered valve. When the engine is not running, a spring keeps the tapered valve seated against the valve housing (**Figure 10-22**). During idle or deceleration, the high intake manifold vacuum moves the tapered valve upward against the spring tension. Under this condition, there is a small opening between the tapered valve and the PCV valve housing (**Figure 10-23**). Since the engine is not under heavy load during idle or deceleration operation, blow-by gases are minimal, and the small PCV valve opening is adequate to move the blow-by gases out of the crankcase.

Intake manifold vacuum is lower during part-throttle operation than during idle operation. Under this condition, the spring moves the tapered valve downward to increase the opening between this valve and the PCV valve housing (**Figure 10-24**). Since engine load is higher at part-throttle operation than at idle operation, blow-by gases are increased. The larger opening between the tapered valve and the PCV valve housing allows all the blow-by gases to be drawn into the intake manifold.

When the engine is operating under heavy load conditions with a wide throttle opening, the decrease in intake manifold vacuum allows the spring to move the tapered valve farther downward in the PCV valve (**Figure 10-25**). This action provides a larger opening

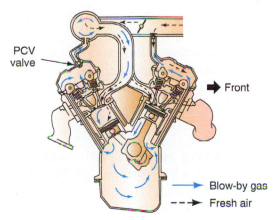

PCV valve

Front

→ Blow-by gas
--→ Fresh air

Figure 10-21 Operation of PCV system.

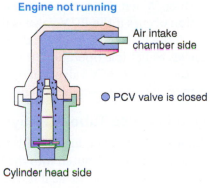

Engine not running

Air intake chamber side

● PCV valve is closed

Cylinder head side

Figure 10-22 PCV valve position with the engine not running.

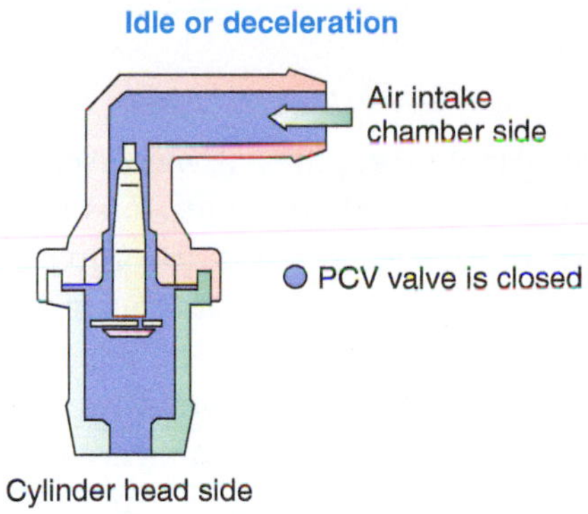

Idle or deceleration

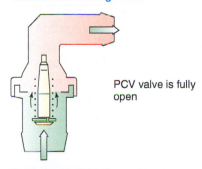

Air intake chamber side

● PCV valve is closed

Cylinder head side

Figure 10-23 PCV valve position during idle or deceleration.

Normal operation

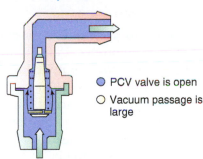

● PCV valve is open
○ Vacuum passage is large

Figure 10-24 PCV valve position during part-throttle operation.

Acceleration or high load

PCV valve is fully open

Figure 10-25 PCV valve position during hard acceleration or heavy load.

between the tapered valve and the PCV valve housing. Since higher engine load results in more blow-by gases, the larger PCV valve opening is necessary to allow these gases to flow through the valve into the intake manifold.

When worn rings or scored cylinders allow excessive blow-by gases into the crankcase, the PCV valve opening may not be large enough to allow these gases to flow into the intake manifold. Under this condition, the blow-by gases create a pressure in the crankcase, and some of these gases are forced through the clean air hose and filter into the air

An orifice is a small opening inserted in a passage that limits the amount of vacuum, in this case, applied to the crankcase.

cleaner. When this action occurs, there is oil in the PCV filter and air cleaner. This same action occurs if the PCV valve is restricted or plugged.

If the PCV valve sticks in the wide-open position, excessive airflow through the valve causes rough idle operation. If a backfire occurs in the intake manifold, the tapered valve is seated in the PCV valve as if the engine is not running. This action prevents the backfire from entering the engine where it could cause an explosion.

Fixed Orifice Tube PCV System

Some engines are equipped with a PCV system that does not use a PCV valve. Rather, the blow-by gases are routed into the intake manifold through a fixed orifice tube. The basic system works the same as if it had a valve except that the system is regulated only by the vacuum on the orifice. The size of the orifice limits the amount of blow-by flow into the intake. The engine's air-fuel system is calibrated for this calibrated air leak. Since the action of the PCV allows unmetered air into the intake, the air-fuel system must be set for this amount of extra air.

Shop Manual
Chapter 10, page 538

IGNITION CONTROL SYSTEMS

Spark control systems have been in use since the earliest gasoline engines. It was discovered that the proper timing of the ignition spark helped reduce exhaust emissions and develop more power output. Incorrect timing affects the combustion process. Incomplete combustion results in HC emissions. High CO emissions can also result from incorrect ignition timing. Advanced timing can also increase the production of NO_x. When timing is too far advanced, combustion temperatures rise. For every 1 degree of over advance, the temperature increases by 125°F (52°C). Today's vehicles have advanced timing control based on inputs to the PCM. The PCM knows the position and speed of the crankshaft and the camshaft at all times.

In many late-model vehicles with variable valve timing (VVT), the EGR valve has been eliminated. The VVT system can be configured so that the amount of valve timing overlap will leave a portion of the exhaust gas in the cylinder to provide for NO_x control. Since the NO_x control requirements vary on different engines, there are several different systems with various controls to provide these functions.

Knock Sensor and Knock Sensor Module

Many engines with an EFI have a knock sensor or sensors. The knock sensors may be mounted in the block, cylinder head, or intake manifold. A piezoelectric sensing element is mounted in the knock sensor, and a resistor is connected parallel to this sensing element (**Figure 10-26**). When the engine detonates, a vibration occurs in the engine. The piezoelectric sensing element changes this vibration to an analog voltage, and this signal is sent to the knock sensor module, which is generally built into the ECM (**Figure 10-27**). When the PCM receives this signal, it reduces the spark advance incrementally to prevent detonation.

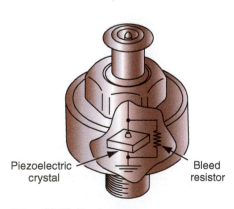

Figure 10-26 Knock sensor.

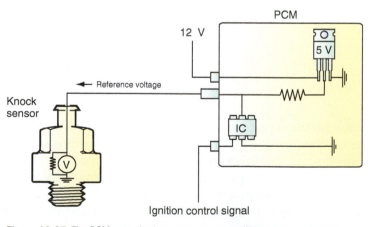

Figure 10-27 The PCM uses the knock sensor to modify timing and prevent spark knock.

EGR SYSTEMS

Exhaust gas recirculation systems reduce the amount of oxides of nitrogen emitted. The EGR system dilutes the air-fuel mixture with controlled amounts of exhaust gas. Since exhaust gas does not burn, this reduces the peak combustion temperatures. At lower combustion temperatures, very little of the nitrogen in the air combines with oxygen to form NO_x. Most of the nitrogen is simply carried out with the exhaust gases. For drivability, it is desirable to have the EGR valve opening (and the amount of gas flow) proportional to the throttle opening. Drivability is also improved by shutting off the EGR when the engine is started up cold, at idle, and at full throttle.

EGR systems have to be monitored by the OBD II system for operation. Exhaust is added into the intake charge to lower the combustion chamber temperature. When the EGR valve opens, the exhaust gas mixes with the incoming air-fuel mixture and then enters the intake manifold. The effect is to dilute or lean-out the mixture so that it still burns completely but with a reduction in combustion chamber temperatures.

The vacuum-controlled EGR is opened by vacuum and is closed by a spring. The vacuum used on modern EGR valve systems is computer controlled (**Figure 10-28**). The duty cycle of the vacuum solenoid can be changed to regulate the opening of the valve. The TPS (throttle position sensor), CTS (coolant temperature sensor), and MAP (manifold absolute pressure) sensor are all monitored to determine if and how much EGR is needed to limit NO_x production. Manifold vacuum that is pulse width modulated (PWM) through a solenoid called the **electronic vacuum regulator valve (EVRV)** is used to open the EGR valve. The PCM on OBD II vehicles must also determine that the EGR valve has opened when commanded and that EGR is flowing into the intake manifold. One strategy to do this monitors the MAP sensor to sense a change in the intake manifold pressure. Other valves may use temperature sensors or dedicated vacuum sensors to perform this function. Make sure to check with service literature before starting work on any system.

Figure 10-29 shows a Ford EGR system diagram. The ECM can command the amount of EGR needed depending on engine temperature, throttle position, and engine load. The PCM can then use this information to verify EGR valve opening by monitoring the MAP sensor signal. When EGR is commanded, the MAP increases, and/or modifies the amount of opening to match the engine's needs.

Shop Manual
Chapter 10, page 544

The term "drivability" is commonly used to describe the ability of an engine to move a vehicle under the different driving conditions.

Shop Manual
Chapter 10, page 547

The ECM is the powertrain control module.

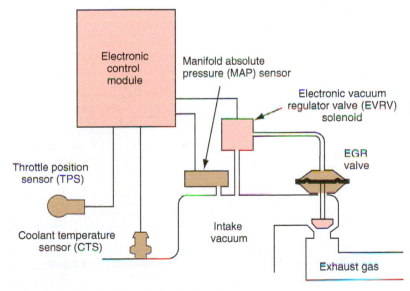

Figure 10-28 A computer-controlled EGR system.

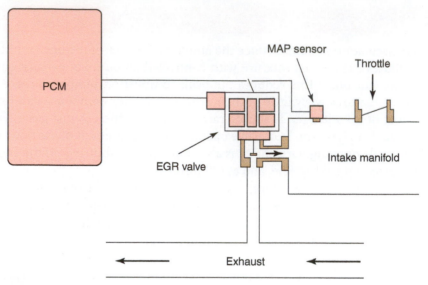

Figure 10-29 EEGR System.

Types of EGR Valves

Digital EGR Valve. A digital EGR valve contains up to three electric solenoids that are operated directly by the PCM (**Figure 10-30**). Each solenoid contains a movable plunger with a tapered tip that seats in an orifice. When any solenoid is energized, the plunger is lifted and exhaust gas is allowed to recirculate through the orifice into the intake manifold. The solenoids and orifices are of different sizes. The PCM can operate one, two, or three solenoids to supply the amount of exhaust recirculation required to provide optimum control of NO_x emissions.

Linear EGR Valve. The linear EGR valve contains a single electric solenoid that is operated by the PCM. A tapered pintle is positioned on the end of the solenoid plunger. When the solenoid is energized, the plunger and tapered valve are lifted and exhaust gas is allowed to recirculate into the intake manifold (**Figure 10-31**). The EGR valve contains an EGR valve position (EVP) sensor, which is a linear potentiometer. The signal from this

Shop Manual
Chapter 10, page 550

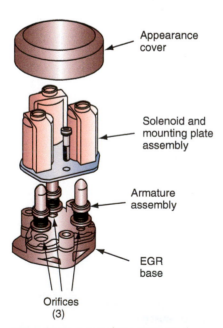

Figure 10-30 A digital EGR valve with three solenoids.

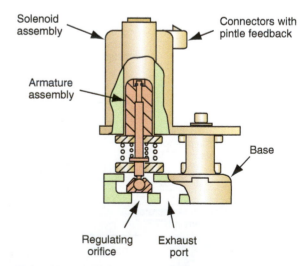

Figure 10-31 Linear EGR valve operation.

sensor varies from approximately 1 volt with the EGR valve closed to 4.5 volts with the valve wide open.

The PCM pulses the EGR solenoid winding on and off with a pulse width modulation principle to provide accurate control of the plunger and EGR flow. The EVP sensor acts as a feedback signal to the PCM to inform the PCM if the commanded valve position was achieved.

Electronic EGR Controls

These various EGR system controls represent some of the common controls currently used by automobile manufacturers. Control devices used in the various systems might have different labels but actually complete the same function within the EGR system. Engines with electronic engine control systems control EGR action in many different ways. The most common of these follow.

Shop Manual
Chapter 10, page 551

EGR Vacuum Regulator (EVR). In many EGR systems the PCM operates a normally closed EGR vacuum regulator solenoid (**Figure 10-32**), which supplies vacuum to the EGR valve. If the EVR solenoid is not energized, the solenoid plunger tip is seated in the vacuum passage and shuts off vacuum to the EGR valve. When the EVR solenoid plunger is shutting off vacuum to the EGR valve, any vacuum in the EGR valve and hose is vented through the EVR solenoid to prevent vacuum from being locked in the system, which could hold the EGR valve open.

When the PCM inputs indicate that the EGR valve should be open, the PCM provides a ground for the EVR solenoid winding. This action moves the solenoid plunger and opens the vacuum passage through the solenoid to the EGR valve. In some systems, the PCM pulses the EVR on and off to supply the precise vacuum and EGR valve opening required by the engine.

The PCM uses inputs such as engine temperature, throttle position, and vehicle speed to operate the EGR valve. The PCM will not energize the EVR solenoid if the engine coolant temperature is below a preset value. The EGR valve is opened by the PCM when the vehicle is operating at normal temperature in the cruising speed range. When the vehicle is operating at low speed or near wide-open throttle, the PCM does not open the EGR valve. If the EGR valve is open while the engine is idling or operating at low rpm, engine operation is erratic.

EGR System with Pressure Feedback Electronic (PFE) Sensor. Some EGR systems have a PFE sensor. These systems have an orifice located in the exhaust passage below the EGR valve. A small pipe connected from this orifice chamber supplies exhaust pressure to the PFE sensor (**Figure 10-33**).

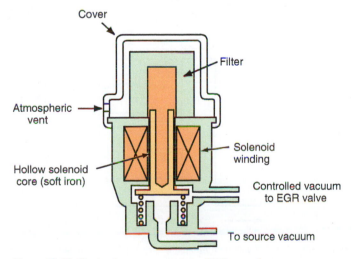

Figure 10-32 Electronic vacuum regulator (EVR) operation.

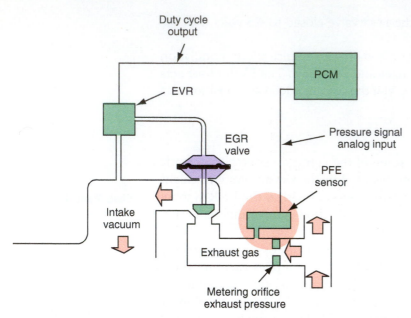

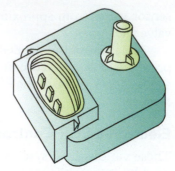

Figure 10-34 PFE sensor.

Figure 10-33 PFE sensor and related circuit.

The PFE sensor changes the exhaust pressure signal to a voltage signal that is sent to the PCM. Three wires are connected from the PFE sensor to the PCM (**Figure 10-34**). These wires include ground, 5-volt reference, and signal wires. The exhaust pressure in the orifice chamber is proportional to the EGR valve flow. The PFE signal informs the PCM regarding the amount of EGR flow, and the PCM compares this signal to the EGR flow requested by the input signals. If there is some difference between the actual EGR flow indicated by the PFE signal and the requested EGR flow, the PCM makes the necessary correction to the EVR output signal.

In some EGR systems, two pipes supply exhaust pressure from above and below the orifice under the EGR valve to the **Delta pressure feedback EGR (DPFE)** sensor (**Figure 10-35**).

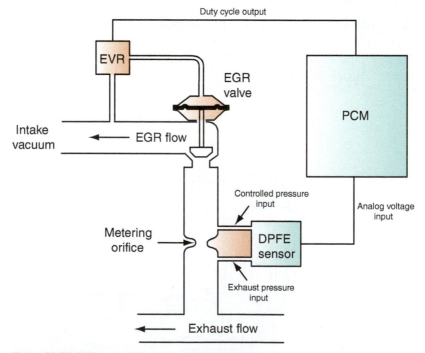

Figure 10-35 Differential PFE with dual exhaust pressure pipes connected above and below the EGR orifice to the DPFE.

EGR System Module Components

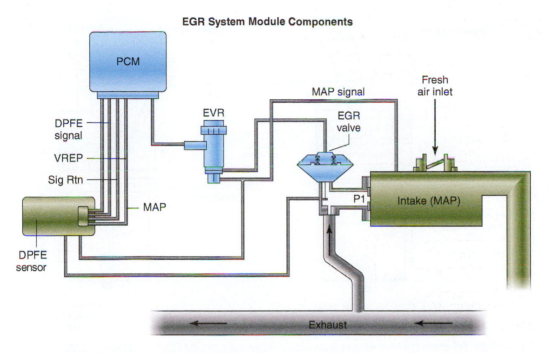

Figure 10-36 ESM EGR system is similar to the older DPFE systems but compares the difference between the pressure at the EGR inlet and vacuum (MAP) signal to verify EGR operation.

ESM EGR Operation

A variation on the DPFE EGR system is the EGR System Module (ESM) system. The DPFE sensor compares the vacuum pressure in the intake manifold to the pressure at the inlet of the EGR (**Figure 10-36**). When the valve is closed, the pressures at the EGR valve port and the manifold vacuum would be the same. When the EGR valve is opened, the pressure at the EGR port would rise due to the pressure in the exhaust, and the MAP pressure would also rise to a higher point than that with the valve closed.

VVT Action That Replaces Conventional EGR

Most late-model vehicles do not use an EGR valve. The EGR valve has been replaced by manipulating the intake valve opening by use of variable valve timing (**Figure 10-37**). Variable valve timing allows the intake valve opening to be advanced to the point that the

Figure 10-37 The camshaft can change its relationship to that of cam sprocket. This allows the intake valves to open and close early or late.

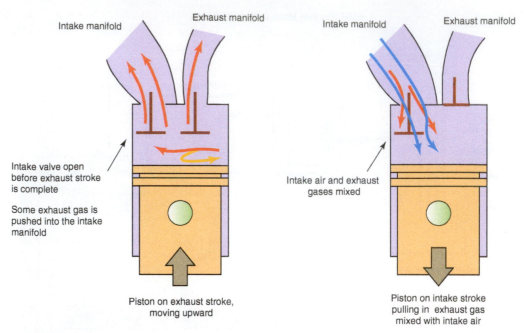

Intake manifold   Exhaust manifold   Intake manifold   Exhaust manifold

Intake valve open
before exhaust stroke
is complete

Some exhaust gas is
pushed into the intake
manifold

Intake air and exhaust
gases mixed

Piston on exhaust stroke,
moving upward

Piston on intake stroke
pulling in exhaust gas
mixed with intake air

Figure 10-38 VVT action can allow exhaust gases into the intake and make the EGR valve and its passages obsolete.

valve opens very early. It actually opens during the exhaust stroke. As the piston is on the way up during the exhaust stroke, some of the exhaust gas is actually pushed into the intake manifold. This exhaust gas cools momentarily, then reenters the combustion chamber during the actual intake stroke (**Figure 10-38**).

POST-COMBUSTION SYSTEMS

Shop Manual
Chapter 10, page 555

Post-combustion emission control devices clean up the exhaust after the fuel has been burned but before the gases exit the vehicle's tailpipe. An excellent example of this is the catalytic converter. A converter is one of the most effective emission control devices on a vehicle for reducing HC, CO, and NO_x.

CATALYTIC CONVERTERS

Catalytic converters are often referred to as "Cats."

Catalytic converters are the most effective devices for controlling exhaust emissions. Until 1975, carmakers had done a somewhat effective job of controlling emissions by the use of other systems—auxiliary air injection systems, exhaust gas recirculation systems, and positive crankcase ventilation. But controlling emissions with these systems alone also meant lean mixtures and exotic ignition timing that often severely penalized power and fuel economy. When catalytic converters were introduced, much of the emission control could be taken out of the engine and moved into the exhaust system. This change allowed manufacturers to retune the engine for better performance and improved fuel economy.

Many changes will be necessary in vehicles, engines, and emission systems to meet emission standards. Approximately 90 percent of HC emissions occur before the catalytic converter is hot enough to provide for proper HC oxidation. Another post-combustion system that has been used in the past is the secondary air or air injection system. This system is used to help reduce cold start emissions. The AIR injects air into the exhaust manifolds for the first few seconds of cold engine operation. The burning of excess HC in the exhaust manifold helps get the catalytic converter hot enough that it can burn excessive HC on its own. With the advent of DFI, one strategy that can help heat the converter is the injection of fuel late during the power stroke. The high pressures will burn the fuel rapidly, and the excess heat can warm the converter quickly.

A catalytic converter contains a ceramic element coated with a catalyst. A catalyst is something that causes a chemical reaction without being part of the reaction. A catalytic converter causes a chemical change to take place in the passing exhaust gases. Most of the harmful gases are changed to harmless gases.

Three different materials are used as the catalyst in automotive converters: platinum, palladium, and rhodium. Platinum and palladium are the oxidizing elements of a converter. When HC and CO are exposed to heated surfaces covered with platinum and palladium, a chemical reaction takes place. The HC and CO are combined with oxygen to become H_2O and CO_2. Rhodium is a reducing catalyst. When NO_x is exposed to hot rhodium, oxygen is removed and NO_x becomes just N. The removal of oxygen is called reduction, which is why rhodium is a reducing catalyst. A major factor in the production of acid rain and converter degradation is sulfur in fuel. **Tier 3** emission standards have reduced sulfur in fuel from around 80 ppm to 10 ppm, resulting in longer converter life and fewer hydrogen sulfide emissions.

Catalytic converters contain all three catalysts; reduce HC, CO, and NO_x; and are called three-way converters (**Figure 10-39**). Three-way converters have the oxidizing catalysts in part of the container and the reducing catalyst in the other (**Figure 10-40**).

Shop Manual
Chapter 10, page 556

SECONDARY AIR INJECTION

One of the earliest methods used to reduce the amount of hydrocarbons and carbon monoxide in the exhaust was to force fresh air into the exhaust system after combustion. This additional fresh air causes further oxidation and burning of the unburned hydrocarbons and carbon monoxide. The process is much like blowing on a dwindling fire. Oxygen in the air combines with the HC and CO to continue the burning that reduces the HC and CO concentrations, while adding enough heat to help the catalytic converter get up to temperature faster. Modern AIR systems run for only 2 minutes or less on a cold engine, just to help reduce cold start emissions until the converter can warm up to operating temperature.

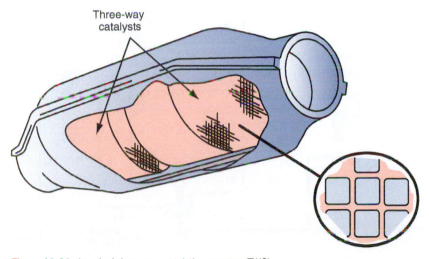

Three-way catalysts

Figure 10-39 A typical three-way catalytic converter (TWC).

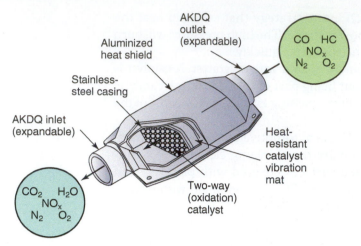

Figure 10-40 The action of a three-way catalytic converter.

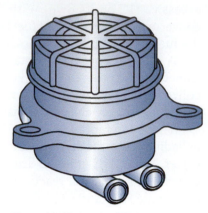

Figure 10-41 Modern AIR systems use an electric pump.

Shop Manual
Chapter 10, page 556

The **air diverter (AIRD)** valve directs the air to the exhaust manifold or to atmosphere.

Most modern vehicles that use secondary air have an air pump driven by an electric motor (**Figure 10-41**). In a commonly used system, air flows from the pump to a pair of diverter valves, which direct the air either to the atmosphere or to the exhaust manifold (**Figure 10-42**).

Some vehicles have used an electric AIR pump system and associated parts (**Figure 10-43**). The pump is activated by a solid state relay, which is controlled by the PCM. The air injection bypass solenoid is also controlled by the PCM. The bypass solenoid directs vacuum to the **air diverter (AIRD)** valves, which opens the normally closed valves (**Figure 10-44**). Air from the AIR system helps warm up the catalytic converter and oxygen sensors by burning HC and CO in the exhaust system. Once the catalytic converter is warm, the PCM shuts the pump down.

Bypass Mode

In the bypass mode, secondary air is vented to the atmosphere. Secondary air may be vented or bypassed due to a fuel-rich condition, which the onboard computer recognizes as a problem in the system or during deceleration.

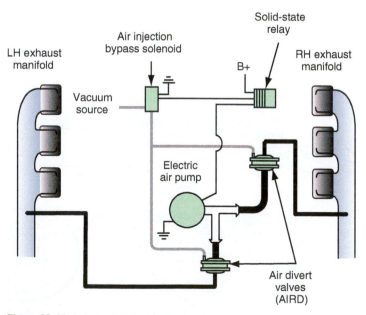

Figure 10-42 Late-model air system.

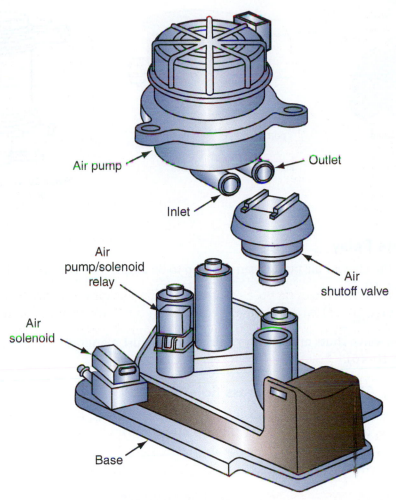

Figure 10-43 Electric AIR pump and components.

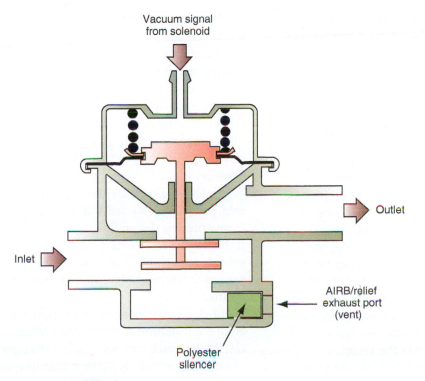

Figure 10-44 AIRD valve.

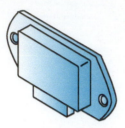

Figure 10-45 Solid state relay.

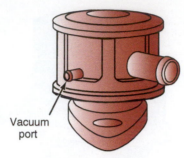

Vacuum port

Figure 10-46 Diverter valve.

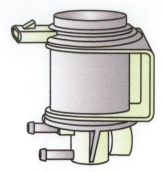

Figure 10-47 Secondary AIR injection bypass solenoid.

Solid State Relay

Given that the electric air pumps can draw up to 35 amps, they require a relay for operation (**Figure 10-45**).

Diverter Valve

The diverter valve shuts off the flow of air to the exhaust manifolds when the air pump is off (**Figure 10-46**).

Secondary AIR Injection Bypass Solenoid

The PCM uses the secondary AIR injection bypass solenoid to control vacuum to the diverter valve, thus closing the valve (**Figure 10-47**).

SUMMARY

- Unburned hydrocarbons, carbon monoxide, and oxides of nitrogen are three types of emissions controlled in gasoline engines.
- HC emissions are unburned gasoline released by the engine because of incomplete combustion.
- CO emissions are a by-product of combustion.
- CO emissions are caused by a rich air-fuel ratio.
- Oxides of nitrogen (NO_x) are formed when combustion temperatures reach more than 2,500°F (1,371°C).
- Pre-combustion control systems prevent emissions from being created in the engine either during or before the combustion cycle. Post-combustion control systems clean up exhaust gases after the fuel has been burned. The evaporative control system traps fuel vapors that would normally escape from the fuel and carburetor into the air.
- The PCV system removes blow-by gases from the crankcase and recirculates them to the engine intake. The PCV system benefits the vehicle's drivability by eliminating harmful crankcase gases, reducing air pollution, and promoting fuel economy.

- With the engine running at idle speed, the high intake manifold vacuum moves the PCV valve toward the closed position.
- During part-throttle operation, the intake manifold vacuum decreases and the PCV valve spring moves the valve toward the open position.
- As the throttle approaches the wide-open position, intake manifold vacuum decreases and the spring moves the PCV valve further toward the open position.
- When the engine backfires into the intake manifold, the PCV valve seats and prevents the backfire from entering the engine.
- If the engine has excessive blow-by or the PCV valve is restricted, crankcase pressure forces crankcase gases through the clean air hose into the air cleaner.
- A knock sensor changes a vibration caused by engine detonation to an analog voltage signal. The electronic spark control module changes the analog detonation sensor signal to a digital signal and sends this signal to the PCM.

- An evaporative emission control (EVAP) system stores vapors from the fuel tank in a charcoal canister until certain engine operating conditions are present. When the proper conditions are present, fuel vapors are purged from the charcoal canister into the intake manifold.
- A digital EGR valve has up to three electric solenoids operated by the PCM.
- A linear EGR valve contains an electric solenoid that is operated by the PCM with a pulse width modulation (PWM) signal.

- A pressure feedback electronic (PFE) sensor sends a voltage signal to the PCM in relation to the exhaust pressure under the EGR valve.
- Many engines use variable valve timing to replace the conventional EGR valve.
- Many secondary air injection systems pump air into the exhaust ports during engine warm-up and deliver air to the catalytic converters with the engine at normal operating temperature knock.
- Tier 3 emissions begin with the 2017 model year.

REVIEW QUESTIONS

Short-Answer Essay

1. Explain why a small PCV valve opening is adequate at idle speed.
2. Describe how photochemical smog is produced.
3. Explain how a knock sensor operates.
4. Describe the operation of a digital EGR valve.
5. Explain why a secondary air injection system pumps air into the exhaust ports during engine warm-up.
6. Describe the causes of high HC emissions.
7. Describe CO emissions in relation to air-fuel ratio.
8. Name the three main emissions being controlled in gasoline engines.
9. Give the basic parts of an Inspection/ Maintenance (I/M) program.
10. List three ways the PCV system benefits the vehicle's drivability.

Fill-in-the Blanks

1. Positive crankcase ventilation systems that route crankcase vapors back to the engine's _____ were developed and incorporated into 1961 cars and light trucks sold in California.

2. The knock sensor module changes the _____ voltage signal to a(n) _____ voltage signal and sends this signal to the PCM.

3. The components necessary to form photochemical smog are _____ and _____ exposed to _____ in stagnant air.

4. _____ emission control systems clean up the exhaust gases after the fuel has been burned.

5. Oxides of nitrogen (NO_x) are formed when combustion temperatures reach more than _____ °F.

6. The EGR valve has been replaced by manipulating the intake valve opening by use of _____.

7. Evaporative emissions from fuel tanks and evaporative systems are a source of _____ emissions.

8. CO emissions are a good indicator of a(n) _____ air-fuel ratio.

9. HC emissions are caused largely by _____ fuel.

10. A(n) _____ emission system stores vapors from the fuel tank in a charcoal canister until certain engine operating conditions are present.

Multiple Choice

1. *Technician A* says that CARB is an abbreviation for *California Air Resources Board*.
   Technician B says that PMVI is to inspect a vehicle's emission controls twice a year.
   Who is correct?

   A. Technician A   C. Both technicians

   B. Technician B   D. Neither technician

2. *Technician A* says that most manufacturers started adding controls to reduce oxides of nitrogen (NO_x) as early as 1960.

   Technician B says that the first catalytic converters were designed to oxidize CO and HC.

   Who is correct?

   A. Technician A
   B. Technician B
   C. Both technicians
   D. Neither technician

3. *Technician A* says that HCs result from unburned fuel and from incomplete combustion.

   Technician B says that CO is formed with a mixture that is too rich.

   Who is correct?

   A. Technician A
   B. Technician B
   C. Both technicians
   D. Neither technician

4. *Technician A* says that NO_x is formed at temperatures over 2,500°F.

   Technician B says that NO_x is formed when an engine overcools, due to a stuck open thermostat.

   Who is correct?

   A. Technician A
   B. Technician B
   C. Both technicians
   D. Neither technician

5. *Technician A* says that if gasoline is allowed to evaporate into the air, it becomes a significant source of CO emissions.

   Technician B says that before evaporative emission (EVAP) controls, 20 percent of the total emissions from a vehicle occurred while it was sitting in the driveway.

   Who is correct?

   A. Technician A
   B. Technician B
   C. Both technicians
   D. Neither technician

6. *Technician A* says that canister purging occurs only when conditions are right.

   Technician B says that modern purge valves are computer controlled.

   Who is correct?

   A. Technician A
   B. Technician B
   C. Both technicians
   D. Neither technician

7. All of the following are pre-combustion emission controls *except*:

   A. catalytic converter
   B. PCV system
   C. timing control systems
   D. EGR systems

8. *Technician A* says that lower compression ratios bring down NO_x emissions.

   Technician B says that lower compression ratios improve performance and fuel mileage.

   Who is correct?

   A. Technician A
   B. Technician B
   C. Both technicians
   D. Neither technician

9. *Technician A* says that PCV systems are used to relieve engine blow-by pressure.

   Technician B says that PCV systems also control sludge.

   Who is correct?

   A. Technician A
   B. Technician B
   C. Both technicians
   D. Neither technician

10. *Technician A* says that EGR systems reduce the amount of HC emission.

    Technician B says that EGR systems use controlled amounts of exhaust gases to cool combustion temperatures.

    Who is correct?

    A. Technician A
    B. Technician B
    C. Both technicians
    D. Neither technician

CHAPTER 11
COMPUTER OUTPUTS AND NETWORKS

Upon completion and review of this chapter, you should be able to:

- Describe high-side and low-side drivers.
- Explain how computer output drivers operate most output actuators.
- Describe quad drivers and output driver modules.
- Explain duty cycle.
- Explain pulse width modulation (PWM).
- Explain the output driver's role in the operation of fuel injectors.
- Describe computer control of the idle air control (IAC) system.
- Describe the powertrain control module (PCM) output role in the operation of the exhaust gas recirculating (EGR) valve.

- Describe computer output control of electronic spark timing (EST).
- Provide a brief overview of electronic transmission control by the PCM.
- Describe vehicle computer networking.
- Explain a serial data bus.
- Describe UART, J1850, KWP2000, and ISO 9141 communications.
- Explain the operation of controller area networking (CAN).
- Explain the two types of vehicle module wiring configurations.
- Describe peer-to-peer and master-to-slave communication.
- Examine vehicle network module function and PCM relationship.

Terms To Know

A/C sunload temperature sensor

Ambient air temperature sensor

Body control module (BCM)

Controller area network (CAN)

Customization

Driver door module (DDM)

Drivers

Duty cycle

Electronic automatic temperature control (EATC)

Electronic crash sensor (ECS)

Electronic spark timing (EST)

Front passenger door module (FPDM)

High-side driver

In-car temperature sensor

Lighting control module (LCM)

Loop network

Low-side driver

Masters

Output driver modules

Output drivers

Peak-and-hold injectors

Power door module (PDM)

Pulse width modulation (PWM)

Quad drivers

Serial data bus

Slaves

Stand-alone modules

Star

Vehicle theft security system (VTSS)

INTRODUCTION

The modern PCM must have a host of both input and output devices to function. You have already studied computer inputs. In this chapter we take an in-depth look at computer outputs and the latest in computer networking. Vehicle wiring and computer integration have undergone a revolution by using smart modules placed strategically throughout the vehicle that have the ability to communicate and make decisions. Networking is a way for individual modules to share information between themselves. Computer networking language has dramatically changed the way vehicles communicate with scan tools and with modules on the vehicle's own serial bus.

BASIC CIRCUIT CONTROL

Any circuit can be activated in one of two basic ways: by completing the ground path or completing the path to power. The computer takes advantage of both of these methods of activating or completing a circuit (**Figure 11-1**). Given that the computer only knows on and off, or digital signals, ways have to be developed to allow the PCM to communicate its commands in this manner.

DRIVERS

High-side drivers apply power to a circuit.
Low-side drivers apply ground to a circuit.

The PCM uses transistorized switches called **drivers** to control circuits or devices. If the computer activates a component or circuit by completing a ground, it is referred to as a **low-side driver**. **Figure 11-2** shows a typical wiring diagram for the instrument cluster warning lamps that are controlled by the PCM by supplying the ground for the bulbs when the conditions are right. Notice that this particular wiring diagram does not show a simplified representation of the complex circuitry inside the PCM, but because the power is supplied continually through the fuse, a technician will know this is how the circuit works. An example of a PCM providing power to a circuit is shown in **Figure 11-3**. When the computer provides power for a circuit, it is known as a **high-side driver**. The PCM turns the fuel pump relay coil on; the relay then closes and provides power to the fuel pump. The use of a relay prevents the PCM from handling large amounts of current flow. You may notice in Figure 11-3 that a small switch to B+ is a simplified representation of a computer driver. Some figures may show a transistor that actually represents a driver inside the PCM, as in **Figure 11-4**.

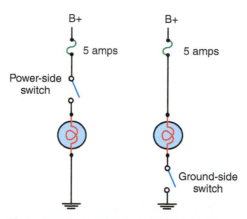

Figure 11.1 A circuit can be switched from either the power side or ground side.

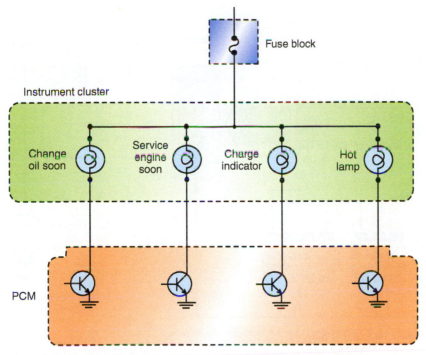

Figure 11-2 These warning lamps are turned on by the PCM by supplying ground.

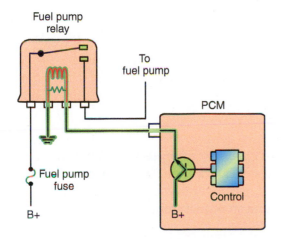

Figure 11-3 This fuel pump relay coil is activated by the PCM supplying power.

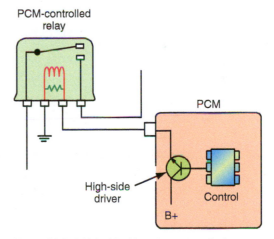

Figure 11-4 A high-side driver shown as a single transistor inside the PCM.

The computer uses **quad drivers** or **output driver modules** to activate a circuit or actuator. Quad drivers are assembled into groups of four (**Figure 11-5**). A quad driver has four independently controllable outputs that generally operate by completing a ground to the controlled circuit. Since the early 1990s, quad drivers have had electronic circuit breakers to protect themselves in the event of over-amperage. Prior to that time, many PCMs were replaced due to the **output drivers** burning out. The quad drivers were damaged by components such as solenoids or relay coils shorting internally and lowering the controlled circuit resistance. With lower resistance, amperage levels climbed too high and damaged components. Modern quad drivers are able to set trouble codes and reset after the problem component or circuit has been replaced. The problem with quad drivers has

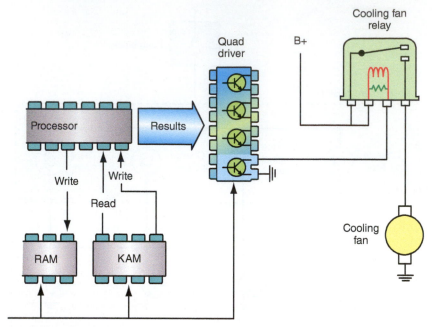

Figure 11-5 Output drivers on the computer usually supply a ground for the actuator solenoids and relays.

been that all four outputs that are usually controlled by a quad driver shut down when any of the drivers are affected.

> **AUTHOR'S NOTE** Whether shown as a simple switch or transistor, the driver is actually much more complex. Because the technician is concerned only with function, the internal circuitry does not have to be a true representation.

> **AUTHOR'S NOTE** Even though the internal circuitry can protect itself, it does not mean that it is immune to damage. For instance, if a high-current driver, such as that for a fuel injector, is shorted to ground, the driver may be destroyed. Be careful doing diagnostic tests and never ground an output from the PCM.

Output driver modules are being used today, which control up to seven independent outputs in much the same manner as the older quad driver, with the exception that the output driver module can shut down one output in case of trouble and leave the others operational. Additionally, more outputs are being controlled from the power sides of the circuit because of the greater diagnostic capability.

DUTY CYCLE

Duty cycle is a percentage of a fixed cycle. If a cycle was one second, a 50-percent duty cycle would mean the circuit was on for 0.5 second.

Duty cycle is the application of power or ground for a percentage of a fixed cycle. The computer can turn a circuit on and off rapidly. It can also vary the amount of time that the circuit is held off or on. Duty cycle is an example. The circuit can be held on for varying percentages of a cycle. The cycle time remains constant at 10 Hz, as shown in **Figure 11-6**,

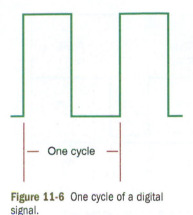

Figure 11-6 One cycle of a digital signal.

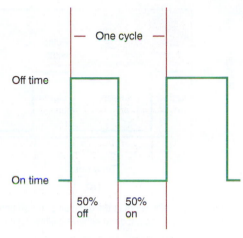

Figure 11-7 A 50-percent duty cycle.

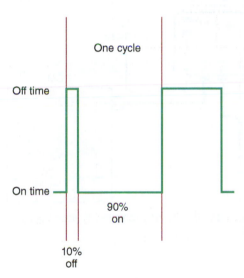

Figure 11-8 A 10-percent duty cycle lasts 10 percent of a complete cycle.

Figure 11-7, and **Figure 11-8**. A 100-percent duty cycle would be a circuit that was on 100 percent of the time. A 0 percent duty cycle would be a circuit that was off 100 percent of the time.

PULSE WIDTH MODULATION

Another form of computer control is **pulse width modulation (PWM)**. PWM is the duration of on time, but is not connected to a specific cycle time. The on time of a pulse width modulated signal is usually measured in milliseconds (ms). The fuel injector control is a good example of PWM, but there are many uses, such as lamp brightness control (**Figure 11-9**). PWM is turning a circuit on for a period of time, usually measured in milliseconds, but could be longer such as in a cooling fan relay.

Most of the devices currently in use are information center displays and electromechanical devices such as relays, motors, switches, and solenoids. Usually, the PCM circuit avoids controlling large amounts of current directly, such as a cooling fan. One way to control a large current with a smaller one is by using a relay. Using relays, the PCM can control such circuits as the cooling fan relay (**Figure 11-10**).

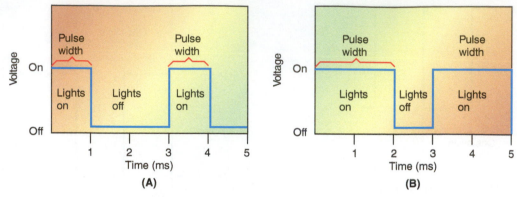

Figure 11-9 Pulse width is the duration of on-off time: (A) pulse width modulation to achieve dimmer lights, (B) pulse width modulation to achieve brighter lights.

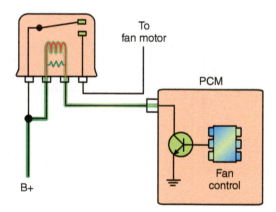

Figure 11-10 A fan relay controlled by grounding the relay coil.

COMPUTER OUTPUTS

Shop Manual
Chapter 11, page 588

Computers all rely on inputs to determine engine or transmission conditions. The computer processes the input information and then acts on the inputs according to its programming. This information is also covered in more detail in other chapters in the text and is provided here as a brief overview during the discussion of computer output control.

Fuel Injectors

Fuel injectors are controlled by a ground-side driver in the PCM. The pulse width of the injector on time is changed based on inputs from the oxygen sensor, MAP; coolant temperature sensor, MAF; and others for optimum fuel control (**Figure 11-11**). The basic fuel control concept is based on the input from the oxygen sensor or A/F sensor; the computer makes a correction and then checks the oxygen sensor or A/F sensor to see if the correction was effective and so forth, in an ongoing process of fuel control (**Figure 11-12**). This process is referred to as closed loop.

Peak-and-hold injectors are opened quickly by a high current and then held open by a lower current.

Fuel injectors can also be **peak-and-hold injectors** where a relatively high current is applied to the injector. This high initial current quickly opens the pintle and then is reduced to a lower level because it takes less current to hold the injector open after it

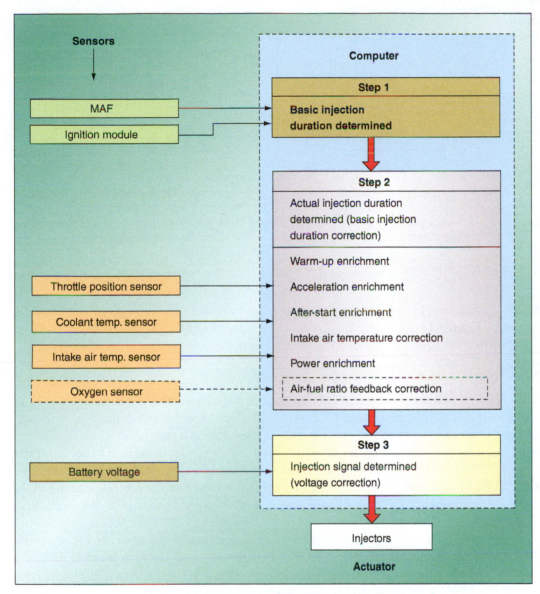

Figure 11-11 Injector "on" time is determined by sensor inputs and the computer's programming.

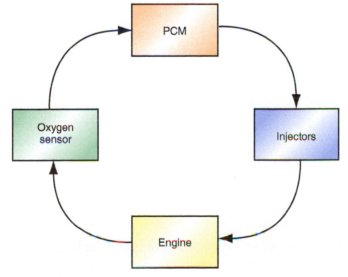

Figure 11-12 The oxygen sensor reports on the amount of oxygen in the exhaust, the PCM adds or subtracts fuel as needed, and then the process starts over again.

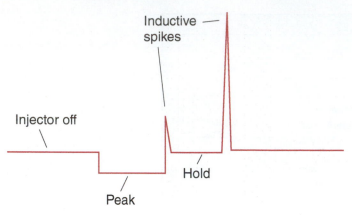

Figure 11-13 A peak-and-hold injector waveform. It takes less energy to hold an injector open than to open it initially. The higher current opens the injector faster.

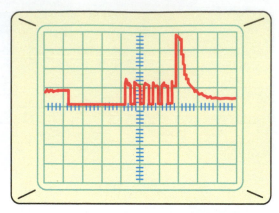

Figure 11-14 A pulse width modulated injector waveform.

Direct fuel injectors are of the peak-and-hold variety. Some vehicles use peak-and-hold injectors that are supplied both power and ground via the engine control module (ECM).

"The amount of fuel that a vehicle's ECM/PCM can add without setting a code for a fuel mixture problem varies between vehicle types."

"Short-term fuel trim" is abbreviated "STFT." "Long-term fuel trim" is abbreviated "LTFT."

Shop Manual
Chapter 11, page 592

is opened (**Figure 11-13**). Injectors can also be operated by PWM, as in the example shown in **Figure 11-14** where the injector is turned on and then pulsed rapidly before being turned off.

Fuel Trim

If the PCM sees that the long-term air-fuel ratio is not maintained at 14.7:1, the PCM can add or subtract fuel from its basic air-fuel calculation to correct this condition. This is in part to compensate for wear on components as they age, or it may point to a fuel delivery problem. The pulse width of the injectors is changed to add or subtract fuel from the mixture. There are two commonly used names in relation to fuel delivery: short-term fuel trim (STFT) and long-term fuel trim (LTFT). STFT is based on the feedback to the PCM of the oxygen sensor. STFTs can change very fast, as they are a reaction to the short-term fuel delivery.

LTFT is based on a long-term trend of adding or subtracting fuel to the base calculation that is programmed into the PCM. For example, a small vacuum leak would cause the STFT to go in the positive direction (adding fuel) and if the trend of adding fuel continued, the LTFT would start adding fuel as a positive number. If the STFT returned back to near zero, no more action would be taken. But the LTFT would stay positive until the leak was repaired. Normally the PCM can add or subtract about 20 percent of the basic LTFT needs before setting a diagnostic trouble code. STFT can add and subtract much larger percentages of fuel than LTFT. STFTs are reset when the key is turned off. LTFTs are retained in memory for use on the next key cycle.

Idle Air Control

The IAC motor works by varying the amount of airflow past the throttle plate as shown in **Figure 11-15**. This air is detected by the oxygen sensor, which increases fuel delivery and idle speed. The idle air control is a reversible stepper motor that is controlled by the PCM (**Figure 11-16**). The PCM also makes small corrections in idle speed by altering the ignition timing. The PCM uses the RPM signal as feedback for the system.

Throttle Actuator Control

The throttle actuator control (TAC) system is an output from the PCM that controls cold idle speed, normal idle speed, cruise control, and traction control (**Figure 11-17**). Instead of an accelerator cable, a reversible DC motor is used to control the throttle operation.

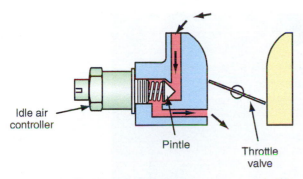

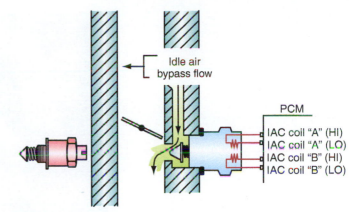

Figure 11-15 The idle is controlled by varying the amount of air around the throttle plate with an IAC valve.

Figure 11-16 The PCM controls the IAC valve by using a reversible stepper motor.

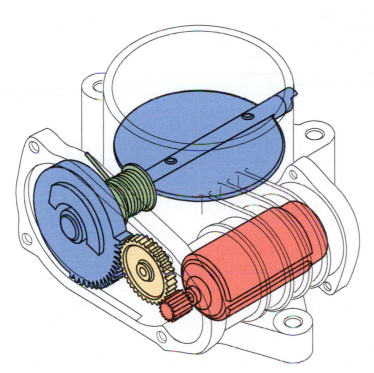

Figure 11-17 A typical throttle actuator control (TAC).

The output is based on in the formation received by the PCM from the engine coolant, vehicle speed, cruise control, and traction control. The PCM also calculates engine performance information to place the throttle at the optimum location for power, fuel economy, and emissions.

Exhaust Gas Recirculating Valve

As we elaborated in Chapter 10, the EGR valve in **Figure 11-18** is a linear EGR valve. The linear EGR valve is controlled by the PCM by PWM. The valve is a normally closed pintle opened by a solenoid. The digital EGR comprises three solenoids that all flow different amounts of EGR and can be opened or closed independently. The combination of open and closed solenoids permits seven different flow rates. In the case of the digital EGR valve, the solenoids are turned either on or off and are not modulated.

Figure 11-18 A linear EGR valve.

Variable Valve Timing

An important addition to the computer output function over the last few years has been variable valve timing (VVT). For many years, automotive engineers have been able to tailor ignition timing to the needs of the vehicle, now the valve timing can be modified as well. The camshaft profile that determines valve timing and duration had been a compromise between acceptable idle, acceleration, and fuel economy. With VVT, the PCM can modify the cam timing for a good idle, good acceleration, and fuel economy over a broad range of engine operating conditions. Many vehicles are replacing the function of the EGR valve by VVT systems. The VVT can leave some of the exhaust gas in the cylinder by the percentage of intake and exhaust valve overlap.

Electronic Spark Timing

One of the most important outputs of the ECM is the spark timing. Spark timing is one of the key elements for a successful and efficient combustion affecting both fuel economy and emissions. The ECM is equipped to make timing decisions by having all the important information needed, such as coolant temperature, EGR added, engine load, and mass airflow. The ECM knows if the engine starts to spark knock because it has a knock sensor to give that information. The ECM sends a digital signal called **electronic spark timing (EST)** to the ignition module that controls the ignition timing.

Transmission Control

Shared information used for both engine and transmission operation cut down on the number of components used, which increases reliability and reduces costs.

Electronically shifted transmissions have several solenoids and control valves governed by the ECM, as shown in **Figure 11-19**. The ECM has information from the engine sensors such as engine load, vehicle speed, and the throttle position, and this information is shared with the transmission control module (TCM). Shift solenoids are opened or closed by the computer based on design. Many transmissions have pulse width modulated solenoids that are used to control line pressure and torque converter clutch apply.

COMPUTER NETWORKING

Computer networking is the way computers communicate. An example of this would be the antilock brake module. The PCM and the ABS computer both need vehicle speed information. Rather than using two individual sensors and the associated wiring or using one sensor and running individual wiring to both computers, it would be much easier for the PCM and the ABS computer to share information over the **serial data bus**. Imagine

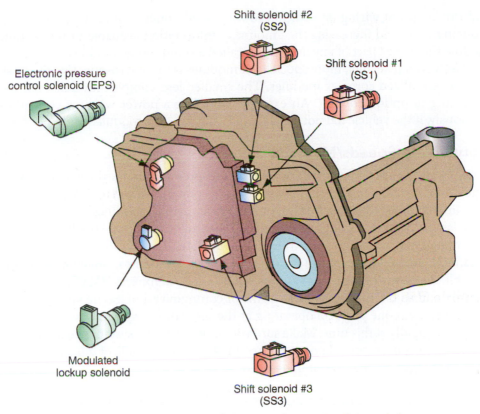

Figure 11-19 Some of the electronically controlled actuators for an automatic transmission.

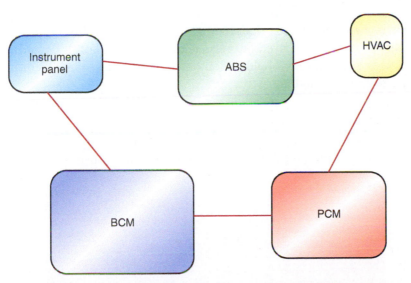

Figure 11-20 Networked modules share information with each other over the serial data bus.

a network where many individual modules share information with each other as needed. Serial data is digitally encoded information that computers can share together. On your home personal computer, this is much like sharing information on an office intranet or worldwide over the Internet. Late-model vehicles have their own network of computers, or modules (**Figure 11-20**). Some modules still stand alone, some are capable of communicating and working with each other, and others take commands from another module. Networking shares information and results in cutting down the number of connectors

A module that is not part of any network is called a standalone module.

Shop Manual
Chapter 11, page 593

and the length of wiring, in addition to the added bonus of increased reliability, while reducing costs and increasing the amount of information available to the technician. A module that is not part of any network is called a stand-alone module.

Networking smaller, more specialized modules is more cost-effective than using one or two centralized complex modules. The smaller, less-complex modules can be added according to option content. An example would be a power window control module. Obviously, if the vehicle did not have power windows, this module would not be needed.

Networking Speeds/Methods

Networking speeds and methods depend on the intended use for the information. OBD II regulations have limited the number of types of communication to three that can be used for connection to a diagnostic scan tool and/or individual modules.

Differences in communication methods are one reason aftermarket scan tools have to rely on different cartridges for domestic, European, and Asian vehicles.

Class B, ISO 9141, SCP, Class 2, and KWP2000 all were allowed methods of powertrain communication under OBD II until 2008, at which time **controller area network (CAN)** became the mandated protocol for emissions-related modules. This means that it is still possible that some of the other systems, such as entertainment, HVAC, security, stability control, and so on, may have other forms of communication. Some vehicles can and do have more than one protocol operating at the same time, especially as changes are happening so rapidly at this time. Make sure you understand the system you are working with before performing repairs. Look at **Figure 11-21** for an example of the different programming languages that have been in use in recent years.

Class A or UART Communication/Local Interconnect Network

Since the early 1980s, a communication language known as Universal Asynchronous Receive and Transmit (UART) has been used in vehicle communication between onboard modules and diagnostic tools. UART is the oldest network protocol in use. UART is also called Class A and is still used in some applications as a communication protocol, although it is relatively low speed at 8,192 bits/second. UART accomplishes communication by toggling 5 volts to ground at a fixed bit pulse width (**Figure 11-22**). Local Interconnect Network (LIN) is a low-speed supplemental database that uses UART communication along with up to 15 slave modules.

Manufacturer	Post 2008 Communications network Non-Emissions related	Post 2008 Emissions Communication
GM	CAN KWP2000 Class 2	CAN
FORD	SCP CAN	CAN
HONDA	ISO 9141 CAN	CAN
VW/AUDI	CAN KWP2000 Class 2	CAN
CHRYSLER	J1850 10.4 CAN	CAN

Figure 11-21 Emissions-related communications mandated as CAN protocol after 2008, but non-emissions-related protocols can be used for any other purpose, such as power windows or seats.

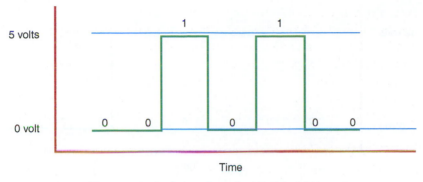

Figure 11-22 UART communication is based on a fixed bit pulse width.

Class B/J1850/SCP

With a speed of 10,400 bits/second, Class B (also called J1850) communication is faster than UART. Ford has been using a Class B communication standard they call Standard Corporate Protocol (SCP) along with an ISO 9141 network, which will be discussed later. Modules that are configured on the SCP allow communication between individual modules in addition to interfacing with a scan tool. GM's Class B communication is called Class 2. Class 2 communications are different from UART in that instead of toggling a voltage from high to low and using 5 volts, it toggles a voltage from 0 to 7 volts, and the signal can be sent in long or short pulse widths (**Figure 11-23**).

ISO 9141/KWP2000

ISO 9141 is similar to UART in that it is a bit-by-bit-style communication method instead of a pulse width communication, and instead of working on 5 volts and 0 volt, it works on 12 volts and 0 volt (**Figure 11-24**). ISO 9141 requires a scan tool that asks a module for the information needed and the module responds. Some Ford vehicles use ISO 9141 for communication with the **electronic crash sensor (ECS)** in addition to the Class B SCP described earlier. DaimlerChrysler and the Asian automakers have used ISO 9141 on many vehicles. ISO 9141 is not a communication language in the sense that it links separate modules, but it is used to allow a scan tool to communicate with a module and is used for diagnostics only. When the vehicle network is operating in normal mode, it may be using different communication rates and methods. European automakers are using Key Word Protocol 2000 (KWP2000) that is similar to the Japanese ISO 9141 in that it is used for diagnostics only, but it can be used to connect a scan tool to more than one module.

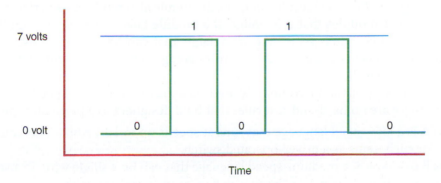

Figure 11-23 Class 2 communication has different pulse times.

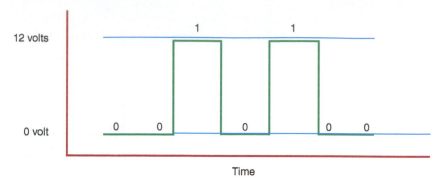

Figure 11-24 ISO communication.

Controller Area Network

GM calls their CAN protocol General Motors Local Area Network (GMLAN).

Shop Manual Chapter 11, page 595

Starting in the 2008 model year, the only communication standard used for emissions-related modules is CAN (controller area network) or CAN J2284/ISO 15765-4, but it was used by many manufacturers sooner.

CAN is a universal communication method or "language" used that allows sharing of technology around the world. CAN was developed by Bosch in 1984 for the automotive industry and was first used on some Mercedes-Benz models in the early 1990s. Since then, CAN has found its way on many devices and has become the world leader in networking languages. CAN technology is faster, more reliable, more cost-effective, and more flexible, because it allows the use of generic modules that are available from several manufacturers. The trend with CAN is to use more modules, for example, a vehicle may use a TCM and an engine control computer in place of the powertrain control module (PCM). This allows the manufacturer to save money on vehicles that use a manual transmission because the TCM can be eliminated. CAN allows the individual modules to power down whenever the module is not needed, which helps conserve power. A module can wait for a signal, power up and perform a function, and then go back into a state of rest. The CAN language contains the rules to be followed by modules when they communicate. The individual modules are sometimes called nodes. There are two types of messages: diagnostic and normal. Normal messages are those that go between modules; diagnostic messages are those that are communicated to a scan tool. Each CAN message contains the following information:

When more than one protocol is used on a vehicle communication network, shared information is translated though a "gateway" module.

- A signal informs the other modules that a message is following.
- The priority of the following message is communicated. A low-priority message will have to wait on a higher-priority message.
- The size of the message.
- Message reliability.
- Other modules acknowledge the message.

Each module in CAN receives every communication, but the module responds only to the messages with the correct heading. Each module also sends out a periodic message telling the other modules that it is online. If a module fails to "check in" with the other modules, a trouble code is set. There are defaults that can be used for most modules if they fail. This prevents the entire system from going down in the event that one of the modules should fail.

CAN currently has some commonly used versions referred to as "speeds" or "classes," although there are no hard-and-fast rules that bind designers to a particular speed.

- Class A is a low-speed language that runs at speeds up to 10.4 Kbps. This could be used for such items as a driver-operated switch.
- Class B or CAN-B, a medium-speed language that can be a single wire. Its use for large amounts of information speed is not an issue. It can communicate at speeds up to 83 Kbps.

- Class C or CAN-C, referred to as high-speed CAN, can run as fast as 1 Mbps. This data bus is used for "real-time" engine performance, antilock brake, and active suspension control. The wiring for CAN-C is a twisted pair to help prevent electrical interference. The bus wires are connected by a termination resistor at the end, which helps prevent "reflected" messages coming over the bus.

Figure 11-25 shows a typical high-speed CAN electrical signal. The "idle" voltage is about 2.5 volts. The modules look for the difference in voltage levels between CAN+ and CAN−. The positive voltage goes to about 3.5 volts and the negative down to 1.3 volts.

As stated previously, the CAN communication standard relates only to emission-control modules, so some vehicles will use CAN along with a "gateway" module to communicate with other protocols as shown in **Figure 11-26**. The gateway module can translate and share serial data over the other buses and diagnostic scan tools as well. The LIN modules are low-priority, low-speed body control devices that do not need sophisticated communications.

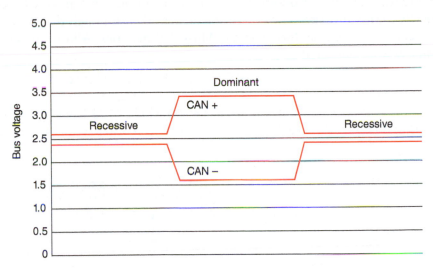

Figure 11-25 CAN-C high-speed messages are transmitted as a voltage differential.

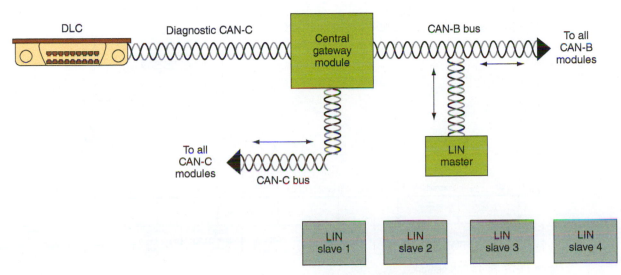

Figure 11-26 A gateway module can be used to communicate with modules of different protocols.

Manufacturer-Specific Versions of CAN

GM. GM's version of CAN is GMLAN. GMLAN has three speeds: high, medium, and low. The high-speed GMLAN is used for real-time communications, such as the stability control, antilock brakes, automatic transmission, cell phone, and navigation. The medium-speed GMLAN is for communications between modules. The low-speed GMLAN is for the radio, some of the air bag systems, heated seats, and instrument cluster gauges. GM is still using some Class 2 and UART devices, but not for engine performance (OBD II) diagnostics. CAN was the only approved OBDII diagnostics protocol after 2008. The BCM is the gateway module for the GMLAN system. The gateway module allows networks of different speeds to communicate with one another and with the scan tool.

Ford. Ford's network has a high-speed CAN and medium-speed CAN, and infotainment CAN or I-CAN (**Figure 11-27**). HS-CAN and I-CAN are for real-time events and runs at 500 Kbps. The MS-CAN system runs at 125 Kbps. The MS-CAN is for bus messages and general information. I-CAN or Infotainment CAN is 500 Kbps speed network for such things as the radio, power running boards, front display panels, backup cameras, and navigation. The HS-CAN is used for antilock brakes (ABS), restraints control module (RCM), and the PCM among others. The instrument cluster (IPC) and the **body control module (BCM)** are the gateway modules for the network, although the I-CAN modules communicate with the scan tool over the MS-CAN network. All of the networks can communicate with one another.

Honda. Honda's CAN network is called F-CAN and B-CAN (**Figure 11-28**). The F-CAN is short for *fast CAN*. Fast can is used for real-time data transfer, up to 500 Kbps speeds. B-CAN is for body CAN. B-CAN that runs at the slower 125 Kbps speed is used for modules that do not require real-time speed, as with the other systems mentioned. The

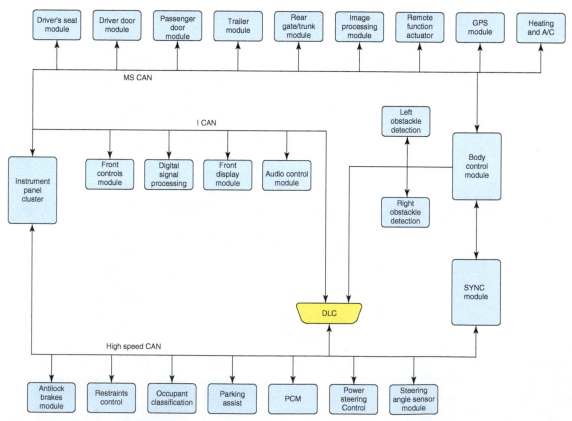

Figure 11-27 Ford's CAN network.

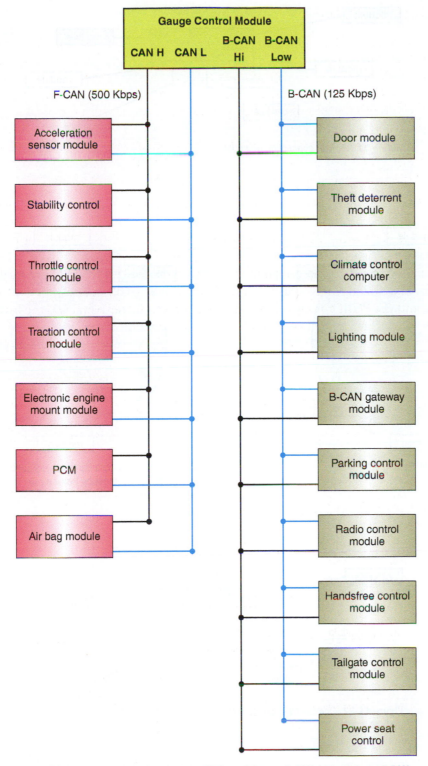

Figure 11-28 Honda CAN network. Note which modules are B-CAN and which are F-CAN.

gateway controller for the F-CAN network in the Honda system is the gauge control module (instrument cluster), although the multiplex integrated control unit (MICU) is the gateway module for the body control units such as heated seats, power window control, hatch control, and so on. The MICU communicates with the scan tool and the other modules on the K-line.

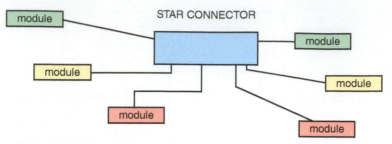

Figure 11-29 Star connection of modules.

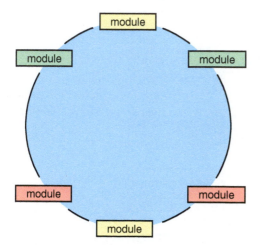

Figure 11-30 Loop network connections of modules.

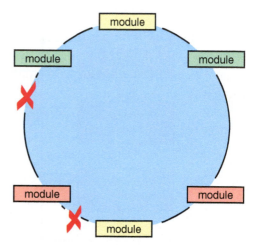

Figure 11-31 Two faults are resumed to isolate a module from the loop network.

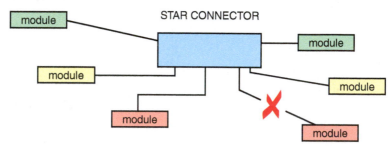

Figure 11-32 One fault will isolate a module on a star-style network, but only one module is affected.

Network Wiring

There are two primary ways to wire the serial bus in a vehicle network: star and loop. The **star** network is shown in **Figure 11-29**. The star network uses star connectors as junction blocks to join the serial bus. In the **loop network** (**Figure 11-30**), the modules are connected in series. The advantage of the loop network is that there has to be two opens to affect the network (**Figure 11-31**). It only takes one open to affect the star network, but only the module on the affected line goes down (**Figure 11-32**).

Some manufacturers call the star connector a splice pack.

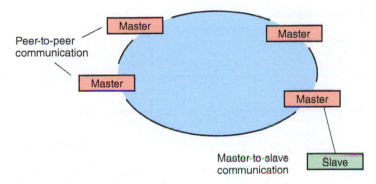

Figure 11-33 Peer-to-peer and master-to-slave communication. Slaves cannot initiate communication on the serial bus.

Peer-to-Peer and Master-to-Slave Communication

Some modules on the serial bus can initiate communication with other modules on the serial data bus. These modules are called **masters**. The master modules can communicate with each other and are called peers. Communication between master modules is called peer-to-peer communication. Some modules are referred to as **slaves**. Slaves just listen to instruction from the master and relay confirmation that instructions have been carried out (**Figure 11-33**).

LIN modules are often referred to as **slave** modules.

Modules of the Vehicle Network

We will look at some of the more common computer modules in recent use. The use of modules can allow for **customization** of the module's function. This section will give you an idea of how the modules work together and why they are used. As a note, to function properly, most of the various body control modules have to be reprogrammed when replaced. Some modules have to be asked for their information that is collected and used to reprogram the replacement module. Once again, there is vast variety in the number, placement, naming, and use of various body control modules, and technologies are changing rapidly. This explanation of body control modules does not cover every available make and function, but tries to familiarize you with the vehicle networking concept. Make sure that you have the factory service information available to you to actually begin diagnosis of these systems.

Lighting Control Module

For this example we will look at the function of the **lighting control module (LCM)** from Ford. This lighting control module takes care of the warning chimes, interior and exterior lighting, and theft deterrent. The lighting control module has a battery saver function that will disrupt power to the courtesy lamps and/or headlamps after 10 minutes. Turning the ignition switch to the run or accessory position or activation of the theft deterrent will prevent the battery saver by disabling the timers and reinstating power to the lamps. The time the lights stay on after leaving the vehicle, the courtesy lamps turning on when the doors are unlocked, and so on are a few of the customizable functions.

Driver Door Module

The **driver door module (DMM)**–controlled functions include the power windows, power door locks, remote keyless entry, and others, depending on the make and model of the vehicle. See **Figure 11-34** for a block diagram example of module operation for a typical GM product. The DDM receives a signal from the driver to open the passenger window. The DDM sends a serial message over the network to all the modules that the window is to be rolled down. Because it is addressed only to the **front passenger door module (FPDM)**, it reads the message and then rolls the window down.

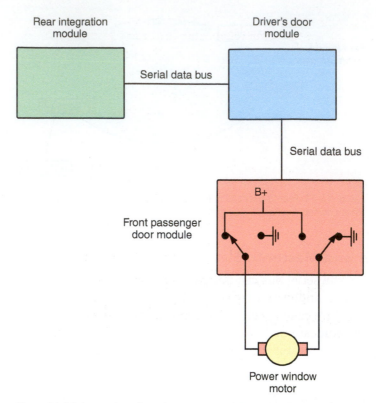

Figure 11-34 Interaction of rear integration module, driver's door module, and front passenger door module of a typical GM product.

The FPDM is a slave module that can receive messages and send acknowledgment that a command has been executed, but it has no access to the main serial bus. You may notice that the FPDM has the ability to provide either power or ground for the circuit to operate.

Heating and Air Conditioning

The electronic heating and air-conditioning system has been in vehicles for some time. To provide the most efficient and comfortable climate control, these systems have also moved toward integration into the body control system (**Figure 11-35**). The system described originated on a Ford product, but most manufacturer systems are similar. The climate control assembly is used to input the desired temperature, which is sent to the **electronic automatic temperature control (EATC)** module. The EATC then knows what course of action to take based on the **ambient air temperature sensor**, the **in-car temperature sensor**, and the **A/C sunload temperature sensor**. The temperature sensors report directly to the EATC, and the A/C compressor clutch is controlled from the PCM based on evaporator discharge temperatures by way of an evaporator temperature sensor. The PCM also has the last word on whether the A/C will be allowed to operate when the EATC requests; for instance, if the engine temperature is too high, the PCM may prevent A/C clutch operation. The EATC automatically controls the temperature control actuator, the outlet position, and blower motor settings. The temperature sensors and sunload sensor give the EATC a basis for action. For example, if the difference in the inside and outside temperatures is great, then the EATC anticipates the required action by moving the temperature door to full heat after the engine has reached the operating temperature. If the selected vehicle temperature is close to the outside air temperature, then the blower speed may be set to low, and very little regulation of heating and cooling may be required.

The **ambient air temperature sensor** measures the outside air temperature and is usually placed near the front grill.

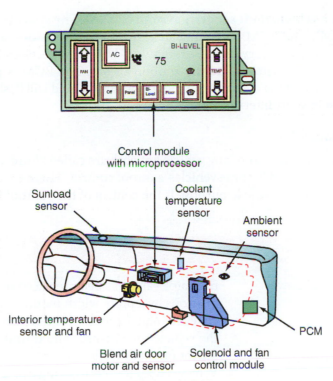

Figure 11-35 Typical automatic air-conditioning system components.

Anti-theft Systems

Ford uses an anti-theft system called passive anti-theft system (PATS), or SecuriLock uses a specially encoded key that must be programmed into the system (**Figure 11-36**). A PATS transceiver module located inside the steering column reads the key's identification and sends it to the PCM. If there is a problem with the system, the theft indicator will either stay on steady or flash rapidly.

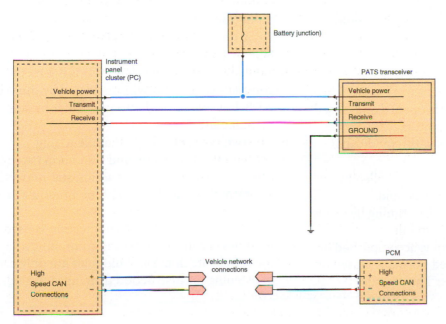

Figure 11-36 Passive Anti-Theft System (PATS) diagram.

Chrysler uses a **vehicle theft security system (VTSS)**. The system uses a body control module (BCM). The BCM communicates with the **power door module (PDM)**, which also has the keyless entry receiver functions. The BCM receives inputs from the door lock and door-ajar switches and the ignition switch. The BCM is programmed to interpret a vehicle break-in or activation of the "panic button" and will flash the headlamps and blow the horn in short intervals.

Stand-Alone Modules

Some modules that are not part of any vehicle network are called **stand-alone modules**. An example would be found in some vehicles' sunroof control. There are a set of switches connected to a module whose sole purpose is the control of the sunroof function.

SUMMARY

- High-side drivers control the application of power to a circuit.
- Low-side drivers provide a ground to operate a circuit.
- Output drivers are grouped into quad drivers and output driver modules.
- Quad drivers are capable of controlling four separate outputs, but a fault on one controlled circuit causes the entire driver to shut down all of its controlled circuits.
- Output driver modules are capable of controlling up to seven separate outputs. If there is a problem on one output, then only the output actuator or circuit at fault is shut down.
- Duty cycle is the application of power or ground for a percentage of a fixed cycle.
- Pulse width modulation is the activation of a circuit for varying amounts of time.
- Fuel injectors are controlled by varying the pulse width of the injector on time according to the vehicle's fuel control needs.
- The IAC system is controlled by varying the amount of air through a bypass passage located in the throttle body. The passage is opened and closed by a PCM-controlled stepper motor.
- The linear EGR valve is a pulse width modulated solenoid.
- Vacuum-operated EGR valves are controlled by a pulse width modulated vacuum solenoid.
- The PCM controls electronic spark timing by using a digital output to the ignition module.
- Electronic transmission control is accomplished by the PCM by grounding selected normally closed or normally open solenoid valves. The pressure

regulator valve has been replaced on some transmissions by a pulse width modulated solenoid. The governor and throttle valve have been replaced by the vehicle speed sensor and throttle position sensor on many transmissions.
- Vehicle computer networking uses many modules to control various vehicle functions. These modules have the ability to communicate with each other over the serial data bus.
- Local Interconnect Network (LIN) is used by slave modules on the data bus.
- UART, J1850, KWP2000, and ISO 9141 are different computer networking languages. Some manufacturers use more than one method in the same vehicle; some use different methods depending on the vehicle make and model.
- UART is the oldest and slowest type of vehicle computer communication, still in use by some LIN modules.
- J1850 communication is called Class 2 by GM and SCP by Ford.
- Controller area network (CAN) is the method of communication that was mandated by the year 2008, but was phased into use earlier by some manufacturers.
- The two types of vehicle module wiring configurations are loop and star. Loop networks use modules connected in series; star networks use a common connection point for a module's serial data line.
- Peer-to-peer module communication is used for modules capable of initiating communication on the serial data line. Slave modules are only capable of listening for instruction from master modules.

REVIEW QUESTIONS

Short Answer Essay

1. Describe high-side output drivers.

2. Explain the difference between quad drivers and output driver modules.

3. Describe duty cycle.

4. Describe pulse width modulation.

5. Describe the advantages of using a computer network.

6. Briefly describe J1850 communications.

7. Briefly describe CAN communications.

8. Describe the EATC system.

9. Describe the difference between peer-to-peer and master-to-slave module communication.

10. Explain the differences between the star and loop network configurations.

Fill-in-the Blanks

1. The PCM uses transistorized switches called _____ to control circuits or devices.

2. Low-side drivers apply_____ to a circuit.

3. Fuel injectors are controlled by a(n)_____ driver in the PCM.

4. A module that is not part of any network is called a(n)_____ module.

5. Duty cycle is the application of_____ or_____ for a percentage of a fixed cycle.

6. _____ is the oldest and slowest type of vehicle computer communication.

7. _____ is a way for individual modules to share information between themselves.

8. _____ module takes care of the warning chimes, interior and exterior lighting, and theft deterrent.

9. _____ communication is called Class 2 by GM and SCP by Ford.

10. GM's version of CAN is_____.

Multiple Choice

1. *Technician A* says that a circuit can be turned on by switching power to the circuit.
   Technician B says that a circuit can be switched on by supplying ground to the circuit.
   Who is correct?
   A. Technician A   C. Both technicians
   B. Technician B   D. Neither technician

2. *Technician A* says that a low-side driver switches power to activate a device.
   Technician B says that an output driver can control up to five devices independently.
   Who is correct?
   A. Technician A   C. Both technicians
   B. Technician B   D. Neither technician

3. *Technician A* says that a quad driver has four independently controlled outputs.
   Technician B says that modern quad drivers do not have a self-resetting circuit breaker in case a controlled device is shorted.
   Who is correct?
   A. Technician A   C. Both technicians
   B. Technician B   D. Neither technician

4. *Technician A* says that Ford's J1850 standard of communication is called DLP.
   Technician B says that GM's J1850 standard of communication is called Class 2.
   Who is correct?
   A. Technician A   C. Both technicians
   B. Technician B   D. Neither technician

5. *Technician A* says that duty cycle is a percentage of time in a fixed cycle that a device is on.
   Technician B says that if a full cycle were 1 second in length, a 50 percent duty cycle means that the device was on for 0.5 second.
   Who is correct?
   A. Technician A   C. Both technicians
   B. Technician B   D. Neither technician

6. While discussing pulse width modulation (PWM), *Technician A* says that pulse width is measured in a length of time, not a percentage of a cycle. *Technician B* says that the cooling fan relay is usually controlled by the PCM through PWM. Who is correct?

A. Technician A
B. Technician B
C. Both technicians
D. Neither technician

7. *Technician A* says that long-term fuel trim is constantly changing for driving conditions. *Technician B* says that the PCM can add or subtract about 10 percent of the base fuel trim before setting a trouble code. Who is correct?

A. Technician A
B. Technician B
C. Both technicians
D. Neither technician

8. *Technician A* says that governors and throttle valve cables used in earlier transmissions have been replaced by the vehicle speed sensor and TP sensor in modern, electrically shifted transmissions.

Technician B says that pulse width modulated solenoids also control line pressure. Who is correct?

A. Technician A
B. Technician B
C. Both technicians
D. Neither technician

9. *Technician A* says that the communications protocol called UART has been around since the early 1980s. *Technician B* says that CAN was the standard for all emissions-related modules starting in 2010. Who is correct?

A. Technician A
B. Technician B
C. Both technicians
D. Neither technician

10. All CAN protocol messages contain all of the following information *except*:

A. the size of the message
B. message reliability
C. other modules acknowledge the message
D. the speed of the message

CHAPTER 12

ON-BOARD DIAGNOSTIC (OBD II) AND COMPUTER SYSTEMS

Upon completion and review of this chapter, you should be able to:

- Describe the primary provisions of OBD II.
- Explain the requirements to illuminate the malfunction indicator light (MIL) in an OBD II system.
- Briefly describe the monitored systems in an OBD II system.
- Describe the main hardware differences between an OBD II system and other systems.
- Describe an OBD II warm-up cycle.
- Explain trip and drive cycle in an OBD II system.
- Describe how engine misfire is detected in an OBD II system.
- Describe the differences between type A and type B misfires.

- Describe the purpose of having two oxygen sensors in an exhaust system.
- List five mandates in OBD II.
- Briefly describe what the comprehensive component monitor looks at.
- Describe briefly the design of a microprocessor chip, and state the basic purpose of this chip.
- Explain briefly how the microprocessor stores and retrieves information.
- Describe the purpose of random access memory (RAM).
- Explain the terms "volatile" and "nonvolatile memory." Describe the purpose of a read only memory (ROM).

Terms To Know

Adaptive strategy

Calibration

Catalyst efficiency monitor

Comprehensive component monitor (CCM)

Cylinder misfire monitoring

Drive cycle

Electronically erasable programmable read only memory (EEPROM)

Enable criteria

Failure record

Flashing

Freeze frame

Fuel system monitor

Heated oxygen sensor

Information retrieval

Information storage

Keep alive memory (KAM)

Long-term fuel trim (long-term FT)

Microprocessor

Parameter identification (PID)

Programmable read only memory (PROM)

Random access memory (RAM)

Read only memory (ROM)

Reflashing

Short-term fuel trim (short-term FT)

Snapshot

Trip

Type A misfire

Type B misfire

INTRODUCTION

Technological advances are being made every year to keep pace with more stringent emission laws and consumer demands. These advancements include computer-controlled engine systems that help provide more power, better performance, lower emissions, and better fuel economy. From one year to the next, changes are common among all manufacturers. Because of this, it would be impossible to memorize all engine control systems and test procedures on every vehicle ever built. It is essential to verify the VIN when performing any service on the vehicle, and always use the appropriate service information. Subtle changes in components or system design can cause changes in diagnostic procedures. Costly mistakes have been made failing to verify the vehicle information before the diagnosis or repair begins. The onboard diagnostic capabilities found in vehicles equipped with computerized engine controls provide the technician with a greater troubleshooting advantage. Once the technician understands the vehicle's operating system and testing routines, the computer can be used to reveal the source of most performance problems. A typical engine control system is shown in **Figure 12-1**.

> OBD II was mandated to be implemented in 1996, but some manufacturers did so earlier.

The On-Board Diagnostic II (OBD II) systems were developed in response to the federal government's and the state of California's emission control system monitoring standards for all automotive manufacturers. The main goal of OBD II was to detect when engine or system wear or when component failure caused exhaust emissions to increase by 50 percent or more. Additionally, monitoring and protecting the catalytic converter is one of the major priorities of an OBD II system. OBD II also called for standard service procedures without the use of dedicated special tools.

- A universal diagnostic test connector, known as the data link connector (DLC), with dedicated pin assignments.
- A standard location for the DLC. It must be under the dash on the driver's side of the vehicle and must be visible (SAE standard J1962).
- A standard list of diagnostic trouble codes (DTCs) (SAEs standard J2012).
- A standard communication protocol (SAE standard J2411) (CAN after 2008).
- The use of common scan tools on all vehicle makes and models (SAE standard J1979).
- Common diagnostic test modes (SAE standard J2190).
- Vehicle identification must be automatically transmitted to the scan tool.
- Stored trouble codes must be able to be cleared from the computer's memory with the scan tool.
- The ability to record and store in memory a freeze frame of the operating conditions that existed when a fault occurred.
- The ability to store a code whenever something goes wrong and affects exhaust emissions.
- A standard glossary of terms, acronyms, and definitions must be used for all components in the electronic control systems (SAE standard J1930).

System Functions

Most electronic engine control systems work in a similar way regardless of the manufacturer. All systems require multiple input signals to indicate engine load, speed, and engine temperatures used to determine fuel and spark requirements. The vehicle's computer is programmed to execute specific commands based on the input data. For example, the computer commands control fuel injectors, spark ignition timing, emission devices, and actuators or other system components to obtain optimal performance and emissions.

Some of the design features of computerized engine management are as follows:

> The vehicle's computer is known by several names, such as electronic control module, engine control module (ECM), or powertrain control module (PCM), as referenced throughout this book.

- Maintain air-fuel ratios as close to 14.7:1 to allow optimal catalytic converter operation and improve fuel economy.
- Accurately control critical emission control devices such as EGR and carbon canister/evaporative control system.

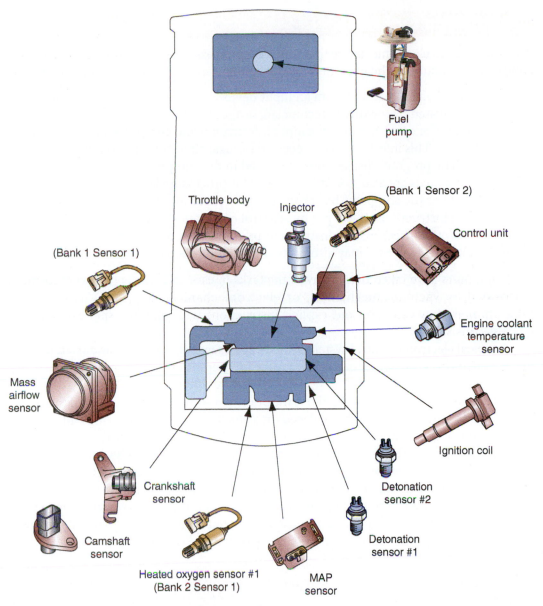

Figure 12-1 Components of a typical engine control system.

- Precisely control ignition timing under all driving conditions adjusted for climate and fuel grade.
- Provide more efficient cold engine operation with lower emissions and faster warm-ups.

AUTHOR'S NOTE When beginning a diagnostic procedure, do not skip steps or attempt shortcuts such as assuming all the basic items will check out fine. It is critical to have proper power supply voltage and system ground circuits that are sound. Multiple symptoms and drivability problems that appear complex may require the simplest of repairs, such as cleaning and tightening a loose ground wire. A recent report from an OEM indicated that less than 2 percent of the PCMs returned for warranty claims were actually verified to be faulty. This means 98 percent of those returned were misdiagnosed and had no problems.

COMPUTER OPERATION

A computer is an electronic device that stores and processes data. It is also capable of operating other computers. The operation of a computer is divided into four basic functions:

1. *Input.* A voltage signal sent from an input device. The device can be a sensor or a button activated by the driver, technician, or mechanical part.
2. *Processing.* The computer uses the input information and compares it to programmed instructions. This information is processed by logic circuits in the computer.
3. *Storage.* The program instructions are stored in the computer's memory. Some of the input signals are also stored in memory for processing later.
4. *Output.* After the computer has processed the sensor input and checked its programmed instructions, it will put out control commands to various output devices. These output devices may be instrument panel displays or output actuators. The output of one computer may also be an input to another computer.

Computers have taken over many of the tasks in cars and trucks that were formerly performed by vacuum, mechanical, or electromechanical devices. When properly programmed, they can carry out explicit instructions with blinding speed and almost flawless consistency.

A typical electronic control system is made up of sensors, actuators, and related wiring that is tied into a central processor called a **microprocessor** or a microcomputer (a smaller version of a computer).

The central processing unit (CPU) is the brain of a computer. The CPU is constructed of thousands of transistors that are placed on a small chip. The CPU brings information into and out of the computer's memory. The input information is processed in the CPU and checked against the program in the computer's memory. The CPU also checks the memory for any other information regarding programmed parameters. The information obtained by the CPU can be altered according to the instructions of the program. The CPU may be ordered to make logic decisions on the information received. Once these decisions or calculations are made, the CPU sends out commands to make the required corrections or adjustments to the system being controlled (**Figure 12-2**).

> A **microprocessor** is the central processing unit in a computer that controls and calculates data.

MICROPROCESSORS

Design

The microprocessor is the calculating and decision-making chip in a computer. Thousands of miniature transistors and diodes are contained in the microprocessor. These transistors act as electronic switches that are either on or off. The components in the microprocessor are etched on an integrated circuit (IC) that is small enough to fit on a fingertip. The silicon chip containing the IC is mounted in a flat, rectangular, protective box. Metal connecting pins extend from each side of the microprocessor container. These pins connect the microprocessor to the circuit board in the computer.

The microprocessor is supported by various memory chips that store information and help the microprocessor in making decisions. These memory chips are similar in appearance to the microprocessor chip. The function of the memory chips will be explained later in this chapter.

> The microprocessor chip is the calculating and decision-making chip in a computer.

Program

A program is a group of instructions that is followed by the microprocessor. The program guides the microprocessor in decision making. For example, the program may inform the microprocessor when sensor information should be retrieved and then tells the microprocessor how to process this information. Finally, the program will guide the

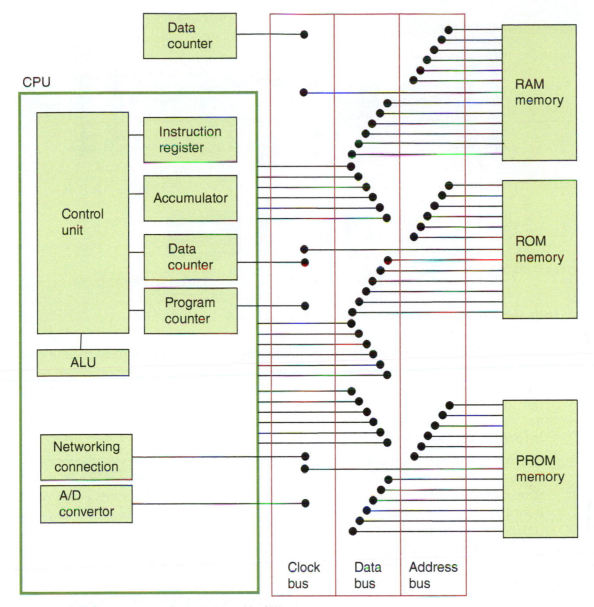

Figure 12-2 Major components of a computer and its CPU.

microprocessor regarding the activation of output control devices such as relays and solenoids. The various memories contain the programs and other vehicle data that the microprocessor refers to as it performs calculations. As the microprocessor performs calculations and makes decisions, it works with the memories in the following ways:

1. The microprocessor can read information from the memories.
2. The microprocessor can write new information into the memories.

Information Storage

The memories contain many different locations. These **information storage** locations may be compared to file folders in a filing cabinet, and each location contains one piece of information. An address is assigned to each memory location. This address may be compared to the lettering or numbering arrangement on file folders. Each address is written in a binary code, and these codes are numbered sequentially beginning with 0.

While the engine is running, the computer receives a large quantity of information from a number of sensors. The computer may not be able to process all this information

Information storage refers to processing and storing a large quantity of information received from a number of sensors.

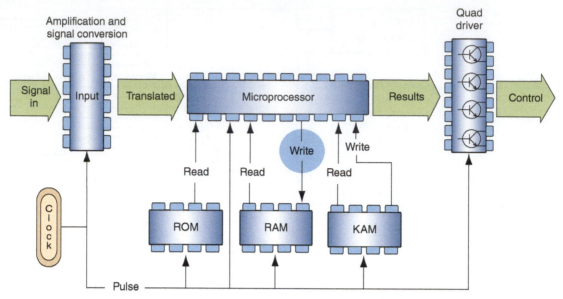

Figure 12-3 Information storage.

immediately. In some instances, the computer may receive sensor inputs that the computer requires to make a number of decisions. In these cases, the microprocessor writes information into memory by specifying a memory address and sending information to this address (**Figure 12-3**).

Information Retrieval

> **Information retrieval** refers to requesting information that has been stored by a microprocessor.

When stored information is required, the microprocessor specifies the stored information address and requests the information. This is known as **information retrieval**. When stored information is requested from a specific address, the memory sends a copy of this information to the microprocessor. However, the original stored information is still retained in the memory address.

The memories store information regarding the ideal air-fuel ratios for various operating conditions. The sensors inform the computer about the engine and vehicle operating conditions. The microprocessor reads the ideal air-fuel ratio information from memory and compares this information with the sensor inputs. After this comparison, the microprocessor makes the necessary decision and operates the injectors to provide the exact air-fuel ratio required by the engine.

TYPES OF COMPUTER MEMORIES

A computer's memory holds the programs and other data, such as vehicle calibrations the microprocessor refers to in performing calculations. To the CPU, the program is a set of instructions or procedures that it must follow. Included in the program is information that tells the microprocessor when to retrieve input (based on temperature, time, and so on), how to process the input, and what to do with it once it has been processed. The microprocessor works with memory in two ways: it can read information from memory or change information in memory by writing in or storing new information.

Random Access Memory

Information that requires temporary storage is sent from the microprocessor to the **random access memory (RAM)**. The information stored in the RAM is subject to change. Since the sensor input information changes frequently in relation to various

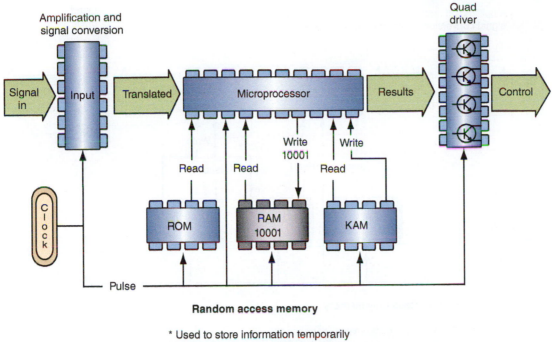

Figure 12-4 Random access memory (RAM).

operating conditions, this information is stored in the RAM (**Figure 12-4**). The microprocessor may write the results of calculations and other changeable data into the RAM. The microprocessor can write information into the RAM, read information from the RAM, and erase RAM information.

The term "random access" indicates that the microprocessor can retrieve information from any RAM address in any order. If the RAM has a volatile memory, each time the ignition switch is turned off, the information stored in the RAM is erased. RAMs may also be designed with a nonvolatile memory. This type of RAM retains information when the ignition switch is turned off. Some types of nonvolatile RAM can even retain memory when the battery power is disconnected. If the RAM has a volatile memory, new information will be written into the RAM when the engine is restarted. Many parts of the RAM are nonvolatile, meaning that the information is not erased when the vehicle is turned off.

> The microprocessor may write information into a **random access memory (RAM)** chip and read information from this chip.

> A **RAM** with a volatile memory erases stored information when the ignition switch is turned off.

Read Only Memory

The microprocessor can read information from the **read only memory (ROM)**, but information cannot be written into the ROM by the microprocessor, and the microprocessor cannot erase ROM information (**Figure 12-5**). During the chip manufacturing process, information is programmed into the ROM. This information is not erased even if the battery terminals are disconnected.

The ROM contains look up tables that contain information about how a vehicle should perform. For example, the lookup table would contain the ideal manifold vacuum under various engine operating conditions. The microprocessor compares the sensor inputs to the ideal vacuum in this table. If the sensor inputs indicate that the actual manifold vacuum is different from the ideal vacuum in lookup tables, the microprocessor will take some appropriate action.

The ROM also contains calibration tables regarding specific engine, transaxle or transmission, and differential specifications.

> The computer can only read information from a **read only memory (ROM)**.

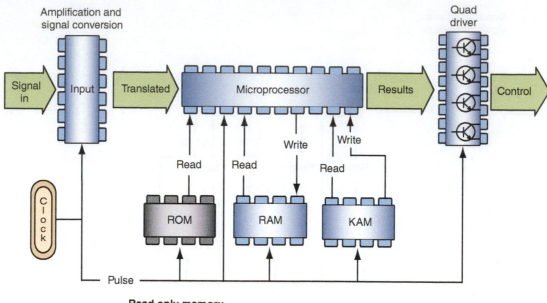

Read only memory

* Used to store information permanently
* Microprocessor cannot write to ROM
* Microprocessor can read from ROM
* Information is retained even when system power
  is disconnected

Figure 12-5 Read only memory (ROM).

Shop Manual
Chapter 12, page 608

Programmable Read Only Memory

Early automotive computers had a removable **programmable read only memory (PROM)**, which is serviced separately from the computer. The PROM contains specific programs such as the spark advance program, which is designed for the specific requirements of each vehicle. For example, this spark advance program varies with different transmissions or rear axle ratios.

OBD II computers are equipped with an **electronically erasable programmable read only memory (EEPROM)**. This type of memory chip can be reprogrammed via a J2534 pass through programming device.

Computers with **electronically erasable programmable read only memory (EEPROM)** have a memory chip that is easily programmable.

If an OBD II computer is replaced, the operating instructions for the vehicle must be programmed into the computer before use. The replacement PCM is much like a new personal computer that has to have software installed before use. Information stored in the PCM has to be tailored to the exact vehicle being serviced according to the VIN. This programming is called **flashing** the PCM. The elimination of the replaceable PROM chip was part of the OBD II legislation to prevent the use of aftermarket high-performance PROMs. There is an advantage to not using the replaceable PROM in that many updates for engine performance concerns can be performed during service by changing the **calibration** of the PCM without the need for having replacement parts. This calibration change is called **reflashing** the PCM. When servicing OBD II vehicles, always check for service bulletins that will alert the technician to any available service updates, including reflashing information.

Flashing the PCM is the initial programming of a new PCM. **Reflashing** is the reprogramming of a PCM based on service bulletins or procedures.

Keep Alive Memory

The **keep alive memory (KAM)** has characteristics similar to the RAM. For example, the microprocessor can read and write information to and from the KAM and it can erase KAM information (**Figure 12-6**). However, the KAM retains information when the

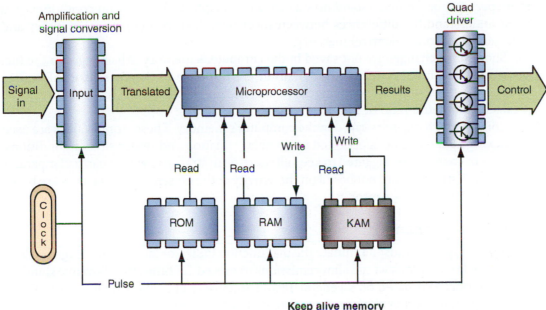

Figure 12-6 Keep alive memory (KAM).

* Used to store information temporarily
* Adaptive strategies use KAM
* Microprocessor can write to KAM
* Microprocessor can read from KAM
* Information is retained even when the system is turned off

ignition switch is turned off. When the battery power is disconnected from the computer, the KAM memory is erased. The KAM is used for adaptive strategies that are explained in the next section.

> The microprocessor can write information into and read information from a keep alive memory (KAM).

ADAPTIVE STRATEGY

Operation

If a computer has **adaptive strategy** capabilities, the computer can actually learn from past experience. For example, the normal voltage input range from a throttle position sensor (TPS) may be 0.6 volt to 4.5 volts. If a worn TPS sends a 0.3-volt signal to the computer, the microprocessor interprets this signal as an indication of component wear. The microprocessor stores this altered calibration in the KAM. The microprocessor now refers to this new calibration during calculations; thus, normal engine performance is maintained. If a sensor output is erratic or considerably out of range, the computer may ignore this input. When a computer has adaptive strategy, a short learning period is necessary:

> **Adaptive strategy** allows the computer to adapt to minor defects in the inputs and outputs.

1. After the battery has been disconnected
2. When a computer system component has been replaced or disconnected
3. On a new vehicle

During this learning period, the engine may surge, idle fast, or have a loss of power. The average learning period lasts for 5 miles of driving.

Most adaptive strategies have two parts: **short-term fuel trim (STFT)** and **long-term fuel trim (LTFT)**. Short-term strategies are those immediately enacted by the computer to overcome a change in operation. These changes are temporary. Long-term strategies are based on the feedback about the short-term strategies. These changes are

> **Short-term fuel trim** adds or subtracts fuel to correct for small or temporary changes.

> **Long-term fuel trim** Moves slowly to keep short-term fuel trim near zero.

more permanent. To understand how a computer adapts itself to certain conditions, you must understand the differences between short-term fuel trim (adaptive) memory and long-term fuel trim (adaptive) memory.

Short-term changes are not saved in the computer's memory. All changes to the fuel system happen immediately and occur in direct response to the O_2 sensor/AF sensor and/ or other sensors. These changes are also designed to keep the fuel mixture at the ideal stoichiometric ratio of 14.7:1.

Long-term changes are saved in the computer's memory. These stored values are used the next time the engine is operated in a similar situation and under similar conditions. Long-term changes are triggered to keep all short-term strategies within particular parameters. These strategies are not based on the activity of the O_2 sensor; rather, they are based on the results of the O_2 sensor.

OBD II Components

The OBD II systems must illuminate the malfunction indicator lamp (MIL) (**Figure 12-7**) if the vehicle conditions would allow emissions to exceed 1.5 times the allowable standard for that model year based on a Federal Test Procedure (FTP). The FTP is an established testing program used by the government to certify new vehicles as compliant with emissions regulations. When a component or strategy failure allows emissions to exceed this FTP level, the MIL is illuminated to inform the driver of a problem, and a diagnostic trouble code is stored in the PCM.

> The oxygen sensor behind the catalytic converter is often called the downstream O_2 sensor.

🖋️📖 A BIT OF HISTORY

Many years ago, before OBD II, vehicles typically had only one oxygen sensor to inform the computer of changes in the mixture. Once the oxygen sensor began to degrade, the computer would add fuel to make the oxygen sensor read correctly, eventually the engine would be running very rich, but the oxygen sensor would report a normal fuel mixture. The problem was that with only one sensor to report mixture, the mixture went out of bounds when that sensor failed.

Oxygen Sensor Locations

OBD vehicles use at least two **heated oxygen sensors**, or an oxygen sensor and an air-fuel ratio sensor (A/F sensor). If the A/F sensor is used, it is in the pre-catalyst position. **Figure 12-8** details the numbering system for multiple heated oxygen sensors. Note that

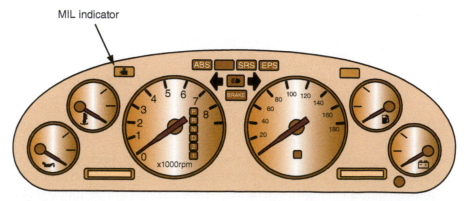

Figure 12-7 A standard MIL.

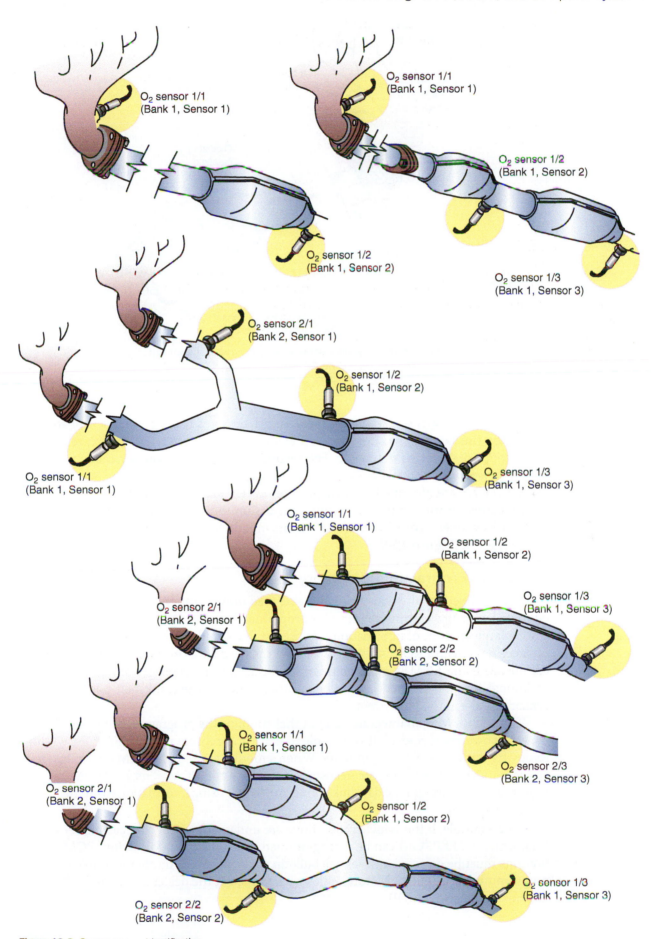

Figure 12-8 Oxygen sensor identification.

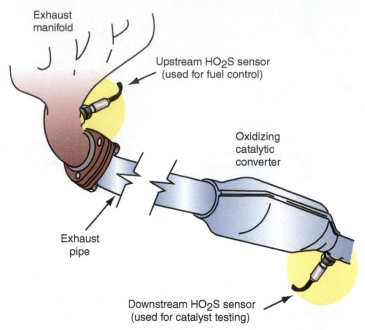

Exhaust manifold

Upstream HO$_2$S sensor
(used for fuel control)

Oxidizing catalytic converter

Exhaust pipe

Downstream HO$_2$S sensor
(used for catalyst testing)

Figure 12-9 OBD II system with heated oxygen sensors in the exhaust manifold and after the catalytic converter (pre- and post-catalyst).

regarding the sensors on V-type engines, bank 1 sensor 1 is the oxygen sensor (or A/F sensor) on the side that has the number 1 cylinder. Bank 2 sensor 1 is the oxygen sensor on the opposite bank as number 1. The second sensor on the number 1 side would be bank 1 sensor 2, and so forth.

The HO$_2$S are identified by their position in relation to the number 1 cylinder and their location relative to the converters. Sensors on the same side of the vehicle as the number 1 cylinder have a 1 prefix, and sensors mounted downstream from the converter have a 2 suffix (**Figure 12-9**).

Computer Programming/Memory

A **heated oxygen sensor** is electrically heated to accurately sense exhaust gases regardless of operating temperature.

Instead of a fixed, unalterable programmable read only memory (PROM), the PCMs are equipped with an electronically erasable programmable read only memory (EEPROM) to store a large amount of information. The EEPROM is soldered into the PCM and is not replaceable. The EEPROM stores data without the need for a continuing source of electrical power. If the ECM is replaced, then it will have to be reprogrammed. Sometimes additional components may have to be relearned, such as the crankshaft position (CKP) sensor and the anti-theft system.

The EEPROM is an integrated circuit that contains the program used by the PCM to provide powertrain control. It is possible to erase and reprogram the EEPROM without removing this chip from the computer. When a modification to the PCM operating strategy is required, it is no longer necessary to replace the PCM. The EEPROM may be reprogrammed through the data link connector (DLC) using information from the manufacturer's website.

For example, if the vehicle calibrations are updated for a specific car model sold in California, the EEPROM can be reprogrammed with updated information. PCM recalibrations must be directed by a service bulletin or recall letter. Another example is "flash" programming on General Motors vehicles performed at the dealership with the multiple diagnostic interface (MDI).

MONITORS

1. Catalyst efficiency monitor
2. Engine misfire monitor
3. Fuel system monitor
4. Heated exhaust gas oxygen sensor monitor
5. Exhaust gas recirculation (EGR) monitor
6. Evaporative (EVAP) system monitor
7. Secondary air injection (AIR) monitor
8. Thermostat monitor
9. PCV monitor
10. Comprehensive component monitor

> **AUTHOR'S NOTE** Even though OBD II systems can all be accessed through the "generic" scan tool mode, most technicians prefer the "enhanced" data available on specialized scan tools. Generic scan tools tend to run slower and do not have access to manufacturer specific codes for emissions or access to systems such as ABS.

The computer has a series of monitors that were mandated through OBD II legislation. A monitor watches a particular system for faults and degraded performance. The monitors in use at this time are discussed in the following paragraphs.

Shop Manual
Chapter 12, page 626

Catalyst Efficiency Monitor

OBD II vehicles use a minimum of two oxygen sensors (HO_2S) as a **catalyst efficiency monitor**. One of these is used for feedback to the PCM for fuel control, and the other, located at the rear of the catalytic converter, gives an indication of the efficiency of the converter. If the converter is operating properly, the signal from the pre-catalyst O_2 will have oscillations, while the post-catalyst O_2 will be relatively flat (**Figure 12-10**). Once the signal from the rear sensor approaches that of the front sensor, the MIL comes on and a DTC is set.

The two oxygen sensors are both heated to bring them into operation sooner, but the heater is not turned on until some warming of the exhaust after a cold start. The internal heater is not turned on until the engine coolant temperature (ECT) sensor signal indicates a warmed-up engine. This action prevents cracking of the ceramic. Gold-plated pins and sockets are used in the HO_2S to help prevent corrosion.

The **catalyst efficiency monitor** uses two oxygen sensors (pre- and post-converter) to determine how well the catalytic converter is treating incoming exhaust gas. This is accomplished by comparing the pre-catalyst and post-catalyst O_2 sensor voltages.

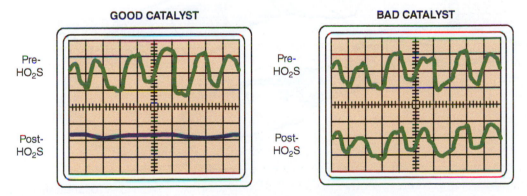

Figure 12-10 Oxygen sensor signals for good and bad catalytic converters.

A catalytic converter stores oxygen during lean engine operation and gives up this stored oxygen during rich operation to burn up excessive hydrocarbons. Catalytic converter efficiency is measured by monitoring the oxygen storage capacity of the converter during closed-loop operation.

When the catalytic converter is storing oxygen properly, the downstream HO_2S sensors provide low-frequency voltage signals. If the catalytic converter is not storing oxygen properly, the voltage signal frequency increases on the downstream HO_2S sensors until the frequency of the downstream HO_2S sensors approaches the frequency of the upstream HO_2S sensors (**Figure 12-10**). When the downstream HO_2S voltage signals reach a certain frequency, a DTC is set in the PCM memory. If the fault occurs on three drive cycles, the MIL light is illuminated. The placement of the HO_2S sensors' sensors are shown in **Figure 12-11**.

Cataltyst Efficiency Monitor with AF Ratio Sensor

Vehicles that use an A/F sensor instead of an HO_2S will use the A/F sensor in the pre-catalyst position and still use the HO_2S in the post-catalyst position. The catalyst efficiency monitor is still measuring the efficiency of the converter using the post-converter HO_2S sensor regardless of whether a HO_2S sensor or A/F sensor is used in the pre-catalyst location. Even though the A/F ratio sensor does not toggle as the oxygen sensor does, the computer still knows what the frequency of the post-catalyst should be according to current conditions in the exhaust, and can make the determination of the catalyst health.

Shop Manual
Chapter 12, page 630

Misfire Monitor

If a cylinder misfires, unburned hydrocarbons (HCs) are exhausted from the cylinder, and these excessive HC emissions enter the catalytic converter. When the catalytic converter changes these excessive HC emissions to carbon dioxide and water, the catalytic converter is overheated and the honeycomb in the converter may melt together into a solid mass. If this occurs, the converter is no longer efficient in reducing emissions. For this reason, most computers are designed to shut down a misfiring cylinder's fuel injector. The injector will

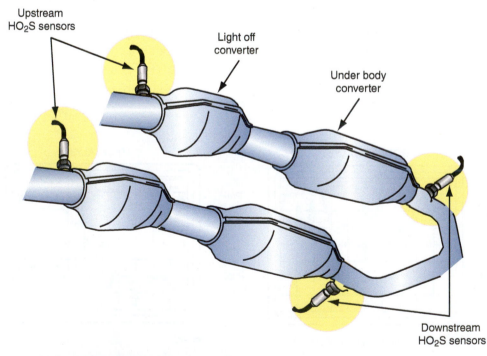

Figure 12-11 Catalyst efficiency monitoring system.

stay off until the next key cycle even if the miss no longer occurs. Additionally, a misfiring cylinder may not only put raw fuel into the exhaust stream, it also allows air from the cylinder to enter the exhaust stream. Remember that an oxygen or A/F sensor only "sees" oxygen, the raw fuel (HC) is not detected by the O_2 or A/F sensor. What it can react to is the presence of the oxygen, which can be interpreted as a lean condition, and actually add more fuel. To prevent this from happening, most late-model vehicles will go into open loop and ignore the oxygen sensor signal during an engine miss, to help protect the converter.

Cylinder misfire monitoring requires measuring the contribution of each cylinder to engine power. The misfire monitoring system uses a highly accurate crankshaft angle measurement to measure the crankshaft acceleration each time a cylinder fires. A high data rate crankshaft sensor is required for this function (**Figure 12-12**). The PCM monitors the crankshaft acceleration time for each cylinder firing. If a cylinder is contributing normal power, a specific crankshaft acceleration time occurs. When a cylinder misfires, the cylinder does not contribute to engine power, and crankshaft acceleration for that cylinder is slowed.

Most OBD II systems allow a random misfire rate of about 2 percent before a misfire is flagged as a fault. It is important to note that this monitor looks at only the speed of acceleration of the crankshaft during a cylinder's firing stroke. It cannot determine if the problem is fuel-, ignition-, or mechanical-related. Misfire is categorized as type A or type B. **Type A misfire** could cause immediate catalyst damage. **Type B misfire** could cause emissions of 1.5 times the design standard and could cause an inspection/maintenance (I/M) failure. When there is a type A misfire, the MIL will flash. If there is a type B misfire, the MIL will turn on but will not flash.

The misfire monitoring sequence includes an adaptive feature compensating for variations in engine characteristics caused by manufacturing tolerances and component wear (**Figure 12-13**). It also has the adaptive capability to allow vibration at different engine speeds and loads. When an individual cylinder's contribution to engine speed falls below a certain threshold, the misfire monitoring sequence calculates the vibration, tolerance, and load factors before setting a misfire code.

Cylinder misfire monitoring uses a complex strategy to calculate crankshaft speed acceleration as each cylinder fires. If a misfire occurs, the relative speed of the crankshaft momentarily slows and resumes when the next cylinder fires.

Type A Misfires

Type A and type B engine misfires are detected by the misfire monitor. When detecting a type A misfire, the monitor checks cylinder misfiring over a 200 rpm period. If cylinder misfires are between 2 and 20 percent, the monitor considers the misfiring to be excessive.

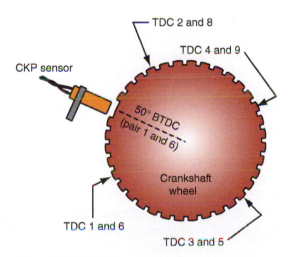

Figure 12-12 High data rate crankshaft sensor used for misfire detection.

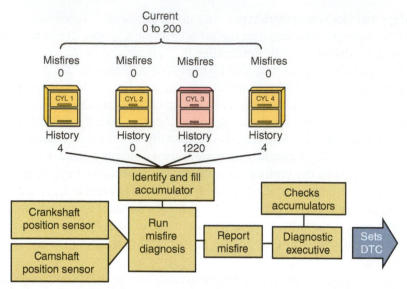

Figure 12-13 Engine misfire file system.

Some manufacturers use the ABS wheel speed sensors to detect rough roads and use this information to modify the misfire code sensitivity.

Shop Manual
Chapter 12, page 634

Under this condition, the PCM may shut off the fuel to the misfiring cylinder or cylinders to limit catalytic converter heat. The PCM may turn off two injectors at the same time on misfiring cylinders. When the engine is operating under heavy load, the PCM will not turn off the injectors on misfiring cylinders.

If the misfire monitor detects a type A cylinder misfire and the PCM does not shut off the injector or injectors, the MIL light begins flashing. When the misfire monitor detects a type A cylinder misfire and the PCM shuts off the injector or injectors, the MIL is illuminated continually.

A misfire means a lack of combustion in at least one cylinder for at least one combustion event. A misfire pumps unburned fuels through the exhaust. Although the converter can handle an occasional sample of raw fuel, too much fuel to the converter can overheat and destroy it. A type A misfire sets a trouble code on the first failure.

Type B Misfires

To detect a type B cylinder misfire, the misfire monitor checks cylinder misfiring over a 1,000 rpm period. A type B misfire occurs if cylinder misfiring exceeds 2 to 3 percent during this 1,000 rpm period. This amount of cylinder misfiring may not overheat the catalytic converter, but it may cause excessive emission levels. When a type B misfire is detected, a pending DTC is set in the PCM memory. A type B misfire can illuminate the MIL on the second nonconsecutive failure if conditions are similar to the first.

Fuel System Monitoring

The **fuel system monitor** checks short-term FT and long-term FT while the PCM is operating in closed loop. Fuel trim is a feedback system that looks at the present state of this oxygen sensor compared to the desired range. When a fuel system problem causes the PCM to make large fuel trim corrections for an excessive time period, the fuel system monitor sets a DTC and illuminates the MIL if the fault occurs on two consecutive drive cycles. The fuel system monitor operates continually when the PCM is in closed loop. Like the misfire monitor, the fuel system monitor can illuminate the MIL on a second nonconsecutive failure under similar conditions.

Heated Oxygen Sensor Monitor

Shop Manual
Chapter 12, page 644

The system also monitors lean-to-rich and rich-to-lean time responses. This test can pick up a lazy O_2 sensor that cannot switch fast enough to keep proper control of the air-fuel mixture in the system. These sensors are the heated type, and the amount of time before activity of the sensor signal is present is an indication of whether it is functional or not. Some systems use current flow to indicate if the heater is working or not.

All of the system's HO_2S sensors are monitored once per drive cycle, but the heated oxygen sensor monitor provides separate tests for the upstream and downstream sensors. The heated oxygen sensor monitor checks the voltage signal frequency of the upstream HO_2S. Excessive time between signal voltage frequencies indicates a faulty sensor. At certain times, the heated oxygen sensor monitor varies the fuel delivery and checks for HO_2S sensor's response. A slow response in the sensor voltage signal frequency indicates a faulty sensor. The sensor signal is also monitored for excessive voltage.

The heated oxygen sensor monitor also checks the frequency of the rear HO_2S sensor signals and checks these sensor signals for excessively high voltage. If the monitor does not detect signal voltage frequency within a specific range, the rear HO_2S sensors are considered faulty. The heated oxygen sensor monitor will command the PCM to vary the air-fuel ratio to check the rear HO_2S sensor's response.

On many V-type engines, there are two primary or upstream O_2 sensors, one for each cylinder bank. These individual sensors report the air-fuel mixture and combustion for the cylinders on that side of the engine.

A/F Ratio Sensors

Just as with the heated oxygen sensors, the A/F ratio sensors are also monitored by the PCM for performance, lack of activity, and opens, grounds, or shorts in the wiring to them from the PCM.

EGR System Monitoring

Although many vehicles no longer have an EGR valve, the exhaust gas recirculation (EGR) monitors use several different strategies to determine if the system is operating properly. Some systems look at the manifold absolute pressure (MAP) signal, energize the EGR valve, and look for corresponding change in vacuum levels. Most electronically controlled EGR systems use a pintle position sensor to determine the actual position of the EGR valve compared to the commanded value from the PCM.

Shop Manual
Chapter 12, page 617

The EGR systems from Ford may contain a delta pressure feedback EGR (DPFE) sensor. The latest DPFE design subtracts the MAP pressure from that of the pressure measured at the EGR valve to determine EGR flow.

With the engine idling and the EGR valve closed, the PCM checks for EGR flow again by comparing MAP to the pressure at the EGR valve fitting. When the EGR valve is closed and there is no EGR flow, the pressure should be the same at both areas. If the pressure is different at these two hoses, the EGR valve is stuck open.

The PCM commands the EGR valve to open and then checks the pressure at the two exhaust hoses connected to the DPFE sensor. With the EGR valve open and EGR flow through the orifice, there should be higher pressure at the EGR valve fitting to the DPFE sensor (**Figure 12-14**).

The PCM checks the EGR flow by checking the DPFE signal value against an expected DPFE value for the engine operating conditions at steady throttle within a specific rpm range. If a fault is detected in any of the EGR monitor tests, a DTC is set in the PCM memory. If the fault occurs during two consecutive drive cycles, the MIL is illuminated. The EGR monitor operates once per OBD II trip.

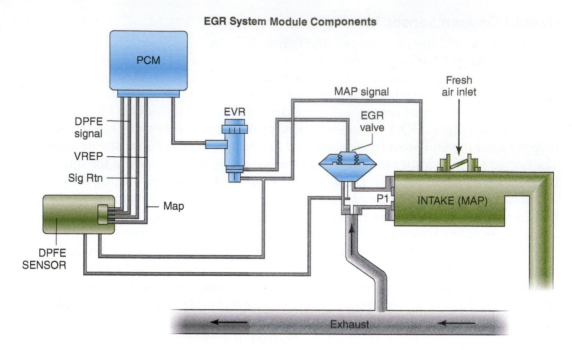

Figure 12-14 EGR system with a delta pressure feedback sensor.

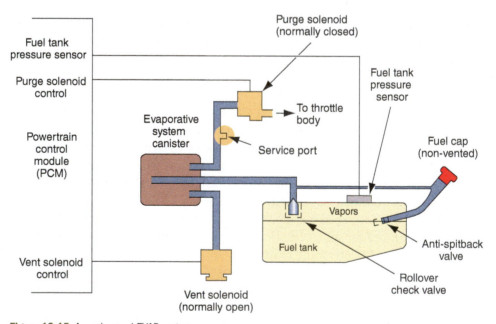

Figure 12-15 An enhanced EVAP system.

Enhanced EVAP System

Since 1996, enhanced EVAP system vehicles have an enhanced evaporative system monitor (**Figure 12-15**). This system detects leaks and restrictions in the EVAP system. A newly designed fuel tank filler cap is used on this system. In these enhanced systems, an evaporative system leak or a missing fuel tank cap will cause the MIL to turn on. This enhanced system has to be able to detect leaks as small as 0.020 in diameter. To accomplish this, the system uses a fuel tank pressure sensor, a vent solenoid, and a purge solenoid. These three devices allow the PCM to run pressure and vacuum checks for leaks in the system.

(See Chapter 12 for more details.) A new fuel cap that was easier to install was added to prevent setting a code from a loose fuel cap. Some vehicles can also illuminate a "check fuel cap" if the cap is loose. Additionally, since 1998, vehicles have to capture refueling vapors in a system called onboard refueling vapor recovery (ORVR). The ORVR is not monitored by OBD II.

SECONDARY AIR INJECTION SYSTEM MONITOR

Some manufacturers turned to secondary air injection (AIR) to reduce cold start emissions. Cold start emissions make up a large share of the vehicle emissions, because they occur before the converter is hot enough to reduce emissions, and the oxygen sensor or A/F sensor is hot enough to accurately report emissions. In recent years, the AIR system is being replaced by functions of DFI such as injecting fuel during the compression stroke while cold, and by the VVT system varying valve timing events when cold. The AIR system operation can be verified by turning the AIR system on to inject air upstream of the oxygen sensor while monitoring its signal. Current designs inject air into the exhaust manifold when the engine is in open loop. This extra air helps heat the converter for proper operation. The pump runs for up to 2 minutes after a cold start.

The AIR system is monitored with passive and active tests. During the passive test, the voltage of the pre-catalyst HO_2S is monitored from start-up to closed-loop operation. The AIR pump is normally on during this time. Once the HO_2S is warm enough to produce a voltage signal, the voltage should be low if the AIR pump is delivering air to the exhaust manifold. The secondary AIR monitor will indicate a pass if the HO_2S voltage is low at this time. The passive test also looks for a higher HO_2S voltage when the AIR flow to the exhaust manifold is turned off by the PCM. When the AIR system passes the passive test, no further testing is done. If the AIR system fails the passive test or if the test is inconclusive, the AIR monitor in the PCM proceeds with the active test.

During the active test, the PCM cycles the AIR flow to the exhaust manifold on and off during closed-loop operation and monitors the pre-catalyst HO_2S voltage and the short-term FT value. When the AIR flow to the exhaust manifold is turned on, the sensor's voltage should decrease and the short-term FT should indicate a leaner condition. The secondary AIR system monitor illuminates the MIL and stores a DTC in the PCM's memory if the AIR system fails the active test on two consecutive trips.

Some Ford vehicles have an electric air pump system controlled by a solid-state relay. The relay is operated by a signal from the PCM. An air-injection bypass solenoid is also operated by the PCM. This solenoid supplies vacuum to dual air diverter valves (**Figure 12-16**).

When the engine is started, the PCM signals the relay to start the air pump. This module supplies the high current required for air pump operation. The air pump may provide a 10-second delay in pump operation after the engine is started. The PCM also energizes the air-injection bypass solenoid. When this solenoid is energized, it supplies vacuum to the dual air diverter valves. This action opens the normally closed air diverter valves. Air from the pump is now delivered to the exhaust manifold. The purpose of the air pump is to oxidize HC and CO in the exhaust manifolds from 20 to 120 seconds after the engine has started and until the catalytic converter is working properly. The length of time the air pump is operating depends on the temperature of the engine. Once the catalyst is warmed up, the PCM signals the relay to shut down the air pump. The PCM also de-energizes the air-injection bypass solenoid, which allows the air diverter valves to close.

The PCM monitors the relay and the air pump to determine if secondary air is present. This PCM monitor for the air pump system functions once per drive cycle. When a

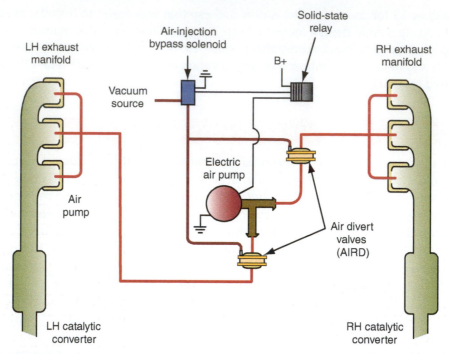

Figure 12-16 An electric air pump system.

malfunction occurs in the air pump system on two consecutive drive cycles, a DTC is stored and the MIL is turned on. If the malfunction corrects itself, the MIL is turned off after three consecutive drive cycles in which the fault is not present.

Thermostat Monitor

The thermostat monitor was added in 2002, after the initial OBD II legislation. Since vehicles cause more pollution when they are cold, not to mention the fact that the vehicle may not get hot enough to boil off unburned fuel and water from the oil in the crankcase, A timer is set when the engine is started after a period of sitting to cool down. The PCM faces the intake air temperature (IAT) sensor when the vehicle is started and decides how long it should take for the coolant to reach normal operating temperature. If it does not reach this temperature, the PCM will set a code on the second consecutive occurrence (type B code).

PCV Monitor

The PCV system monitor was phased in starting in 2002 and required for vehicles built after the 2004 model year. This monitor has the ability to monitor PCV hoses for disconnection. GM has built their PCV into the intake manifold. Since this eliminated the hoses, it also eliminated the need for a separate system. Ford's strategy for the monitor was to make the hoses large enough that the vehicle would not run with the PCV hoses disconnected. The exception is some of Eco-Boost engines. These systems are using a fresh air inlet hose equipped with a pressure sensor to monitor the PCV system for malfunctions, such as disconnected hoses in the system. If the fresh air hose became disconnected, the vehicle would discharge blow by gases into the atmosphere.

Comprehensive Component Monitor

The system looks at any electronic input that could affect emissions. The strategy is to look for opens and shorts or input signal values that are out of the normal range. It also looks to see if the actuators have their intended effect on the system and to monitor other abnormalities.

Shop Manual
Chapter 12, page 641

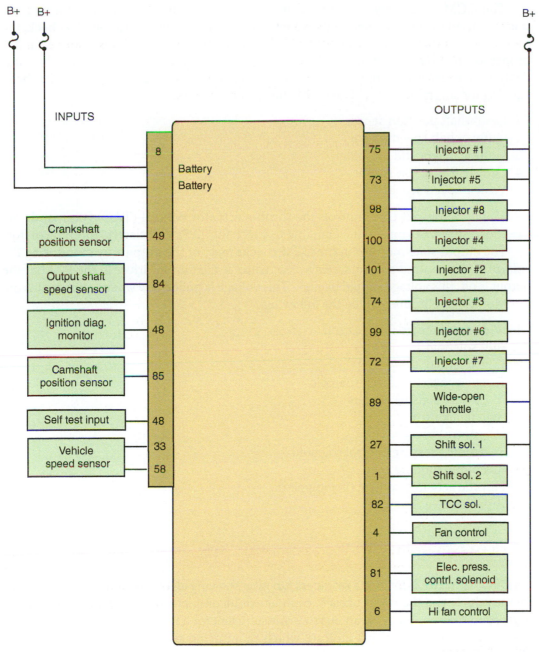

Figure 12-17 Examples of some of the components monitored by the CCM.

The **comprehensive component monitor (CCM)** uses two strategies to monitor inputs and two strategies to monitor outputs. One strategy for monitoring inputs involves checking certain inputs for electrical defects and out-of-range values by checking the input signals at the analog-to-digital converter. The input signals monitored in this way are illustrated in **Figure 12-17**:

1. Mass airflow sensor
2. Manual lever position sensor
3. Throttle position sensor
4. Accelerator pedal position sensor
5. Engine coolant temperature sensor
6. Intake air temperature sensor
7. Knock sensors
8. System wiring

The **comprehensive component monitor (CCM)** strategy determines potential electronic inputs that could affect emission levels.

The CCM checks signal inputs that do not have their own monitor. One of the methods used to check components is a rationality check. During a rationality check, the monitor uses other sensor readings and calculations to determine if a sensor reading is proper for the present conditions. An example of this would be an ECT sensor that suddenly goes from a warm engine reading to a cold engine reading, after the engine has been running for several minutes. The CCM checks these inputs:

1. Crankshaft position sensor
2. Output shaft speed sensor
3. Ignition diagnostic monitor
4. Camshaft position sensor
5. Vehicle speed sensor

The PCM output that controls the throttle actuator control (TAC) commands the throttle to sweep during engine cranking to check the TPSs and electric motor actuator.

The output state monitor in the CCM checks most of the outputs by monitoring the voltage of each output solenoid, relay, or actuator at the output driver in the PCM. If the output is off, this voltage should be high. This voltage is pulled low when the output is on.

Monitored outputs include the following:

1. Wide-open throttle A/C cutoff
2. Shift solenoid 1
3. Shift solenoid 2
4. Torque converter clutch solenoid
5. HO_2S (A/F sensor) heaters
6. High fan control
7. Fan control
8. Electronic pressure control solenoid
9. Fuel pump module
10. Throttle actuator control (if equipped)
11. Variable valve timing

Shop Manual
Chapter 12, page 641

SYSTEM READINESS MODE

All OBD II scan tools include a readiness function showing all of the monitoring sequences on the vehicle and the status of each, complete or incomplete. If vehicle travel time, operating conditions, or other parameters were insufficient for a monitoring sequence to complete a test, the scanner will indicate which monitoring sequence is not yet complete (**Figure 12-18**).

Warm-Up Cycle

OBD II standards define a warm-up cycle as a period of vehicle operation, after the engine had been turned off, in which the coolant rises by at least 40°F (4°C) and reaches at least 160°F (71°C). Most DTCs are automatically erased after 40 warm-up cycles if the fault is not detected during that time. Some manufacturers retain erased DTCs in a flagged condition. This can be useful if the technician notices a pattern of component failure, all of which may be related to a single intermittent cause like low fuel pressure.

A **trip** in an OBD II system refers to starting and driving the vehicle until a monitor is completed.

OBD II TRIP

The OBD II **trip** consists of an engine start following an engine off period, with enough vehicle travel to allow for a particular monitor to complete.

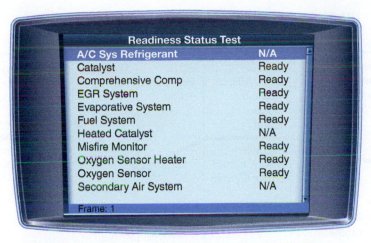

Figure 12-18 The available monitors have all run and passed for this vehicle.

The definition of a trip depends on the monitor involved. For instance, the fuel system monitor will run once the following criteria are met:

1. Closed loop
2. Coolant temperature between 20°F (−7°C) and 257°F (125°C)
3. Intake air temperature between 20°F (−7°C) and 293°F (145°C)
4. Mass airflow between 1 and 500 grams per second
5. Barometric pressure greater than 10.7 psi
6. Engine speed between 400 rpm and 5,700 rpm
7. Vehicle speed less than 80 mph (129 km/h)
8. Manifold absolute pressure between 2.2 and 12.3 psi

Every vehicle may have a slightly different set of criteria; this is just an example. The monitor will run once when these criteria are met while driving (**Figure 12-19**).

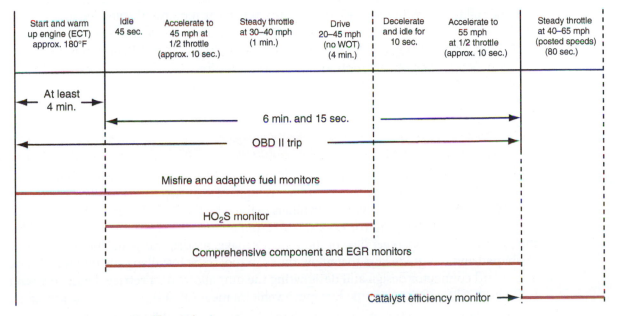

Figure 12-19 Chart showing different trips for various monitors.

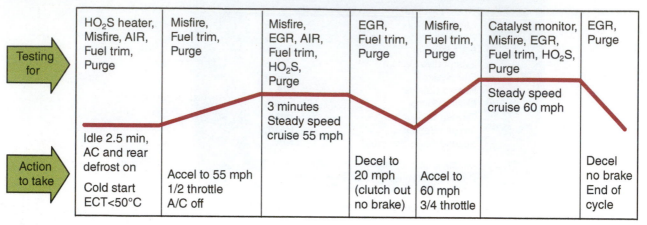

Figure 12-20 An OBD drive cycle.

DRIVE CYCLE

A **drive cycle** begins at engine start-up and lasts until all the monitors are completed.

An OBD II **drive cycle** (**Figure 12-20**) consists of an engine start and vehicle operation that brings the vehicle into closed loop and includes whatever specific operating conditions are necessary either to initiate and complete a specific monitoring sequence or to verify a symptom or a repair. A monitoring sequence is an operational strategy designed to test the operation of a specific system, function, or component.

To complete a drive cycle, all trip monitors must be completed. The catalyst monitor must be completed after the other monitors in a drive cycle. A steady throttle opening between 40 mph and 60 mph (64 km/h and 96 km/h) for 80 seconds is required to complete the catalyst efficiency monitor.

AUTHOR'S NOTE It is important to note that if a vehicle has never been driven under all the conditions necessary to run a particular monitor, the vehicle may rarely complete a drive cycle. An example of this is the catalyst efficiency monitor that requires a steady cruise speed at 40 mph to 60 mph. A vehicle driven in the city may not see this condition for long periods of time.

TEST CONNECTOR

Shop Manual
Chapter 12, page 634

OBD II and the Society of Automotive Engineers (SAE) standards require the DLC to be mounted in the passenger compartment out of sight of vehicle passengers. The DLC must be a 16-terminal connector with 9 terminals defined by the SAE (**Figure 12-21**).

The standard DLC (**Figure 12-22**) is a 16-pin connector. The same pins are used for the same information, regardless of the vehicle's make, model, and year. The connector is D-shaped and has guide keys that allow the scan tool to be installed only one way. Using a standard connector design and designating the pins allows data retrieval with any scan tool designed for OBD II. Some European vehicles meet OBD II standards by providing the designated DLC along with their own connector for their own scan tool.

Cavity	General assignment
1	Ignition control
2	BUS (+) SCP
3	Discretionary (not used)
4	Chassis ground
5	Signal ground
6	CAN high
7	K line of ISO 9141
8	Discretionary (not used)
9	Discretionary (not used)
10	BUS (−) SCP
11	Discretionary (not used)
12	Discretionary (not used)
13	FEPS (flash EEPROM)
14	CAN low
15	L line of ISO 9141
16	Battery power

Figure 12-21 DLC terminal identification.

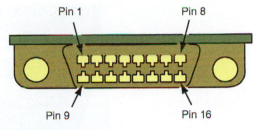

Pin 1   Pin 8
Pin 9   Pin 16

Figure 12-22 A 16-pin DLC.

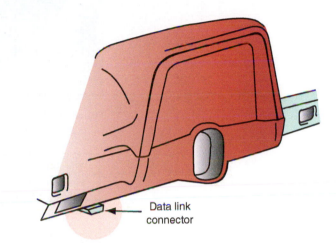

Data link connector

Figure 12-23 Designated location for the DLC.

The DLC must be easily accessible while sitting in the driver's seat (**Figure 12-23**). All DLCs must be located somewhere between the left end of the instrument panel and the centerline of the vehicle. The connector cannot be located on the center console or in a cupholder on the console. The DLC cannot be located any lower than the bottom of the steering wheel and should be visible to a crouched technician entering the vehicle from the driver's side of the vehicle. The DLC cannot be hidden behind panels and must be accessible without tools. Any generic scan tool can be connected to the DLC and can access the diagnostic data stream.

The connector pins are arranged in two rows and are numbered consecutively. Seven of the sixteen pins have been assigned by the OBD II standard. The remaining nine pins can be used by the individual manufacturers to meet their needs and desires.

When a vehicle has a 16-pin DLC, this does not necessarily mean that the vehicle is equipped with OBD II.

MALFUNCTION INDICATOR LAMP OPERATION

An OBD II system continuously monitors the entire emissions system, switches on an MIL if something goes wrong, and stores a fault code in the PCM when it detects a problem. The codes are well defined and can lead a technician to the problem. A scan tool must be used to access and interpret emission-related DTCs regardless of the make and model of the vehicle (**Figure 12-24**).

Generic codes begin with P0; manufacturer-specific codes begin with a P1. P2XXX codes are again generic codes, P3XXX can be either manufacturer specific or generic. If a generic scan tool is used, the manufacturer-specific codes cannot be read. Manufacturer-specific codes can be read only on an "enhanced" scan tool.

Many of the trouble codes from an OBD II system will mean the same thing regardless of manufacturer. These are the generic codes. However, some of the codes will pertain

Shop Manual
Chapter 12, page 632

Figure 12-24 Scan tool connected to the DLC.

The SAE J2012 standards specify that all DTCs will have a five-digit alphanumeric numbering and lettering system. The following prefixes indicate the general area to which the DTC belongs:

1. P — power train
2. B — body
3. C — chassis
4. U — network codes

The first number in the DTC indicates who is responsible for the DTC definition.

1. 0 — SAE
2. 1 — manufacturer

The third digit in the DTC indicates the subgroup to which the DTC belongs. The possible subgroups are as follows:

0 — Total system
1 — Air-fuel control
2 — Air-fuel control
3 — Ignition system misfire
4 — Auxiliary emission controls
5 — Idle speed control
6 — PCM and I/O
7 — Transmission
8 — Non-EEC powertrain

The fourth and fifth digits indicate the specific area where the trouble exists. Code P1711 has this interpretation:

P — Powertrain DTC
1 — Manufacturer-defined code
7 — Transmission subgroup
11 — Transmission oil temperature (TOT) sensor and related circuit

Figure 12-25 OBD II DTC interpretation chart.

only to a particular system or will mean something different with each system, they are the manufacturer-specific codes. The DTC is a five-character code with both letters and numbers (**Figure 12-25**). This is called the alphanumeric system. The first character of the code is a letter. This defines the system where the code was set. Currently, there are three possible first character codes: B for body, C for chassis, and P for powertrain. The U-codes are designated for use as network problem codes.

The second character is a number. This defines the code as being a mandated code or a special manufacturer code. This number will be a 0, 1, 2, or 3. A 0 code means that the fault is defined or mandated by OBD II. A "1" code means that the code is manufacturer specific. Codes with P2 are generic codes, and P3 can be either generic or manufacturer specific.

The third through fifth characters are numbers. These describe the fault. The third character of a powertrain code tells you where the fault occurred. The remaining two characters describe the exact condition that set the code.

There are two main types of OBD II diagnostic trouble codes (DTCs), type A and type B. There are also type C and type D codes that are not emissions related. The type C codes can illuminate a lamp other than the MIL on some models; type D codes do not usually illuminate any lamp but can be used by the technician to aid troubleshooting. In this discussion, we will be concerned with type A and type B codes. In the OBD II monitors previously discussed, the vehicle runs tests called monitors whenever the **enable criteria** for the test is correct. A type A DTC will illuminate the MIL on the first failure. Type B codes are "pending" on the first failure, and a MIL is illuminated on the second *consecutive* failure unless the pending DTC is for fuel trim or misfire. The fuel trim and misfire codes can be set on a second *nonconsecutive* failure. Additionally, the catalyst monitor must fail *three consecutive* tests before illuminating the MIL. If the MIL flashes, a type A misfire is indicated. A flashing MIL means that the catalytic converter could be damaged due to raw fuel being burned inside the converter.

Enable Criteria

For misfire and fuel monitors, if the fault does not occur on three consecutive drive cycles under similar conditions, the MIL is turned off. The system defines similar conditions as follows:

1. Engine speed within 375 rpm compared to when the fault was detected.
2. Engine load within 10 percent compared to when the fault was detected.
3. Engine warm-up state or coolant temperature must match the temperature when the fault was detected.

For the catalyst efficiency, HO_2S, EGR, and comprehensive component monitors, the MIL is turned off if the same fault does not reappear for three consecutive drive cycles. When the fault is no longer present and the MIL is turned off, the DTC is erased after 40 engine warm-up cycles. A technician may use a scan tester to erase DTCs immediately.

A pending DTC is a code representing a fault that has occurred but not enough times to illuminate the MIL. Some scan testers are capable of reading pending DTCs with the continuous DTCs.

TEST MODES

All OBD II systems have the same basic test modes. These test modes must be accessible with an OBD II scan tool.

Mode 1 is the **parameter identification (PID)** mode. It allows access to certain data values, analog and digital inputs and outputs, calculated values, and system status information. Some of the PID values will be manufacturer specific, while others are common to all vehicles.

Mode 2 is the **freeze-frame** data access mode. This mode permits access to emission-related data values from specific generic PIDs. These values represent the operating

"P0"xxx codes are mandated.

"P1"xxx codes are manufacturer specific.

Enable criteria are conditions that must be met for the PCM to run a system monitor. For example, the HO_2S monitor is run only during 3 minutes of steady-state cruise at 55 mph.

Test modes are available only in "generic mode."

Freeze frame refers to a function of a scan tool that captures a particular set of data when a trouble code is set.

conditions at the time the fault was recognized and logged into memory as a DTC. Once a DTC and a set of freeze-frame data are stored in memory, they will stay in memory even if other emission-related DTCs are stored. The number of these sets of freeze frames that can be stored are limited.

One type of failure is an exception to this rule—misfire. Fuel system misfires will overwrite any other type of data except for other fuel system misfire data. This data can be removed only with a scan tool. When a scan tool is used to erase a DTC, it automatically erases all freeze-frame data associated with the events that lead to that DTC.

Mode 3 permits scan tools to obtain stored DTCs. The information is transmitted from the PCM to the scan tool following an OBD II mode 3 request. Either the DTC or its descriptive text, or both will be displayed on the scan tool.

The PCM reset mode, mode 4, allows the scan tool to clear all emission-related diagnostic information from its memory. Once the PCM has been reset, the PCM stores an inspection maintenance readiness code until all OBD II system monitors or components have been tested to satisfy an OBD trip cycle without any other faults occurring. Specific conditions must be met before the requirements for a trip are satisfied.

Mode 5 is the oxygen sensor monitoring test. This mode gives the oxygen sensor fault limits and the actual oxygen sensor outputs during the test cycle. The test cycle includes specific operating conditions that must be met to complete the test. This information helps determine the effectiveness of the catalytic converter. It is important to note that some A/F sensors report in mode 6.

Mode 6 is the onboard test of noncontinuously monitored systems. The PCM will indicate the pass/fail status of each monitored system in one trip. Using this mode, the technician can determine that the monitor has run and passed or failed without having the system fail twice and set the MIL. The technician can also tell how closely the monitor passed or failed. The mode can be helpful, but the material has to be decoded from the hexadecimal system. There is more on mode 6 in the Shop Manual. The systems that are monitored include the following:

- Thermostat
- Evaporative system
- EGR system
- Secondary AIR system
- Catalyst
- Oxygen sensor heater
- Oxygen sensor PCV system
- Heated catalyst

Mode 7 is the monitoring of test results for continuously monitored systems. A two-trip DTC may be displayed here after the first failure as a pending code, but the information may not be accurate until confirmed by another failure. If the vehicle shows a failure here after one trip, the technician should make sure that the vehicle is OK before returning it to the customer.

Mode 8 is the request for control of an onboard system test or component. In this mode, the technician can use the scan tool to control the PCM to test a system and its related components. Presently, mode 8 is most commonly used for testing the EVAP system. The EVAP leak test is conducted in this mode because the PCM has set conditions to enable the technician to test the EVAP system for leaks.

Mode 9 (if supported) allows the PCM to identify:

- The vehicle's identification number (VIN)
- The PCM's calibration ID and calibration verification

Shop Manual
Chapter 12, page 638

VMode 10 Permanent Codes. Mode 10 was added a few years ago. The name is a bit misleading, as the codes are not really permanent, but they can only be erased by the PCM/ECM. The intent is to keep a repair technician from clearing a code and then only driving the vehicle far enough to get the readiness code to reset. These codes cannot be cleared by using a scan tool or disconnecting the battery.

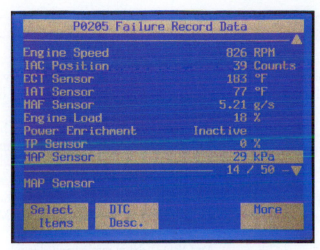

Figure 12-26 A Tech 2 showing a failure record from a random misfire code P0300.

Freeze-Frame Data

This information can be used by the technician to see if the latest calibration has been programmed into the PCM.

The basic advantage of the freeze-frame feature is the ability to look at the existing conditions when a code was set. The freeze frame is set automatically by the PCM. This will be especially valuable for diagnosing intermittent problems. Whenever a code is set, a record of all related activities will be stored in memory. This allows the technician to look at the action of sensors and actuators when the code was set and can help identify the cause of the problem. Only one freeze frame is available. Some manufacturers such as GM have provided more than one freeze frame known as **failure records**. An example is shown in **Figure 12-26**. A freeze frame should not be confused with a **snapshot**. Most scan tools are capable of a snapshot, which is triggered by a technician. The snapshot is a series of frames of data (movie) taken during a drivability problem that can be analyzed in the shop.

Snapshot refers to a short "movie" of several frames of data that can be triggered by the technician with a scan tool.

AUTHOR'S NOTE OBD II systems provide an outstanding opportunity to simultaneously perform multiple tests on engine systems. The root cause of most drivability problems can likely be observed in the OBD II data. It is the interrelationship of the data that will reveal a potential problem area. For example, consider what occurs during a snap acceleration test. The TPS voltage quickly rises, the MAF also rises as a result of an increased volume of intake air, and the rpm will increase along with the voltage from the oxygen sensor, indicating a rich condition. If enrichment occurs, the injectors respond to the PCM's command based on the other sensors' inputs. This is a very direct, albeit basic, test that confirms the fundamental operation of the PCM. With additional experience, a technician can superimpose other interrelated data to help with even the most complex diagnosis.

SUMMARY

- An OBD II system has many monitors to check system operation, and the MIL is illuminated if vehicle emissions exceed 1.5 times the allowable standard for that model year.

- According to the guidelines of OBD II, all vehicles must have a universal diagnostic test connector, a standard location for the DLC, a standard list of diagnostic trouble codes, a standard communication protocol, common use of scan tools on all vehicle makes and models, common diagnostic test modes, the ability to record and store in memory a snapshot of the operating conditions that existed when a fault occurred, and a standard glossary of terms, acronyms, and definitions that must be used for all components in the electronic control systems.

- Monitors included in OBD II are catalyst efficiency, engine misfire, fuel system, heated exhaust gas oxygen sensor, EGR, EVAP, PCV, secondary air injection, thermostat, and comprehensive component monitors.

- OBD II vehicles use a minimum of two oxygen sensors. One sensor is located in front (upstream) of the catalytic converter and is used for feedback to the PCM for fuel control, and the other sensor is located behind (downstream) the converter and gives an indication of the efficiency of the converter. Many vehicles use an A/F sensor in the pre-catalyst position.

- Cylinder misfire monitoring requires measuring the contribution of each cylinder to engine power.

- Fuel system monitoring checks short-term fuel trim and long-term fuel trim while the PCM is operating in closed loop.

- The heated oxygen sensor monitor checks lean-to-rich and rich-to-lean time responses.

- The EGR monitor uses several different strategies to determine if the system is operating properly.

- The EVAP system monitor measures the ability of the fuel tank to hold pressure and vacuum. The monitor also checks the operation of the vacuum and purge valves.

- The AIR system monitor operates by turning the AIR system on to inject air upstream of the oxygen sensor while monitoring its signal.

- The comprehensive monitor looks at any electronic input that could affect emissions that does not have its own monitor, such as the TPS, MAF sensor, and so on.

- An OBD II drive cycle includes whatever specific operating conditions are necessary to initiate and complete all of the vehicle's monitors. The completed drive cycle can be used to verify a symptom or verify a repair.

- The DLC must be a 16-terminal connector with 9 terminals defined by the SAE.

- An OBD II system continuously monitors the entire emissions system and stores a fault code in the PCM when it detects a problem.

- Each manufacturer must use the same protocol between the PCM and its sensors and actuators. The same protocol must be used to send diagnostic information to the scan tool through the DLC.

- All OBD II systems have the same basic test modes.

- One of the mandated capabilities is the freeze-frame or snapshot feature, which gives the ability to record data from all of its sensors and actuators at a time when the system turns the MIL on.

- Compared to previous systems, the main difference in an OBD II system is in the software contained in the PCM.

- An EEPROM may be erased and reprogrammed with the proper equipment without removing the chip.

REVIEW QUESTIONS

Short Answer Essay

1. Describe a computer input.

2. Describe mode 8.

3. Describe an OBD II warm-up cycle.

4. Explain a drive cycle in an OBD II system.

5. Describe how engine misfire is detected in an OBD II system.

6. Describe the differences between a type A misfire and a type B misfire.

7. Describe the purpose of having two oxygen sensors in an exhaust system.

8. Explain the advantages of the freeze-frame feature of OBD II.

9. Briefly describe what the comprehensive component monitor looks at.

10. Briefly describe the EVAP system monitor.

Fill-in-the-Blanks

1. The main goal of OBD II was to detect when engine wear or component failure caused exhaust emissions to increase by_____ percent or more.

2. All OBD II scan tools include a_____ function showing all of the monitoring sequences on the vehicle and the status of each, complete or incomplete.

3. The_____ checks short-term fuel trim (short-term FT) and long-term fuel trim (long-term FT) while the PCM is operating in closed loop.

4. Catalytic converter efficiency is measured by monitoring the_____ _____ capacity of the converter during closed-loop operation.

5. A type B misfire occurs if cylinder misfiring exceeds_____ to_____ percent during a 1,000 rpm period.

6. The thermostat monitor was added in_____.

7. The AIR system operation can be verified by turning the AIR system on to inject air_____ of the oxygen sensor while monitoring its signal.

8. Mode 7 is the monitoring of test results for _____ systems.

9. Mode 5 is the_____ monitoring test.

10. A_____ DTC is a code representing a fault that has occurred but not enough times to illuminate the MIL.

Multiple Choice

1. While discussing OBD II vehicles,
   Technician A says that OBD II vehicles must have a universal test connector (DLC).
   Technician B says that the DLC must be under the dash on the driver's side.
   Who is correct?
   A. Technician A   C. Both technicians
   B. Technician B   D. Neither technician

2. While discussing a computerized engine control system,
   Technician A says that its job is to maintain an air-fuel ratio as close to 14.7:1 as possible.
   Technician B says that it must accurately provide ignition timing under all driving conditions.
   Who is correct?
   A. Technician A   C. Both technicians
   B. Technician B   D. Neither technician

3. All of the following are parts of a computer's operation *except*:
   A. input.   C. processing.
   B. determination.   D. output.

4. *Technician A* says that the computer memory that can be written into by the PCM is RAM.
   Technician B says that nonvolatile RAM is erased when the ignition is switched off.
   Who is correct?
   A. Technician A   C. Both technicians
   B. Technician B   D. Neither technician

5. *Technician A* says that a short-learning period may be necessary on a vehicle with adaptive strategy after the battery is disconnected.
   Technician B says that the learning period is also needed after some computer system parts are replaced.
   Who is correct?
   A. Technician A   C. Both technicians
   B. Technician B   D. Neither technician

6. *Technician A* says that the elimination of the replaceable PROM chip was part of the OBD II legislation to prevent the use of aftermarket high-performance PROMs.
   Technician B says that many updates for engine performance concerns can be performed during service by changing the PROM chip.
   Who is correct?
   A. Technician A   C. Both technicians
   B. Technician B   D. Neither technician

7. *Technician A* says that the misfire monitor is designed to prevent catalytic converter damage by excessive amounts of NO_x.

    Technician B says that the misfire monitor uses a highly accurate crankshaft sensor to detect engine misfires.

    Who is correct?

    A. Technician A   C. Both technicians

    B. Technician B   D. Neither technician

8. *Technician A* says that to complete a warm-up cycle, the engine temperature must reach at least 160°F (71°C).

    Technician B says that the engine must be shut down long enough that the coolant must rise at least 40°F (4°C) to complete a warm-up cycle.

    Who is correct?

    A. Technician A   C. Both technicians

    B. Technician B   D. Neither technician

9. All of the following are mandates of OBD II *except*:

    A. a data link connector (DLC) with dedicated pin assignments.

    B. a standard list of diagnostic trouble codes.

    C. vehicle identification number (VIN) having to be automatically transmitted to the scan tool.

    D. stored vehicle codes having to be erased by disconnecting the battery.

10. While discussing the numbering of HO_2S,

    Technician A says that they are identified by their position in relation to the number 1 cylinder and their location relative to the converters.

    Technician B says that HO_2S sensors on the same side of the vehicle as the number 1 cylinder have a 1 prefix, and sensors mounted downstream from the converter have a 1 suffix.

    Who is correct?

    A. Technician A   C. Both technicians

    B. Technician B   D. Neither technician

CHAPTER 13
RELATED SYSTEMS

Upon completion and review of this chapter, you should be able to:

- List the basic systems that make up an automobile, and name their major components and functions.
- Describe the various clutch components and their functions.
- Explain the design characteristics of the gears used in manual transmission and transaxles.
- Explain the fundamentals of torque multiplication and overdrive.
- Explain the basic design and operation of standard and lockup torque converters.
- Explain the function and operation of a differential and drive axles.
- Describe how an automotive air-conditioning system operates.

- Understand how cruise or speed control operates and the differences between various systems.
- Consider the use and value of engine cooling fans.
- Identify the major components of a typical drum brake, and describe its functions.
- Describe the components of a hydraulic brake system and its operation.
- Briefly describe the operation of drum and disc brakes.
- Describe the main purposes of the steering and suspension systems.

Terms To Know

A/C pressure switch	Input speed sensor	Thrust angle
Band	Multiple-disc clutch	Toe
Camber	Output speed sensor	Torque converter
Caster	Pilot bushing	Torque converter clutch
Clutch disc	Pressure control solenoid	Transmission control module (TCM)
Clutch fork	Radial ply	Transmission fluid pressure (TFP) switch
Clutch pack	Shift adapt	
Clutch release bearing	Shift schedules	Turbine
Coupling point	Shift solenoid valves	Valve body
Cycling clutch system	Static balance	Wheel shimmy
Dynamic balance	Stator	Wheel tramp
Final drive gear	Steering gear	
Flywheel	Steering linkage	
Impeller	TCC control PWM solenoid	

INTRODUCTION

Although the main emphasis of this book is on the engine and its systems, other systems affect the way a vehicle or engine runs or appears to run. The engine is the power source for a car or truck; poor drivability can result from problems other than the engine. To better understand this, let's define drivability. Good drivability requires the following conditions:

1. The engine must start quickly.
2. The engine must idle smoothly.
3. The engine's idle speed must be constant to prevent stalling or racing of the engine.
4. The engine must accelerate smoothly and without hesitation.
5. The engine must run smoothly at all speeds.
6. Normal amounts of fuel should be used by the engine.
7. Exhaust emissions should be at a minimum, and there should be no noticeable smoke from the tailpipe.

Most of these conditions concern only the engine. Some are affected by other systems. For example, if a car has a flat tire, it will not accelerate smoothly or quickly. This may seem ridiculously simple, but you would be surprised at how often simple issues are the cause of drivability problems.

In this chapter, we will look at the components and systems that affect drivability. Most of these systems have the purpose of moving the car down the road comfortably and safely.

The engine provides the power to drive the wheels of the vehicle. All automobile engines, both gasoline and diesel, are classified as internal combustion engines because the combustion or burning that creates energy takes place inside the engine. The combustion process is the burning of an air and fuel mixture. As a result of combustion, large amounts of pressure are generated in the engine. This pressure or energy is used to power the car. The engine must be built strong enough to hold the pressure and temperatures formed by combustion.

The drivetrain is made up of all the components that transfer power from the engine to the driving wheels of the vehicle. The exact components used in a vehicle's drivetrain depend on whether the vehicle is equipped with rear-wheel drive, front-wheel drive, or four-wheel drive (**Figure 13-1**).

Power flow through the drivetrain of a rear-wheel-drive vehicle passes through the clutch or torque converter, manual or automatic transmission, and the driveshaft. Then it goes through the rear differential, through the rear-driving axles, and onto the rear wheels.

Power flow through the drivetrain of a front-wheel-drive vehicle passes through the clutch or torque converter, through the manual or automatic transmission, and then through a front differential, the driving axles, and onto the front wheels.

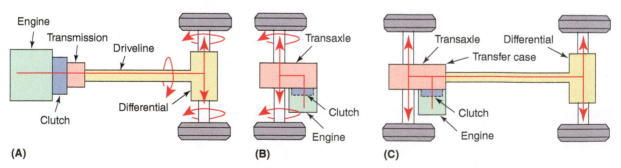

Figure 13-1 Typical drivetrains (A) rear-wheel drive, (B) front-wheel drive, and (C) four-wheel drive.

Four-wheel-drive, or all-wheel-drive, vehicles combine features of both rear- and front-wheel-drive systems so that power can be delivered to all wheels either on a permanent or on-demand basis.

Shop Manual
Chapter 13, page 660

CLUTCHES

Clutches are used on vehicles with manual transmissions or transaxles. The clutch is used to mechanically connect the engine's flywheel to the transmission or transaxle input shaft (**Figure 13-2**). It does this through the use of a special friction plate that is splined to the input shaft. When the clutch is engaged, the friction plate contacts the flywheel, transferring power through the plate to the input shaft.

When stopping, starting, and shifting from one gear to the next, the clutch is disengaged by pushing down on the clutch pedal. This moves the clutch plate away from the flywheel. Power flow stops at the pressure plate. The driver can then shift gears without damaging the transmission or transaxle. Allowing the clutch pedal to come up re-engages the clutch. This allows power to flow from the engine through the transmission.

All manual transmissions require a clutch to engage or disengage the transmission. If the vehicle had no clutch and the engine was always connected to the transmission, the engine would stop every time the vehicle was brought to a stop. The clutch allows the engine to idle while the vehicle is stopped. It also allows for easy shifting between gears.

The clutch engages the transmission gradually by allowing a certain amount of slippage between the input and the output shafts on the clutch. The basic principle of engaging a clutch is demonstrated in **Figure 13-3**. The flywheel and the pressure plate are the drive or driving members of the clutch. The driven member splined to the transmission input shaft is the **clutch disc**, also called the friction disc. As long as the clutch is disengaged (clutch pedal depressed), the drive members turn independently of the driven member and the engine is disconnected from the transmission. However, when the clutch is engaged (clutch pedal released), the pressure plate moves in the direction of the arrows and the clutch disc is bound between the two revolving drive members and forced to turn at the same speed.

The **clutch disc** is the friction material component that couples the flywheel to the pressure plate disc.

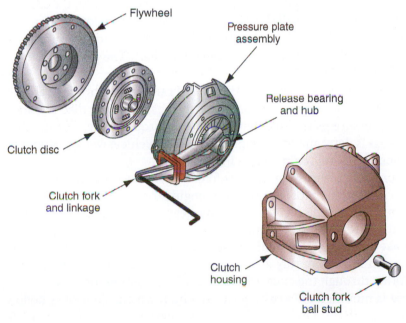

Figure 13-2 Parts of a clutch assembly.

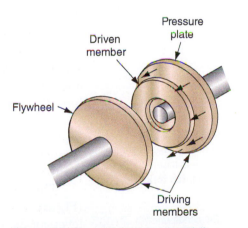

Figure 13-3 When the clutch is engaged, the driven member is squeezed between the two driving members. The transmission is connected to the driven member.

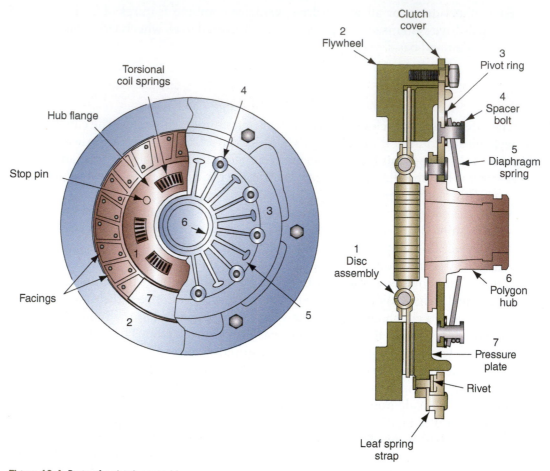

Figure 13-4 Parts of a clutch assembly.

The **flywheel** is the main driving member of the clutch. The rear surface of the flywheel is a friction surface machined very flat to ensure smooth clutch engagement.

A bore in the center of the flywheel and crankshaft holds the **pilot bushing**, which supports the front end of the transmission input shaft and maintains alignment with the engine's crankshaft. Sometimes a ball or roller needle bearing is used instead of a pilot bushing.

The clutch disc receives the driving motion from the flywheel and pressure plate assembly and transmits that motion to the transmission input shaft. The parts of a clutch disc are shown in **Figure 13-4**.

The clutch disc is designed to absorb such things as crankshaft vibration, abrupt clutch engagement, and driveline shock. Torsional coil springs allow the disc to rotate slightly in relation to the pressure plate while they absorb the torque forces. The number and tension of these springs is determined by engine torque and vehicle weight. Stop pins limit this torsional movement to approximately ⅜ inch.

The purpose of the pressure plate assembly is twofold. First, it must squeeze the clutch disc onto the flywheel with sufficient force to transmit engine torque efficiently. Second, it must move away from the clutch disc so the clutch disc can stop rotating even though the flywheel and pressure plate continue to rotate.

The **clutch release bearing** is usually a sealed, prelubricated ball bearing (**Figure 13-5**). Its function is to smoothly and quietly move the pressure plate release levers or diaphragm spring through the engagement and disengagement process.

The release bearing is mounted on a casting called a hub, which slides on a hollow shaft at the front of the transmission housing. This hollow shaft, shown in **Figure 13-6**, is part of the transmission bearing retainer. Note also that the clutch slave cylinder is part of the release bearing assembly on this unit.

The **flywheel** is the counterbalance of the engine's crankshaft that is the driving component of the clutch disc.

The **pilot bushing**, or bearing, is located in the recess in the end of the crankshaft. The input shaft of the transmission is fitted into this component.

The pressure plate levers ride on the clutch release bearing, which is mounted to the **clutch fork**.

The **clutch release bearing** is commonly referred to as the throw-out bearing.

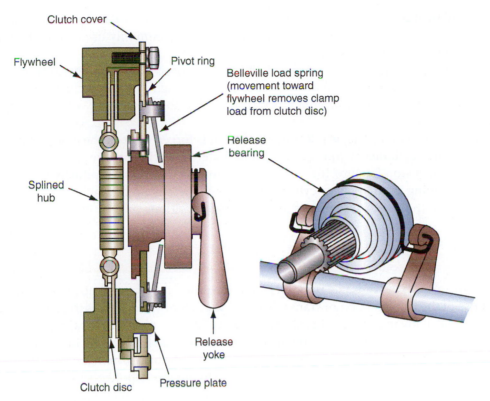

Figure 13-5 Typical clutch-release bearing.

To disengage the clutch, the release bearing is moved forward on its shaft by the **clutch fork**. As the release bearing contacts the release levers or diaphragm spring of the pressure plate assembly, it begins to rotate with the rotating pressure plate assembly. As the release bearing continues forward, the clutch disc is disengaged from the pressure plate and flywheel.

To engage the clutch, the release bearing slides to the rear of the shaft. The pressure plate moves forward and traps the clutch disc against the flywheel to transmit engine torque to the transmission input shaft. Once the clutch is fully engaged, the release bearing is normally stationary.

Usually, the clutch assembly is controlled by a hydraulic system (**Figure 13-7**). In the hydraulic clutch linkage system, hydraulic (liquid) pressure transmits motion from one sealed cylinder to another through a hydraulic line. In addition, the hydraulic pressure

Many vehicles have combined the clutch release bearing with the clutch slave cylinder. This arrangement eliminated the clutch fork.

The **clutch fork** is the pivotal unit connecting the pedal linkage to the pressure plate.

Figure 13-6 This clutch release bearing also includes the slave cylinder. This eliminated the need for the clutch linkage mechanism.

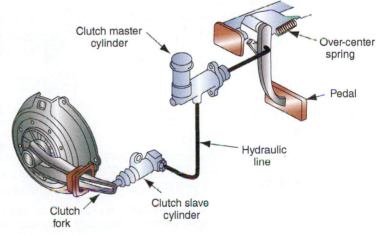

Figure 13-7 Typical hydraulic clutch linkage arrangement.

developed by the master cylinder decreases required pedal effort and provides a precise method of controlling clutch operation. Brake fluid is commonly used as the hydraulic fluid in hydraulic clutch systems.

When the clutch pedal is depressed, the movement of the piston develops hydraulic pressure that is displaced from the master cylinder through a tube into the slave cylinder. The slave cylinder piston movement is transmitted to the clutch fork, which disengages the clutch.

When the clutch pedal is released, the piston is forced back to the engaged position by the master cylinder piston return spring. External springs move the slave cylinder pushrod and piston back to the engaged position. Fluid pressure returns through the hydraulic tubing to the master cylinder assembly. There is no hydraulic pressure in the system when the clutch assembly is in the engaged position.

MANUAL TRANSMISSIONS

Shop Manual
Chapter 13, page 662

The transmission or transaxle is a vital link in the powertrain of any modern vehicle. The purpose of the transmission or transaxle is to use gears of various sizes to give the engine a mechanical advantage over the driving wheels. During normal operating conditions, power from the engine is transferred through the engaged clutch to the input shaft of the transmission or transaxle. Gears in the transmission or transaxle housing alter the torque and speed of this power input before passing it on to other components in the powertrain. Without the mechanical advantage the gearing provides, an engine can generate only limited torque at low speeds. Without sufficient torque, moving a vehicle from a standing start would be impossible.

In any engine, the crankshaft always rotates in the same direction. If the engine transmitted its power directly to the drive axles, the wheels could be driven only in one direction. Instead, the transmission or transaxle provides the gearing needed to reverse direction so the vehicle can be driven backward.

Vehicles propelled by the rear wheels normally use a transmission. Transmission gearing is located within an aluminum casting called the transmission case assembly (**Figure 13-8**). The transmission case assembly is attached to the rear of the engine, which

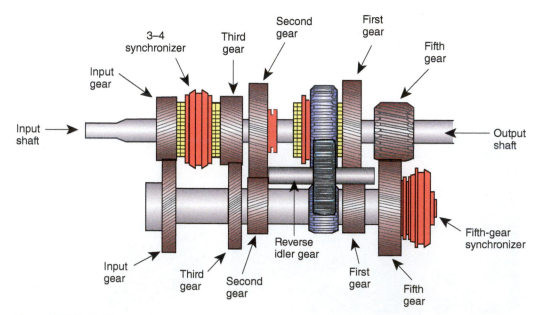

Figure 13-8 Typical five-speed manual transmission with the mainshaft (speed) gears, countershaft cluster gears, and shaft forks visible through the case cutaway.

is normally located in the front of the vehicle. A driveshaft links the output shaft of the transmission with the differential and drive axles located in a separate housing at the rear of the vehicle (**Figure 13-9**). The differential splits the driveline power and redirects it to the two rear drive axles, which then pass it onto the wheels. For many years, rear-wheel-drive systems were the conventional method of propelling a vehicle.

Front-wheel-drive vehicles are propelled by the front wheels. For this reason, they must use a drive design different from that of a rear-wheel-drive vehicle. The transaxle is the special power transfer unit commonly used on front-wheel-drive vehicles. A transaxle combines the transmission gearing, differential, and drive axle connections into a single-case aluminum housing located in front of the vehicle (**Figure 13-10**). This design offers many advantages. One major advantage is the good traction on slippery roads due to the weight of the powertrain components being directly over the driving axles of the vehicle.

Four-wheel-drive vehicles typically use a transmission and transfer case. The transfer case mounts on the side or back of the transmission. A chain or gear drive inside the transfer case receives power from the transmission and transfers it to two separate

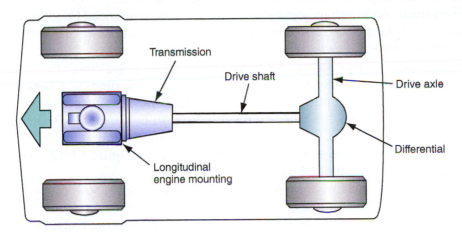

Front-engine, rear-wheel drive

Figure 13-9 Location of typical rear wheel powertrain components.

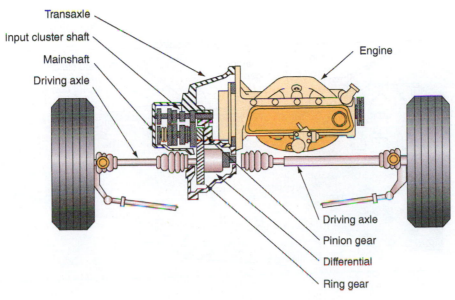

Figure 13-10 Location of typical front-wheel drive powertrain components.

driveshafts. One driveshaft connects to a differential on the front drive axle. The other driveshaft connects to a differential on the rear drive axle.

Most manual transmissions and transaxles are constant mesh, fully synchronized units. Constant mesh means that regardless of the vehicle being stationary or moving, the gears within the unit are constantly in mesh. Fully synchronized means that the unit uses a mechanism of brass rings and clutches to bring rotating shafts and gears to the same speed before shifts occur. This promotes smooth shifting. In a vehicle equipped with a six-speed manual shaft transmission or transaxle, all six forward gears are synchronized. Reverse gearing may or may not be synchronized, depending on the type of transmission or transaxle.

TORQUE CONVERTERS

An automatic transmission eliminates the use of a mechanical clutch and shift lever. In place of a clutch, it uses a fluid coupling called a **torque converter** to transfer power from the engine's flywheel to the transmission input shaft. The torque converter allows for smooth transfer of power at all engine speeds (**Figure 13-11**).

The torque converter operates through hydraulic force provided by automatic transmission fluid. The torque converter changes or multiplies the twisting motion of the engine crankshaft and directs it through the transmission.

The torque converter automatically engages and disengages power from the engine to the transmission in relation to engine rpm. With the engine running at the correct idle speed, there is not enough fluid flow for power transfer through the torque converter. As engine speed is increased, the added fluid flow creates sufficient force to transmit engine power through the torque converter assembly to the transmission.

A standard torque converter consists of three elements (**Figure 13-12**): the impeller, the stator assembly, and the turbine.

The **impeller** assembly is the input (drive) member. It receives power from the engine. The **turbine** is the output (driven) member. It is applied to the input shaft of the transmission and to the turbine shaft assembly. The **stator** assembly is the reaction member or torque multiplier. The stator is supported on a roller race, which operates as an overrunning clutch and permits the stator to rotate freely in one direction and lock up in the opposite direction.

> A torque converter's impeller is also called the pump assembly.

> **Shop Manual**
> Chapter 13, page 665

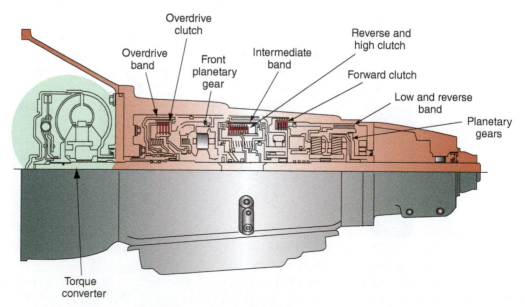

Figure 13-11 Typical torque converter and automatic transmission.

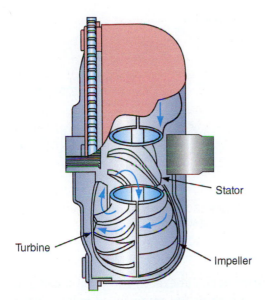

Figure 13-12 A torque converter's major internal parts are its impeller, turbine, and stator.

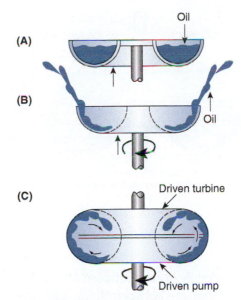

Figure 13-13 Fluid travels inside the torque converter: (A) fluid at rest in impeller/pump, (B) fluid thrown up and outward by spinning pump, and (C) fluid flow harnessed by turbine and redirected back into the pump.

Transmission oil is used as the medium to transfer energy in the torque converter. **Figure 13-13A** illustrates the torque converter impeller or pump at rest. Figure 13-13B shows it being driven. As the pump impeller rotates, centrifugal force throws the oil outward and upward due to the curved shape of the impeller housing.

The faster the impeller rotates, the greater the centrifugal force becomes. In Figure 13-13B, the oil is simply flying out of the housing and is not producing any work. To harness some of this energy, the turbine assembly is mounted on top of the impeller (Figure 13-13C). Now the oil thrown outward and upward from the impeller strikes the curved vanes of the turbine, causing the turbine to rotate. An oil pump driven by the converter shell and the engine continually delivers oil under pressure into the torque converter through a hollow shaft at the center axis of the rotating torque converter assembly. A seal prevents the loss of fluid from the system.

With the transmission in gear and the engine at idle, the vehicle can be held stationary by applying the brakes. Because the impeller is driven by engine speed, it turns slowly, creating little centrifugal force within the torque converter. Therefore, little or no power is transferred to the transmission.

When the throttle is opened, engine speed, impeller speed, and the amount of centrifugal force generated in the torque converter increase dramatically. Oil is then directed against the turbine blades, which transfer power to the turbine shaft and transmission.

Types of Oil Flow

Two types of oil flow take place inside the torque converter: rotary and vortex flow (**Figure 13-14**). Rotary oil flow is the oil flow around the circumference of the torque converter caused by the rotation of the torque converter on its axis. Vortex oil flow is the oil flow occurring from the impeller to the turbine and back to the impeller.

Figure 13-15 also shows the oil flow pattern as the speed of the turbine approaches the speed of the impeller. This is known as the **coupling point**. The turbine and the impeller are running at essentially the same speed. Coupling speed in most torque converters

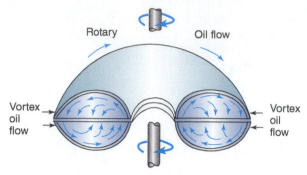

Figure 13-14 Rotary and vortex oil flow in the torque converter.

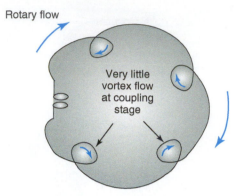

Figure 13-15 Rotary flow is at its greatest at the coupling stage.

generally occurs between 40 and 45 mph. They cannot run at exactly the same speed due to slippage between them. The only way they can turn at the same speed is by using a lockup clutch to mechanically tie them together. Torque converter multiplication can only occur when the impeller is rotating faster than the turbine.

As the vehicle begins to move, the stator stays in its stationary or locked position because of the difference between the impeller and turbine speeds.

As the vehicle road speed increases, turbine speed increases until it approaches impeller speed. Oil exiting the turbine vanes strikes the back face of the stator, causing the stator to rotate in the same direction as the turbine and impeller.

If the vehicle slows, engine speed also slows along with turbine speed. This decrease in turbine speed allows the oil flow to change direction. It now strikes the front face of the stator vanes, halting the turning stator and attempting to rotate it in the opposite direction.

As this happens, the stator is locked in position. In a stationary position, the stator now redirects the oil exiting the turbine so that torque is again multiplied.

Lockup Torque Converters

A lockup torque converter eliminates the 10 percent slip that takes place between the impeller and turbine at the coupling stage of operation. The engagement of a clutch between the engine crankshaft and the turbine assembly has the advantage of improving fuel economy and reducing torque converter operational heat and engine speed.

The lockup torque converter clutch assembly is controlled by the powertrain control module (PCM). When the computer receives electronic signals from the different sensors confirming the requirements for lockup have been met, lockup clutch engagement begins. These sensors include an engine coolant sensor, vehicle speed sensor, engine vacuum sensor, and throttle position sensor.

The system operates in the following manner. The engine operates for more than 5 minutes, and the engine coolant temperature sensor reports 150°F (66°C). Engagement could take place if all the other sensors agree. However, the vehicle operates in congested traffic at speeds varying from 15 mph to 35 mph and the converter clutch engagement speed is approximately 40 mph. Under these operating conditions, the vehicle speed sensor reports that the vehicle speed is too low for clutch converter engagement. In addition, the throttle position sensor reports to the computer the unsteady up-and-down movement of the throttle. The computer interprets this as a reason not to engage the converter clutch. The brake switch also opens periodically. Thus, the computer does not energize the clutch solenoid to engage the converter clutch.

If the vehicle breaks out of congested traffic and is traveling at a steady higher speed, the speed sensor reports that the vehicle is at a speed higher than the converter clutch engagement speed. The throttle position sensor reports that the throttle is in a steady position in favor of engagement, and the brake switch is closed because the driver's foot is not on the brake pedal. When the computer scans all the sensors and determines that all and the converter clutch is engaged for lockup operation.

AUTOMATIC TRANSMISSIONS

Many rear-wheel-drive and four-wheel-drive vehicles are equipped with automatic transmissions. Automatic transaxles, which combine an automatic transmission and final drive assembly in a single unit, are used on front-wheel-drive, all-wheel-drive, and some rear-wheel-drive vehicles (**Figure 13-16**).

Shop Manual
Chapter 13, page 663

An automatic transmission or transaxle selects gear ratios according to engine speed, powertrain load, vehicle speed, and other operating factors. Little effort is needed on the part of the driver because both upshifts and downshifts occur automatically. A driver-operated clutch is not needed to change gears, and the vehicle can be brought to a stop without shifting to neutral. This is a great convenience, particularly in stop-and-go traffic. The driver can also manually select a lower forward gear, reverse, neutral, or park. Depending on the forward range selected, the transmission can provide engine braking during deceleration.

Until recently, all automatic transmissions were controlled by hydraulics. However, new systems now feature computer-controlled operation of the torque converter and transmission. Based on the input data supplied by electronic sensors and switches, the computer sets the torque converter's operating mode, controls the transmission's shifting sequence, and regulates transmission oil pressure.

Most automatic transmissions rely on planetary gear sets to transfer power and multiply engine torque to the drive axle. Compound gear sets combine two simple planetary gear sets so that load can be spread over a greater number of teeth for strength as well as to obtain the largest number of gear ratios possible in a compact area.

 A BIT OF HISTORY

During the 1950s through the 1960s, many vehicles with automatic transmissions had the gear selector placed in a variety of locations. Many had the more traditional steering column-mounted selector lever, while others had floor-mounted shifters. Instead of a shift lever, other unique designs incorporated dash-mounted push buttons, one button for each selectable gear.

A simple planetary gear set consists of three parts: a sun gear, a carrier with planetary pinions mounted to it, and an internally toothed ring gear or annulus. The sun gear is located in the center of the assembly (**Figure 13-17**). It can be either a spur or helical gear design. It meshes with the teeth of the planetary pinion gears. Planetary pinion gears are small gears fitted into a framework called the planetary carrier. The planetary carrier is designed with a shaft for each of the planetary pinion gears.

The planetary pinions surround the sun gear's center axis, and they themselves are surrounded by the annulus or ring gear, which is the largest part of the simple gear set. The ring gear acts like a band to hold the entire gear set together and provide great strength to the unit.

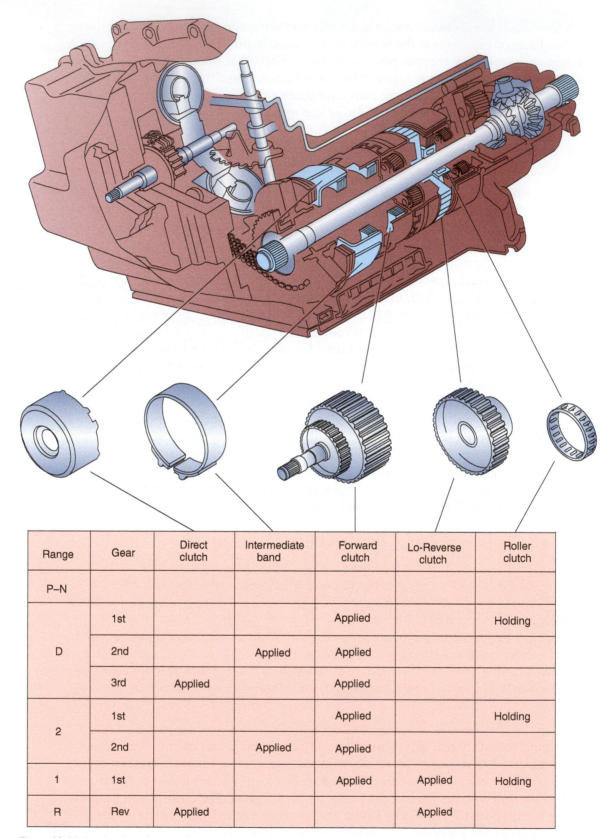

Range	Gear	Direct clutch	Intermediate band	Forward clutch	Lo-Reverse clutch	Roller clutch
P–N						
D	1st			Applied		Holding
	2nd		Applied	Applied		
	3rd	Applied		Applied		
2	1st			Applied		Holding
	2nd		Applied	Applied		
1	1st			Applied	Applied	Holding
R	Rev	Applied			Applied	

Figure 13-16 Interior view of a typical transaxle.

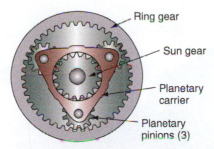

Figure 13-17 Planetary gear configuration is similar to the solar system, with the sun gear surrounded by the planetary pinion gears. The ring gear surrounds the complete gear set.

Sun Gear	Carrier	Ring Gear	Speed	Torque	Direction
1. Input	Output	Held	Maximum reduction	Increase	Same as input
2. Held	Output	Input	Minimum reduction	Increase	Same as input
3. Output	Input	Held	Maximum increase	Reduction	Same as input
4. Held	Input	Output	Minimum increase	Reduction	Same as input
5. Input	Held	Output	Reduction	Increase	Reverse of input
6. Output	Held	Input	Increase	Reduction	Reverse of input
7. When any two members are held together, speed and direction are the same as input. Direct 1:1 drive occurs.					
8. When no member is held or locked together, output cannot occur. The result is a neutral condition.					

Figure 13-18 Laws of simple planetary gears.

The planetary pinion gears are called planetary pinions for short.

Any one of the three members can be used as the driving or input member. At the same time, another member might be kept from rotating and thus becomes the held or stationary member. The third member then becomes the driven or output member. Depending on which member is the driver, which is held, and which is driven, either a torque increase or a speed increase is produced by the planetary gear set. Output direction can also be reversed through various combinations.

Figure 13-18 summarizes the basic laws of simple planetary gears. It indicates the resultant speed, torque, and direction of the various combinations available. Also, remember that when an external-to-external gear tooth set is in mesh, there is a change in the direction of rotation at the output. When an external gear tooth is in mesh with an internal gear, the output rotation for both gears is the same.

Planetary Gear Controls

Certain parts of the planetary gear train must be held, while others must be driven to provide the needed torque multiplication and direction for vehicle operation. *Planetary gear controls* is the general term used to describe transmission bands, servos, and clutches.

A **band** is a braking assembly positioned around a stationary or rotating drum. The band brings a drum to a stop by wrapping itself around the drum and holding it. The band is hydraulically applied by a servo assembly. Connected to the drum is a member of the planetary gear train. The purpose of a band is to control the planetary gear train by holding the drum and connecting planetary gear member stationary.

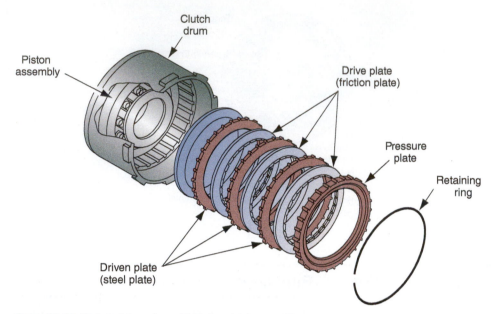

Figure 13-19 Exploded view of a multiple-disc clutch assembly.

In contrast to a band, which can hold only a planetary gear member, transmission clutches, either overrunning or multiple-disc, are capable of both holding and driving members. A **multiple-disc clutch** uses a series of friction discs to transmit torque or apply braking force. The discs have internal teeth that are sized and shaped to mesh with splines on the clutch assembly hub. In turn, this hub is connected to a planetary gear train component so gear-set members receive the desired braking or transfer force when the clutch is applied or released.

Multiple-disc clutches have a large drum-shaped housing that can be either a separate casting or part of the existing transmission housing (**Figure 13-19**). This drum housing holds all other clutch components: the cylinder, hub, piston, piston return springs, seals, pressure plate, **clutch pack**, and snap rings.

> A **multiple-disc clutch** can be referred to as clutch packs.

> A **clutch pack** contains plain steel and friction discs.

Hydraulic Systems

A hydraulic system uses a liquid to perform work. In an automatic transmission, this liquid is automatic transmission fluid (ATF). An automatic transmission uses ATF pressure to control the action of the planetary gear sets through clutches, bands, valves, and gears. This fluid pressure is regulated and directed to change gears automatically using various pressure regulators and control valves.

The **valve body** can be best understood as the control center of an automatic transmission (**Figure 13-20**). The purpose of the valve body is to control the application of gear upshifts and downshifts commanded by the PCM/TCM at the appropriate times.

> The **valve body** controls various valves and solenoids in order to respond to driving and load conditions.

Many very precisely machined holes are in the valve body to accommodate the various valves. The purpose of a valve is to start, stop, or direct and regulate fluid flow. The movement of the valves engages and disengages the gears. The choice of gear is determined by the driver and the placement of the shift lever.

Electronic Controls

Shifting in an automatic transmission is controlled by a hydraulic system through the PCM or TCM. All late-model automatic transmissions use electronics to control shifting with the hydraulic system. In a hydraulic system, an intricate network of valves and other components use hydraulic pressure to control the operation of planetary gear sets. These gear sets generate as many as nine forward speeds, neutral, park, and reverse gears

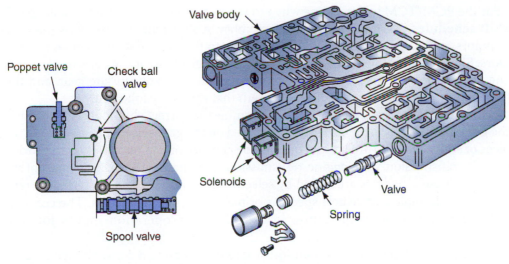

Figure 13-20 Typical valve body assembly.

normally found on automatic transmissions. Electronic shifting systems use electric solenoids to control shifting mechanisms. Electronic shifting is precise and can be varied to suit certain operating conditions. Electronic control is superior because information about the engine, fuel, ignition, vacuum, and operating temperature is fed into the computer so that shifting and lockup are closely monitored to take place at exactly the right time.

When a computer controls the shifting of the automatic transmission, input signals for engine and road speed, manifold vacuum, engine operating temperature, gear selection, throttle position, and other factors are fed to the computer. The computer produces output signals, which in turn operate electrical solenoid valves. Instead of hydraulic pressures and springs, the motion of the solenoids controls the position of the valves. When activated, a solenoid moves its valve to control fluid pressure in a valve body (**Figure 13-21**).

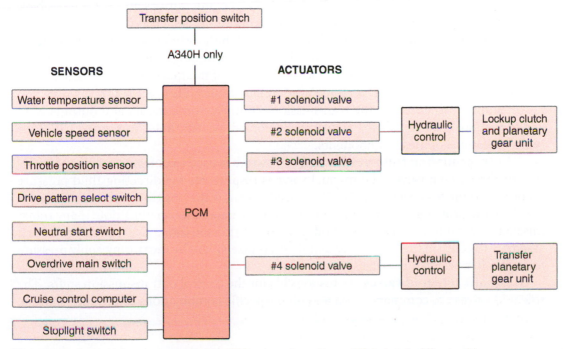

Figure 13-21 Some of the sensors used by the PCM to determine optimum shift timing via shift solenoids.

Shift schedules are the
programmed shift points
in the computer's
memory.

For the PCM/TCM to determine when to start a gear change, it must be able to refer to **shift schedules** that it has stored in its memory. A shift schedule contains the actual shift points to be used by the computer according to the input data it receives from the sensors. Shift schedule logic chooses the proper shift schedule for the current conditions of the transmission. It uses the shift schedule to select the appropriate gear, and then determines the correct shift schedule or pattern that should be followed.

The first input a computer looks at to determine the correct shift logic is the position of the gear shift lever. All shift schedules are based on the gear selected by the driver. The choices of shift schedules are limited by the type and size of engine that is coupled to the automatic transmission. Each engine/transmission combination has a different set of shift schedules. These schedules are coded by selector lever position and current gear range and use throttle angle and vehicle speed as primary determining factors. The computer also looks at different temperature, load, and engine operation inputs for more information.

The shift schedules set the conditions that need to be met for a change in gears. Because the computer frequently reviews the input information, it can make quick adjustments to the schedule if needed and as needed. The result of the computer's processing of this information and commanding outcomes according to a logical program is optimum shifting of the automatic transmission. This results in improved fuel economy and overall performance.

The electronic control systems used by the manufacturers differ with the various transmission models and the engines they are attached to. The components in each system and the overall operation of the system also vary with the different transmissions; however, all operate in a similar fashion and use the same parts.

Electronic Transmission Controls, Inputs, and Outputs

Transmissions can be controlled by a separate computer called a **transmission control module (TCM)**, or the transmission control can be part of the function of the PCM. In either case, the transmission and engine share both information and sensors with other components to make shifting and drivability as smooth and efficient as possible. Additionally, the technician must be aware that a problem with an engine sensor may carry over into transmission operation. Electronic engine sensors and computers have taken the place of their vacuum and mechanical counterparts such as mechanical breaker points, vacuum advance, distributor flyweights, and thermostatic vacuum control valves. Transmissions have also benefited from advances in technology since electronic controls have replaced several traditional components. Three of the most basic mechanical controls that were in use have all been replaced by electronics (**Figure 13-22**).

In addition to updating the transmissions controls, some inputs and outputs were added that are basically for the transmission control system. Refer to **Figure 13-23** for typical locations of these components.

The **transmission fluid pressure (TFP) switch** is used to determine the exact position of the manual valve. The manual valve is responsible for directing fluid pressure according to the gear selected. The PCM needs to know the position of the manual valve to follow the proper shift program. For example, in manual second position, many transmissions start out and stay in second gear. Some transmissions use a pressure switch assembly to confirm that a commanded shift has taken place by measuring apply pressure for a gear in a selected passage.

The **shift solenoid valves** are responsible for the control of transmission shifts. The solenoid valves are commanded on and off in specific combinations to produce shifting.

Component	Function	Replaced by
Vacuum modulator	**Engine load** Transmission shifting was altered according to manifold vacuum level.	**MAP, MAF, TPS** PCM can alter shifting via shift solenoids according to the load it sees.
Governor	**Vehicle speed** The governor used weights that moved an internal valve that changed hydraulic pressure into a speed signal used to time shifts according to vehicle speed.	**Vehicle speed sensor** The PCM knows the precise vehicle speed and can alter shifting based accordingly.
Shift (TV) cable	**Throttle position (load)** A cable was used to change position of a valve that altered hydraulic pressure to form a load (throttle position) signal.	**Throttle position sensor** The cable has been replaced by the TPS signal to the PCM.

Figure 13-22 Older hydraulic controls and their electronic replacements.

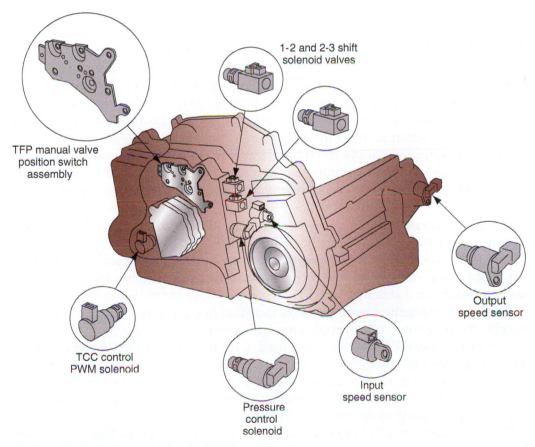

Figure 13-23 Electrical components on a 4T40-E transaxle.

SHIFT SOLENOID OPERATION CHART					
Transaxle range selector lever position	Powertrain control module Gear commanded	Eng braking	AX4N solenoids		
			SS1	SS2	SS3
P/N[a]	P/N	NO	OFF[b]	ON[b]	OFF
R (Reverse)	R	YES	OFF	ON	OFF
Overdrive	1	NO	OFF	ON	OFF
	2	NO	OFF	OFF	OFF
	3	NO	ON	OFF	ON
	4	YES	ON	ON	ON
D (Drive)	1	NO	OFF	ON	OFF
	2	NO	OFF	OFF	OFF
	3	YES	ON	OFF	OFF
Manual 1	2[c]	YES	OFF	ON	OFF
	3[c]	YES	OFF	OFF	OFF
		YES	ON	OFF	OFF

a When transmission fluid temperature is below 50°, then SS1 = OFF, SS2 = ON, SS3 = ON to prevent cold creep.

b Not contributing to power flow.

c When a manual pull-in occurs above calibrated speed, the transaxle will downshift from the higher gear until the vehicle speed drops below this calibrated speed.

Figure 13-24 Shift solenoid reference chart for Ford's AX4N transmission.

Figure 13-24 shows a typical application chart for Ford's AX4N transmission. Shift solenoids made shift timing much easier to accomplish than the previous hydraulically controlled shift valves. Shift timing involves not only shifting the transmission at the proper time according to torque, but also the turning off and on of components at the precise moment needed. Two gears cannot be applied at the same time. The timing of components is much like runners in a relay race; one member hands off to the next. This timing is easier to accomplish electronically than hydraulically. The electronically shifted transmission has on average one-half the number of valves that its totally hydraulic counterpart used. Also, should a common concern with transmission shifting occur, many times a software correction made by reprogramming the PCM with a manufacturer's update can relieve the problem.

The **pressure control solenoid** inside the transmission controls the line pressure of the transmission according to commands from the PCM. The control of transmission pressure allows the PCM to better adapt itself to driver habits and wear. It is also called **shift adapt**. If the PCM senses a shift that is occurring too slowly, it can increase transmission pressure for that particular shift and then return to normal for the next shift. The PCM can also alter ignition timing and transmission line pressure to make a shift occur smoothly if, for instance, a band applies too aggressively.

The **TCC control PWM solenoid** is used to control apply of the **torque converter clutch**. Many older torque converter clutches (TCC) applied noticeably especially during driving from 40 mph to 50 mph. Instead of applying the clutch either on or off all at once, the TCC control PWM solenoid feeds the apply solenoid with pulse width modulated fluid pressure. This is done to cushion apply of the torque converter clutch by gradually increasing the duty cycle of the PWM solenoid until the TCC is fully applied. The TCC can also be turned off in the same manner.

The **input speed sensor** is used to monitor the speed of the transmission's input shaft. This measurement is used along with the **output speed sensor** to calculate the gear range of the transmission. The PCM takes the two measurements and confirms that the gear commanded has actually applied and is not slipping. This information can also be used to measure how long it takes to complete a particular shift and use the information to help adapt shift pressure to conditions in the transmission or driver habits. If the gear range calculated is incorrect, the PCM can set an invalid gear ratio code. The vehicle speed sensor can also be used as the output shaft speed sensor in some applications.

DRIVESHAFT

Driveshafts are used on rear-wheel-drive vehicles and four-wheel-drive vehicles. They connect the output shaft of the transmission with the gearing in the rear axle housing (**Figure 13-25**) on rear-wheel-drive vehicles. They are also used to connect the output shaft to the front and rear drive axles on a four-wheel-drive vehicle.

Shop Manual
Chapter 13, page 670

A driveshaft consists of a hollow drive or propeller shaft that is connected to the transmission and drive axle differential by universal joints. The driveshaft is nothing more than an extension of the transmission output shaft. The driveshaft transfers engine torque from the transmission to the rear driving axle.

The universal joint (or U-joint) allows two rotating shafts to operate at a slight angle to each other. Although simple in appearance (**Figure 13-26**), the universal joint is more intricate than it seems. This is because its natural action is to speed up and slow down twice in each revolution when operating at an angle. The rate of speed varies depending upon the steepness of the U-joint angle.

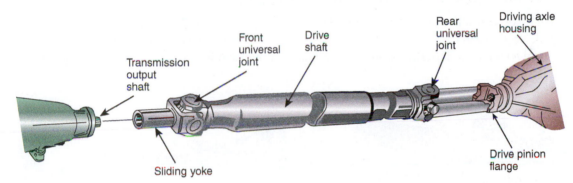

Figure 13-25 Output power from the transmission is connected to the differential in the drive axle housing by a driveshaft.

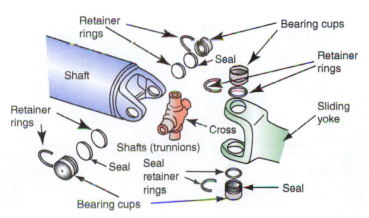

Figure 13-26 Exploded view of Cardan universal joint.

The driveshafts on front-wheel-drive vehicles are correctly called axle or half shafts.

The universal joint operating angle is derived by taking the difference between the transmission installation angle and the drive shaft installation angle. For instance, in **Figure 13-27**, a true horizontal centerline is drawn through the transmission. To align better with the rear axle, the transmission is installed at an angle 5 degrees off the true horizontal line. The drive shaft installation angle is 8 degrees off the true horizontal. The difference between the transmission installation angle and the drive shaft installation angle is 3 degrees. Therefore, the universal joint operating angle is 3 degrees. This 3-degree operating angle should be okay, but always consult the manufacturer's specifications to be sure. When the universal joint is operating at an angle, the driven yoke speeds up and slows down twice during each driveshaft revolution. This acceleration and deceleration of the universal joint is known as speed variation or fluctuation.

The speed changes are not normally visible during rotation. They might be felt as torsional vibrations due to improper installation, steep or unequal operating angles, or high speeds.

Speed variations must be canceled at exactly the same point in driveshaft rotation. The two driving yokes at opposite ends of the driveshaft must be at the same point of rotation.

Vibrations can be reduced by using canceling angles (**Figure 13-28**). Carefully examine the illustration and note that the operating angle at the front of the driveshaft is offset by the one at the rear of the driveshaft. When the front universal joint accelerates causing a vibration, the rear universal joint decelerates, causing a vibration. The vibrations created by the two joints oppose and dampen the vibrations from one to the other. The use of canceling angles provides a smoother driveshaft operation.

Front-Wheel-Drive Constant Velocity Joints

The driving axles of a front-wheel-drive vehicle operate at higher angles than a rear-wheel-drive driveshaft and have to be able to move with the suspension (**Figure 13-29**). A universal joint actually speeds up and slows down while turning. This speeding up and

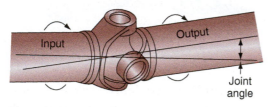

Figure 13-27 U-joint action.

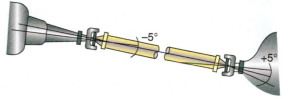

Figure 13-28 Canceling angles reduce vibrations.

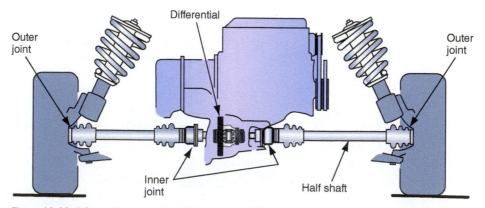

Figure 13-29 A front drive axle with half shafts and CV joints.

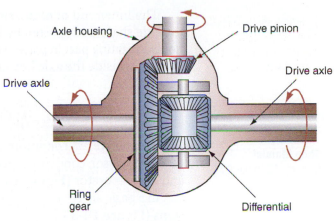

Figure 13-31 Rear differential components.

Figure 13-30 A typical front-wheel-drive CV joint.

slowing down is more dramatic as the angle increases. A constant velocity (CV) joint such as found in a front-wheel-drive vehicle (**Figure 13-30**) allows a constant speed during rotation.

Differential

On rear-wheel-drive vehicles, the driveshaft turns perpendicular to the forward motion of the vehicle. The differential gearing in the rear axle housing is designed to turn the direction of the power so it can be used to drive the wheels of the vehicle. The power flows into the differential where it changes direction, then to the rear axles and wheels (**Figure 13-31**).

   The differential also performs two other important jobs. It multiplies the torque of the power it receives from the driveshaft by providing a final gear reduction. Also, it divides this power between the left and right driving axles and wheels in such a way that a differential wheel speed is possible. This means one wheel can turn faster than the other when going around turns. All vehicles use a differential to provide an additional gear reduction (torque increase) above and beyond what the transmission or transaxle gearing can produce. This is known as the **final drive gear**. In a transmission-equipped vehicle, the differential gearing is located in the rear axle housing. However, in a transaxle, the final reduction is produced by the final drive gears housed in the transaxle case.

The **final drive gear** is provided by the differential.

Driving Axles

Driving axles are solid steel shafts that transfer differential torque to the driving wheels. A separate axle shaft is used for each driving wheel. The driving axles and part of the differential are enclosed in an axle housing that protects and supports these parts.

   Each driving axle is connected to the side gears in the differential. The inner or differential ends of the axles are splined to fit into the side gears. As the side gears are turned, the axles to which they are splined turn at the same speed.

   At their outer or wheel ends, the axles are attached to the driving wheels. For attachment to a wheel, the outer end of each axle has a flange mounted to it. A flange is a rim for attaching one part to another part. Studs are attached to hold the wheel in place against the flange. Studs are threaded shafts resembling bolts without heads. One end of the stud is screwed or pressed into the flange. The wheel fits over the studs, and a nut, called the lug nut, is tightened over the open end of the stud. This holds the wheel in place.

The inner end of each axle is supported by the differential carrier. The outer end of the axle shaft is supported by a bearing inside the axle housing. A bearing supports and holds a rotating part in place. This bearing, called the axle bearing, allows the axle to rotate smoothly inside the axle housing.

HEATING AND AIR CONDITIONING

Shop Manual
Chapter 13, page 671

An automotive air-conditioning (A/C) system is a closed pressurized system. It consists of a compressor, condenser, receiver/dryer or accumulator, expansion valve or orifice tube, and an evaporator (**Figure 13-32**).

In a basic air-conditioning system, the heat is absorbed and transferred in the following steps (**Figure 13-33**):

1. Refrigerant leaves the compressor as a high-pressure, high-temperature vapor.
2. By removing heat via the condenser, the vapor becomes a high-pressure, high-temperature liquid.
3. Moisture and contaminants are removed by the receiver/dryer where the cleaned refrigerant is stored until it is needed.
4. The expansion valve converts the high-pressure liquid into a low-pressure liquid by controlling its flow into the evaporator.
5. Heat is absorbed from the air inside the passenger compartment by the low-pressure, low-temperature refrigerant, causing the liquid to vaporize.
6. The refrigerant returns to the compressor as a low-pressure vapor.

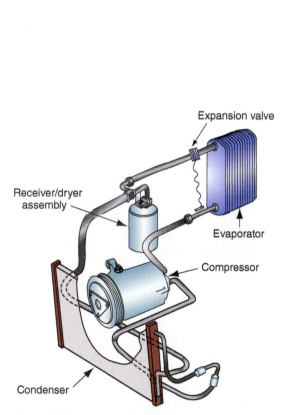

Figure 13-32 An air-conditioning system using an expansion valve.

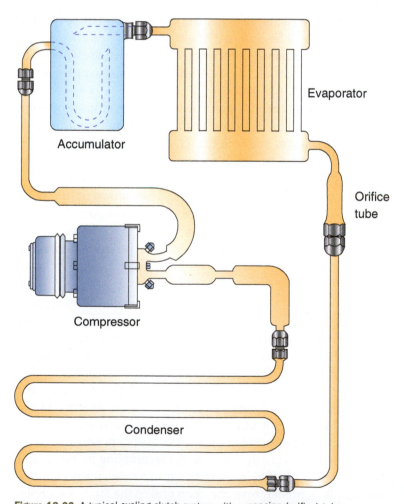

Figure 13-33 A typical cycling clutch system with expansion (orifice) tube.

The compressor is the heart of the automotive air-conditioning system. It separates the high-pressure and low-pressure sides of the system. The primary purpose of the unit is to draw the low-pressure vapor from the evaporator and compress this vapor into high-temperature, high-pressure vapor. This action results in the refrigerant having a higher temperature than surrounding air and enables the condenser to condense the vapor back to a liquid. The secondary purpose of the compressor is to circulate or pump the refrigerant through the condenser under the different pressures required for proper operation.

In a **cycling clutch system**, the compressor is run intermittently through controlling the application and release of its clutch by a thermostatic switch. The thermostatic switch senses the evaporator's outlet air temperature through a capillary tube that is part of the switch assembly. With a high sensing temperature, the thermostatic switch is closed and the compressor clutch is energized. As the evaporator outlet temperature drops to a preset level, the thermostatic switch opens the circuit to the compressor clutch. The compressor then ceases to operate until such time as the evaporator temperature rises above the switch setting. The term *cycling clutch* is derived from this on/off operation.

> In a **cycling clutch system**, the compressor controls refrigerant flow by turning on and off as determined by pressure.

When the temperature of the evaporator approaches the freezing point (or the low setting of the switch), the thermostatic switch opens the circuit and disengages the compressor clutch. The compressor remains inoperative until the evaporator temperature rises to the preset temperature, at which time the switch closes and compressor operation resumes.

Automatic Air Conditioning

Automatic air conditioning (A/C) is in wide use today, and these systems have also felt the effect of technology. The A/C system has many of its own inputs and outputs, along with sharing some of the familiar inputs of the engine control system (**Figure 13-34** and **Figure 13-35**).

The **A/C pressure switch(s)** monitors the pressure of refrigerant in the A/C system. There are usually two on each vehicle, a high-pressure switch and a low-pressure switch. These switches can tell the PCM when the refrigerant pressure is too low to allow

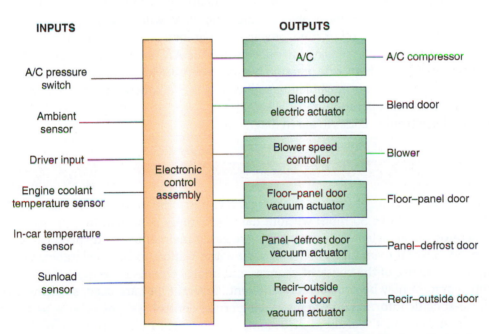

Figure 13-34 Many vehicles use computers to monitor the operation of the air-conditioning system, and some inputs are shared with the PCM.

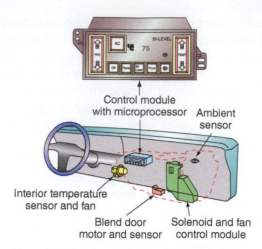

Figure 13-35 A/C control module works to maintain a comfortable interior automatically with the help of several inputs.

operation of the A/C or that the pressure is too high and the A/C system must be shut down to prevent damage.

Air-Conditioning Sensors

The ambient temperature sensor is used to inform the A/C system controller of the outside air temperature. The ambient temperature sensor, combined with the in-car temperature sensor, allows the computer to calculate the load required. The A/C system then knows what position it must place the mode door's blower motor speed and determine whether cooling or heat is required to achieve the desired temperature in the shortest possible time. The sun load sensor allows the A/C controller to compensate for the effects of the sun on heating in the winter and cooling in the summer.

The engine coolant temperature sensor is used by the PCM and the A/C controller. Some systems do not start the heater blower motor until there is heat available. Also, if the engine temperature is too high, the PCM will not allow the A/C to come on, because it would make the condition worse. Further, the PCM will usually shut the A/C down under wide-open throttle or high-power steering loads. The cooling fan is also PCM controlled and generally runs on low speed when the A/C system pressures are moderate and high speed if the high-side pressure begins to climb. Most cooling fans are shut down over 35 mph because air movement is reduced by the fan operating above that speed. To help monitor their operation, most automatic A/C systems can be accessed with a scan tool and have trouble codes to aid in diagnosis.

CRUISE CONTROL SYSTEMS

Shop Manual
Chapter 13, page 672

Cruise or speed control systems are designed to allow the driver to maintain a constant speed (usually above 30 mph) without having to apply continual foot pressure on the accelerator pedal. Selected cruise speeds are easily maintained, and speed can easily be changed. Several override systems also allow the vehicle to be accelerated, slowed, or stopped. Because of the constant changes and improvements in technology, each cruise control system may be considerably different. Several types are used, so always consult vehicle-specific information when servicing cruise control systems.

When engaged, the cruise control components set the throttle position to the desired speed. The speed is maintained unless heavy loads and steep hills interfere. The cruise control is disengaged whenever the brake pedal is depressed. The common speed or cruise control system components function in the following manner:

- The cruise control switch is located on the end of the turn signal or near the center or sides of the steering wheel. There are usually several functions on the switch, including off/on, resume, and engage buttons.
- Electronic throttle control vehicles (TAC equipped) do not require a stepper motor, as there is already one in use for the throttle control. In either case, the cruise control module is usually built into the PCM and uses the vehicle speed sensor signal along with others to maintain vehicle speed.
- Depressing the brake or clutch will disengage the system.
- Many cruise control systems are accessible through the vehicle's PCM and may have diagnostic trouble codes and switch tests to aid in diagnosis. A typical block diagram for a modern cruise control is illustrated in **Figure 13-36**.

The cruise control unit uses vehicle speed sensor (VSS) information from the PCM to sense vehicle speed.

BRAKE SYSTEMS

Automobiles are stopped by activating the brake system (**Figure 13-37**). Brakes, which are located at each wheel, use friction to slow and stop the automobile.

The brakes are activated when the vehicle operator depresses a brake pedal. The brake pedal is connected to a plunger in a *master cylinder*, which is filled with hydraulic fluid. When the brake pedal is depressed, a force is put onto the hydraulic fluid in the master cylinder. The force is increased by the master cylinder and transferred through brake hoses and lines to the four brake assemblies.

Shop Manual
Chapter 13, page 672

Two types of brakes are used on automobiles: disc brakes and drum brakes. Many automobiles use a combination of the two types: disc brakes at the front wheels and drum brakes at the rear wheels, although rear disc brakes are fast becoming more popular.

Most vehicles have power-assisted brakes. A brake booster typically uses manifold vacuum to increase the pressure applied to the plunger in the master cylinder. This lessens the amount of pressure that must be applied to the brake pedal by the operator and increases the responsiveness of the brake system.

The brake system is designed to slow and halt the motion of a vehicle. To do that, various components within a hydraulic brake system must convert the momentum of the vehicle into heat. They do so by using friction.

As the brakes on a moving automobile are actuated, rough-textured pads or shoes are pressed against rotating parts of the vehicle—either rotors or drums. The kinetic energy,

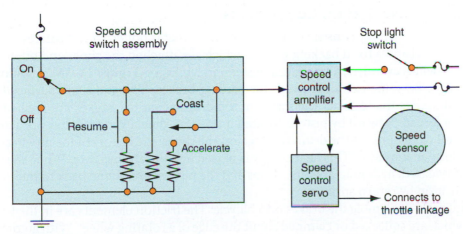

Figure 13-36 Electronic cruise control uses an electronic control module (controller) to operate a servo that controls the position of the throttle.

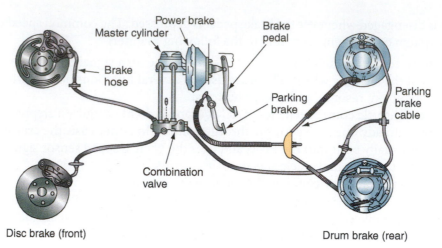

Disc brake (front)

Drum brake (rear)

Figure 13-37 Many automobiles feature a combination of drum and disc brakes.

> Disc brake rotors are commonly called discs.

or momentum, of the vehicle is then converted into heat energy by the kinetic friction of rubbing surfaces, and the car or truck slows down.

When the vehicle comes to a stop, it is held in place by static friction. The friction between the surfaces of the brakes as well as the friction between the tires and the road resist any movement. To overcome the static friction that holds the car motionless, the brakes are released. The heat energy of combustion in the engine crankcase is converted into kinetic energy by the transmission and drivetrain, and the vehicle moves.

Static friction also plays an important part in controlling a moving vehicle. The rotating tires grip the road, and the static friction between these two surfaces enables the driver to control the speed and direction of the car. When the brakes are applied, the kinetic friction of the rubbing brake components slows the rotation of the tires. This increases the static friction between the tires and the road, decreasing the motion of the car. If the kinetic or sliding friction of the brake components overcomes the static friction between the tires and road, the wheels lock up and the car begins to skid. Static friction then exists between the components in the brakes and kinetic friction between the skidding tires and the road—the car is out of control. Obviously, the most effective braking effort is achieved just below the friction levels that result in wheel lockup. This is the role antilock braking systems play in modern vehicles. By electronically pumping the brakes on and off many times each second, antilock brake systems keep kinetic friction below the static friction between the tires and road.

Hydraulic Brake System Components

A drum brake assembly consists of a cast-iron drum that is bolted to and rotates with the vehicle's wheel and a fixed backing plate to which are attached the shoes and other components—wheel cylinders, automatic adjusters, and linkages (**Figure 13-38**). Additionally, there might be some extra hardware for parking brakes. The shoes are surfaced with frictional linings that contact the inside of the drum when the brakes are applied. The shoes are forced outward by pistons located inside the wheel cylinder. They are actuated by hydraulic pressure. As the drum rubs against the shoes, the energy of the moving drum is transformed into heat. This heat energy is passed into the atmosphere. When the brake pedal is released, hydraulic pressure drops and the pistons are pulled back to their unapplied position by return springs.

Disc brakes resemble the brakes on a bicycle: The friction elements are in the form of pads, which are squeezed or clamped about the edge of a rotating wheel. With automotive disc brakes, this wheel is a *rotor* inboard of the vehicle wheel (**Figure 13-39**). The rotor

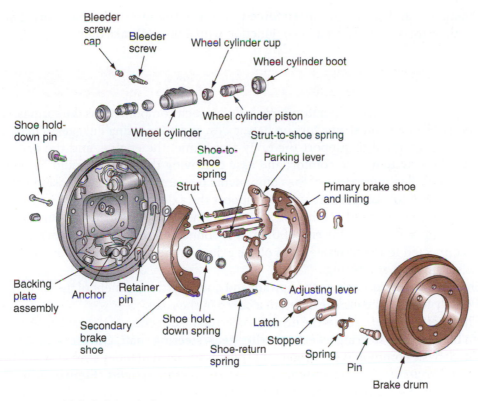

Figure 13-38 Typical drum brake.

Figure 13-39 A disc brake assembly.

is made of cast iron. Since the pads clamp against both sides of it, both sides are machined smooth. The pads are attached to metal shoes, which are actuated by pistons, the same as with drum brakes. The pistons are contained within a caliper assembly, a housing that wraps around the edge of the rotor. The caliper is kept from rotating by way of bolts holding it to the car's suspension framework.

The caliper is a housing containing the pistons and related seals, springs, and boots as well as the cylinders and fluid passages necessary to force the friction linings or pads against the rotor. The caliper resembles a hand in the way it wraps around the edge of the rotor. It is attached to the steering knuckle.

Unlike shoes in a drum brake, the pads act perpendicular to the rotation of the disc when the brakes are applied. This effect is different from that produced in a brake drum, where frictional drag actually pulls the shoe into the drum. Disc brakes are said to be

Some brake discs are manufactured with two separate discs joined together by a finned center section. These discs are called ventilated brake rotors.

Shop Manual
Chapter 13, page 674

The **steering gear** changes rotary movement to lateral movement. The **steering linkage** connects the wheels to the gear box.

non-energized and so require more force to achieve the same braking effort. For this reason, they are ordinarily used in conjunction with a power brake unit.

SUSPENSION AND STEERING SYSTEMS

The suspension system on the automobile includes such components as the springs, shock absorbers, MacPherson struts, torsion bars, axles, and connecting linkages. These components are designed to support the body and frame, the engine, and the drivelines. Without these systems, the comfort and ease of driving the vehicle would be reduced. **Figure 13-40** illustrates some of the components that are used in a suspension system.

Springs and torsion bars are used to support the axles of the vehicle. The two types of springs commonly used are the coil spring and the leaf spring. Torsion bars are made of long spring steel rods. One end of the rod is connected to the frame, while the other end is connected to the movable parts of the axles. As the axles move up and down, the rod twists and acts as a spring.

Shock absorbers slow down the upward and downward movement of the springs. This is necessary to limit the car's reaction to a bump in the road.

The steering system allows the driver to control the direction of the vehicle. A steering system includes the steering wheel, **steering gear**, steering shaft, and **steering linkage**.

Three basic common types of steering systems are used in today's vehicles: *rack-and-pinion* and *recirculating ball* systems and *electric steering systems* (**Figure 13-41**). The rack-and-pinion system is commonly used in passenger cars and trucks. The recirculating ball system is normally used only on heavy-duty vehicles. Most modern vehicles are using electric steering racks, or electrically driven steering assist columns.

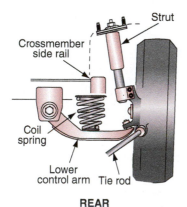

REAR

Rack-and-pinion steering linkage

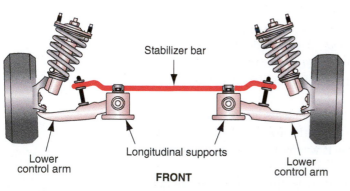

FRONT

Figure 13-40 Typical front and rear suspension systems.

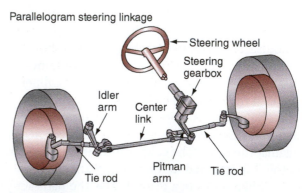

Parallelogram steering linkage

Figure 13-41 Common steering systems.

Steering gears provide a gear reduction to make changing the direction of the wheels easier. On all but a few subcompact and compact car models, the steering gear is also power assisted to ease the effort of turning the wheels. In a power-assisted system, a pump provides hydraulic fluid under pressure to the steering gear. A spool valve directs fluid to one side or the other of the steering gear to assist the operator in turning the wheels. Electric steering has been developed to take part of the load off the engine (power steering pump) and use an electrical steering system. Although the advantages of electric steering are numerous, a few of the advantages are as follows:

- It is not affected by engine stalling.
- There is no hydraulic fluid.
- There is no pump to drive.
- It is more compact.
- It works with hybrid vehicles as well as conventional cars.
- It is more adaptable to vehicle stability control systems.

> All new GM vehicles, with the exception of heavy-duty trucks, are now equipped with electric steering.

Principles of Wheel Alignment

Wheel alignment allows the wheels to roll without scuffing, dragging, or slipping on different types of road conditions. This gives greater safety in driving, easier steering, longer tire life, reduction in fuel consumption, and less strain on the parts that make up the front end of the vehicle.

There is a multitude of angles and specifications that the automotive manufacturers must consider when designing a car. The multiple functions of the suspension system complicate things a great deal for design engineers. They must take into account more than basic geometry. Durability, maintenance, tire wear, available space, and production cost are all critical elements. Most elements contain a degree of compromise to satisfy the minimum requirements of each.

Most technicians do not need to be concerned with all of this. All they need to do is restore the vehicle to the condition the design engineer specified. However, to do this, the technician must be totally familiar with the purpose of basic alignment angles.

The alignment angles in the vehicle are designed to properly locate the vehicle's weight on moving parts and to facilitate steering. If these angles are incorrect, the vehicle is misaligned. The effects of misalignment are given in **Figure 13-42**. It is important to remember that alignment angles, when specified in the text, are those specific angles that should exist when the system is being measured under a given set of conditions. During regular performance, these angles change as the traveling surface and vehicle driving forces change.

Caster is the angle of the steering axis of a wheel from the vertical, as viewed from the side of the vehicle. The forward or rearward tilt from the vertical line (**Figure 13-43**) illustrates caster. Caster is the first angle adjusted during an alignment. Tilting the wheel forward is negative caster. Tilting backward is positive caster.

Caster is designed to provide steering stability. The caster angle for each wheel on an axle should be equal. Unequal caster angles cause the vehicle to steer toward the side with less caster. Too much negative caster can cause the vehicle to have sensitive steering at high speeds. The vehicle might wander as a result of negative caster.

Camber is the angle represented by the tilt of either the front or rear wheels inward or outward from the vertical as viewed from the front of the car (**Figure 13-44**). Camber is designed into the vehicle to compensate for road crown, passenger weight, and vehicle weight. Camber is usually set equally for each wheel. Equal camber means each wheel is tilted outward or inward the same amount. Unequal camber causes tire wear and causes the vehicle to steer toward the side that is more positive.

Problem	Effect
Incorrect camber setting	Tire wear Ball joint/wheel bearing wear Pull to side of most positive/least negative camber
Too much positive caster	Hard steering Excessive road shock Wheel shimmy
Too much negative caster	Wander Weave Instability at high speeds
Unequal caster	Pull to side most negative/least positive caster
Incorrect SAI	Instability Poor return Pull to side of lesser inclination Hard steering
Incorrect toe setting	Tire wear
Incorrect turning radius	Tire wear Squeal in turns

Figure 13-42 Effects of incorrect alignment.

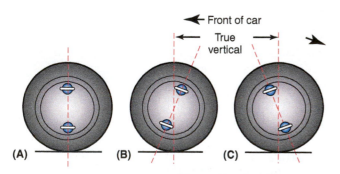

Figure 13-43 Three types of caster: (A) zero, (B) positive, and (C) negative.

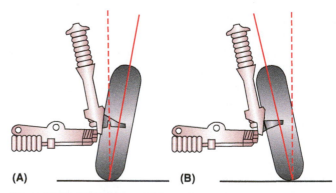

Figure 13-44 (A) Positive and (B) negative camber.

Toe is the distance comparison between the leading edge and trailing edge of the front tires. If the leading edge distance is less, then there is toe in. If it is greater, there is toe out (**Figure 13-45**). Toe is critical as a tire-wearing angle. Wheels that do not track straight ahead have to drag as they travel forward. Excessive toe measurements (in or out) cause a saw tooth edge on the tread surface from dragging the tire sideways.

A main consideration in any alignment is to make sure the vehicle runs straight down the road with the rear tires tracking directly behind the front tires when the steering wheel is in the straight-ahead position. The geometric centerline of the vehicle should be parallel to the road direction. This is the case when rear toe is parallel to the vehicle's geometric

Figure 13-45 Typical rear toe condition.

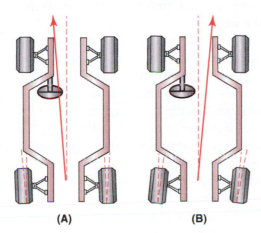

(A)   **(B)**

Figure 13-46 (A) Left and (B) right thrust direction.

centerline in the straight-ahead position. If rear toe does not parallel the vehicle centerline, a thrust direction to the left or right is created (**Figure 13-46**). This difference of rear toe from the geometric centerline is called the **thrust angle**. The vehicle tends to travel in the direction of the thrust line rather than straight ahead.

Steering axle inclination (SAI) locates the vehicle weight to the inside or outside of the vertical centerline of the tire. The SAI is the angle between true vertical and a line drawn between the steering pivots as viewed from the front of the vehicle. It is an engineering angle designed to project the weight of the vehicle to the road surface for stability. The SAI helps the vehicle's steering system return to straight ahead after a turn.

Turning radius or cornering angle is the amount of toe out present in turns. As a car goes around a corner, the inside tire must travel in a smaller radius circle than the outside tire. This is accomplished by designing the steering geometry to turn the inside wheel sharper than the outside wheel. The result can be seen as toe out in turns. This eliminates tire scrubbing on the road surface by keeping the tires pointed in the direction they have to move.

All vehicles are built around a geometric centerline that runs through the center of the chassis from the back to the front. The thrust line is the direction the rear axle would travel if unaffected by the front wheels. This condition is also called tracking. An ideal alignment has all four wheels parallel with the centerline, making the thrust line parallel with the centerline. However, the rear-wheel thrust line of a vehicle might not always be parallel to the actual centerline of the vehicle, so the angle of the thrust line must be checked first.

Correct tracking refers to a situation with all suspension and wheels in their correct location and condition and aligned so that the rear wheels follow directly behind the front wheels while moving in a straight line. For this to occur all the wheels must be parallel with one another, and axle and spindle lines must be at 90-degree angles to the vehicle centerline. Simply stated, all four wheels should form a perfect rectangle.

Rear Alignment

A car with a perfect front alignment can still experience poor handling and premature tire wear—particularly on front-wheel-drive cars and cars with independent rear suspensions—if the rear suspension is misaligned. Approximately 80 percent of today's vehicles not only have front-end alignment specifications, but also require rear-wheel alignment.

Like front camber, rear camber affects both tire wear and handling. The ideal situation is to have zero running camber on all four wheels to keep the tread in full contact with the road for optimum traction and handling. Camber is not a static angle. It changes as the suspension moves up and down. Camber also changes as the vehicle is loaded and the suspension sags under the weight.

Besides wearing the tires unevenly across the tread, uneven side-to-side camber (as when one wheel leans in and the other does not) creates a steering pull just like it does when the camber readings on the front wheels do not match. It is like leaning during a turn on a bicycle. A vehicle always pulls toward a wheel with the most positive camber. If the mismatch is at the rear wheels, the rear axle pulls toward the side with the greatest amount of positive camber. If the rear axle pulls to the right, the front of the car drifts to the left, resulting in a steering pull even though the front wheels may be perfectly aligned.

Rear toe, like front toe, is a critical tire wear angle. If toed in or toed out, the rear tires scuff just like the front ones. Either condition can also contribute to steering instability as well as reduced braking effectiveness. (Keep this in mind with antilock brake systems.)

Like camber, rear toe is not a static alignment angle. It changes as the suspension goes through jounce and rebound. It also changes in response to rolling resistance and the application of engine torque. With four-wheel-drive vehicles, the front wheels tend to toe in under power while the rear wheels toe out in response to rolling resistance and suspension compliance. With rear-wheel-drive vehicles, the opposite happens: The front wheels toe out while the rear wheels on an independent suspension try to toe in as they push the vehicle ahead.

WHEELS AND TIRES

Shop Manual
Chapter 13, page 677

A vehicle's tires, wheels, and suspension and steering systems provide the contact between the driver and the road. They allow the driver to safely maneuver the vehicle in all types of conditions as well as ride in comfort and security. Tire design has improved dramatically during the past few years. Modern tires require increased attention to achieve their full potential of extended service and correct ride control. Tire wear that is uneven or premature is usually a good indicator of problems in the steering and suspension system. Tires become not only a good diagnostic aid to a technician, but can also be clear evidence to the customer for the need to service the front end.

The primary purpose of tires is to provide traction. Tires also help the suspension absorb road shocks, but this is a side benefit. They must perform under a variety of conditions. The road might be wet or dry; paved with asphalt, concrete, or gravel; or there might not be a road at all. The car might be traveling slowly on a straight road or moving quickly through curves or over hills. All of these conditions call for special requirements that must be present, at least to some degree, in all tires.

In addition to providing good traction, tires are also designed to carry the weight of the vehicle, to withstand side thrust over varying speeds and conditions, and to transfer braking and driving torque to the road.

Many different designs of tires are available today (**Figure 13-47**), the most common of which is the **radial ply**. Radial-ply tires have body cords that extend from bead to bead at an angle of about 90 degrees "radial" to the tire circumferential centerline, plus two or more layers of relatively inflexible belts under the tread. This construction of various

Radial ply is a common tire construction.

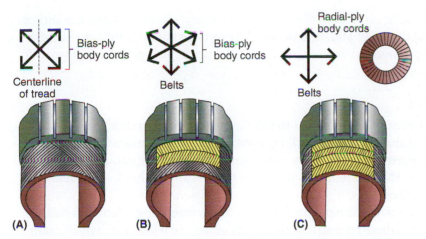

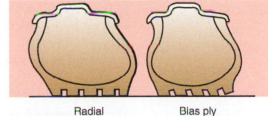

Figure 13-48 A radial tire's highly flexible sidewalls give maximum tread contact area during fast, hard turning.

Figure 13-47 Three types of tire construction: (A) bias ply, (B) bias belted, and (C) radial ply.

combinations of rayon, nylon, fiberglass, and steel gives greater strength to the tread area and flexibility to the sidewall (**Figure 13-48**). The belts restrict tread motion during contact with the road, thus improving tread life and traction. Radial-ply tires also offer greater fuel economy, increased skid resistance, and more positive braking.

Although the newer synthetics are being used more frequently in radial tires, steel is still the most popular belt material. Bias ply and belted bias are available in all cord materials mentioned earlier, except aramid and Kevlar. Non-radial belts are usually of the same material as the sidewalls.

Combining Tire Types

As a general rule, tires should be replaced with the same size designation or an approved optional size as recommended by the auto or tire manufacturer. In addition to following the vehicle manufacturer's recommendations for tire size, type, inflation pressures, and rotation patterns, the following points should be observed:

1. Never mix size or construction types on the same axle.
2. Tires on the same axle should be of approximately equal tread depth.
3. All tires on station wagons and all other vehicles used for trailer towing should be of the same size, type, and load rating.
4. New tires should be installed in pairs on the same axle. When replacing only one tire, it should be paired with the tire having the most tread to equalize braking traction.
5. Snow tires should be of a size and type equivalent to the other tires on the vehicle. Otherwise, the safety and handling of the vehicle might be adversely affected.

Tire Care

To maximize tire performance, inspect for signs of improper inflation and uneven wear, which can indicate a need for balancing, rotation, or front suspension alignment. Tires should also be checked frequently for cuts, stone bruises, abrasions, blisters, and for objects that might have become imbedded in the tread. More frequent inspections are recommended when rapid or extreme temperature changes occur or where road surfaces are rough or occasionally littered with debris.

A properly inflated tire gives the best tire life, riding comfort, handling stability, and gas mileage for normal driving conditions. Too little air pressure can result in tire squeal, hard steering, excessive tire heat, abnormal tire wear, and increased fuel consumption by as much as 10 percent. An under-inflated tire shows maximum wear on the outside edges of the tread. There is little or no wear in the center. Conversely, an overinflated tire shows its wear in the center of the tread and little wear on the outside edges. A higher tire inflation pressure than recommended can cause a hard ride, tire bruising, and rapid wear at the center of the tire.

Tire/Wheel Balance

Proper front-end alignment allows the tires to roll straight without excessive tread wear. The wheels can go out of alignment from striking raised objects or potholes. Misalignment subjects the tires to uneven and/or irregular wear. An out-of-balance condition can also cause increased wear on the ball joints as well as deterioration of shock absorbers and other suspension components.

Should an inspection show uneven or irregular tire wear, wheel alignment and balance service is a must. Wheel balancing distributes weights along the wheel rim, which counteract heavy spots in the wheels and tires and allow them to roll smoothly without vibration. There are two types of wheel imbalance: static and dynamic.

Static balance is the equal distribution of weight around the wheel. Wheels that are statically unbalanced cause a bouncing action called **wheel tramp** (**Figure 13-49**). This condition eventually causes uneven tire wear. As the name implies, static balance is balancing a wheel at rest. This is done by adding a compensating weight. Static balance is achieved when the wheel does not rotate by itself regardless of the position in which it is placed on its axis. A statically unbalanced wheel tends to rotate by itself until the heavy portion is down.

Dynamic balance is the equal distribution of weight on each side of the centerline. When the tire spins, there is no tendency for the assembly to move from side to side. Wheels that are dynamically unbalanced can cause **wheel shimmy** and a wear pattern (**Figure 13-50**). Simply stated, dynamic balance is balancing a wheel in motion. Once a wheel starts to rotate and is in motion, the static weights try to reach the true plane of

> Computerized wheel balancers have a setting called "static balance," which places all the wheel weights on the inside of the rim. This is necessary for some customized wheels that do not have a lip to hold a wheel weight, or chrome wheels that might suffer damage to their finish.

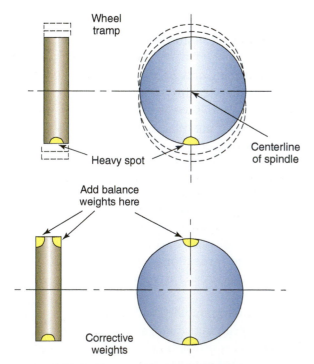

Figure 13-49 Static unbalance causes wheel tramp.

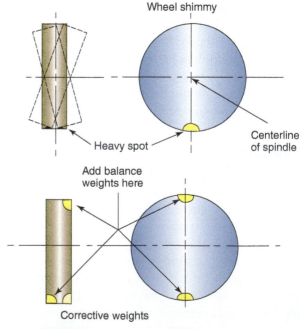

Figure 13-50 Dynamic unbalance causes wheel shimmy.

rotation of the wheel because of the action of centrifugal force. In an attempt to reach the true plane of rotation when there is an imbalance, the static weights force the spindle to one side.

At 180 degrees of wheel rotation, static weights kick the spindle in the opposite direction. The resultant side thrusts cause the wheel assembly to wobble or wiggle. When severe enough, as already mentioned, it causes vibration and front-wheel shimmy.

SUMMARY

- The clutch, located between the transmission and the engine, provides a mechanical coupling between the engine flywheel and the transmission's input shaft. All manual transmissions and transaxles require a clutch.

- The flywheel, an important part of the engine, is also the main driving member of the clutch.

- The clutch disc receives the driving motion from the flywheel and pressure plate assembly and transmits that motion to the transmission input shaft.

- The twofold purpose of the pressure plate assembly is to squeeze the clutch disc onto the flywheel and to move away from the clutch disc so that the disc can stop rotating.

- The clutch release bearing, also called a throw-out bearing, smoothly and quietly moves the pressure plate release levers or diaphragm spring through the engagement and disengagement processes.

- The clutch fork moves the release bearing and hub back and forth. It is controlled by the clutch pedal and linkage.

- Clutch linkage can be mechanical or hydraulic. Mechanical linkage is divided into two types: shaft and lever linkage and cable linkage.

- A transmission or transaxle uses meshed gears of various sizes to give the engine a mechanical advantage over its driving wheels.

- Transaxles contain the gear train plus the differential gearing needed to produce the final gear ratios. Transaxles are commonly used on front-wheel-drive vehicles.

- Transmissions are normally used on rear-wheel-drive vehicles.

- All vehicles use a gear set in a differential to provide additional gear reduction (torque increase) above and beyond what the transmission or transaxle gearing can produce. Modern automatic transmissions use a computer to match the demand for acceleration with engine speed, wheel speed, and load conditions. It then chooses the proper gear ratio and if necessary, initiates a gear change.

- The torque converter is a fluid clutch used to transfer engine torque from the engine to the transmission. It automatically engages and disengages power transfer from the engine to the transmission in relation to engine rpm. It consists of three elements: the impeller (input), turbine (output), and stator (torque multiplier). Two types of oil flow take place inside the torque converter: rotary and vortex flow. An overrunning clutch keeps the stator assembly from rotating in one direction and permits overrunning when turned in the opposite direction.

- A lockup torque converter eliminates the 10 percent slip that takes place between the impeller and turbine at the coupling stage of operation. There are two types: centrifugal lockup clutch and the more popular piston lockup clutch.

- Planetary gear sets transfer power and generate torque from the engine to the drive axle. Compound gear sets combine two simple planetary gear sets so that load can be spread over a greater number of teeth for strength and also to obtain the largest number of gear ratios possible in a compact area. A simple planetary gear set consists of a sun gear, a carrier with planetary pinions mounted to it, and an internally toothed ring gear.

- Planetary gear controls include transmission bands, servos, and clutches. A band is a braking assembly positioned around a drum. There are two types: single wrap and double wrap.

- Simple and compound servos are used to engage bands. Transmission clutches, either overrunning or multiple-disc, are capable of both holding and driving members.

- The valve body is the control center of the automatic transmission. It is made of two or three main parts. Internally, the valve body has many fluid passages called worn tracks. A shift schedule contains the actual shift points to be used by the computer according to the input data it receives from the sensors. Its logic chooses the proper shift schedule for the current conditions of the transmission.

- Front-wheel-drive axles transfer engine torque generally from the transaxle differential to the front wheels.

- Constant velocity (CV) joints provide the necessary transfer of uniform torque and a constant speed while operating through a wide range of angles.

- A differential is a geared mechanism located between the driving axles of a vehicle. Its job is to direct power flow to the driving axles. Differentials are used in all types of powertrains.

- The major components of an air-conditioning system are compressor, condenser, receiver/dryer or accumulator, expansion valve or orifice tube, and evaporator.

- The compressor is the heart of an automotive air-conditioning system. It separates the high-pressure and low-pressure sides of the system. The primary purpose of the unit is to draw the low-pressure vapor from the evaporator and compress this vapor into high-temperature, high-pressure vapor. This action results in the refrigerant having a higher temperature than surrounding air, enabling the condenser to condense the vapor back to liquid.

- The secondary purpose of the compressor is to circulate or pump the refrigerant through the condenser under the different pressures required for proper operation. The compressor is located in the engine compartment.

- Among the many controls used to monitor and maintain the compressor during its operational cycle within various systems are the pressure relief valve, the low- and the high-pressure cutout switches, the ambient temperature switch, and the thermostatic switch. Each of these represents the most common protective control devices designed to ensure safe and reliable operation of the compressor.

- Cruise control is used to mechanically or electronically control the position of the throttle during highway operation. These systems help the operator maintain a constant speed without having to apply foot pressure to the accelerator pedal. A cruise control switch is used to engage or disengage the system.

- Electronic cruise control systems use several additional parts. These include the electronic control module, a vehicle speed sensor (VSS), a clutch switch, and a brake switch.

- The four factors that determine a vehicle's braking power are pressure, which is provided by the hydraulic system; coefficient of friction, which represents the frictional relationship between pads and rotors or shoes and drums and is engineered to ensure optimum performance; frictional contact surface, meaning that bigger brakes stop a car more quickly than smaller brakes; and head dissipation, which is necessary to prevent brake fade.

- The drum is mounted to the wheel hub. When the brakes are applied, a wheel cylinder uses hydraulic power to press two brake shoes against the inside surface of the drum. The resulting friction between the shoe's lining and drum slows the drum and wheel. Caster is the angle of the steering axis of a wheel from the vertical, as viewed from the side of the vehicle. Tilting the wheel forward is negative caster. Tilting backward is positive.

- Camber is the angle represented by the tilt of either the front or rear wheels inward or outward from the vertical as viewed from the front of the car.

- Toe is the distance comparison between the leading edge and trailing edge of the front tires. If the edge distance is less, then there is toe in. If it is greater, there is toe out.

- The primary purpose of tires is to provide traction. They also are designed to carry the weight of the vehicle, to withstand side thrust over varying speeds and conditions, to transfer braking and driving torque to the road, and to absorb much of the rock shock from surface irregularities.

- There are two types of wheel balancing: static balance and dynamic balance.

REVIEW QUESTIONS

Short Answer Essays

1. Explain the operation of the manual transmission clutch.

2. Explain the terms *constant mesh* and *fully synchronized*.

3. Define final drive gear.

4. What component in an electronic cruise control system is used to monitor or sense vehicle speed?

5. Define dynamic and static wheel balance.

6. Briefly describe how an automatic transmission knows when to shift gears.

7. Explain the primary purpose of the torsional coil springs in the transmission's clutch disc.

8. Describe the purpose of a brake wheel cylinder.

9. Explain what happens during the coupling point of conventional torque converter operation.

10. Explain what is necessary for torque converter clutch lockup engagement to take place.

Fill-in-the-Blanks

1. When the brake pedal is depressed, a force is put onto the hydraulic fluid in the _____.

2. A vehicle always pulls toward a wheel with the most _____ camber.

3. The cruise control on a throttle-by-wire system does not require a(n) _____.

4. The primary purpose of tires is to provide _____.

5. A driveline consists of a hollow drive or propeller shaft that is connected to the transmission and drive axle differential by _____.

6. The clutch is used to mechanically connect the engine's flywheel to the transmission or transaxle _____.

7. The pressure control solenoid inside the transmission controls the _____ pressure of the transmission according to commands from the PCM.

8. When the vehicle comes to a stop, it is held in place by _____.

9. An automatic transmission eliminates the use of a mechanical clutch and shift lever. In place of a clutch, it uses a fluid coupling called a(n) _____.

10. The _____ sensor allows the A/C controller to compensate for the effects of the sun on heating in the winter and cooling in the summer.

Multiple Choice

1. *Technician A* says that the clutch disc is also called the friction disc.
Technician B says that the clutch disc is splined to the transmission input shaft.
Who is correct?
   A. Technician A
   B. Technician B
   C. Both technicians
   D. Neither technician

2. *Technician A* says that an automatic transmission uses a torque converter instead of a clutch to transfer power from the engine to the transmission.
Technician B says that the torque converter allows the vehicle to stop and idle without releasing a clutch.
Who is correct?
   A. Technician A
   B. Technician B
   C. Both technicians
   D. Neither technician

3. All of the following are parts of the torque converter *except*:
   A. impeller.
   B. roller gear.
   C. turbine.
   D. stator.

4. *Technician A* says that most automatic transmissions rely on planetary gears to transfer and multiply engine torque.
Technician B says that the planetary gear set consists of four parts.
Who is correct?
   A. Technician A
   B. Technician B
   C. Both technicians
   D. Neither technician

5. *Technician A* says that a transmission *band* is hydraulically applied by a multiple-disc clutch.
Technician B says that for the PCM/TCM to determine when to start a gear change, it must be able to refer to shift schedules that it has stored in its memory.
Who is correct?
   A. Technician A
   B. Technician B
   C. Both technicians
   D. Neither technician

6. *Technician A* says that the driveline connects the transmission output shaft to the gearing in the rear axle housing.
Technician B says that a driveline consists of a solid drive or propeller shaft connected to the transmission and differential by U-joints.
Who is correct?
   A. Technician A
   B. Technician B
   C. Both technicians
   D. Neither technician

7. *Technician A* says that U-joints speed up and slow down as they rotate.
   Technician B says that CV joints run at a constant rate of speed.
   Who is correct?
   A. Technician A
   B. Technician B
   C. Both technicians
   D. Neither technician

8. *Technician A* says that the differential divides the power between the left and right axles.
   Technician B says that the differential allows the rear wheels to turn at different speeds when going around corners.
   Who is correct?
   A. Technician A
   B. Technician B
   C. Both technicians
   D. Neither technician

9. While discussing the air-conditioning system, *Technician A* says that the expansion valve converts high-pressure liquid into a low-pressure liquid.
   Technician B says that the refrigerant enters the compressor as a low-pressure liquid from the evaporator.
   Who is correct?
   A. Technician A
   B. Technician B
   C. Both technicians
   D. Neither technician

10. *Technician A* says that a common steering system is the rack and pinion.
    Technician B says that another common system is the recirculating ball type.
    Who is correct?
    A. Technician A
    B. Technician B
    C. Both technicians
    D. Neither technician

Note: **Terms are highlighted in bold**, followed by Spanish translation in color.

Accelerator pedal position (APP) sensor The APP sensor tells the exact location of the accelerator pedal.

Sensor de posición del pedal del acelerador El sensor de posición del pedal del acelerador (APP, por sus siglas en inglés) indica la ubicación exacta del pedal del acelerador.

Acid A substance that reacts with a basic substance and produces salts, generally corrosive and dangerous.

Ácido Una sustancia que reacciona con una sustancia base y produce los sales, generalmente corrosivos y peligrosos.

A/C pressure switch An electrical device used to control or protect the A/C system based on pressures of refrigerant.

Interruptor de presión A/C Un dispositivo eléctrico usado para controlar o cuidar los sistemas del A/C basado en las presiones de los fluidos refrigerantes.

A/C sunload temperature sensor A sensor that measures the amount of heating in the vehicle interior due to sunlight. Used on automatic A/C systems.

Sensor de la temperatura A/C cargada por sol Un sensor que mide la cantidad de calentamiento en el interior del vehículo que se debe al sol. Se usa en los sistemas automáticos de A/C.

Actuator A control device that delivers mechanical action in response to an electrical signal.

Actuador Un dispositivo de control que suministra una acción mecánica en respuesta a una señal eléctrica.

Adaptive strategy The ability of a computer to adapt to certain defects in the system.

Estrategia de adaptación Capacidad de una computadora para adaptarse a ciertas fallas en el sistema.

After top dead center (ATDC) Refers to the position of the piston while it is moving away from its upmost position.

Después del punto muerto superior Se refiere a la posición del pistón mientras se está moviendo después de su punto más alto en su carrera hacia la posición inferior.

Air cleaner ducts Ducts connected from the air intake source to the air cleaner.

Conductos del filtro de aire Conductos conectados desde la fuente del aire aspirado hasta el filtro de aire.

Air-cooled cooling system A cooling system that provides engine cooling by passing air over the outside of the engine.

Sistema de enfriamiento enfriado por aire Sistema de enfriamiento que enfría el motor al pasar aire sobre la parte exterior del mismo.

Air diverter (AIRD) A valve that directs the air to the exhaust manifold or to the outside air.

Derivador de aire (AIRD) Una válvula que dirige el aire al colector de escape o al convertidor catalítico.

Air door A butterfly-type valve in the heated air inlet system, usually positioned in the air cleaner snorkel.

Puerta de ventilación Válvula tipo mariposa en el sistema de admisión de aire calentado, que normalmente se coloca en el tubo de respiración del filtro de aire.

Air door vacuum motor diaphragm A vacuum diaphragm that opens and closes the air door.

Diafragma del motor de vacío de la puerta de ventilación Diafragma de vacío que abre y cierra la puerta de ventilación.

Air-fuel ratio The ratio of the amount of air to the amount of fuel entering the cylinders.

Relación de aire y combustible Relación entre la cantidad de aire y la cantidad de combustible que entra en los cilindros.

Air injection reactor (AIR) A system that induces additional air into the exhaust to continue burning exhaust gases.

Reactor de inyección de aire (AIR) Un sistema que induce aire additional al escape para continuar quemando los gases del escape.

Alternating current An electric current that reverses its direction at regularly recurring intervals.

Corriente Alterna Una corriente eléctrica que reserva su dirección a intervalos de repetición regular.

Ambient air temperature sensor A sensor that measures the temperature of the outside air. Used on automatic A/C systems.

Sensor de temperaturas del ambiente Un sensor que mide la temperatura del aire exterior. Se usa en los sistemas automáticos de A/C.

American Petroleum Institute (API) A technical society that has developed service ratings for motor oil.

Instituto de Petroleo Americano (API) Una sociedad de technología que ha desarrollado las especificaciones del aceite de motor.

Ammeter A device used to measure electrical current.

Amperímetro Un dispositivo para medir la corriente eléctrica.

Ampere A measurement for the amount of electron movement or current flow.

Amperio Medida de la cantidad del movimiento de electrones o del flujo de corriente.

Amp-hour rating A battery rating indicating the amount of amperes a battery will deliver over a longer time period with a small electrical load.

Clasificación de amperios-horas Clasificación de una batería que indica la cantidad de amperios que la misma podrá generar durante un mayor espacio de tiempo con una pequeña carga eléctrica.

Amplitude The difference between the highest and lowest voltage in a waveform signal.

Amplitud La diferencia entre el voltaje más alto y el voltaje más bajo en una señal en forma de onda.

Analog A nondigital measuring method that uses a needle to indicate readings. A typical dashboard gauge with a moving needle is an analog instrument.

Análogo Un método de medir en forma no digital, que usa una aguja para indicar la lectura. Un reloj en el tablero del vehículo con una aguja que se mueve es un instrumento análogo.

Analog/digital conversion The process of changing analog voltage signals to digital signals.

Conversión analógica/digital Proceso de convertir señales de tensión analógicas en señales digitales.

Analog voltage signal A voltage signal that is usually produced by input sensors and is continuously variable within a certain voltage range.

Señal de tensión analógica Señal de tensión que normalmente es producida por sensores de entrada y que varía de modo continuo en proporción a cierto margen de tensión.

Anti drainback valve A valve that prevents oil drainback from components in the cylinder heads into the lubrication system.

Válvula que evita la filtración del aceite Válvula que evita que el aceite de los componentes en las culatas de los cilindros se filtre a través del sistema de lubrificación.

Antifoaming agents Oil additives that prevent oil foaming.

Agentes antiespumantes Aditivos de aceite que evitan que éste se espume.

Antiknock Refers to a fuel's ability to resist self-ignition.

Antidetonante Se refiere a la habilidad de un combustible de resistir la tendencia de que se incinere por si mismo.

API American Petroleum Institute. An organization that sets standards for petroleum-based products, such as engine oil.

API Instituto Americano del petróleo. Una organización que establece estándares para productos basados en petróleo.

Atmospheric pressure The pressure exerted on the earth by the atmosphere.

Presión atmosférica Presión que la atmósfera ejerce sobre la tierra.

Atom The smallest particle of an element.

Átomo La partícula más pequeña de un elemento.

Atomization The process of breaking a liquid up into small particles or droplets.

Atomización El proceso de romper un líquido adentro de partículas pequeñas o gotas.

Automatic shutdown relay A computer-controlled relay that opens and closes the circuit to the fuel pump, coil primary winding, and other components on Chrysler products.

Relé de parada automática Relé controlado por computadora que abre y cierra el circuito que conduce hacia la bomba del combustible, el bobinado primario, y otros componentes de productos fabricados por la Chrysler.

Backpressure Pressure created by restriction in an exhaust system.

Contrapresión La presión creada por la restricción en un sistema de escape.

Ballast resistor A resistor connected in series between the ignition switch and the coil positive primary terminal on some DI systems.

Resistor de compensación Resistor conectado en serie entre el botón conmutador de encendido y el borne primario positivo de la bobina en algunos sistemas de encendido con distribuidor.

Band A holding member for an automatic transmission.

Banda Un miembro de retención de la transmisión.

Bar A bar equals 100 kPa or 14.5 psi. It can also be used in terms of pressure measurements for oil and fuel.

Bar Un bar equivale a 100 kPa o a 14.5 psi. También se puede utilizar en mediciones de la presión de aceite y de combustible.

Barometric pressure switch A sensor or signal circuit that sends a varying frequency signal to the processor relating actual barometric pressure.

Interruptor para la presión barométrica Un sensor o su circuito de señal que envía una señal de frecuencia variable al procesador relacionándole la presión barométrica actual.

Base A substance that reacts with acids to form salts.

Base Una sustancia que reacciona con los ácidos para formar los sales.

Base timing Base or initial timing is timing without computer advance added. Base timing is adjustable on some vehicles.

Sincronización de base La sincronización inicial o de base es el valor de la sincronización sin ningún avance de la computadora. En algunos vehículos, el tiempo de encendido base es ajustable.

Basic timing The initial timing specified by the manufacturer.

Tiempo básico El tiempo original especificado por el fabricante.

Baud rate The rate at which a PCM is able to transfer and receive data. Baud rate is measured in bits per second.

Velocidad Baud La velocidad a la cual el PCM (módulo de control de la potencia del motor) es capaz de transferir y recibir data. La velocidad Baud es medida en mordidas por segundo.

Bearing crush Bearing inserts are slightly longer than the connecting rod bore in which they are mounted. This design crushes the bearing slightly when the rod bolts are tightened to provide improved contact between the bearing insert and the connecting rod bore.

Quiebra de cojinete Piezas insertas de cojinetes que son ligeramente más largas que el calibre de las bielas en las que van montadas. Este diseño quebranta un poco el cojinete cuando se aprietan los pernos de la biela a fin de proporcionar un mejor contacto entre la pieza inserta del cojinete y el calibre de la biela.

Bearing inserts Circular bearings mounted between the connecting rod bore and the crankshaft journal.

Piezas insertas de cojinetes Cojinetes circulares montados entre el calibre de la biela y el gorrón del cigüeñal.

Bearing spread A bearing design where the bearing curvature is slightly larger than the bore in which it is mounted.

Extensión del cojinete Cojinete diseñado de forma tal que su curvatura es un poco más grande que el calibre en el que está montado.

Before top dead center (BTDC) Refers to the position of the piston while it is moving toward its upmost position.

Antes del punto muerto superior Se refiere a la posición del pistón mientras se está moviendo hasta su punto más alto.

Belt-driven cooling fan A cooling fan driven by a belt from the crankshaft.

Ventilador de enfriamiento accionado por correa Ventilador de enfriamiento accionado por una correa desde el cigüeñal.

Binary coding A group of numbers assigned to digital voltage signals.

Código binario Grupo de números asignados a las señales de tensión digitales.

Blow-by Compression and exhaust gases that blow by the piston rings and enter into the engine's crankcase.

Fuga en la cámara de la combustión Compresión y gases de escape que se escapan atraves de los anillos del pistón y entran adentro de la cacerola del aceite.

Body Control Module (BCM) A controller (computer) that is responsible for controlling vehicle body functions.

Módulo de control del Chasis (BCM) Un controlador (de computadora) que se encarga de controlar las funciones del chasis del vehículo.

Boost pressure The amount of pressure in the intake manifold created by a turbocharger or supercharger.

Presión de sobrealimentación Cantidad de presión en el colector de aspiración producida por un turbocompresor o un compresor.

Bore The diameter of a hole, commonly used to describe the dimensions of an engine.

Diámetro del cilindro El diámetro del orificio, comúnmente usado para describir las dimensiones de un motor.

Bore and stroke The diameter of the cylinder bore and the length of the piston stroke.

Calibre y carrera El diámetro del calibre del cilindro y el largo de la carrera del pistón

Bottom dead center (BDC) Piston position at the bottom of the cylinder.

Punto muerto inferior La posición del pistón en la parte inferior del cilindro.

Burn time The length of the spark line while the spark plug is firing, measured in milliseconds.

Duración del encendido Espacio de tiempo que la línea de chispas de la bujía permanece encendida, medido en milisegundos.

Bypass Bypass voltage is sent from the ECM to tell the module to switch to computer-based timing.

Derivación El voltaje de derivación se envía desde el módulo de control del motor (ECM, por sus siglas en inglés) para indicar que es necesario cambiar a la sincronización basada en la computadora.

Bypass oil filter An engine oil filter in which a portion of the flow from the pump circulates through the filter.

Filtro del aceite de paso Filtro del aceite de un motor en el que parte del flujo de la bomba circula a través del filtro.

Calibration A set of instructions programmed into the PCM or contained on a PROM that tailors the PCM to the vehicle.

Calibración Un conjunto de instrucciones programados en el PCM or contenidos en un PROM que adapta el PCM al vehículo.

California Air Research Board (CARB) The part of the California government that monitors vehicle emissions and sets standards for the state to reduce vehicle emissions.

Junta de recursos del aire de California La parte del gobierno de California que revisa las emisiones de los vehículos y establece los estándares del estado para reducir las emisiones de los vehículos.

Cam Elliptical circles or raised parts of a shaft.

Leva Los círculos elípticos o las partes alzadas de un árbol.

Camber The attitude of a wheel and tire assembly when viewed from the front of a car. If it leans outward, away from the car at the top, the wheel is said to have positive camber. If it leans inward, it is said to have negative camber.

Comba La aptitud de un ensamblaje de rueda y el neumático cuando se observa desde la parte delantera del vehículo. Si se inclina hacia la parte de afuera del vehículo en la parte superior, es dicho que la rueda tiene comba positiva. Si se inclina hacia la parte de adentro del vehículo en la parte superior, es dicho que la rueda tiene comba negativa.

Cam-ground piston A piston with the skirt designed in the shape of a cam rather than being perfectly round.

Pistón de leva excéntrica Pistón con una faldilla diseñada en forma de leva en vez de ser perfectamente redonda.

Cams They are raised sections of a shaft that have high spots called lobes.

Levas Son secciones elevadas del eje que poseen puntos salientes llamados lóbulos.

Camshaft The component in the engine that opens and closes the valves.

Árbol de levas Componente en el motor que abre y cierra las válvulas.

Camshaft lobes Rotating high points on the camshaft that open the valves.

Lóbulos del árbol de levas Puntos altos giratorios en el árbol de levas que abren las válvulas.

Camshaft reference sensor A sensor that sends a voltage signal to the ignition module; this signal is often used for injector sequencing.

Sensor de referencia del árbol de levas Sensor que le envía una señal de tensión al módulo del encendido; dicha señal se utiliza con frecuencia para el ordenamiento del inyector.

Canister purge valve The valve that controls the release of fuel fumes from the charcoal canister into the intake.

Válvula para la purga del canasto La válvula que controla la liberación de los vapores del combustible desde el canasto de carbón hacia adentro del múltiple de escape.

Carbon dioxide (CO₂) A gas that is a by-product of the combustion process.

Bióxido de carbono (CO2) Gas que es un producto derivado del proceso de combustión.

Carbon monoxide (CO) A gas formed as a by-product of the combustion process in the engine cylinders. This gas is very dangerous or deadly to the human body in high concentrations.

Monóxido de carbono (CO) Gas que es un producto derivado del proceso de combustión en los cilindros del motor. Este gas es muy peligroso y en altas concentraciones podría ocasionar la muerte.

Caster Angle formed between the kingpin axis and a vertical axis as viewed from the side of the vehicle. Caster is considered positive when the top of the kingpin axis is behind the vertical axis.

Inclinación del eje delantero El ángulo formado entre el axis del eje pivote de la dirección y un axis vertical según es observado desde el lado del vehículo. La inclinación del eje es considerada positiva cuando la parte superior del axis del eje pivote de la dirección está detrás del axis vertical.

Catalyst efficiency monitor A monitor in OBD II systems that checks the oxygen content of the exhaust before it enters the catalytic converter and after it leaves the converter.

Monitor de la eficiencia del catalítico Un monitor en los sistemas OBD II (diagnostico abordo del vehículo II) que chequea el contenido del oxigeno en el escape antes de que entre en el convertidor catalítico y después que sale del convertidor.

Catalytic converter A device installed in a vehicle's exhaust system that changes undesired gases into harmless gases. An emission control device.

Convertidor catalítico Un dispositivo instalado en el sistema de escape de un vehículo que cambia los gases indeseables a gases no dañinos. Un dispositivo de control para las emisiones.

Cell group The collection of positive and negative plates in one cell of a battery. Each group typically provides 2.1 volts.

Grupo de celdas La colección de los plato negativos y positivos en una celda de la batería. Cada grupo típicamente provee 2.1 volteos.

Cellular radiator core A type of core made from many small interconnected cells.

Núcleo del radiador celular Tipo de núcleo compuesto de muchas células conectadas entre sí.

Central port injection (CPI) A system that delivers a steady stream of pressurized fuel into the intake manifold.

Inyección por orificio central (CPI) Un sistema que entrega un flujo fijo de combustible bajo presión a la toma de aire.

Cetane A rating used to classify diesel fuel that refers to the volatility of the fuel.

Cetano Una relación usada para clasificar el combustible diesel, se refiere a la votálidad del combustible.

Charcoal canister An emission control device that collects gasoline vapors and prevents them from entering into the atmosphere.

Canasto de carbón Un dispositivo para el control de las emisiones que colecta los vapores de gasolina y los previene de que entren adentro de la atmósfera.

Circuit opening relay A relay that opens and closes the fuel pump circuit on some Toyota products.

Relé para la apertura del circuito Relé que abre y cierra el circuito de la bomba del combustible en algunos productos fabricados por Toyota.

Clear flood mode A computer operating mode that supplies a leaner air-fuel ratio if an engine becomes flooded. This mode is entered by holding the throttle wide open while cranking the engine.

Modo para disminuir una inundación Modo de funcionamiento de una computadora que suminista una relación de aire y combustible más pobre si el motor se inunda. Este modo se activa manteniendo la mariposa abierta de par en par mientras se arranca el motor.

Closed loop A computer system operating mode in which the computer uses the oxygen sensor to help control the air-fuel ratio.

Bucle cerrado Modo de funcionamiento de una computadora en el que se utiliza el sensor de oxígeno para ayudar a controlar la relación de aire y combustible.

Clutch disc The part of a clutch that receives the driving motion from the flywheel and pressure plate assembly and transmits that motion to the transmission input shaft.

Disco del embrague La parte de un embrague que recibe el movimiento desde el volante del motor, el ensamblaje del plato de presión y transmite ese movimiento al eje de entrada de la transmisión.

Clutch fork A forked lever that moves the clutch release bearing and hub back and forth.

Tenedor del embrague Una palanca en configuración a un tenedor que libera el balero de liberación del embrague y el cubo hacia adelante y hacia atrás.

Clutch pack A common term for the multiple-disc clutch pack used to hold a member of the planetary gearset in an automatic transmission.

Páguete de embrague Un termino común para un embrague con múltiple discos, usados para detener un miembro del juego de los engranes planetarios en una transmisión automática.

Clutch pedal position (CPP) switch The clutch pedal position (CPP) switch informs the computer that there is no load on the engine, as well as prevents the engine from starting unless the clutch is depressed.

Posición del pedal de embrague El interruptor de posición del pedal de embrague (CPP, por sus siglas en inglés) le informa a la computadora que no hay carga en el motor e impide que este se encienda a menos que el embrague esté bajo.

Clutch release bearing A sealed, prelubricated ball bearing that moves the pressure plate release levers or diaphragm spring through the engagement and disengagement of the clutch.

Balero de liberación del embrague Un balero prelubricado y sellado que mueve la palanca de liberación del plato de presión, resortes del diafragma atraves del enganche y desenganche del embrague.

CNG Compressed Natural Gas. An alternative fuel source for engines.

CNG Gas Natural Comprimido. Una fuente de combustible alternativo para los motores.

Coefficient of drag (Cd) The total resistance to motion caused by friction between a moving vehicle and the air.

Coeficiencia de resistencia (Cd) La resistencia total al movimiento causada por la fricción entre un vehículo en movimiento y el aire.

Coil-near-plug (CNP) The single coil mounted directly to the spark plug is commonly referred to as a coil-on-plug (COP) system, while the single coil connected to a spark plug with a spark plug wire is referred to as a coil-near-plug (CNP).

Bobina cerca de bujía Una bobina que está montada directamente en la bujía por lo general se denomina sistema de bobina sobre bujía (COP, por sus siglas en inglés), mientras que una bobina que está conectada a una bujía mediante un cable se denomina bobina cerca de bujía (CNP, por sus siglas en inglés).

Coil-on-plug (COP) The single coil mounted directly to the spark plug is commonly referred to as a coil-on-plug (COP) system, while the single coil connected to a spark plug with a spark plug wire is referred to as a coil-near-plug (CNP).

Bobina sobre bujía Una bobina que está montada directamente en la bujía por lo general se denomina sistema de bobina sobre bujía (COP, por sus siglas en inglés), mientras que una bobina que está conectada a una bujía mediante un cable se denomina bobina cerca de bujía (CNP, por sus siglas en inglés).

Coil pack A common term for the ignition coil assembly in a distributorless ignition system.

Juego de bobinas Un termino común usado para el ensamblaje de la bobina de la ignición en un sistema de ignición sin distribuidor.

Coil sequencing Firing coils on an EI system to match the engine firing order.

Ordenamiento de bobina Bobinas del encendido en un sistema de encendido electrónico para equilibrar la secuencia del encendido del motor.

Cold cranking ampere rating A battery rating indicating the amount of amperes a battery will deliver at a specific temperature.

Clasificación de amperios de arranque en frío Clasificación de una batería que indica la cantidad de amperios que la misma podrá generar a una temperatura específica.

Cold spark plug A spark plug designed so that the electrodes operate at lower temperatures.

Bujía en frío Bujía diseñada para que los electrodos funcionen a temperaturas más bajas.

Combustion The process whereby the proper mixture of fuel and oxygen is exposed to high temperature or spark and causes the material or fuel to burn.

Combustión El proceso por el cual la mezcla correcta del combustible y el oxígeno se expone a una temperatura alta o una chispa para causar que se quema el combustible o la material.

Combustion chamber The area formed in the cylinder head when the piston is at TDC.

Cámara de combustión El área formada adentro de la cabeza del cilindro cuando el pistón está en el TDC (punto muerto superior).

Compound A material with two or more types of atoms.

Compuesto Material que tiene dos o más tipos de átomos.

Comprehensive component monitor One of the monitor systems in an OBD II system that checks the activity and efficiency of several systems and major components.

Componente de monitor comprensivo Uno de los monitores en el sistema OBD II (diagnostico abordo del vehículo II) que chequea las actividades y eficiencias de varios sistemas y componentes mayores.

Compressed natural gas (CNG) A petroleum-based pressurized fuel.

Gas natural comprimido (CNG) Un combustible a base de petroleo bajo presión.

Compressibility The ability of a material to become smaller when pressure is applied.

Compresibilidad Capacidad de un material para empequeñecerse al aplicársele presión.

Compression Placing a gas under pressure raising its temperature.

Compresión Poniendo un gas bajo presión subiendo su temperatura.

Compression ignition (CI) An engine in which the air-fuel mixture is ignited by the heat of compression.

Encendido por compresión Motor en el que la mezcla de aire y combustible se enciende por medio del calor de compresión.

Compression ratio The ratio of the volume in the cylinder above the piston when the piston is at bottom dead center to the volume in the cylinder above the piston when the piston is at top dead center.

Relación de compresión La relación del volumen en el cilindro encima del pistón cuando el pistón está en el punto muerto inferior, comparado con el volumen en la parte superior del cilindro, cuando el pistón está en el punto muerto superior.

Compression rings Rings mounted near the top of the piston to seal the compression and combustion gases in the cylinder.

Anillos de compresión Anillos montados cerca de la parte superior del pistón para atrapar los gases de compresión y de combustión dentro del cilindro.

Compression sense ignition A waste spark ignition which determines which cylinder is on compression by measuring the amount of secondary ignition voltage, thereby eliminating the need for a cam sensor.

Encendido por compresion Encendido por chispa perdida que establece que cilindro esta en compresion al medir la intensidad del voltaje secundario y, en consecuencia, eliminando la necesidad de un sensor de camara.

Compressor wheel A vaned wheel mounted in the air intake and connected to one end of the turbocharger shaft.

Rueda compresora Rueda con paletas montada en el aire aspirado y conectada a un extremo del árbol turbocompresor.

Computer-controlled dwell (CCD) The length of time that a coil primary stays on using computer control.

Parada de movimiento controlado por computadora (CCD) La cantidad de tiempo que esta encendida la bobina primaria usando control por computadora.

Conduction A method of heat transfer in which heat from a warmer object is transferred to a cooler object.

Conducción Método de transferencia de calor en el que el calor de un objeto más tibio se transfiere a un objeto más frío.

Conductor An element with one, two, or three valence electrons that easily conducts electric current.

Conductor Elemento con uno, dos o tres electrones de valencia que conduce una corriente eléctrica fácilmente.

Connecting rod Steel rod that connects the crankshaft to the piston, able to pivot on both ends that enables a reciprocating motion to become a circular motion.

Biela Una varilla de acero que conecta el cigueñal al piston, capáz de pivotear en ambas extremidades que permite que un movimiento recíproco se convierte en un movimiento circular.

Connecting rod bore The circular opening in the lower end of the connecting rod that retains the connecting rod bearings.

Calibre de biela Abertura circular en el extremo inferior de la biela que retiene los cojinetes de la biela.

Connecting rod eye A circular opening in the top of the connecting rod in which the piston pin is located.

Ojal de biela Abertura circular en la parte superior de la biela en la que está ubicado el pistón.

Controller Area Network or Computer Area Network (CAN) Starting in the 2008 model year, the universal communication method or "language" used for the sharing of information between modules on a vehicle network or with scan tools.

Sistema Red de área de Controladores o Sistema de área de Computadora (CAN) Comenzando con el modelo del año 2008, el metodo de comunicación universal utilizado para compartir la información entre los módulos en un sistema red del vehículo o con las herramientas exploradoras.

Controllers A term used to describe electronic devices that control the activity of various actuators.

Controladores Un termino usado para describir dispositivos electrónicos que controlan las actividades de varios actuadores.

Convection A method of heat transfer in which heated atoms or molecules rise upward and cooler atoms or molecules sink downward where they are heated.

Convección Método de transferencia de calor en el que los átomos o moléculas calentados ascienden y los átomos o moléculas más fríos descienden a donde son calentados.

Coolant A mixture of antifreeze and water.

Fluido refrigerante Una mezcla de anticongelante y el agua.

Coolant bypass passage A coolant passage or hose through which coolant is circulated when the thermostat is closed.

Tubo para el paso del refrigerante Tubo o manguera por el que circula el refrigerante cuando el termóstato está cerrado.

Coolant recovery system A container that catches any coolant that comes out the radiator overflow hose and returns this coolant to the radiator.

Sistema para la recuperación del refrigerante Recipiente que atrapa el refrigerante que se escapa de la manguera de rebose del radiador y lo devuelve al radiador.

Cooling system Radiator, hoses, thermostat, heater core, and coolant designed to dissipate excess heat from the engine.

Sistema de enfriamiento El radiador, las mangueras, el termostáto, el núcleo de calor y el líquido de enfriamiento diseñado a dispersar el calor excesivo del motor.

Core plugs Metal plugs in a cast component, such as the engine block, which seal openings required by the casting tools.

Tapones para núcleo Tapones de metal en un componente fundido, como por ejemplo el bloque de un motor, que sellan las aberturas requeridas por las herramientas de fundición.

Corporate average fuel economy (CAFE) Fleet average fuel economy standards imposed on car manufacturers by the U.S. Congress.

Economía promedio de combustible para fabricantes Normas sobre la economía promedio de combustible para flotas de vehículos que el Congreso de los Estados Unidos les impone a los fabricantes de automóviles.

Corrosion and rust inhibitors Oil additives that neutralize acids in the oil that cause corrosion of engine components.

Inhibidores de corrosión y oxidación Aditivos de aceite que neutralizan los ácidos en el aceite que pueden provocar la corrosión de los componentes del motor.

Corrosive A material that dissolves metals or burns the skin.

Corrosivo Material que disuelve metales o que quema la piel.

Counterweights Heavy weights that are part of the crankshaft and help to provide proper balance.

Contrapesos Pesos pesados que forman parte del cigüeñal y que ayudan a proporcionar un equilibrio adecuado.

Coupling point This is the period of time when the speed of a torque converter's turbine approaches the speed of the impeller. At this point, torque multiplication is minimized.

Punto de acoplación Este es el periodo de tiempo cuando la velocidad de la turbina del convertidor de torque se acerca a la velocidad del impulsor. En este punto la multiplicación de torque es minimizada.

Crankshaft A rotating component mounted in the lower side of the block that changes vertical piston motion to rotary motion.

Cigüeñal Componente giratorio montado en la parte inferior del bloque que convierte el movimiento vertical del pistón en movimiento giratorio.

Crankshaft position sensor A sensor that helps the PCM determine crankshaft position and speed necessary for proper spark and fuel delivery.

Sensor de posición del cigüeñal Un sensor que ayuda al PCM en determinar la posición y la velocidad del cigüeñal esencial para la entrega de chispa y combustible adecuado.

Crankshaft timing sensor A pickup assembly that produces a voltage signal used for ignition triggering in an EI system.

Sensor de regulación del cigüeñal Conjunto de captación que produce una señal de tensión utilizada para el arranque del encendido en un sistema de encendido electrónico.

Crank throw The distance from the crankshaft main bearing centerline to the connecting rod journal centerline. The stroke of any engine is the crank throw.

Tiro del cigüeñal La distancia central desde el cojinete principal del cigüeñal a la línea central del muñón de la biela. La carrera de cualquier motor es el tiro del cigüeñal.

Crossflow radiator A radiator in which the tanks are positioned on each side of the core and the coolant flows horizontally through the core.

Radiador de flujo transversal Radiador en el que los tanques se colocan a cada lado del núcleo y donde el refrigerante fluye horizontalmente a través de éste.

Crude oil Petroleum products that are unrefined. This term is normally used to describe oil that has just been pumped from an oil well.

Aceite crudo Productos de petróleo que no están refinados. Este termino es normalmente usado para describir el aceite que se han acabado de ser bombeado desde un pozo de petróleo.

Current The movement of electricity.

Corriente El movimiento de la electricidad.

Customization Refers to the ability of the customer to customize certain accessories such as door locks, power sliding doors, etc., to act according to a set routine (such as all doors lock when the vehicle is put into gear).

Personalización Refiere a la habilidad del cliente a personalizar ciertos acesorios tal como los cerraduras de puerta, las puertas corredizas de poder, etc. a funcionar según una rutina predeterminada (como todas las puertas se cierran cuando el vehículo se pone en marcha).

Cycle One complete set of changes in a recurring signal.

Ciclo Un juego de cambio completo en una señal recurrente.

Cycling clutch A type of air-conditioning system in which system pressure is controlled by turning the compressor clutch on and off.

Embrague de ciclo Un sistema de tipo de aire acondicionado en cual la presión del sistema es controlada apagando y encendiendo el embrague del compresor.

Cylinder head Casting that seals the top of the direct drive: a one to one gear ratio.

Cabeza del cilindro Una pieza moldeada que sella la parte superior de la toma directa: una relación de embrague uno a uno.

Cylinder misfire monitoring Calculates crankshaft speed acceleration as each cylinder fires.

Regulación del fallo del encendido del cilindro Calcula la acceleración de la velocidad del cigüeñal al disparar cada cilíndro.

Cylinders Holes bored into the engine block to accept pistons.

Cilindros Los hoyos que se han maquinado en el bloque del motor para aceptar los pistones.

Data link connector (DLC) A computer system connector to which the computer supplies data for diagnostic purposes.

Conector de enlace de datos Conector de computadora al que ésta suministra datos para propósitos diagnósticos.

Deep skirt block An engine with the block skirt extending well below the bottom of the cylinders.

Faldilla profunda de bloque Motor con una faldilla de bloque que se extiende hasta más abajo de la parte inferior de los cilindros.

Delta pressure feedback EGR (DPFE) sensor It relays a difference in pressure between the exhaust and intake side of a metering orifice to the PCM.

Sensor de retroalimentación de presión diferencial EGR El sensor de retroalimentación de presión diferencial EGR (DPFE, por sus siglas en inglés) transmite una diferencia de presión entre el escape y la admisión de un orificio medidor al módulo de control del tren de potencia (PCM, por sus siglas en inglés).

Detergents and dispersants Oil additives that help the oil to clean carbon particles off engine components and disperse large carbon particles in the oil.

Detergentes y dispersores Aditivos de aceite que ayudan al aceite a remover las partículas de carbón de los componentes del motor y a dispersar las partículas grandes de carbón en el aceite.

Detonation Abnormal combustion. Refers to the ignition of the air-fuel mixture inside the combustion chamber prior to the firing of a spark plug.

Detonación Combustión anormal. Se refiere a la ignición de la mezcla de aire/combustible adentro de la cámara de combustión antes de que la chispa de la bujía ocurra.

Diesel A compression-ignited engine.

Diesel Un motor que hace su combustión con su compresión.

Diesel oxidation converter (DOC) The diesel oxidation converter works similarly to the gasoline catalytic converter and oxidizes HC and CO and converts them to H_2O and CO_2.

Convertidor de oxidación diésel El convertidor de oxidación diésel (DOC, por sus siglas en inglés) funciona de un modo similar al convertidor catalítico de gasolina y oxida el HC y el CO para convertirlos en H2O y en CO2.

Diesel particulate emissions Diesel exhaust emissions that consist mainly of small carbon particles.

Emisiones de partículas de diesel Emisiones del escape de diesel que consisten principalmente en pequeñas partículas de carbón.

Diesel particulate filter (DPF) Diesel engines also produce particulate emissions, which are small carbon particles, often called soot. Emissions of particulates have come under scrutiny because they have been listed as possibly contributing to lung cancer. A ceramic monolith inside the diesel particulate converter reduces particulate emissions.

Filtro antipartículas diésel Los motores diésel también producen emisiones de partículas, es decir, pequeñas partículas de carbono que a menudo se denominan "hollín". Las emisiones de partículas se encuentran en estudio porque han sido mencionadas como un posible factor de riesgo de cáncer de pulmón. Un monolito de cerámica situado dentro del convertidor de partículas diésel reduce las emisiones de partículas.

Digital Normally refers to a signal that is either on or off. Digital signals are required by most automotive computers.

Digital Normalmente se refiere a una señal que no está ni apagada ni encendida. Las señales digitales son requeridas por la mayorías de las computadoras automotrices.

Digital EGR valve An EGR valve that contains a computer-operated solenoid, or solenoids.

Válvula EGR digital Una válvula EGR que contiene un solenoide o solenoides accionados por computadora.

Digital voltage signal A voltage signal that is either on or off, high or low.

Señal de tensión digital Una señal de tensión que está encendida o apagada, o que es alta o baja.

Diode A one-way flow valve for electricity.

Diodo Válvula de flujo de una vía para la electricidad.

Direct current An electrical current that flows in one direction.

Corriente Directa Una corriente eléctrica que fluye solamente en una dirección.

Direct drive A 1:1 gear ratio, meaning that neither torque nor speed were gained or lost in the transfer of power.

Accionamiento directo Una relación de 1:1, lo que significa que ni el par ni la velocidad se gana o se pierde en la transferencia del poder.

Direct fuel injection (DFI) Injects the fuel directly into the combustion chamber.

Inyección directa de combustible El combustible se inyecta directamente en la cámara de combustión.

Direct injection A fuel injection system in which fuel is injected directly into the combustion chamber instead of the intake manifold.

Toma directa Una relación de embrague de 1:1, significando que ni la torsión ni la velocidad se gana o pierda en la transferencia de fuerza.

Dispatch sheets Contain appointment records in the shop.

Hojas de servicios Contienen información sobre las reparaciones de vehículos que se llevan a cabo en el taller mecánico.

Displacement The volume of each cylinder multiplied by the number of cylinders, usually expressed in liters, cubic centimeters, or cubic inches.

Desplazamiento Volumen de cada cilindro multiplicado por el número de cilindros, y expresado normalmente en litros, centímetros cúbicos, o pulgadas cúbicas.

Display scope pattern A scope pattern in which the voltage traces from the cylinders are displayed one after the other across the screen.

Modelo de radio de visualización Modelo de radio en el que las pequeñas cantidades de tensión provenientes de los cilindros se proyectan una tras otra a través de la pantalla.

Distributor cap and rotor A mechanical device that distributes the ignition sparks from the secondary coil winding to the spark plugs.

Tapa y rotor del distribuidor Dispositivo mecánico que distrubuye las chispas del encendido desde el bobinado secundario hasta las bujías.

Distributor ignition (DI) system SAE J1930 terminology for an ignition system with a distributor.

Sistema de encendido con distribuidor Término utilizado por la SAE J1930 para referirse a un sistema de encendido que tiene un distribuidor.

Distributorless ignition system (DIS) An ignition system that does not use a distributor for spark distribution; rather, spark distribution is controlled by the vehicle's computer.

Sistema de ignición sin distribuidor Un sistema de ignición que no usa un distribuidor para la chispa de la ignición, envés la distribución de la chispa es controlada por la computadora del vehículo.

Diverter valve Part of a typical air injection system. This valve allows the air from the air pump to be diverted into the atmosphere rather than the exhaust to prevent backfiring.

Válvula de diversión Una parte típica del sistema de inyección de aire. Esta válvula permite que el aire desde la bomba de aire sea divertido adentro de la atmósfera envés de enviarlo adentro del sistema de escape para prevenir una contra explosión.

Downflow radiator A radiator in which the tanks are positioned on the top and bottom of the core, and the coolant flows vertically through the core.

Radiador con flujo descendente Radiador en el que los tanques se colocan sobre las partes superior e inferior del núcleo y donde el refrigerante fluye verticalmente a través de éste.

Drive cycle A specified set of driving conditions. The drive cycle is important to adaptive strategies of a computer. The proper drive cycle is also required of OBD II systems to complete monitor tests on certain systems.

Ciclo de ejecución Un juego especifico de condiciones de manejar. Estas condiciones son importantes para la estrategia adaptativa de la computadora. El ciclo de manejo apropiado es también requerido para el sistema OBD II (diagnóstico abordo del vehículo II) para competir con las pruebas del monitor en algunos sistemas.

Driver door module (DMM) A computer module that is part of the vehicle network and located in the driver's door that is responsible for power window and door lock operation and customization features.

Módulo de puerta conductor (DMM) Un módulo de computadora que es parte del sistema red del vehículo y que se encuentra en la puerta del conductor siendo responsable de las operaciones de las ventanas de poder y las cerraduras de las puertas y las características personalizadas.

Drivers Electronic "switches" that a module (computer) uses to control actuators.

Accionador Los "interruptores" que usa un módulo (de computadors) para controlar los acutadores.

Dry-type cylinder sleeve A cylinder sleeve installed in the block so that the coolant does not contact the outside area of the sleeve.

Manguito de cilindro tipo seco Manguito de cilindro instalado en el bloque para que el refrigerante no entre en contacto con la parte exterior del mismo.

Dry-type intake manifold An intake manifold without coolant circulation through the manifold.

Colector de aspiración tipo seco Colector de aspiración a través del cual no circula refrigerante.

Dual overhead camshaft (DOHC) An engine with two camshafts positioned above each cylinder head.

Árbol de levas superpuesto doble Motor con dos árboles de levas colocados sobre cada cada una de las culatas de los cilindros.

Duty cycle On-time to off-time ratio, as measured in a percentage of pulse width or degrees of dwell.

Ciclo de duración La relación de apagado y encendido, según es medido en porcentaje de la amplitud del pulso o grados Dwell (tiempo en que los puntos están cerrados medidos en grados).

Dwell The amount of time the current is flowing through a circuit. Most often, this term is applied to ignition systems.

Tiempo en que los puntos están cerrados medidos en grados La cantidad de tiempo que la corriente está fluyendo atraves de un circuito. Más común, este termino es aplicado a sistemas de ignición.

Dwell time Coil primary on time.

Tiempo de breve parada La bobina primaria en tiempo.

Dynamic balance Refers to the balance of a wheel and tire assembly when it is in motion.

Balanceo dinámico Se refiere al balance del ensamblaje de una llanta y un neumático cuando está en moción.

Efficiency The relationship between the amount of energy put into the engine and the amount available from the engine.

Rendimiento La relación entre la cantidad de la energía introducido al motor y la cantidad disponible del motor

EGR vacuum regulator (EVR) A vacuum solenoid that is pulsed on and off by the computer to supply the proper amount of vacuum to the EGR valve.

Regulador de vacío EGR Solenoide de vacío que es encendido y apagado por la computadora para suministrarle una cantidad adecuada de vacío a la válvula EGR.

Electric-drive cooling fan A cooling fan driven by an electric motor.

Ventilador de enfriamiento accionado eléctricamente Ventilador de enfriamiento accionado por un motor eléctrico.

Electricity The release of energy as electrons jump from the orbit of one atom to another.

Electricidad La descarga de energía que ocurre cuando los electrones brincan de la órbita de un átomo a otra.

Electrolyte A sulfuric acid and water solution in automotive batteries.

Electrolito Solución de ácido sulfúrico y agua en baterías de automóviles.

Electromagnet A magnet created by the flow of electricity through a coil of wire.

Electroimán Imán producido por el flujo de electricidad a través de un alambre.

Electromagnetic induction The process of inducing a voltage by moving a conductor through a magnetic field or vice versa.

Inducción electromagnética Proceso de inducir una tensión moviendo un conductor a través de un campo magnético o viceversa.

Electron A negatively charged particle that orbits around the nucleus of an atom.

Electrón Partícula de carga negativa que orbita alrededor del núcleo de un átomo.

Electronic automatic temperature control (EATC) Heating and A/C controller that maintains a preset temperature and blower speed without further action by the driver.

Control electrónico de temperatura automático (EATC) El controlador de calentura y A/C que mantiene una temperatura predeterminada y la velocidad del ventilador sín que el conductor toma ningúna acción.

Electronic control unit (ECU) An electronic device that opens and closes the primary ignition circuit.

Unidad de control electrónico Dispositivo electrónico que abre y cierra el circuito primario de encendido.

Electronic crash sensor (ECS) A controller that determines when the air bags should deploy based on crash severity and angle of impact.

Sensor electrónico de impacto (ECS) Un controlador que determine cuando deben desplegarse las bolsas de aire basado en la severidad y el ángulo del impacto.

Electronic data system Service bulletins and other information accessed on the Internet.

Sistema electrónico de datos Publicaciones y otra información referente al servicio mecánico; dicha información se guarda en discos compactos.

Electronic engine control V (EEC V) A term applied to a fifth-generation Ford computer system.

Sistema de control electrónico del motor V Término aplicado a una computadora de cuarta generación de la Ford.

Electronic fuel injection (EFI) A generic term applied to various fuel injection systems.

Inyección electrónica de combustible Término general aplicado a varios sistemas de inyección de combustible.

Electronic ignition (EI) system SAE J1930 terminology for an ignition system without a distributor.

Sistema de encendido electrónico Término utilizado por la SAE J1930 para referirse a un sistema de encendido que no tiene distribuidor.

Electronic spark timing (EST) A term applied to computer-controlled spark advance on some ignition systems.

Regulación electrónica de chispas Término aplicado al avance de chispas controlado por computadora en algunos sistemas de encendido.

Electronic vacuum regulator valve (EVRV) Manifold vacuum that is pulse width modulated (PWM) through a solenoid called the **electronic vacuum regulator valve (EVRV)** is used to open the EGR valve.

Válvula electrónica reguladora de vacío El vacío del múltiple, que está modulado por ancho de pulso a través de un solenoide llamado **válvula electrónica reguladora de vacío (EVRV, por sus siglas en inglés)**, se utiliza para abrir la válvula para recirculación de los gases de escape (EGR, por sus siglas en inglés).

Electronically erasable programmable read only memory (EEPROM) A computer chip that may be easily erased and reprogrammed with special equipment.

Memoria de solo lectura borrable y programable electrónicamente (EEPROM) Pastilla de memoria que puede borrarse fácilmente y reprogramarse con equipo especial.

Element A liquid, solid, or gas containing only one type of atom.

Elemento Líquido, sólido o gas que contiene sólo un tipo de átomo.

Emission control system Components that make up a broad array of emission control devices on a vehicle.

Sistema de control de emisión Los componentes que incluyen un gran surtido de dispositivos de control de emisiones en un vehículo.

Enable criteria Conditions that must be met before the PCM can initiate a test of a component or system. These criteria are specific to the sensor or system being tested.

Criterio de aplicación Las condiciones que deben cumplirse antes de que el PCM pueda iniciar una prueba de un componente o un sistema. Estos criterios son específicos al sensor o al sistema que esta bajo prueba.

Energy The ability to do work.

Energía Capacidad para realizar un trabajo.

Energy conserving oil Engine oil that is specially formulated to reduce power losses from friction.

Aceite de conservación de energía Aceite de motor que es especialmente formulado para reducir la perdida de fuerza debido a la fricción.

Engine block Solid casting into which the cylinders are bored, also contains the necessary oil and coolant passages.

Bloque del motor Una pieza moldeada dentro de la cual se maquinan los cilíndros, tambien tiene los pasajes necesarios para el aceite y el líquido de enfriamiento.

Engine coolant temperature (ECT) sensor An input sensor that sends a voltage signal to the computer in relation to coolant temperature.

Sensor de la temperatura del refrigerante del motor Sensor de entrada que le envía una señal de tensión a la computadora referente a la temperatura del refrigerante.

Engine off natural vacuum (EONV) An EVAP monitoring system that runs pressure and vacuum tests based on the natural volatility of the fuel as the fuel warms up and cools off after driving.

Vacío natural de motor apagado (EONV) Un sistema monitor del EVAP que efectúa las pruebas de presión y de vacío basado en la volatilidad natural del combustible cuando éste se calienta o se enfría después de la marcha.

Enhanced EVAP An EVAP system used on OBD II in which the PCM conducts several tests to determine if the system is operational and there are no leaks.

EVAP intensificado Un sistema de EVAP usado en un OBDII en el cual el PCM efectúa varias pruebas para determinar si esta operacional el sistema y si no hay fugas.

Environmental Protection Agency (EPA) The federal government agency in charge of air and water quality in the United States.

Agencia para la Protección del Medio Ambiente (EPA) Agencia del gobierno federal que tiene a su cargo todos los aspectos relacionados a la calidad del agua y del aire en los Estados Unidos.

Ethanol A grain alcohol used as a gasoline additive.

Etanol Un alcohol de cereales que se usa como aditivo de gasolina.

Ethylene glycol The basic chemical used in automotive antifreeze.

Glicol de etileno Producto químico básico utilizado en anticongelantes de automóviles.

Evaporative emission control systems (EVAP) Emission systems that control evaporative emissions from the fuel tank.

Sistemas de control de emisiones de evaporación Sistemas de emisión que controlan las emisiones de evaporación del tanque del combustible.

Exhaust The by-product of combustion.

Emisión de vapor Un derivado de la combustión.

Exhaust gas recirculation (EGR) valve A valve that dilutes the intake charge with spent exhaust in order to cool combustion chamber temperatures.

Válvula de recirculación de gases del escape (EGR): Válvula que diluye la carga de entrada con los gases de escape ya utilizados para disminuir las temperaturas de la cámara de combustión.

Exhaust gas recirculation valve position (EVP) sensor A sensor that sends a voltage signal to the computer in relation to EGR valve position.

Sensor de la posición de la válvula de recirculación del gas del escape Sensor que le envía una señal de tensión a la computadora referente a la posición de la válvula EGR.

Exhaust gas temperature sensor An input sensor that sends a voltage signal to the computer in relation to exhaust gas temperature.

Sensor de la temperatura del gas del escape Sensor de entrada que le envía una señal de tensión a la computadora referente a la temperatura del gas del escape.

Exhaust headers An exhaust manifold designed to prevent interference between the exhaust pulses from the cylinders.

Colectores del escape Un múltiple de escape diseñado para prevenir la interferencia entre los impulses de escape de los cilíndros.

Exhaust valves Vales that open in the exhaust stroke of a 4-cycle engine.

Válvulas del escape Las válvulas que abren en la carrera de escape de un motor de 4 ciclos.

Expansion tank Part of the cooling system that allows high-pressure antifreeze to escape the system. This tank allows the antifreeze to leave and relieve the system of some pressure without spilling the antifreeze onto the ground or into the atmosphere.

Tanque de expansión Parte del sistema de enfriamiento que permite que el anticongelante de alta presión se escape del sistema. Este tanque permite que el anticongelante se salga y libere la presión del sistema sin derramar el anticongelante en el piso o en la atmósfera.

Extreme pressure resistance An oil additive that helps prevent high pressure from squeezing the oil out of the engine bearings.

Resistencia a presión extrema Aditivo de aceite que ayuda a evitar que la alta presión fuerce el aceite fuera de los cojinetes del motor.

Failure record Some manufacturers such as GM have provided more than one freeze frame known as "failure records." See **freeze frame**.

Registro de falla Algunos fabricantes, como GM, han proporcionado varias imágenes congeladas, que se denominan "registros de fallas". Consulte **imagen congelada**.

Fast-start EI system An EI system with the capability to start firing with less crankshaft rotation compared to other systems.

Sistema de encendido electrónico de arranque rápido Sistema de encendido electrónico que, a diferencia de otros sistemas, tiene la capacidad de encenderse con menor giro del cigüeñal.

Federal Test Procedure A transient-speed mass sampling emissions test conducted on a loaded dynamometer. This is the test used by manufacturers to certify vehicles before they can be sold.

Procedimiento de la Prueba Federal Una prueba de la velocidad transitoria para la prueba de la masa en un dinamómetro cargado. Esta es la prueba usada por los fabricantes para certificar los vehículos antes de que ellos puedan ser vendidos.

Feedback A term used to describe a PCM's ability to control the activity of an actuator in response to input from sensors.

Retroalimentación Un termino usado para describir la habilidad del PCM (módulo de control de la potencia del motor) de controlar la actividad de un actuador en respuesta a la información de los sensores de entra.

Field of flux The lines of force that are emitted by a magnet. The strength of this field decreases as it moves away from the magnet.

Campo del flujo La líneas de fuerza que son emitidas por un magneto. La fuerza de este campo disminuye según se mueve hacia afuera del magneto.

Filler pipe A pipe through which fuel flows when filling the fuel tank.

Tubo de llenado Tubo a través del cual fluye el combustible al llenarse el tanque del combustible.

Filler pipe vent hose A hose connected from the top of the fuel tank to the outer end of the filler pipe.

Manguera de alivio del tubo de llenado Manguera conectada desde la parte superior del tanque del combustible hasta el extremo exterior del tubo de llenado.

Final drive gear The final set of gears for the powertrain. These are typically fixed-torque multiplication gears and are in the differential assembly.

Engrane de conducción final El juego final del engrane para el tren de fuerza transmitida. Estos son los engranes para la multiplicación de torque/par e torsión aplicados fijamente y están en el ensamblaje del diferencial.

Firing line The vertical line on a secondary scope pattern. The top of this line indicates the voltage required to start the spark plug firing.

Línea del encendido Línea vertical en los modelos de radio secundarios. La parte superior de esta línea indica la tensión que se requiere para encender la bujía.

Firing order The order in which the cylinders of an engine are on their power stroke.

Orden del encendido El orden en el cual los cilindros de un motor están en su carrera de fuerza.

Flame front The outer edge of the immense heat buildup in a cylinder during combustion.

Frente de la flema La parte exterior del inmenso calor acumulado en un cilindro durante la combustión.

Flash codes Codes that could readily be obtained without the use of a scan tool, usually be flashing a lamp or LED on the dash or PCM itself. Usually available only on OBD I vehicles.

Códigos de destello Los códigos que se pueden obtener fácilmente sín ayuda de una herramienta explorador, normalmente destella un luz o un LED en el panel de instumentos o en el mismo PCM. Suele ser disponible sólo en los vehículos OBD I.

Flashing A term that is used to describe programming an OBD II PCM.

Destello Un término que se usa para describir la programación de un OBDII PCM.

Flat-rate manual A manual that lists suggested times for each repair operation on various vehicles.

Manual de tarifa única Manual que especifica la duración aproximada de una reparación según los diferentes modelos de vehículos.

Flathead or side valve An engine with the valves mounted in the block.

Válvula lateral o de cabeza plana Motor con las válvulas montadas en el bloque.

Flexplate or flywheel The component that is bolted to the crankshaft and connects the crankshaft to the torque converter and supports the ring gear.

Volante de motor Componente que se emperna al cigüeñal, lo conecta al convertidor de torsión y apoya la corona.

Flywheel A heavy wheel attached to an engine's crankshaft used to absorb crankshaft vibrations, mount the clutch assembly, and add inertia to the rotation of the crankshaft.

Volante del motor Una rueda pesada acoplada al cigüeñal del motor usado para absorber las vibraciones del cigüeñal, montada en el ensamblaje del embrague y agregarle inercia a la rotación del cigüeñal.

Four-stroke cycle An engine in which a complete cycle requires four piston strokes.

Ciclo de cuatro tiempos Motor en el que un ciclo completo requiere cuatro carreras del pistón.

Four-valve (4-V) cylinder head A cylinder head with four valves per cylinder.

Culata del cilindro con cuatro válvulas Culata del cilindro que tiene cuatro válvulas por cilindro.

Free-wheeling or spin-free engine An engine in which the valves do not contact the pistons if the timing chain or belt slips or breaks.

Motor de rueda libre o giro libre Motor en el que las válvulas no entran en contacto con los pistones si se rompen o se deslizan la cadena o la correa de regulación del encendido.

Freeze frame A feature of some scan tools and a requirement of OBD II. This feature takes a snapshot of the operating conditions present when the PCM sets a diagnostic trouble code.

Marco congelado Una característica de algunas herramientas exploratorias y un equipo de OBD II (diagnóstico abordo del vehículo II). Esta característica toma una foto de las condiciones de operaciones presente cuando el PCM (módulo de control de la potencia del motor) establece un código de diagnostico de problema.

Frequency The number of complete cycles that occur in a specific period of time.

Frecuencia El número de ciclos completos que ocurren en un periodo especifico de tiempo.

Friction The resistance to motion when one object is moved over another object.

Fricción Resistencia al movimiento cuando un objeto se mueve sobre otro.

Friction modifiers Oil additives that reduce friction in the oil and between moving components that contact each other to improve fuel economy.

Modificadores de fricción Aditivos de aceite que disminuyen la fricción en el aceite y entre componentes en movimiento en contacto el uno con el otro, a fin de mejorar la capacidad para economizar combustible.

Front passenger door module (FPDM) A computer module that is part of the vehicle network and located in the front passenger door that is responsible for power window and door lock operation and customization features.

Módulo de la puerta delantera del pasajero (FPDM) Un módulo de computadora que es parte del sistema red del vehículo y que se encuentra en la puerta del pasajero responsable de las operaciones de las ventanas de poder y las cerraduras de las puertas y las características personalizadas.

Fuel induction system Refers to the method and form of fuel delivery, such as throttle body injection or port fuel injection. Can also refer to a turbo or supercharging.

Sistema de inducción de combustible Refiera al método y forma de la entrega de combustible, tal como la inyección por reguladora o la inyección de compuerta de combustible. Puede referirse tambien al turbo o a la sobrealimentación.

Fuel pressure regulator A device designed to limit the amount of pressure built-up in a fuel delivery system.

Regulador de presión del combustible Un dispositivo diseñado para limitar la cantidad de acumulación de presión en un sistema de suministro de combustible.

Fuel pump impeller A vaned rotating impeller attached to the fuel pump motor shaft that creates fuel pressure.

Impulsor de la bomba del combustible Impulsor giratorio con paletas fijado al árbol motor de la bomba del combustible que produce la presión del combustible.

Fuel pump module It controls the operation of the fuel pump according to the demand on the system.

Módulo de control de la bomba de combustible Controla el funcionamiento de la bomba de combustible de acuerdo con la demanda del sistema.

Fuel pump relay A computer-operated relay that opens and closes the fuel pump relay.

Relé de la bomba del combustible Relé accionado por computadora que abre y cierra el relé de la bomba del combustible.

Fuel rail A metal or plastic pipe in which the upper end of the injectors is installed in port injection systems.

Carril del combustible Tubo metálico o plástico en el que se monta el extremo superior de los inyectores en sistemas de inyección con lumbrera.

Fuel return line type Fuel system that utilizes a return line.

Tipo con línea de retorno de combustible: Sistema de combustible que utiliza una línea de retorno.

Fuel system Fuel pump, tank, lines and evaporative emission system of a vehicle.

Sistema de combustible La bomba, el tanque, las líneas, y el sistema de emisión evaporativo de combustible de un vehículo.

Fuel system monitor A monitor that checks short- and long-term fuel corrections during closed loop.

Monitor del sistema de combustible Un monitor que verifica las correcciones de corto y largo plazo del combustible durante un circuito cerrado.

Fuel tank level sensor The input that the PCM uses to determine the fuel level in the fuel tank important for EVAP leak detection sensing.

Sensor del nivel del tanque de combustible La entrada que usa el PCM para determinar el nivel del combustible en el tanque que es importante en la detección de fugas del EVAP.

Fuel tank pressure control valve The valve designed to control fuel tank pressure while the vehicle is sitting still. If tank pressure builds too high the valve opens and lets the vapor into the canister.

Válvula de control de presión del tanque de combustible La válvula diseñanda para controlar la presión del tanque de combustible mientras que el vehículo queda imóbil. Si la presión del tanque sube demasiado, abre la válvula para dejar escapar el vapor al bote.

Fuel tank pressure sensor Informs the PCM of the pressure inside the fuel tank for EVAP system tests.

Sensor de presión del tanque de combustible Informa al PCM de la presión dentro del tanque de combustible para las pruebas del sistema EVAP.

Fuel vaporization Changing fuel from liquid to vapor.

Vaporización de combustible Cuando el combustible líquido se convierte en vapor.

Full-flow oil filter An engine oil filter in which all the flow from the pump flows through the filter.

Filtro del aceite de flujo completo Filtro del aceite de un motor a través del cual fluye todo el aceite de la bomba.

Gap The space in which the high-voltage current from the coil flows across.

Separación El espacio a través del cual fluye un corriente de alta voltage de la bobina.

Gear reduction A gear ratio that results in a reduction in speed and increase in torque, usually by a small gear driving a larger one.

Reducción de engranaje Una relación de engranaje que resulta en una reducción de velocidad y un incremento en torsión, normalmente en un engranaje pequeña que acciona una más grande.

Gear reduction starting motor A starting motor with a set of reduction gears between the armature and the drive.

Motor de arranque con desmultiplicación de engranajes Motor de arranque con un juego de engranajes de reducción entre la armadura y el mecanismo de mando.

Gear-type oil pump An oil pump with a meshed pair of gears that create oil pressure and move the oil through the lubrication system.

Bomba del aceite con engranajes Bomba del aceite con un par de engranajes endentados que producen la presión del aceite y lo conducen a través del sistema de lubrificación.

General repair manual A repair manual published by a company that is independent from the car manufacturers. This manual usually includes several years or models of vehicles.

Manual para reparaciones generales Manual de reparación publicado por una compañía no relacionada a los fabricantes de automóviles. En dicho manual se incluyen normalmente varios años o modelos de vehículos.

G-Lader supercharger A special design of supercharger in which the rotors rotate on an eccentric. The primary benefit of this design is the quietness of operation.

Supercargador con el sobre nombre de escalera G Un diseño especial de supercargador, en el cual los rotores giran en un excéntrico. El beneficio primario de este tipo de diseño, es que la operación es callada.

Glitch A momentary disruption in an electrical circuit.

Falla Una interrupción momentánea en un circuito eléctrico.

Glow plug A plug used in diesel engines to warm the intake air to allow for quicker starting and better operation when the engine is cold.

Precalentador para la cámara de combustión Una resistencia eléctrica usada en un motor diesel para calentar el aire de admisión, para permitirlo de que arranque más rápido y que tenga mejor operación cuando el motor está frío.

Grounded circuit An unwanted copper-to-metal connection in an electric circuit.

Circuito puesto a tierra Conexión no deseada entre cobre y metal en un circuito eléctrico.

Halogen and halon fire extinguishers Fire extinguishers that are effective on class B fires, but that give off toxic gases.

Extintores de incendio de halógeno y de halón Extintores de incendio que eficazmente apagan incendios de clase B, pero que emiten gases tóxicos.

Hazard Communication Standard The beginning of the right-to-know laws published by the Occupational Safety and Health Administration (OSHA).

Norma de Comunicación de Riesgo Comienzo de las leyes del derecho a saber; documento publicado por la Dirección para la Seguridad y Salud Industrial (OSHA).

Heat range A rating of a spark plug. This defines the plug's ability to dissipate heat. The hotter the heat range, the more heat it retains.

Rango de calor Una clasificación de una bujía. Esto define la habilidad de las bujías de disipar en calor. Lo más caliente que sea el rango, la mayor cantidad de calor que retiene.

Heated oxygen sensor A sensor that is electrically heated to accurately sense exhaust gases regardless of operating temperature.

Sensor de oxígeno calentado Un sensor que es calentado electrónicamente para detectar correctamente los gases del escape sin hacer caso a la temperatura de operación.

Helmholtz resonator There are so many tubes and containers under the hood going to the air cleaner. These are often called Helmholtz resonators because they are designed to counteract what are called Helmholtz oscillations.

Resonador de Helmholtz Tubos y recipientes situados debajo del capó que se conectan con el purificador de aire. Suelen llamarse resonadores de Helmholtz porque están diseñados para contrarrestar las denominadas oscilaciones de Helmholtz.

Hertz A unit of measurement for counting the number of times an electrical cycle repeats every second. One hertz is one pulse per second.

Hertz Una unidad de medida para contar los números de veces que un ciclo eléctrico se repite cada segundo. Un hertz es un pulso cada segundo.

High Energy Ignition (HEI) The name given to early GM electronic ignition systems.

Sistema de ignición de alta energía El nombre otorgado a los sistemas de inyección de la GM.

High-side drivers *See* feed-side drivers.

Impulsores de lado alto Véase impulsores de lado de alimentación.

Hold-in winding A winding in the starter solenoid that is energized while the solenoid is activated.

Devanado de retención Devanado en el solenoide de arranque que se energiza cuando se activa el solenoide.

Horsepower (hp) A rating that indicates engine power.

Potencia en caballos Clasificación que indica la potencia del motor.

Hot spark plug A spark plug designed with a longer heat path so that the electrodes operate at higher temperatures.

Bujía caliente Bujía diseñada con una trayectoria de calor más larga para que los electrodos funcionen a temperaturas más altas.

Hydrocarbons (HC) Left-over fuel after the combustion process.

Hidrocarburos El combustible restante después del proceso de combustión.

Idle air control (IAC) motor A computer-controlled motor that controls idle speed under all engine operating conditions.

Motor para el control de la marcha lenta con aire Motor controlado por computadora que controla la velocidad de la marcha lenta bajo cualquier condición del funcionamiento del motor.

Ignitable If a material burns when contacted by a spark, flame, or a certain degree of heat, it is ignitable.

Inflamable Si un material se quema al ser expuesto a una chispa, a una llama o a cierto grado de calor, se dice que dicho material es inflamable.

Ignition dwell time The number of degrees that the distributor shaft rotates while the primary circuit is closed prior to firing each cylinder.

Duración de retraso del encendido Número de grados que gira el árbol del distribuidor mientras el circuito primario se cierra antes de encender cada cilindro.

Ignition module An electronic device used to open and close the primary ignition circuit.

Módulo del encendido Dispositivo electrónico utilizado para abrir y cerrar el circuito primario del encendido.

Ignition timing Refers to the time, in relationship to piston movement, when spark ignition begins.

Tiempo de regulación de la ignición Se refiere al tiempo, en relación al movimiento del pistón, cuando la chispa de la ignición comienza.

Impeller The name given to one of the rotor or fin assemblies in a torque converter, turbocharger, and supercharger.

Propulsor El nombre otorgado a uno de los rotores o ensamblajes de las aletas en un convertidor de torque, turbo cargador o supercargador.

Impeller-type water pump A water pump in which a vaned impeller is driven to create pressure and force coolant through the cooling system.

Bomba del agua tipo impulsor Bomba del agua en la que se acciona un impulsor con paletas para producir presión y hacer circular el refrigerante a través del sistema de enfriamiento.

In-car temperature sensor A thermistor that informs the BCM or A/C controller of the in-car temperature. Used on automatic A/C systems only.

Sensor de temperatura interior Un termistor que informa al PCM o al controlador de A/C de la temperatura dentro del vehículo. Se usa solamente en los sistemas automáticas de A/C.

Induction The spontaneous creation of an electrical current in a conductor as the conductor passes through a magnetic field or a magnetic field passes across the conductor.

Inducción La creación espontanea de una corriente eléctrica en un conductor según el conductor atraviesa un campo magnético o un campo magnético atraviesa el conductor.

Inductive reluctance Refers to the opposition to current flow that takes place when a conductor is generating electricity as it is passed by a magnetic field.

Reluctancia inductiva Se refiere a la posición del flujo de la corriente que toma lugar cuando un conductor está generando electricidad según pasa atraves de un campo magnético.

Inertia The tendency of an object at rest to remain at rest or the tendency of an object to remain in motion.

Inercia La tendencia de un objeto inmóvil a permanecer inmóvil, o la tendencia de un objeto a permanecer en movimiento.

Inertia switch A switch used in some fuel pump circuits that opens when it is impacted by moderate collision force.

Conmutador por inercia Conmutador utilizado en algunos circuitos de bombas de combustible que se abre al ser impactado por una colisión moderada.

Information retrieval Refers to the microprocessor retrieving information from some of the computer memories.

Recuperación de información Se refiere a la recuperación de información que lleva a cabo el microprocesador de algunas de las memorias de la computadora.

Information storage Refers to the microprocessor sending information to some of the memories where it is stored for future reference.

Almacenamiento de información Se refiere al envío de información que lleva a cabo el microprocesador hacia algunas de las memorias; es aquí donde la información se guarda para uso futuro.

Infrared analyzer An analyzer used to measure oxygen, carbon monoxide, carbon dioxide, and hydrocarbons in the exhaust.

Analizador infrarrojo Analizador utilizado para medir el oxígeno, el monóxido de carbono, el bióxido de carbono, y los hidrocarburos en el escape.

Initial timing mode (also called base timing) Timing without any advance from the computer.

Modo de regulación de tiempo (tambien regulación de tiempo base) La regulación del tiempo sín avance ninguno de la computadora.

Injector sequencing Opening the injectors in the proper order to match the cylinder firing order.

Ordenamiento del inyector La apertura de los inyectores en una secuencia adecuada para equilibrar la secuencia del encendido del motor.

In-line block An engine block in which the cylinders are arranged in line with each other from the front to the back.

Bloque en línea Bloque de un motor en el que los cilindros están arreglados en relación el uno con el otro de la parte delantera a la trasera.

Input signal amplification The process of increasing input voltage signal power.

Amplificación de la señal de entrada Proceso de aumentar la potencia de tensión de la señal de entrada.

Input speed sensor Measures the speed of the transmission input shaft, used with the output speed sensor by the PCM to calculate gear ratio.

Sensor de velocidad de entrada Mide la velocidad del eje de entrada de la transmisión, usado por el PCM con el sensor de velocidad de salida para calcular la relación del engranaje.

Inspection/Maintenance (I/M) programs Programs that are administered by various states to monitor emission levels on vehicles.

Programas de inspección y mantenimiento Programas administrados por varios estados para controlar los niveles de emisiones provenientes de vehículos.

Insulator An element with five or more valence electrons that does not conduct electric current.

Aislador Elemento con cinco o más electrones de valencia que no conduce corriente eléctrica.

Intake Casting that contains passages for the introduction of the air-fuel mixture into the cylinder heads.

Entrada Pieza moldeada que contiene los pasajes para introducir la mezcla de aire y combustible dentro de las cabezas de los cilíndros.

Intake air temperature (IAT) Sensor *See* manifold air temperature (MAT) sensor.

Temperatura del aire de entrada (IAT) Véase manifold air temperature (MAT) sensor

Intake manifold tuning valve (IMTV) A computer-operated valve that changes the length of the intake manifold air passages.

Válvula de ajuste del colector de aspiración Válvula accionada por computadora que varía el largo de los tubos de aire del colector de aspiración.

Intake valves Valves that open during the intake stroke to allow the air-fuel mixture into the cylinders.

Válvulas de entrada Las válvulas que abren durante la carrera de entrada para permitir entrar la mezcla de aire-combustible dentro de los cilíndros.

Integrated circuit A silicon chip with many components such as diodes and transistors etched on it.

Circuito integrado Pastilla de silicio en la que se han grabado una variedad de componentes, como por ejemplo diodos y transistores.

Integrated circuit (IC) regulator A voltage regulator designed on a silicon chip, which is usually integral with the alternator.

Regulador del circuito integrado Regulador de tensión diseñado en una pastilla de silicio, que normalmente es un complemento del alternador.

Intercooler A device that cools the intake air on a supercharged or turbocharged engine.

Interenfriador Dispositivo que enfría el aire aspirado en un motor de compresor o de turbocompresor.

Intermediate pipe Part of the exhaust system, primarily used to connect different major parts of the system together.

Pipa intermedia Parte del sistema de escape, primariamente usado para acoplar diferentes partes mayores del sistema de escape.

Internal combustion An engine whose combustion occurs inside the engine.

Combustión interna Un motor cuyo combustión ocurre dentro del motor.

Ion sensing ignition An ignition system which uses the spark plug as a combustion sensor by placing a voltage across the spark plug electrode immediately after the spark plug firing event.

Encendido por deteccion De iones sistema de encendido que emplea una bujia como sensor de combustion a través de un voltaje en el electrodo de la bujia, inmediatamente después del evento de encendido de la bujia.

Job estimate A cost estimate for repair work given to the customer.

Estimado del servicio Precio estimado que se le ofrece al cliente sobre la reparación de un vehículo.

Keep alive memory (KAM) The microprocessor can read information from the KAM but it cannot write information into this chip. KAM memory is retained when the ignition switch is turned off, but it is erased when battery power is disconnected from the computer.

Memoria de entretenimiento (KAM) El microprocesador puede leer la información que contiene la KAM, pero no puede escribir información en esta pastilla. La memoria de entretenimiento se retiene cuando se desconecta el botón conmutador de encendido, pero se borra cuando la potencia de la batería se desconecta de la computadora.

Kilopascal Measurement of pressure in the metric system.

Kilopascal Medida de presión en el sistema métrico.

Knock sensor (KS) A sensor that sends a voltage signal to the computer in relation to engine detonation.

Sensor de golpeteo Sensor que le envía una señal de tensión a la computadora referente a la detonación del motor.

Lead (Pb) An element used in paste form on negative battery plates.

Plomo (Pb) Elemento utilizado en forma de pasta en placas negativas de baterías.

Lead peroxide (PbO₂) A chemical paste on positive battery plates in a fully charged battery.

Peróxido de plomo (PbO2) Producto químico en forma de pasta utilizado en placas positivas de baterías en una batería completamente cargada.

Lead sulfate (PbSO₄) A compound on the plates of a discharged battery.

Sulfato de plomo (PbSO4) Compuesto utilizado en las placas de una batería descargada.

Lean A condition where there is a larger amount of air in the air-fuel mixture than would be if the mixture was stoichiometric.

Pobre Una condición donde hay una cantidad mayor de aire en la mezcla de aire combustible que si hubiera una mezcla estequiometrica.

Leak detection pump (LDP) It pressurizes the system and checks for leaks, along with the fuel tank pressure sensor.

Bomba de detección de fugas La bomba de detección de fugas (LDP, por sus siglas en inglés) presuriza el sistema y comprueba si existen fugas, junto con el sensor de presión del tanque de combustible.

L-head design An engine with the valves mounted in the block with all the valves on one side of the cylinders.

Motor de válvulas al costado Motor con las válvulas montadas en el bloque; todas las válvulas se encuentran en un solo lado de los cilindros.

Light emitting diode (LED) A diode that gives off light when it conducts electric current.

Diodo emisor de luz Diodo que emite luz cuando conduce una corriente eléctrica.

Lighting control module (LCM) A module that controls the interior lighting. Function may be customizable by the driver.

Módulo de control de luz (LCM) Unmódulo que controla las luces interiores. Sus funciones se pueden personalizar por el conductor.

Linear EGR valve An EGR valve containing a solenoid that is pulsed on and off by the computer to provide a precise EGR flow.

Válvula EGR lineal Válvula EGR con un solenoide que la computadora enciende y apaga para proporcionar un flujo exacto de EGR.

Linear Hall-effect TP sensor It has the advantage of no contacts to wear or corrode and is even referred to as a noncontact TP sensor.

Sensor lineal de efecto Hall Ofrece la ventaja de carecer de contactos que puedan sufrir desgaste o corrosión; también se lo llama sensor TP sin contacto.

Linear TPS A TPS with a moveable contact on a horizontal or vertical resistance coil.

TPS lineal TPS con un contacto móvil en una bobina de resistencia horizontal o vertical.

Load A term used to describe the weight or force an engine must overcome or an electric device that consumes electricity.

Carga Un termino usado para describir el peso o la fuerza que un motor debe de superar o la electricidad que consume un dispositivo con un motor eléctrico.

Lobe An eccentric rise from a base circle. Typically refers to the shape of a camshaft.

Lóbulo Una elevación desde la base de un circulo. Típicamente se refiere a la figura de un árbol de levas.

Longitudinal engine mounting An engine mounted lengthways in the chassis.

Montaje longitudinal del motor Motor montado longitudinalmente en el chasis.

Long-term adaptive fuel trim Long-term fuel injector pulse width compensation determined by the PCM according to operating conditions. This fuel trim is set to maintain minimum emissions output and is the base point for short-term fuel trim.

Restricción del combustible por tiempo largo Compensación amplia de los pulsos del inyector de combustible determinado por el PCM (módulo de control de la potencia del motor) de acuerdo con las condiciones de operaciones. Esta restricción de combustible es calibrada para mantener las emisiones a un mínimo y es el punto de base para la restricción del combustible por tiempo corto.

Long-term fuel trim (LTFT) The long-term trends in fuel control. Positive numbers mean that fuel is being added; negative numbers mean that fuel is being subtracted from the long-term fuel calculation.

Eficiencia de combusitible de largo plazo (LTFT) La tendencia de control de combustible en largo plazo. Los números positivos quieren decir que se agrega el combustible, los números negativos reflejan que se substrae el combustible de la calculación de largo plazo.

Lookup tables Contained in control computers, these tables contain the specifications and calibrations for the powertrain and its systems.

Tabla de observación Contenido en los controles de computadoras, estas tablas contienen las especificaciones y calibraciones para el tren de fuerza transmitida y sus sistemas.

Loop network A continuous adjust and measure usually describing fuel control. The O_2 sensor measures the mixture; the PCM adjusts mixture based on this input, and then watches the O_2 sensor for the desired change, starting the process over.

Bucle Un ajuste y medida contínuo que suele describe el control del combustible. El sensor de O_2 mide la mezcla; el PCM ajusta la mezcla basado en estos resultados, y luego vigila al sensor de O_2 para efectuar los cambios requeridos, repitiendo el proceso.

Low-maintenance battery A battery designed to minimize electrolyte loss but water can be added to the battery.

Batería de bajo mantenimiento Batería diseñada para disminuir la pérdida de electrolitos, pero a la que se le puede agregar agua.

Low-side drivers Electronic drivers that control a circuit by switching ground on or off to a component. *See also* drivers.

Impulsores del lado bajo Los impulsores electrónicos que controlan un circuito prendiendo o apagando la tierra de un componente. Véase tambien impulsores.

Low-tension rings Piston rings designed with reduced tension on the cylinder walls to decrease friction and improve fuel economy.

Anillos de baja tensión Anillos de pistón diseñados con menos tensión en las paredes de los cilindros a fin de disminuir la fricción y mejorar la capacidad para economizar combustible.

LP-gas Liquefied petroleum gas, often referred to as propane, that burns clean in the engine and can be precisely controlled.

Gas LP Gas de petróleo liquido, comúnmente referido como propano, que quema limpio en el motor y puede ser precisamente controlado.

Magnetic field collapse Refers to magnetic field movement across conductors such as the ignition coil windings.

Derrumbe del campo magnético Se refiere al movimiento del campo magnético a través los conductores, como por ejemplo los bobinados del encendido.

Magnetic sensor A triggering device containing a winding and a permanent magnet.

Sensor magnético Dispositivo accionador que contiene un devanado y un imán permanente.

Main bearing bores Circular openings in the lower side of the engine block that contain the crankshaft main bearings and support the crankshaft.

Calibres de cojinetes principales Aberturas circulares en la parte inferior del bloque del motor que contienen los cojinetes principales del cigüeñal y que apoyan a éste.

Maintenance-free battery A battery to which water cannot be added because there are no filler caps in the top of the case. Special design reduces electrolyte loss.

Batería libre de mantenimiento Batería a la que no se le puede agregar agua ya que no hay tapones de llenado en la parte superior de la caja. Este diseño especial disminuye la pérdida de electrolitos.

Malfunction indicator light (MIL) A light in the instrument panel that is illuminated if a defect occurs in the computer system.

Luz indicadora de funcionamiento defectuoso Luz en el panel de instrumentos que se ilumina si ocurre una falla en la computadora.

Manifold absolute pressure (MAP) sensor An input sensor that sends a voltage signal to the computer in relation to intake manifold vacuum.

Sensor de la presión absoluta del colector Sensor de entrada que le envía una señal de tensión a la computadora referente al vacío del colector de aspiración.

Manufacturer's service manuals Manuals published by the car manufacturer for a specific vehicle.

Manuales de servicio del fabricante Manuales publicados por el fabricante de automóviles para un vehículo particular.

Mass airflow (MAF) sensor An input sensor that sends a voltage signal to the computer in relation to the total volume of air entering the engine.

Sensor del flujo de aire en masa Sensor de entrada que le envía una señal de tensión a la computadora referente al volumen total de aire que entra en el motor.

Mass, weight, and volume Mass is the measurement of an object's inertia. Weight is the measurement of the earth's gravitational pull on an object. Volume is the length, width, and height of a space occupied by an object.

Masa, peso, y volumen La masa es la medida de la inercia de un objeto. El peso es la medida de la fuerza gravitacional de la tierra sobre un objeto. El volumen es el largo, el ancho, y la altura del espacio ocupado por un objeto.

Master module A vehicle network component that is able to issue commands and initiate communications on the vehicle network.

Módulo maestro Un componente del sistema red del vehículo que es capaz de dictar los mandatos e iniciar las comunicaciones en el sistema red del vehículo.

Material Safety Data Sheets (MSDS) Sheets that provide information regarding hazardous materials.

Hojas de datos sobre la seguridad de un material Hojas de información sobre materiales peligrosos.

Maximum secondary coil voltage The maximum voltage that the coil is capable of producing.

Tensión máxima de la bobina secundaria Tensión máxima que la bobina es capaz de generar.

Mechanical efficiency The relationship between the engine power delivered and the power that would be delivered if the engine operated without any power loss.

Rendimiento mecánico Relación entre la potencia generada por el motor y la potencia que podría generarse si el motor funcionase sin pérdida de potencia alguna.

Metering The process of controlling the flow of something. Metering of gasoline is controlled by the time it is allowed to flow, by the size of the opening it is flowing through, or by the pressure causing it to flow.

Regulación El proceso de controlar el flujo de algo. La regulación de la gasolina es controlada por el tiempo que es permitida que fluya, por el tamaño de la apertura que está fluyendo, o por la presión que la está causando que fluya.

Methanol The lightest and simplest of the alcohols; also known as wood alcohol.

Metanol El más liviano y el más simple de los alcoholes; también conocido como alcohol de madera.

Microprocessor The decision-making chip in a computer.

Microprocesador Pastilla sobre la cual están implementadas las funciones aritméticas y lógicas de una computadora.

Microprocessor control unit (MCU) A term applied to some computer-controlled systems.

Unidad de control del microprocesador Término aplicado a algunos sistemas controlados por computadora.

Mid-engine transverse engine mounting An engine mounted crossways near the center of the vehicle.

Montaje transversal central del motor Motor montado transversalmente cerca del centro del vehículo.

Mild hybrid A hybrid vehicle that has an internal combustion engine and electric motor. The internal combustion engine runs to charge the batteries for the electric motor. The electric motor only runs when the internal combustion engine needs assistance.

Híbrido moderado Un vehículo híbrido que tiene un motor de combustión interno y un motor eléctrico. El motor de combustión interno esta en marcha para cargar las baterías para el motor eléctrico. El motor eléctrico sólo funciona cuando el pequeño motor interno de combustión requiere asistencia.

Molecule The smallest particle of a compound.

Molécula La partícula más pequeña de un compuesto.

Momentum An object gains momentum when a force overcomes static energy and moves the object.

Impulso Un objeto cobra impulso cuando una fuerza supera la energía estática y mueve el objeto.

Monolithic-type catalytic converter A catalytic converter in which a honeycomb-type element is coated with the catalyst materials.

Convertidor catalítico tipo monolítico Convertidor catalítico en el que un elemento parecido a un panal está cubierto de materiales catalizadores.

Muffler A part of the exhaust system in which the pulses from the exhaust are dampened and the noise reduced.

Silenciador Una parte del sistema del escape en el cual los pulsos desde el escape son amortiguados y el sonido reducido.

Multiple-disc clutch *See* Clutch pack.

Embrague de discos múltiples Vea clutch pack (páguete de embrague)

Multiport fuel injection (MFI) An electronic fuel injection system in which the injectors are grounded in the computer in pairs or groups of three or four.

Inyección de combustible de paso múltiple Sistema de inyección electrónico de combustible en el que los inyectores se ponen a tierra en la computadora en pares o en grupos de tres o cuatro.

Multipurpose dry chemical fire extinguisher A common type of extinguisher that may be used on various types of fires.

Extinctor de incendio de producto químico seco para aplicaciones múltiples Tipo común de extinctor que puede utilizarse en varios tipos de incendios.

Multiviscosity oil A chemically modified oil that has been tested for viscosity at cold and hot temperatures.

Aceite de viscosidad múltiple Un aceite modificado químicamente que ha sido probado por viscosidad a temperaturas frías y calientes.

Negative pressure A pressure less than atmospheric pressure.

Presión negativa Presión más baja que la presión atmosférica.

Negative temperature coefficient (NTC) A thermistor that reduces its resistance as temperature decreases.

Coeficiente de temperatura negativa (NTC) Un termistor que disminuye su resistencia cuando baja la temperatura.

Neutral drive switch (NDS) A switch that informs the computer regarding gear selector position.

Conmutador de mando neutral Conmutador que le advierte a la computadora sobre la posición del selector de velocidades.

Neutral safety switch A switch connected in the starter circuit that prevents the starting motor from operating except in park or neutral.

Conmutador de seguridad neutral Conmutador conectado en el circuito de arranque que evita que el motor de arranque funcione si el selector de velocidades no se encuentra en las posiciones PARK o NEUTRAL.

Neutron A particle with no electric charge positioned in the nucleus of an atom.

Neutrón Partícula desprovista de carga eléctrica ubicada en el núcleo de un átomo.

Noncompressibility A material with the capability to remain the same size when pressure is applied.

No compresibilidad Un material que tiene la capacidad de retener su tamaño al aplicársele presión.

OBD On-board diagnostics, which refers to the self-diagnostic capabilities of the vehicle's computer control system.

OBD: Diagnóstico de a bordo; hace referencia a las funciones de autodiagnóstico del sistema de control de la computadora del vehículo.

OBD I First generation on-board diagnostics, used from approximately 1979 to 1996.

OBD I Los diagnósticos abordo de primera generación, usados aproximadamente del 1979 al 1996.

OBD II Second generation on-board diagnostics mandated for use since 1996.

OBD II Los diagnósticos abordos de segunda generación usados bajo mandato desde 1996.

Occupational Safety and Health Administration (OSHA) The federal government agency in the United States in charge of safety and healthful working conditions.

Dirección para la Seguridad y Salud Industrial Agencia del gobierno federal en los Estados Unidos que tiene a su cargo el establecimiento de condiciones seguras y saludables en el trabajo.

Octane A rating used to classify gasoline, refers to the volatility of the fuel.

Octano Una clasificación usada para clasificar la gasolina, se refiere a que tan volátil es el combustible.

Octane number A unit of measurement on a scale intended to indicate the tendency of a fuel to detonate or knock.

Número de octano Una unidad de medida en una escala con la intención para indicar la tendencia de un combustible de detonar o producir golpe de chispa.

Ohm A measurement for electrical resistance.

Ohmio Medida de resistencia eléctrica.

Ohmmeter The device used to measure the resistance in an electrical circuit.

Ohm's Law The basic law of electricity.

Ley de Ohm La ley básica de la electricidad.

Ohmímetro Un dispositivo que se usa para medir la resistencia en un circuito eléctrico.

Oil classifications Various oil categories in relation to the type of use.

Clasificaciones de aceite Clases de aceite divididos en diferentes categorías de acuerdo a su utilización.

Oil pump pickup A line from the oil pump to the oil in the pan.

Tubo de la bomba de aceite Una línea desde la bomba de aceite al aceite en el cárter.

Oil rings Usually a single ring positioned below the compression rings on the piston to control the amount of oil on the cylinder wall.

Anillos de aceite Normalmente un solo anillo colocado debajo de los anillos de compresión en el pisión para controlar la cantidad de aceite en la pared del cilindro.

Oil viscosity ratings Various oil categories in relation to the oil's ability to flow.

Clasificaciones de la viscosidad del aceite Clases de aceite divididos en diferentes categorías de acuerdo a su capacidad de flujo.

Oil-wetted, resin-impregnated pleated paper element A type of air cleaner element.

Elemento de papel plegado, impregnado de resina y saturado de aceite Tipo de elemento del filtro de aire.

On-board diagnostics II (OBD II) *See* OBD II.

Diagnóstico instalado en el vehículo Sistema diagnóstico bajo mandato de la Comisión para Recursos del Aire de California (CARB) y otras agencias estadounidenses; dicho sistema enciende la luz MIL si los niveles de emisión superan 1,5 veces el límite especificado. El sistema diagnóstico deberá instalarse en vehículos fabricados a partir del año 1996.

On-board refueling vapor recovery (ORVR) It has been discovered that more HC emissions were released when refueling a vehicle than were released by burning a tank of fuel! This has lead to a redesign of some components in the fuel filler neck and tank assembly to help prevent these emissions.

Sistema de recuperación de vapores en la carga de combustible a bordo Se ha descubierto que son mayores las emisiones de HC al cargar combustible en un vehículo que al quemar un tanque de combustible. Esto ha llevado a modificar el diseño de algunos componentes del montaje del tanque y la boca de llenado de combustible para ayudar a evitar estas emisiones.

Open circuit An unwanted break in an electric circuit.

Circuito abierto Interrupción no deseada en un circuito eléctrico.

Open loop A computer system operating mode in which the computer ignores the oxygen sensor signal as it controls the air-fuel ratio.

Bucle abierto Modo de funcionamiento de una computadora en el que se pasa por alto la señal del sensor de oxígeno mientras se controla la relación de aire y combustible.

Opposed-type block An engine block designed with the cylinders opposite each other.

Bloque tipo opuesto Diseño de bloque de motor en el que los cilindros se encuentran en lados opuestos.

Optical-type pickup A pickup assembly containing a photo diode and a light emitting diode with a slotted plate rotating between these components.

Capacitación tipo óptico Conjunto de capacitación que contiene un fotodiodo y un diodo emisor de luz entre los cuales gira una placa ranurada.

Output driver A transistor in the output area of a control device that is used to turn various output devices off and on.

Ejecutador de salida Un transistor en el área de salida de un dispositivo de control, que es usado para apagar y prender varios dispositivos de ejecución.

Output drivers and actuators Computer system outputs including relays and solenoids.

Excitadores y accionadores de salida Datos producidos por una computadora que incluyen relés y solenoides.

Output driver modules Computer uses this to activate a circuit or actuator.

Módulo controlador de salida Permite a la computadora activar un circuito o accionador.

Output speed sensor Measures the speed of the transmission output shaft, used with the output speed sensor by the PCM to calculate gear ratio.

Sensor de velocidad de salida Mide la velocidad del eje de salida de la transmisión, usado por el PCM con los sensores de la salida de velocidad para calcular la relación de los engranajes.

Overdrive A gear ratio that outputs less torque but more speed than the input, such as a large wheel driving a smaller wheel.

Sobremarcha Una relación de engranaje que produce menos torsión pero más velocidad que la entrada, tal como una rueda grande que acciona una rueda más pequeña.

Overhead cam engine An engine with the camshaft located above or in the cylinder head.

Motor con árbol de levas superpuesto Motor con el árbol de levas ubicado sobre o en la culata del cilindro.

Overhead camshaft (OHC) This camshaft operates the valves directly eliminating rocker arms and pushrods.

Árbol de levas en cabeza (OHC) Este árbol de levas opera las válvulas directamente eliminando el balancín empuja-válvulas y las varillas empujadoras

Overhead valve (OHV) The valve uses rocker arms and pushrods to operate the valves.

Válvula en cabeza (OHV) La válvula usa un balancín empujaválvulas y las varillas empujadoras para operar las válvulas.

Overhead valve engine An engine with the valves positioned in the cylinder head.

Motor con válvulas superpuestas Motor con las válvulas colocadas en la culata del cilindro.

Overrunning clutch drive A starter drive that connects and disconnects the armature and the ring gear.

Mando del embrague de rueda libre Mecanismo de mando de arranque que conecta y desconecta la armadura y la corona.

Oxidation catalyst A catalyst that changes hydrocarbons and carbon monoxide to carbon dioxide and water vapor.

Catalizador del oxidación Un catalizador que cambia los hidrocarburos y el monóxido de carbono al anhídrido carbónico y el vapor de agua.

Oxidation inhibitors Oil additives that reduce sticky tar-like substances in the oil caused by oxidation of the oil when oxygen in the air combines with hot oil.

Inhibidores de oxidación Aditivos de aceite que disminuyen sustancias pegajosas, parecidas al alquitrán. Dichas sustancias se encuentran en el aceite y son el producto de la oxidación del aceite cuando el oxígeno en el aire se mezcla con el aceite caliente.

Oxides of nitrogen (NO_x) A gas formed by the combining of oxygen and nitrogen at high combustion temperatures in the engine cylinders.

Oxidos de nitrógeno (NOx) Gas formado por la combinación de oxígeno y nitrógeno a altas temperaturas de combustión en los cilindros del motor.

Oxidize Adding oxygen to a compound by burning. Refers to catalytic converter operation.

Oxidar Agregando el oxígeno a una compuesta por medio de la combustión. Refiere a la operación del convertidor catalítico.

Oxygen (O_2) A gaseous element that is present in air.

Oxígeno (O_2) Elemento gaseoso presente en el aire.

Oxygen (O_2) sensor An input sensor that sends a voltage signal to the computer in relation to the amount of oxygen in the exhaust stream.

Sensor de oxígeno (O2) Sensor de entrada que le envía una señal de tensión a la computadora referente a la cantidad de oxígeno en el caudal del escape.

Oxygenate A fuel additive that adds oxygen as the fuel burns, notably ethanol.

Oxigenador Aditivo del combustible que agrega oxígeno a medida que se quema combustible, en especial, etanol.

Paraffin A wax substance in the middle distillates of crude oil.

Parafina Una substancia de cera en los destilados intermedios del aceite en crudo.

Parallel circuit In this type of circuit, there is more than one path for the current to follow.

Circuito paralelo En este tipo de circuito, hay más de un camino para que fluya la corriente.

Parallel hybrid A hybrid vehicle that uses both the electric motor and the internal combustion engine to power the drive wheels. The electric motor in the parallel system is small and the internal combustion engine is the primary power source.

Híbrido paralelo Un vehículo híbrido que usa un motor eléctrico y tambien un motor de combustión interno para propulsar las ruedas motrices. El motor eléctrico en el sistema paralelo es pequeño y usa principalmente el motor de combustión internos para proporcionar la fuerza primaria.

Parameter identification (PID) A mode that allows access to certain data.

Identificación de parametros (PID) Un modo que permite acceso a ciertos datos.

Park neutral position (PNP) switch informs the computer regarding gear selector position.

Posición neutral/de estacionamiento Envía a la computadora información relacionada con la posición del selector de cambios.

Particulates Part of the exhaust from a diesel engine. Typically, particulates are caused by the melting of the paraffin present in diesel fuel.

Partículas Parte del escape de un motor diesel. Típicamente estas partículas son causadas por el derretimiento de la parafina presente en el combustible diesel.

Parts requisition An order requesting the necessary parts for a vehicle.

Pedido de piezas Encargo de las piezas necesarias para un vehículo.

Passive anti-theft security system (PATS) An anti-theft system that requires no action from the driver to activate.

Sistema pasiva de seguridad anti-robo (PATS) Un sistema de anti robo que no requiere una acción del conductor para activarse.

Peak-and-hold injectors Fuel injectors that are controlled by applying a relatively high current for a very short time, then reducing the current after the injector is opened. Based on the fact that it is more difficult to open the injectors than it is to keep them open.

Inyectores de pico y retención Los inyectores de combustible que se controlan al aplicarse un corriente relativamente alto por un tiempo muy corto, y luego reduciendo el corriente después de que se abre el inyector. Se base en el hecho de que es más difícil abrir los inyectores que mantenerlos abiertos.

Peak inverse voltage (PIV) The maximum voltage blocked by a diode.

Tensión inversa de picado (PIV) El voltaje máximo que puede ser bloqueado por un diodo.

Pellet-type catalytic converter A converter that contains small pellets coated with catalyst materials.

Convertidor catalítico tipo grano gordo Convertidor que contiene pequeños granos gordos cubiertos de materiales catalizadores.

Periodic motor vehicle inspection (PMVI) *See* Inspection and Maintenance.

Inspección periódico del motor del vehículo (PMVI) Vease Inspection and Maintenance.

Photo diode A diode that produces a voltage signal when light shines on it.

Fotodiodo Diodo que produce una señal de tensión cuando es iluminado por la luz.

Photoelectric sensor A sensor that uses a LED, photo cell, and a slotted disc to measure rotational speeds.

Sensor fotoeléctrico Un sensor que usa un LED (diodo emisor de luz o electroluminiscente,) foto celda y un disco ranurado para medir la velocidad de las rotaciones.

Pickup coil A winding and permanent magnet assembly mounted in the distributor and used for ignition triggering.

Bobina de captación Conjunto de devanado e imán permanente montado en el distribuidor y utilizado para el arranque del encendido.

Piezoresistive Sensors that measure pressure changes.

Piezoresistente Los sensores que miden los cambio de presión.

Pilot bearing A bearing inserted in the flywheel to support and guide the movement of the transmission's input shaft.

Balero piloto Un balero insertado en el volante para soportar y guiar el movimiento del eje de entrada de la transmisión.

Pilot bushing A bearing located in the recess in the end of the crankshaft.

Buje de piloto Un buje ubicado en la muesca en la extremidad del cigüeñal.

Pintle The center pin used to control a fluid passing through a hole; a small pin or pointed shaft used to open or close a passageway.

Pivote central La clavija central usada para controlar un flúido pasando atraves de un orificio; una clavija pequeña o un eje con una punta usado para abrir o cerrar un pasaje.

PIP or Profile Ignition Pickup Usually refers to a Ford product. The crank position sensor signal sent to the PCM.

PIP o Captador Perfíl del Encendido Suele referirse a un producto Ford. El señal del sensor de posición del cigüeñal mandado al PCM.

Piston boss The area around the piston pin bore.

Cubo de pistón El área alrededor del calibre del pasador de pistón.

Piston pin bore The circular opening in the piston in which the piston pin is located.

Calibre del pasador de pistón Abertura circular en el pistón en la que se encuentra el pasador del pistón.

Piston skirt The area of the piston below the rings.

Faldilla del pistón El área del pistón debajo de los anillos.

Piston stroke Piston movement from top dead center to bottom dead center.

Carrera del pistón Movimiento del pistón desde el punto muerto superior hasta el punto muerto inferior.

Pistons The cylindrical casting that moves up and down in the cylinder, powered by combustion.

Pistones El cilíndro moldeado que viaja arriba y abajo en el cilíndro, accionado por la combustión.

Pleated paper fuel filter element The cleaning element in some fuel filters.

Elemento de papel plegado del filtro del combustible Elemento depurador en algunos filtros de combustible.

Points Breaker point ignition systems used a mechanical switch called points riding on lobes on the distributor shaft to open and close the primary coil connection. The points were prone to wear and required frequent replacement.

Platinos Los sistemas de encendido por contacto utilizaban interruptores mecánicos llamados platinos o puntos, situados sobre los lóbulos del eje del distribuidor, para abrir y cerrar la conexión de la bobina principal. Los platinos eran susceptibles al desgaste y debían ser cambiados con frecuencia.

Polyurethane air cleaner element cover A circular polyurethane ring surrounding the air cleaner element for improved cleaning.

Cubierta de poliuretano del elemento del filtro de aire Anillo circular de poliuretano que rodea el elemento del filtro de aire para facilitar la limpieza.

Poppet nozzles Mechanically operated nozzles in the intake ports of a central port injection system.

Toberas de movimiento vertical Toberas accionadas mecánicamente en las lumbreras de aspiración de un sistema de inyección de lumbrera central.

Port fuel injection (PFI) A fuel injection system with an injector mounted in each intake port.

Inyección de combustible de lumbrera Sistema de inyección de combustible que tiene un inyector montado en cada una de las lumbreras de aspiración.

Positive crankcase ventilation (PCV) A system that moves crankcase vapors into the intake manifold rather than having them escape to the atmosphere.

Ventilación positiva del cárter Sistema que conduce los vapores del cárter hacia el colector de aspiración en vez de permitir que los mismos se escapen hacia la atmósfera.

Positive pressure A pressure greater than atmospheric pressure.

Presión positiva Presión más alta que la presión atmosférica.

Positive temperature coefficient (PTC) A thermistor that increases its resistance as temperature increases.

Coeficiente de temperatura positiva (PTC) Un termistor que aumenta su resistencia cuando sube la temperatura.

Post-combustion control A system that cleans up exhaust gas left by burning fuel.

Control de la poscombustión Un sistema que quema el combustible así limpiando el gas del escape

Potentiometer A type of resistor that has three connections.

Potenciómetro Un tipo de resistor que tiene tres conexiones.

Power The measurement of the rate at which work is done.

Potencia Medida de la capacidad a la que se realiza un trabajo.

Power door module (PDM) A networked module that controls the power lock and window function for one particular door.

Módulo de puerta de poder (PDM) Un módulo que es parte de un sistema que controla las operaciones de las cerraduras de poder y la ventana de una puerta específica.

Power module An older term used by Chrysler Corporation for an engine computer that worked with a logic module.

Módulo de potencia Término que antiguamente utilizaba la Chrysler Corporation para referirse a una computadora de motor que funcionaba con un módulo lógico.

Power (watt) rating A battery rating indicating the total amount of electrical power a battery will deliver.

Clasificación de potencia (watio) Clasificación de una batería que indica la cantidad total de energía eléctrica que la misma puede generar.

Power source Generally the battery and/or alternator.

Fuente de poder Generalmente la batería y/o el alternador.

Powertrain control module (PCM) Common term in SAE J1930 terminology for an automotive engine computer.

Módulo del control del tren transmisor de potencia Término utilizado comúnmente por la SAE J1930 para referirse a la computadora del motor de un automóvil.

Pre-combustion control A system that reduces potential emission prior to the air-fuel combustion.

Control del precombustión Un sistema que disminuye la emission potencial antes de la combustión del aire con combustible.

Pre-ignition The ignition of the air-fuel mixture before the spark occurs.

Pre encendido El encendido de la mezcla de aire-combustible antes de que ocurre la chispa.

Pressure A force exerted on a surface area.

Presión Una fuerza aplicada en la superficie de una área.

Pressure control solenoid The solenoid in an automatic transmission that controls transmission line pressure according to commands from the PCM.

Solenoide de control de presión El solenoide en una transmisión automática que controla la presión de la línea de transmisión según los órdenes del PCM.

Pressure feedback electronic (PFE) sensor An input sensor that sends a voltage signal to the computer in relation to the exhaust pressure in a chamber under the EGR valve.

Sensor electróncio de realimentación de presión Sensor de entrada que le envía una señal de tensión a la computadora referente a la presión del escape en una cámara debajo de la válvula EGR.

Pressure regulator A mechanical device that controls fuel pressure in an electronic fuel injection system.

Regulador de presión Dispositivo mecánico que controla la presión del combustible en un sistema de inyección electrónica de combustible.

Pressure transducer A vacuum switching device operated by exhaust pressure that opens and closes the vacuum supply to the EGR valve.

Transconductor de presión Dispositivo de conmutación de vacío accionado por la presión del escape que le abre y cierra el suministro de vacío a la válvula EGR.

Primary circuit The low-voltage circuit of an ignition system.

Circuito primario El circuito de bajo voltaje en un sistema de ignición.

Primary coil winding An ignition coil winding with a few hundred turns of heavy wire.

Bobinado primario Bobinado del encendido enrollado con varios cientos de vueltas de alambre pesado.

Profile ignition pickup The signal sent to the ignition module and PCM concerning crankshaft speed and position.

Captador de perfíl del encendido El señal mandado al módulo del encendido y al PCM con respeto a la velocidad y posición del ciguañal.

Programmable read only memory (PROM) A computer chip containing some of the computer program. This chip is removeable in some computers.

Memoria de solo lectura programable (PROM) Pastilla de memoria que contiene parte del programa de la computadora. Esta pastilla es desmontable en algunas computadoras.

Protocol The language or method used by a PCM to communicate with other computers.

Protocolo El lenguaje o método usado por una PCM (módulo de control de la potencia del motor) para comunicarse con las otras computadoras.

Proton A positively charged particle located in the nucleus of an atom.

Protón Partícula con carga positiva ubicada en el núcleo de un átomo.

Pull-in winding A winding in the starter solenoid that is energized while the starter is engaging.

Devanado energizado Devanado en el solenoide de arranque que se energiza cuando se acopla el motor de arranque.

Pulse A voltage signal that increases from a constant value and then decreases back to its original value.

Pulso Una señal de voltaje que incrementa desde un valor constante y después disminuye de regreso a su valor original.

Pulse rate The number of pulses that takes place over a specific period of time.

Velocidad del pulso Los números de pulso que toman lugar sobre un periodo de tiempo especifico.

Pulse transformer An electrical device that transforms or changes low voltage into high voltage on a pulse basis.

Transformador del pulso Un dispositivo eléctrico que transforma o cambia el voltaje bajo a voltaje alto en una base de pulso.

Pulse width The duration from the beginning to the end of a signal's on time or off time.

Amplitud del pulso La duración desde el principio al final de una señal prendida y apagada.

Pulse width modulation (PWM) The signal that is pulsed for a percentage of a fixed pulse width.

Modulación de la anchura del impulso El señal que pulsa por un porcentaje de una anchura de impulso fijo.

Purge solenoid A purge valve that is opened electrically by the PCM to purge fuel vapors.

Purga del solenoide Una válvula de purga abierta electrónicamente por el PCM para expulsar los vapores del combustible.

Pyrometer A tool used to measure the temperature of something.

Pirómetro Una herramienta usada para medir la temperatura de algo.

Quad drivers A set of four output drivers in a computer.

Conjunto de cuatro accionadores Un grupo de cuatro accionadores de salida en una computadora.

Quick-disconnect fuel line fittings Fuel line fittings that may be disconnected without using a wrench.

Conexiones de la línea del combustible de desmontaje rápido Conexiones de la línea del combustible que se pueden desmontar sin la utilización de una llave de tuerca.

Radial ply A type of tire that has body cords that extend from bead to bead at an angle of about 90 degrees.

Placas radiales Un tipo de neumático que tiene un cuerpo de cordón que se extiende desde un borde al otro borde en un ángulo de 90 grados.

Radiation A method of heat transfer that occurs when heat waves travel through the atmosphere and strike another object.

Radiación Método de transferencia de calor que occure cuando ondas de calor viajan a través de la atmósfera y chocan con otro objeto.

Radiator cap sealing gasket A gasket that seals the radiator cap to the radiator filler neck.

Guarnición de estanqueidad de la tapa del radiador Guarnición que sella herméticamente la tapa del radiador al cuello de llenado del radiador.

Radiator cap vacuum valve A valve that opens and allows coolant to flow from the coolant recovery container to the radiator when the engine coolant temperature decreases. This valve prevents radiator hose collapse.

Válvula de vacío de la tapa del radiador Válvula que se abre para permitir que el refrigerante fluya desde el recipiente de recuperación de refrigerante hasta el radiador cuando la temperatura del refrigerante del motor disminuye. Esta válvua evita que la manguera del radiador se plegue.

Random access memory (RAM) A computer chip that the microprocessor can write information into and retrieve information from.

Memoria de acceso aleatorio (RAM) Pastilla de memoria en la que el microprocesador puede escribir y de la que puede leer información.

Raster scope pattern A scope pattern in which all the voltage traces from the cylinders are displayed one above the other on the screen.

Modelo de radio de rastreo Modelo de radio en el que las pequeñas cantidades de tensión provenientes de los cilindros se proyectan una sobre otra a través de la pantalla.

Reactive The capability of a material to react violently when it comes in contact with another material.

Reactivo Capacidad de un material para producir una reacción violenta cuando entra en contacto con otro material.

Read only memory (ROM) The microprocessor can read information from the ROM chip but it cannot write information into this chip.

Memoria de sólo lectura (ROM) El microprocesador puede leer la información que contiene la pastilla ROM, pero no puede escribir información en la misma.

Recovery tank Part of the cooling system in which coolant spillage or excess is collected to be reused when the system is low.

Tanque de recuperación Parte del sistema de enfriamiento en el cual el derramamiento del anticongelante o exceso es coleccionado para ser usado nuevamente cuando el sistema está bajo.

Recycled oil Waste oil that has been reconditioned to original standards.

Aceite reciclado Residuos de aceite que han sido reacondicionados a su estado original.

Reduced In a catalytic converter, the splitting of oxides of nitrogen into nitrogen and oxygen gas.

Reducido En un convertidor catalítico, la división de los óxidos de nitrógeno a gases de nitrógeno y de oxígeno.

Reducing catalyst The section of the catalytic converter that separates oxygen from nitrogen.

Catalizador de reducción La parte del convertidor catalítico que separa el oxígeno del nitrógeno.

Reference pickup A type of pickup assembly that is often used for ignition triggering.

Captación de referencia Tipo de conjunto de captación que se utiliza con frecuencia para el arranque del encendido.

Reference voltage (Vref) sensors A voltage provided by a voltage regulator to operate potentiometers and other sensors at a constant level.

Voltaje de referencia Un voltaje suministrado por un regulador de voltaje para operar un potenciómetro y otros sensores a un nivel constante.

Reflashing A term used to describe the reprogramming of a PCM, usually to correct a problem repaired by updated software.

Redestello Un término que se usa para describir la reprogramación de un PCM, normalmente para corregir un problema que se puede reparar instalando las programas más nuevas.

Reference A term used by some manufacturers to describe the rpm signal into the PCM.

Referencia Un término usado por algunos fabricantes para describir la señal de rpm en el PCM

Reference low A term used to describe the ground connection between the ignition module and PCM to ensure both are at the same ground plane.

Baja referencia Un término que se usa para describir la conexión a tierra entre el módulo del encendido y el PCM para asegurar que ambos estan en el mismo plano a tierra.

Refueling vapor recovery The act of containing hydrocarbon vapors that could be released during refueling operations.

Recobro de vapor de reaprovisión de combustible El acto de contener los vapores de hidrocarburo que se pueden escapar durante las operaciones de reaprovisión de combustible.

Relay An electrical device used to switch a high-current circuit with a low-current circuit.

Relé Un dispositivo eléctrico usado para cambiar una corriente alta con un circuito de corriente baja.

Reluctor ring A metal ring that is integral with the crankshaft; slots on this ring are used for ignition triggering.

Anillo de reluctancia Anillo de metal que es un complemento del cigüeñal; las ranuras de este anillo se utilizan para el arranque del encendido.

Remote air cleaner An air cleaner mounted separately from the engine.

Filtro remoto de aire Filtro de aire montado lejos del motor.

Repair order An order written by the service writer or shop foreman, detailing the repairs to be completed on a customer's vehicle.

Solicitud de reparación Solicitud que llenan el mecánico o el encargado del taller donde se especifican las reparaciones que se llevarán a cabo en el vehículo de un cliente.

Required secondary coil voltage The secondary coil voltage required to fire the spark plugs with the engine operating at idle or low speed.

Tensión de la bobina secundaria normal requerida Tensión de la bobina secundaria que se requiere para encender las bujías durante el funcionamiento del motor a una velocidad de marcha lenta o baja.

Reserve capacity rating A battery rating that indicates the length of time a battery will deliver current with a 25-ampere load.

Clasificación de capacidad en reserva Clasificación de una batería que indica el espacio de tiempo que una batería podrá generar corriente con una carga de 25 amperios.

Reserve secondary coil voltage The difference between normal required secondary coil voltage and maximum secondary coil voltage.

Tensión de la bobina secundaria en reserva La diferencia entre la tensión de la bobina secundaria normal requerida y la tensión máxima de la bobina secundaria.

Resonator Part of some exhaust systems. Best described as a backup muffler. The resonator reduces the noise level of the exhaust after it has passed through the muffler.

Resonador Una parte de algunos sistemas de escape. Mejor descrito como un silenciador auxiliar. El resonador reduce el nivel del ruido del sistema de escape después de que haya pasado atraves del silenciador.

Resource Conservation and Recovery Act (RCRA) An act that controls hazardous waste disposal in the United States.

Ley de Conservación y Recuperación de Recursos Ley que controla la eliminación de residuos peligrosos en los Estados Unidos.

Returnless fuel system A fuel injection system with the pressure regulator mounted on top of the fuel tank so fuel is returned directly from this regulator to the tank.

Sistema de combustible sin retorno Sistema de inyección de combustible en el que el regulador de presión está montado sobre el tanque del combustible para que se pueda devolver el combustible directamente del regulador al tanque.

Reverse-flow cooling system A cooling system in which the engine coolant flows through the cylinder heads before it flows through the block.

Sistema de enfriamiento de flujo inverso Sistema de enfriamiento en el que el refrigerante del motor fluye a través de las culatas de los cilindros antes de fluir a través del bloque.

Reverse-flow muffler A muffler that reverses the flow of exhaust through the muffler to provide quieter operation.

Silenciador de flujo inverso Silenciador que invierte la dirección del flujo del escape a través del silenciador para proporcionar un funcionamiento con menos ruido.

Rheostat A variable resistor with two leads, an input and an output.

Reóstato Un resistor variable con dos cables, uno de entrada y uno de salida.

Rich A condition in which there is more fuel in the air-fuel mixture than would be if the mixture was stoichiometric.

Rico Una condición en la cual hay más combustible en la mezcla de aire/combustible que abría si la mezcla fuera estequiometrica.

Right-to-know laws Laws that state the workers have a right to know when the materials they use at work are hazardous.

Leyes del derecho a saber Estas leyes establecen que los empleados tienen el derecho a saber cuando los materiales utilizados en su trabajo son peligrosos.

Roots-type A supercharger that uses a pair of rotor vanes to drive the crankshaft.

Talones Un supercargador que usa un par de palas de rotor para impulsar al cigueñal.

Rotary engine An engine design that relies on the movement of a rotor instead of pistons to cycle through the four strokes.

Motor rotatorio Un diseño de motor que depende en el movimiento de un rotor envés de pistones para ciclar atraves de los cuatro tiempos.

Rotary-type throttle position (TP) sensor A potentiometer with a pointer that is rotated by the throttle shaft that informs the PCM of throttle position by varying voltage.

Sensor tipo rotativo de posición de la reguladora (TP) Un potenciómetro con una aguja girado por el eje regulador que informa al PCM de la posición de la reguladora por medio de un voltaje variable.

Rotor-type oil pump An oil pump in which two meshed rotors create pressure to force oil through the lubrication system.

Bomba del aceite tipo rotor Bomba del aceite en la que dos rotores engranados producen una presión para hacer circular el aceite a través del sistema de lubrificación.

SAE J1930 terminology An attempt to standardize electronic terminology in the automotive industry.

Terminología de la SAE J1930 Intento de uniformar los términos electrónicos utilizados en la industria automotriz.

Saturation The point reached when current flowing through a coil or wire has built up the maximum magnetic field.

Saturación El punto alcanzado cuando la corriente fluye atraves de una bobina o alambre que haya llenado el campo magnético a su máximo.

Schmitt trigger A pulse-shaping device.

Gatillo Schmitt Un dispositive que forma los impulsos.

Sealed-terminal battery A battery with the terminals on the side of the case.

Batería con terminales sellados Batería en la que los terminales se encuentran en el lado de la caja.

Secondary air bypass (AIRB) Part of the air injection cycle when air is sent to the atmosphere.

Desvío secundario del aire Parte del ciclo de la inyección de aire cuando el aire es enviado a la atmósfera.

Secondary circuit The part of the ignition system that produces and delivers high voltage to the spark plugs.

Circuito secundario La parte del sistema de la ignición que produce y suministra voltaje alto a las bujías.

Secondary coil winding A coil winding with many turns of fine wire in which the high voltage is induced to fire the spark plugs.

Bobinado secundario Bobinado enrollado con varias vueltas de alambre fino en el que se induce alta tensión para encender las bujías.

Selective catalyst reduction (SCR) It uses a urea solution (diesel exhaust fluid or DEF) to reduce NO_X into N_2 and CO_2. The urea solution is injected into the exhaust stream and collects on the SCR surface. The DEF reacts with the NO_X in the exhaust and reduces NO_X into N_2 and O_2. The DEF is injected into the exhaust stream as needed by NO_X sensors installed in the exhaust stream before and after the SCR.

Reducción catalítica selectiva Utiliza una solución de urea (líquido de escape diésel o DEF, por sus siglas en inglés) para reducir el NO_X a N_2 y CO_2. La solución de urea se inyecta en la corriente de escape y se recolecta en la superficie del sistema de reducción catalítica selectiva (SCR, por sus siglas en inglés). El líquido de escape diésel reacciona con el NO_X en el escape y reduce el NO_X a N_2 y O_2. El líquido de escape diésel se inyecta en la corriente de escape según lo requieran los sensores de NO_X instalados en la corriente antes y después del SCR.

Semiconductor An element with four valence electrons that has unusual characteristics when combined with some other elements.

Semiconductor Elemento con cuatro electrones de valencia que muestra características inusuales al combinarse con algunos otros elementos.

Serial data bus Communication wiring used by various vehicle modules to communicate with each other and diagnostic scan tools.

Autobús de datos en serie El cableado de comunicación usado por varios módulos del vehículo para comunicarse entre ellos y las herramientas exploradoras diagnósticas.

Series circuit An electrical circuit that has two or more resistors connected in line so that the same amount of current must pass through them.

Circuito en serie Un circuito eléctrico que tiene dos o más resistores conectados en línea para que la misma cantidad de corriente deba de pasar através de ellos.

Series hybrid Uses a small internal combustion engine to charge batteries when they become discharged. The electric motor provides power to the drive wheels. The series hybrid uses a complex controller to provide seamless acceleration from the electric motor.

Híbrido en serie Usa un pequeño motor de combustión interno para cargar las baterías cuando éstas se discargan. El motor eléctrico provee la fuerza a las ruedas motrices. El híbrido en serie usa un controlador complejo para proveer la aceleración sín interrupción del motor eléctrico.

Series-parallel circuit An electrical circuit that contains at least one parallel circuit connected in series to a resistor.

Circuito en serie-paralelo Un circuito eléctrico que contiene por lo menos un circuito paralelo conectado en serie al resistor.

Service bulletin Bulletins published by the car manufacturers or independent sources which provide information on service changes and procedures.

Folleto de servicio Folletos publicados por los fabricantes de automóviles o compañías no relacionados a la empresa automotriz que proveen información sobre procedimientos o revisiones acerca del servicio.

Shift adapt PCM learning of driver habits and minor transmission flaws that can be corrected by modifying line pressures and therefore shift speeds, feel, and timing.

Ajuste al cambio de velocidades El PCM aprende los costumbres del conductor y las pequeñas faltas de la transmisión que se pueden corregir modificando las presiones de las líneas, así la velocidad de los cambios de velocidas, la sensación y la regulación.

Shift schedules Programmed shift points in the computer's memory that allow the computer to determine when to start a gear change.

Horario de cambio de velocidades Los puntos de cambio de velocidad programados en la memoria de la computadora que permiten que la computadora determine cuándo comienza un cambio de velocidad.

Shift solenoid valves Valves controlled by shift solenoids that enable the PCM to electrically shift an automatic transmission.

Válvulas de solenoide de cambio de velocidad Las válvulas controladas por los solenoides de cambio de velocidad que permiten que el PCM cambie la velocidad electrónicamente en una transmisión automática.

Shorted circuit An unwanted copper-to-copper connection in an electric circuit.

Cortocircuito Conexión no deseada entre cobre y cobre en un circuito eléctrico.

Short-term adaptive fuel trim Short-term fuel injector pulse width compensation determined by the PCM according to operating conditions. This fuel trim is set to minimize emissions output and represents minor adjustments to the long-term fuel trim strategy.

Tiempo corto para la adaptación de la restricción del combustible Compensación determinada del pulso corto para la amplitud del inyector de combustible de acuerdo con el PCM (módulo de control de la potencia del motor) para las condiciones de operaciones. Esta restricción del combustible es establecida para minimizar las emisiones y representan pequeños ajustes a la restricción del combustible por tiempo largo.

Shutter blades Rotating blades in the distributor that rotate past the pickup assembly.

Aletas enrejadas Aletas giratorias en el distribuidor que sobrepasan el conjunto de captación.

Side-terminal battery A battery with the terminals located on the side of the case.

Batería con terminales laterales Batería en la que los terminales se encuentran en el lado de la caja.

Single overhead camshaft (SOHC) An engine with one camshaft positioned above or in the cylinder head.

Árbol de levas superpuesto sencillo Motor con un árbol de levas colocado sobre o en la culata del cilindro.

Sintered brass fuel filter A type of fuel filter with a sintered brass filtering element.

Filtro del combustible de latón sinterizado Tipo de filtro del combustible que tiene un elemento filtrante de latón sinterizado.

Slant-type block An in-line engine mounted on a slant rather than at the true vertical position.

Bloque tipo inclinado Motor en línea montado de forma oblicua en vez de en una posición completamente vertical.

Slave module A computer module that can signal receipt of a command, but is not able to initiate communication on the serial data bus.

Módulo esclavo Un módulo de computadora que puede señalar que ha recibido un orden, pero que ha recibido un orden, pero que no es capaz de iniciar la comunicación en el autobús de datos en serie.

Slow-start EI system An EI system that may require two crankshaft revolutions before it begins firing the spark plugs.

Sistema de encendido electrónico de arranque lento Sistema de encendido electrónico que podría requerir dos revoluciones del cigüeñal antes de encender las bujías.

Snapshot A feature of OBD II systems that captures all data present when the computer detects a problem and sets a trouble code.

Foto instantánea Una característica del sistema OBD II (diagnóstico abordo del vehículo) que captura toda la data presente cuando la computadora detecta un problema y establece un código de problema.

Society of Automotive Engineers (SAE) A society that established a classification system for oil viscosity ratings.

Sociedad de Ingenieros Automotivos (SAE) Una sociedad que establece el sistema de clasificación de grados de viscosidad del aceite.

Solenoid An electromagnetic device with a moveable core. Primarily used to cause some mechanical action or to act like a switch.

Solenoide Un dispositivo electromagnético con un núcleo removible. Primariamente usado para causar alguna acción mecánica o para actuar como un interruptor.

Spark ignition (SI) An engine in which the air-fuel mixture in the cylinders is ignited by a spark at the spark plug electrodes.

Encendido por chispa Motor en el que la mezcla de aire y combustible en los cilindros se enciende por una chispa en los electrodos de la bujía.

Spark line The horizontal line on a secondary scope pattern that indicates the voltage required to keep the spark plug firing.

Línea de chispa Línea horizontal en un modelo de radio secundario que indica la tensión que se requiere para mantener la bujía encendida.

Spark output A measurement of the voltage of the spark, usually measured in kV.

Producción de chispa Una medida del voltaje de una chispa que suele ser medida en kV.

Spark plug heat range The rate at which a spark plug conducts heat away from the center electrode.

Intervalo de calor de la bujía Velocidad a la que una bujía conduce el calor fuera del centro de un electrodo.

Speed density system A computer-controlled fuel injection system in which the computer calculates the amount of air entering the engine from the engine rpm and manifold absolute pressure (MAP) sensor signals.

Sistema de densidad de velocidad Sistema de inyección de combustible controlado por computadora en el que la misma calcula la cantidad de aire que entra en el motor mediante las señales de las rpm del motor y de los sensores de la presión absoluta del colector.

SPOUT Spark out signal.

SPOUT Señal de chispa saliendo.

Stand-alone modules A module that is not part of any network, but exists for a sole purpose.

Módulos independientes Un módulo que no es parte de ningún sistema red pero que existe para un sólo propósito.

Star network A computer network where the modules are interconnected at a common point. An open serial line would only affect one module.

Sistema estrella Un sistema red de computadoras que son interconectadas en un punto común. Una línea de seria abierta sólo afectaría un módulo.

Static balance Balance at rest, or still balance. It is the equal distribution of weight of the wheel and tire around the axis of rotation such that the wheel assembly has no tendency to rotate by itself regardless of its position.

Balanceo estático Balanceo en descanso o balanceo sin movimiento. Es la distribución igual de peso de la rueda y el neumático alrededor del axis de rotación, tal como el ensamblaje de la rueda no tiene tendencia de girar por si misma sin considerar su posición.

Stator The name given to many automotive parts, such as the stationary windings of an alternator, the third member of a torque converter, and the trigger wheel of an electronic distributor-type ignition system.

Estator El nombre otorgado a una parte automotriz, tal como el embobinado estacionario de un alternador, el tercer miembro de un convertidor de torque y la rueda disparadora de un distribuidor electrónico sistema de inyección electrónico.

Steering gear The device that converts the movement of the steering wheel to the movement of the linkages that turn the wheels.

Caja de la dirección El dispositivo que convierte el movimiento del volante de la dirección al movimiento de la varilla que gira las ruedas.

Steering linkage The arms that connect the steering gear to the wheels.

Varilla de la dirección Los brazos que conectan la caja de la dirección a las ruedas.

Stepper motor An electric motor that moves a pintle or valve horizontally in steps.

Motor por etapas Motor eléctrico que mueve una clavija o válvula horizontalmente por etapas.

Stoichiometric air-fuel ratio The ideal air-fuel ratio at which combustion is most complete.

Relación estequiométrica de aire y combustible Relación ideal de aire y combustible donde la combustión es más completa.

Strategy A plan. Typically refers to the programs of a PCM that ensure low emission levels.

Estrategia Un plan. Típicamente se refiere a los programas del PCM (módulo de control de la potencia del motor) que aseguran niveles bajos de emisiones.

Stroke Normally refers to the movement of an engine's piston from TDC to BDC.

Carrera Normalmente se refiere a los movimientos de los pistones de un motor desde el punto muerto superior al punto muerto inferior.

Sulfuric acid (H_2SO_4) A corrosive acid mixed with water and used in automotive batteries.

Ácido sulfúrico (H2SO4) Ácido sumamente corrosivo mezclado con agua y utilizado en las baterías de automóviles.

Supercharger A device driven by the engine's crankshaft used to force more air into the cylinder during its intake stroke.

Supercargador Un dispositivo conducido por el cigüeñal de los motores usado para forzar más aire adentro de los cilindros durante su carrera de admisión.

Superimposed scope pattern A scope pattern in which the voltage traces from all the cylinders are superimposed on top of each other.

Modelo de radio sobrepuesto Modelo de radio en el que las pequeñas cantidades de tensión provenientes de todos los cilindros se sobreponen una sobre otra.

Surge tank A small tank that acts as a reservoir for the cooling system, which is under pressure.

Cámara de compensación Pequeño tanque que actúa como reserva del sistema de enfriamiento, que está bajo presión.

Synchronizer (SYNC) pickup A pickup assembly that produces a voltage signal for ignition triggering or injector sequencing.

Captación del sincronizador Conjunto de captación que produce una señal de tensión para el arranque del encendido o para el ordenamiento del inyector.

Synthetic oil Oil that is formulated in laboratories.

Aceite sintético Aceite creado en laboratorios.

Tailpipe The part of the exhaust system that allows the exhaust gases to leave the rear of the vehicle and into the atmosphere.

Tubo de escape La parte del sistema de escape que permite que los gases de escape salgan de la parte de atrás del vehículo y adentro de la atmósfera.

TCC control PWM solenoid Torque converter clutch solenoid that is applied using pulse width modulation to cushion the feel of apply and release.

Solenoide PWM de control TCC Un solenoide del embrague del convertidor del par que se acciona usando la modulación de anchura de pulso para suavizar la sensación del aplicación y desembrague.

Theft deterrent computer A computer that controls the theft deterrent system and activates specific warning devices if the vehicle is entered improperly.

Computadora para el sistema anti-robo Computadora que controla el sistema anti-robo y activa dispositivos de advertencia específicos en caso de que alguien intente adueñarse del vehículo indebidamente.

Thermal efficiency The relationship between the engine power output and the heat energy available in the fuel.

Rendimiento térmico Relación entre la salida de la potencia del motor y la energía térmica disponible en el combustible.

Thermistor A resistor that changes values as its temperature changes.

Termistor Un resistor que cambia de valor al cambiar su temperatura.

Thermostatic element A heat-sensitive wax pellet that may be used to open and close the air door in a heated air inlet system.

Elemento termostático Grano gordo de cera sensible al calor que puede utilizarse para abrir y cerrar la puerta de ventilación en un sistema de admisión de aire calentado.

Thermostatic or bimetallic coil A coil made from two different metals with different expansion rates. The coil winds and unwinds as it is heated and cooled.

Bobina termostática o bimetálica Bobina compuesta de dos metales diferentes con diferentes capacidades de expansión. La bobina se enrolla y desenrolla al calentarse y enfriarse.

Thick-film integrated ignition systems (TFI) Multilayering of high circuit element density on the board, allowing more circuitry components to fit in a smaller area.

Sistemas del encendido integrados de placa gruesa (TFI) Poniendo capas múltiples densas en componentes en la placa de soporte, permitiendo que más componentes de circuitos quepan en un área más pequeña.

Three-way catalyst A catalytic converter that reduces three pollutants: carbon monoxide, unburned hydrocarbons, and oxides of nitrogen.

Catalizador triple Convertidor catalítico que disminuye tres sustancias contaminantes: el monóxido de carbono, los hidrocarburos no quemados, y los óxidos de nitrógeno.

Throttle actuator control (TAC) It is used to control the throttle opening by the PCM. It has often been called "throttle by wire."

Control del accionador del acelerador Se utiliza para controlar la abertura del acelerador a través del módulo de control del tren de potencia (PCM, por sus siglas en inglés). A menudo se denomina "acelerador por cable".

Throttle body injection (TBI) A fuel injection system in which the injector, or injectors, is positioned above the throttles.

Inyección del cuerpo de la mariposa Sistema de inyección de combustible en el que el inyector, o los inyectores, están colocados sobre las mariposas.

Thrust angle A line that divides the total toe angle of the rear wheels.

Ángulo del eje en relación al chasis Una línea que divide el ángulo total de la convergencia o la divergencia de las ruedas traseras.

Tier 2 emissions Strict emission control standards made mandatory in 2009 that place vehicles in emission categories called bins, depending on the levels of pollutants they emit.

Emisiones de nivel 2 Son estrictas normas de control de emisiones, obligatorias desde 2009, que clasifican a los vehículos en categorías según los niveles de contaminantes que emiten.

Tier 3 emissions Emissions and fuel mileage standards that are based on the footprint of the vehicle.

Emisiones de Nivel 3 Normas de emisiones y de combustible que se basan en la huella del vehículo.

Timing chain tensioner A mechanical or hydraulic device used to maintain tension on the engine timing chain.

Mecanismo tensor de la cadena de regulación del encendido Dispositivo mecánico o hidráulico utilizado para mantener la tensión en la cadena de regulación del encendido del motor.

Timer core The trigger wheel or armature in some electronic distributors.

Cronometro del núcleo La rueda disparadora o armadura en algunos distribuidores electrónicos.

Toe A wheel geometry concern that deals with the direction the front of the wheels is facing compared to the direction of the rear of the wheels.

Convergencia o divergencia Una geometría que concierne con la dirección de como las ruedas delanteras apuntan comparado con la dirección de las ruedas traseras.

Top dead center (TDC) Piston position at the top of the cylinder.

Punto muerto superior La posición del pistón en la parte superior del cilindro.

Torque A force that does work with a turning force.

Par de torsión Fuerza que realiza trabajo mediante una fuerza de torsión.

Torque converter A fluid coupling that transmits power from the engine to the input shaft of the automatic transmission. It contains a stator to enable the production of mechanical advantage.

Convertidor del par Un acoplador de líquido que transmite la fuerza del motor al eje de entrada de una transmisión automática. Contiene un estator para permitir la producción de una ventaja mecánica.

Torque converter clutch An additional clutch either inside the converter or transmission that enables the turbine to physically lock itself to the impeller, reducing friction and heat.

Embrague del convertidor del par Un embrague adicional que puede estar dentro del convertidor o la transmisión que permite que la turbina se traba al impelador, así reduciendo la fricción y el calor.

Torque converter clutch (TCC) lockup A system to lock the turbine to the front of the converter.

Fijador del embrague del convertidor del par motor Sistema para fijar la turbina en la parte delantera del convertidor.

Toxic Poisonous to the human body.

Tóxico Dañino para el ser humano.

Transducer An electronic device that monitors conditions, primarily pressure conditions, and sends out a voltage signal that represents the changes.

Transductor Un dispositivo electrónico que monitorea las condiciones, primariamente las condiciones de la presión y envía una señal de voltaje de salida que representa los cambios.

Transistor An automatic electronic switch with no moving parts.

Transistor Conmutador electrónico automático desprovisto de piezas móviles.

Transitional low emission vehicle (TLEV) Emission standards required by 1994.

Vehículo transisional de baja emisión Normas sobre emisiones a establecerse antes del año 1994.

Transmission control module (TCM) The module to control the function of the automatic transmission. Most manufacturers have incorporated into the PCM, but not all.

Módulo de control de transmisión (TCM) El módulo para controlar la función de la transmisión automática. La mayoría de los fabricantes lo han incorporado al PCM, pero no todos.

Transmission fluid pressure switch Hydraulically actuated switch inside the transmission that verifies shift occurrence and/or range selection matches the PCM command.

Interruptor de presión del líquido de transmisión Un interruptor actualizado hidráulicamente adentro de la transmisión que verifica un suceso de cambio de velocidad y/o una selección de un serie seleccionado que cumple con un orden del PCM.

Transverse engine mounting An engine mounted crossways.

Montaje transversal del motor Motor montado transversalmente.

Triggering device A sensor or A/C generator that measures the rotation of the crankshaft whose signal is sent to the ignition control module to control ignition and fuel delivery.

Dispositivo gatillo Un sensor o generador A/C que mide la rotación del ciguenal cuyo sensor se manda al módulo de control del encendido para controlar el encendido y la entregada del combustible.

Trip Starting and driving the vehicle until five monitors are completed.

Viaje Encendiendo y manejando el vehículo hasta que se hayan completado los cinco monitores.

Tube and fin radiator core A core with tubes surrounded by fins connected between the radiator tanks.

Núcleo del radiador de tubo y aleta Núcleo en el que los tubos están rodeados por aletas conectas entre los tanques del radiador.

Tuned exhaust headers Headers are connected between the cylinder head exhaust outlets and the exhaust pipe. Each exhaust passage, or pipe, in the header is the same length to improve exhaust flow.

Pipas afinadas de escape Las pipas están conectadas entre las salidas del escape de la culata del cilindro y el tubo de escape. Cada paso o tubo de escape en la pipa es de igual largo para un mejor flujo de escape.

Turbine The output component of the converter that is driven by the impeller.

Turbina El componente de potencia del convertidor que es accionado por el impulsor.

Turbine wheel A vaned wheel mounted in the exhaust system and connected to the opposite end of the turbocharger shaft from the compressor wheel.

Rueda de la turbina Rueda con paletas montada en el sistema de escape y conectada al extremo opuesto del árbol turbocompresor desde la rueda compresora.

Turbo lag The term used to describe the condition that exists when full engine power is needed and the turbocharger is not developing air pressure boost.

Demora del turbocargador Un termino usado para describir la condición que existe cuando la fuerza completa del motor es necesitada y el turbocargador no está desempeñando amplificación para la presión de aire adentro del múltiple de admisión.

Turbocharger A device, driven by exhaust gas flow, that forces more air into the engine's cylinders.

Turbocargador Un dispositivo, conducido por el flujo de los gases de escape, que fuerza más aire adentro de los cilindros del motor.

Two-stroke cycle An engine in which a complete cycle requires two piston strokes.

Ciclo de dos tiempos Motor en el que un ciclo completo requiere dos carreras del pistón.

Two-valve (2-V) cylinder head A cylinder head with two valves per cylinder.

Culata del cilindro de dos válvulas Culata del cilindro que tiene dos válvulas por cilindro.

Type A misfire A misfire that can cause immediate catalyst damage.

Fallo del encendido Tipo A Un fallo del encendido que puede causar daños catalíticos inmediatos.

Type B misfire A misfire that results in one and one half times the emission output standard.

Fallo del encendido Tipo B Un fallo del encendido que resulta en emisiones que excedan el escape normalizado por uno y medio.

Ultra-low emission vehicle (ULEV) Emission standards required since the year 2000.

Vehículo de emisión ultra baja Normas sobre emisiones a establecerse antes del año 2000.

Vacuum A pressure less than atmospheric pressure.

Vacío Presión más baja que la presión atmosférica.

Valence ring The outer ring on an atom.

Anillo de valencia Anillo exterior de un átomo.

Valve body The control center that senses the load on the engine.

Núcleo de la válvula El centro de control que detecta la carga en el motor.

Valve overlap The few degrees of crankshaft rotation when both valves are open with the piston near top dead center on the exhaust stroke.

Solape de la válvula Los pocos grados que gira el cigüeñal cuando ambas válvulas están abiertas y el pistón se encuentra cerca del punto muerto superior durante la carrera de escape.

vapor lock The boiling of fuel inside the fuel system, preventing fuel flow.

Bloqueo por vapor Se produce cuando el combustible bulle dentro del sistema de combustible e impide que este circule.

Vaporization The process in which a liquid changes to a gas or vapor.

Vaporización El proceso en el cual un líquido cambia de un gas a un vapor.

Variable dwell A feature of some electronic distributor-type ignition systems in which dwell time can vary according to engine conditions.

Dwell variable (tiempo en que la bobina se está saturando medido en grados) Una característica de algunos sistemas de ignición de tipos de distribuidores electrónicos en el cual el dwell puede variar de acuerdo con las condiciones del motor.

Variable nozzle turbine (VNT) The variable nozzle turbine unit designed to allow a turbocharger to accelerate quickly, thus reducing lag time.

Turbina con tobera variable La unidad de la turbina con la unidad de la tobera variable diseñada para permitir que un turbocargador acelere rápidamente, reduciendo su tiempo de suministrar presión al múltiple de admisión.

Variable valve timing A system that can adjust valve timing with the engine running for optimum performance.

Ajuste de válvula variable Sistema que puede ajustar la sincronización de la válvula con el motor en funcionamiento para un óptimo rendimiento.

Vehicle speed sensor (VSS) A sensor that is usually mounted in the transmission and sends a voltage signal to the computer in relation to the vehicle speed.

Sensor de la velocidad del vehículo Sensor montado normalmente en la transmisión y que le envía una señal de tensión a la computadora referente a la velocidad del vehículo.

Vehicle theft security system A system that attempts to prevent unauthorized vehicle use. It usually shuts down fuel and/or ignition systems.

Sistema de seguridad de antirobo del vehículo Un sistema que intenta prevenir el uso no autorizado del vehículo. Suele cortar el sistema de combustible o del incendio.

Vent valve It is opened to allow fresh air through the charcoal canister. The vent valve is also PCM controlled. The vent valve can be closed with the purge valve open while the PCM monitors the fuel tank pressure sensor for vacuum.

Válvula de ventilación Se abre para permitir la entrada de aire fresco a través de un filtro de carbón. La válvula de ventilación también se controla mediante el módulo de control del tren de potencia (PCM, por sus siglas en inglés). Puede estar cerrada con la válvula de purga abierta mientras el PCM supervisa el vacío en el sensor de presión del tanque de combustible.

Venturi A narrowing of a passage through which a liquid or gas is flowing.

Venturi Estrechamiento de un paso a través del que fluye un líquido o un gas.

Viscosity A term used to describe the ability of a liquid to flow.

Viscocidad Un termino usado para describir la habilidad de un líquido en fluir.

Viscous fan clutch A cooling fan clutch that drives the fan faster at high temperatures and slower at low temperatures.

Embrague viscoso del ventilador Embrague del ventilador de enfriamiento que acciona el ventilador para que gire más rápido a temperaturas altas y más despacio a temperaturas bajas.

Volt A measurement of electrical pressure difference.

Voltio Medida de una diferencia en presión eléctrica.

Voltage The force developed by the attraction of the electrons to the protons.

Voltaje La fuerza desarrollada por la atracción de los electrones a los protonos.

Voltage-generating devices Electronic devices that generate voltage in response to movement or a change in conditions.

Dispositivo generador de voltaje Dispositivos eléctricos que generan voltaje en respuesta a los movimientos o un cambio en las condiciones.

Voltmeter The device used to measure electromotive force.

Voltímetro Un dispositivo que se usa a medir la fuerza electromotívo.

Volumetric efficiency The relationship between the amount of air actually taken into the engine on the intake stroke compared to the amount of air required to fill the cylinder at atmospheric pressure.

Rendimiento volumétrico Relación entre la cantidad de aire realmente aspirado en el motor durante la carrera de aspiración y la cantidad de aire requerido para llenar el cilindro a una presión atmosférica.

V-type block An engine with two banks of cylinders arranged in a V configuration.

Bloque tipo V Motor con dos bancos de cilindros arreglados en forma de V.

Wankel engine A rotary cycle engine.

Motor Wankel Un motor de ciclo rotatorio.

Waste spark system A term that may be applied to any EI system because each pair of spark plugs fires with one cylinder on the compression stroke and the other cylinder on the exhaust stroke.

Sistema de chispa residual Término que puede aplicarse a cualquier sistema de encendido electrónico ya que cada par de bujías se enciende con un cilindro durante la carrera de compresión y con el otro cilindro durante la carrera de escape.

Waste-gate A valve that allows control of the amount of exhaust flow over the turbo vanes that enables control of boost pressure.

Compuerta de resíduo Una válvula que permite controlar la cantidad del flujo del escape por las aletas del turbo que permite controlar la presión de sobrealimentación.

Waste-gate diaphragm and valve A valve operated by boost pressure that controls the amount of turbocharger boost pressure.

Diafragma y válvula de la compuerta de desagüe Válvula accionada por presión de sobrealimentación que controla la cantidad de presión de sobrealimentación del turbocompresor.

Water (H₂O) A compound containing hydrogen and oxygen.

Agua (H2O) Compuesto que contiene hidrógeno y oxígeno.

Wax pellet A small sealed container filled with wax that expands and contracts when it is heated and cooled to open and close components such as the engine thermostat.

Grano gordo de cera Pequeño recipiente sellado, lleno de cera, que se expande y contrae al calentarse y enfriarse para abrir y cerrar componentes, como por ejemplo el termostato del motor.

Wet-type cylinder sleeve A cylinder sleeve with engine coolant contacting the outside area of the sleeve.

Manguito de cilindro tipo húmedo Manguito de cilindro donde el refrigerante del motor entra en contacto con la parte exterior del mismo.

Wet-type intake manifold An intake manifold with coolant circulated through the manifold.

Colector de aspiración tipo húmedo Colector de aspiración a través del cual fluye el refrigerante.

Wheel shimmy The rapid side-to-side vibration of a wheel/tire assembly.

Abaniqueo de las ruedas Una vibración lateral rápida de la asamblea rueda/neumático.

Wheel tramp The bouncing action of rotating wheels that is caused by static imbalance.

Rebote de la rueda La acción de saltar de una rueda girando, que es causado cuando está fuera de balance estático.

Wide range air/fuel ratio sensor A sensor that measures air/fuel ratio and relays it to the PCM by means of varying current direction and voltage.

Sensor de relacion aire de alto alcance/combustible Sensor que mide la relacion entre el aire y el combustible enviando distintos voltajes y amperes al PCM.

Work The result of applying a force.

Trabajo El resultado de la aplicación de una fuerza.

Workplace hazardous materials information systems (WHMIS) Sheets that provide information regarding hazardous materials.

Sistemas de información relacionados a materiales peligrosos en el lugar de trabajo Hojas de información sobre materiales peligrosos.

Y-pipe Three pipes connected in a Y configuration.

Tubo en Y Tres tubos conectados en forma de Y.

Zener diode A special type of diode that allows current to flow in the desired direction only when certain conditions are met.

Diodo Zener Un tipo especial de diodo que permite que la corriente fluya en la dirección deseada solamente cuando ciertas condiciones son cumplidas.

Zero-emission vehicle (ZEV) Emission standards required on a certain percentage of vehicles beginning in 1998.

Vehículo sin emisiones Normas sobre emisiones a establecerse en cierto número de vehículos a partir del año 1998.

INDEX

Note: page references with *f* notation refer to a figure on that page.